# Historical & Cultural Astronomy

More information about this series at http://www.springer.com/series/15156

Wayne Orchiston

# John Tebbutt

## Rebuilding and Strengthening the Foundations of Australian Astronomy

Springer

Wayne Orchiston
National Astronomical Research Institute of Thailand
Suthep District, Muang, Chiang Mai, Thailand

ISSN 2509-310X ISSN 2509-3118 (electronic)
Historical & Cultural Astronomy
ISBN 978-3-319-44520-5 ISBN 978-3-319-44521-2 (eBook)
DOI 10.1007/978-3-319-44521-2

Library of Congress Control Number: 2016953048



Printed on acid-free paper

This Springer imprint is published by Springer Nature
The registered company is Springer International Publishing AG
The registered company address is: Gewerbestrasse 11, 6330 Cham, Switzerland

# Foreword

Astronomy is still one of the very few sciences in which independent astronomers can play a valuable role and in which their work is welcomed by professional astronomers. There have been many outstanding independent astronomers in astronomical history, and one of the greatest of these is, without doubt, John Tebbutt.

Tebbutt spent his entire life in Australia, and never ventured abroad, but his reputation was truly international. His home-built observatory at Windsor, in New South Wales, was comparatively modest, but he made the best possible use of it, and his work covered many aspects of observational research; he was also the discoverer of two of the brightest comets of the nineteenth century, those of 1861 and 1881. Long before the end of his career, he was accepted as Australia's leading astronomer, which, it has to be admitted, caused a certain amount of jealousy in some quarters, and explains why he did not receive as much Government recognition as he unquestionably deserved (though this has been to some extent put right much more recently; his portrait was to be found on an Australian banknote—an honour certainly unique for an amateur astronomer—and a lunar crater has been named after him).

Up to now, there has been no full-scale biography of this remarkable man. Wayne Orchiston is uniquely qualified to write one, and he has filled a notable gap in the literature. His book gives full details of Tebbutt's life and accomplishments, but it does more than this; it paints a graphic picture of the scientific community of the time, and does not attempt to gloss over Tebbutt's rather strained relationships with some of his contemporaries. At least, nobody ever questioned the quality of Tebbutt's observational work.

Wayne Orchiston has carried out a tremendous amount of research, and one outstanding feature of his book is the very extensive list of references, which will be of great value to future historians of science. The text is also very readable, so that it will be of equal interest to readers who have little or no prior knowledge of the subject.

This is a timely book, which has been long awaited. It merits a place in every astronomical library and will certainly remain the standard work in its field.

Patrick Moore

# Preface

How well I remember that sweltering Saturday back in 1959, not long after my father had scored a new job and we all migrated across 'the ditch', from tiny Lincoln in New Zealand, and settled in Sydney, Australia's largest city. With lunch pack in hand, I took the train from Wollstonecraft to Blacktown, in Sydney's far-western suburbs, and the rickety little rail motor out to Windsor, some 30 miles distant (see Figs. P.1, P.2 and P.3 for Australian, greater Sydney and Windsor localities mentioned in this book). Then came a long hot walk from the station, my destination being a conspicuous cluster of tall trees just east of the town where the weekend before I had caught a fleeting glimpse of a dome-like structure amongst some old trees as my father and I sped through Windsor for the first time by car.

Could this possibly be the famous Windsor Observatory of John Tebbutt? Some years before, under the starry skies of my native New Zealand I had read with great fascination about Tebbutt, and when our family shifted to Sydney at the end of 1958 it gave me an unexpected opportunity to go in search of this remarkable Australian astronomer. But there was no point in raising my hopes unduly, for the Observatory may not have survived the ravages of time—after all, Tebbutt had died in 1916, more than 40 years earlier. These thoughts and others raced through my teenage mind as I walked with mounting excitement through the town, the distant trees growing ever nearer.

At long last, I reached the outskirts of picturesque historic Windsor (Fig. P.3), passed the old courthouse and headed down Pitt Street, with paddocks on either side, punctuated to the right by a line of willows marking South Creek. Upon nearing Palmer Street, I caught sight of the group of ageing trees I had glimpsed the week before, plus a stately old two-storey homestead and a cluster of farm buildings, all nestled on a little hillock, surrounded by the fertile alluvial silts deposited by flood-prone South Creek. As I moved ever closer, I suddenly saw it: one of the supposed farm buildings underwent a visual metamorphism to emerge as a handsome little brick cottage, but surmounted by a green dome. Beside it was another, much smaller, brick building, also topped by a dome (see Fig. P.4). There was no doubt—I had 'discovered' Windsor Observatory. The thrill of that moment will live with me forever!

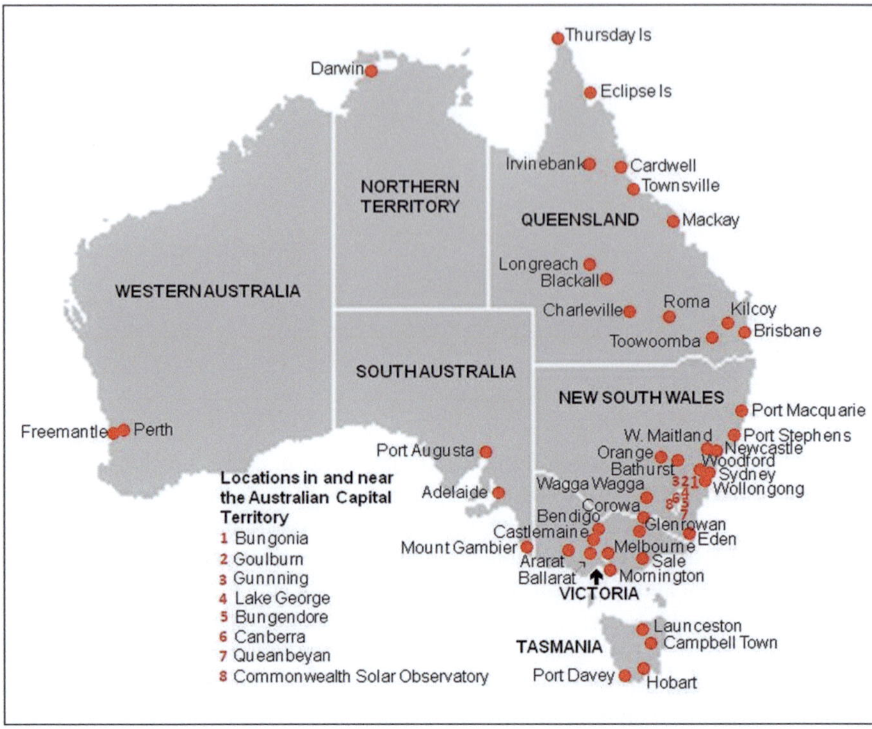


**Fig. P.1** Australian localities mentioned in this book (*Map* Wayne Orchiston)

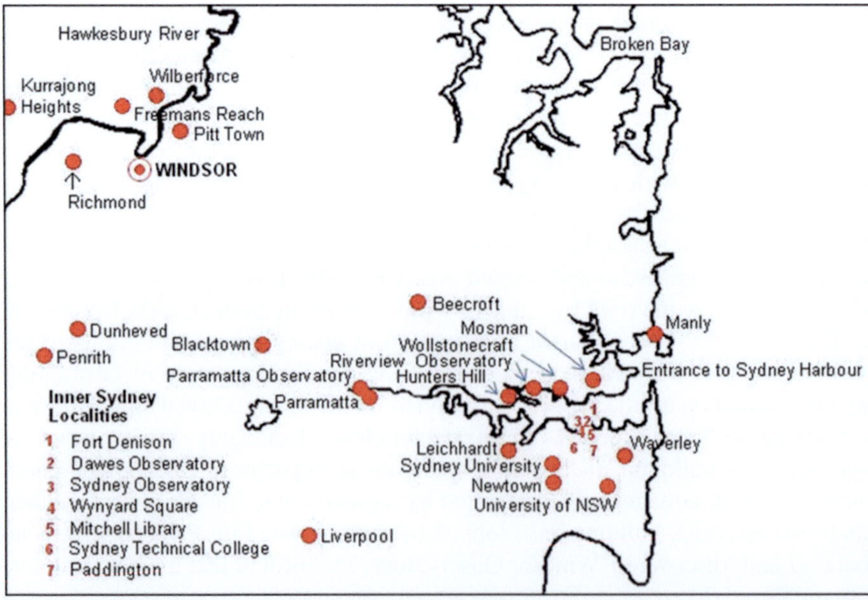


**Fig. P.2** The ‘greater Sydney region’, showing localities and geographical features mentioned in this book (*Map* Wayne Orchiston)

**Fig. P.3** Present-day Windsor, showing the locations of the Railway Station (*1*), the historic Courthouse (*2*), South Creek (*3*), Tebbutt's Homestead and Windsor Observatory (*4*), St. Matthew's Church (*5*) and the Hawkesbury Museum (*6*) (*Map modifications* Wayne Orchiston)

**Fig. P.4** This is very similar to my first view, in 1959, of the Tebbutt Homestead (in the trees left of centre) and assorted 'farm' buildings on the Peninsula Estate; initially, the dome of the 'little cottage' was not obvious because of the large tree directly behind it (*Photograph* Wayne Orchiston)

**Fig. P.5** John Halley Tebbutt (on the *left*) and my brother Peter Orchiston at the John Tebbutt Observatories on 25 April 1992 (ANZAC Day); in those days, many people mistook Peter for my twin brother (*Photograph* Wayne Orchiston)

Thus, began a research project that was to extend, intermittently, over more than half a century and culminated in the writing of this book. In the process, I came to know the owner of the property, Bruce Tebbutt (1904–1963), who was a grandson of the astronomer, and his successor, a great-grandson named John Halley Tebbutt (b. 1942; Fig. P.5).

I also explored the treasure trove of material in Sydney's Mitchell Library, deposited in 1917 by John Tebbutt's son for the benefit of future researchers wishing to flesh out the astronomical skeleton of this greatest of Australian nineteenth century astronomers. In addition to 16 annual reports and several volumes of meteorological observations (all published at his own expense), John Tebbutt left us copies of his published papers, tomes of inwards bound letters, observational notebooks and a plethora of other invaluable archival material (for details, see the next chapter of this book).

During 1960 and 1961, there were numerous weekend sorties to Windsor, sometimes with other members of the North Shore Astronomical Society, and I began cleaning up the smaller observatory building with the intention of reassigning a telescope to it (Fig. P.6). I wrote my first Tebbutt 'paper' at this time (McDonagh and Orchiston 1961), a simple three-page year-by-year compilation of the sorts of astronomical objects and events that Tebbutt had observed. So, it was not really a research paper, or even a review paper, in the strict sense, but it did serve as an illuminating guide for other members of the Society and as a source of inspiration for me.

**Fig. P.6** This July 1973 photograph shows the ivy-covered stand-alone observatory in which the North Shore Astronomical Society planned to instal an 18-in. (46-cm) reflector back in 1960–1961 (*Photograph* Wayne Orchiston)

For reasons that had mainly to do with my relative youth and poverty, unfortunately the Astronomical Society's plan to install Ignace Debono's historic 18-in. (46-cm) reflector (see Orchiston and Bembrick 1995) in the dome never materialised, but this did not dampen my programme of archival research on Tebbutt, and in 1968 I presented a 1 h lecture titled 'Biography of a gentleman astronomer: John Tebbutt, F.R.A.S'. at the 20 March meeting of the New South Wales Branch of the British Astronomical Association in Sydney. A two-page precis of my paper was printed in the Branch's *Bulletin* the following month (Orchiston 1968).

Four years later, Joseph Ashbrook (1972) prepared a general account on Tebbutt for international readers of *Sky & Telescope*, and in 1976 a volume of the *Australian Dictionary of Biography* appeared with an essay on Tebbutt by the late Dr. Harley Wood (1976), Director of Sydney Observatory. In the same year, Ann Moyal's fascinating *Scientists in Nineteenth Century Australia* was published, and Tebbutt was one of those who featured in the short section on 'Astronomical Pioneers' (Moyal 1976). Then, in 1979 and 1980, first Graeme White (Fig. P.7) and then Chris Kimpton published review papers on Tebbutt that specifically targeted Australian professional and non-professional astronomers, respectively.

**Fig. P.7** In 1979, radio astronomer Graeme White was the first person to bring John Tebbutt's achievements to the attention of modern Australian astronomers (*Photograph* Wayne Orchiston)

While these developments were occurring, I was experiencing a self-imposed exile from astronomy as the challenges of Ph.D. research and the resulting job-quest took priority, and it was only in 1980 that I was able to return to my Tebbutt Project. This turned out to be particularly timely since 1981 marked the centennial of Tebbutt's discovery of the Great Comet of 1881, and I was moved to prepare a commemorative paper for the occasion (Orchiston 1981; cf. Orchiston 1999a).

Centennials then proved excellent catalysts for further research. The year 1882 marked Tebbutt's attempt to form Australia's first national astronomical group, and also the construction date of what was to become his largest and most important telescope, and two further commemorative papers were prepared (Orchiston 1982a, b). A summary of the second of these also was presented *in absentia*, and prior to publication, at the 1982 national conference of the Royal Astronomical Society of New Zealand. The late Dr. Frank Bateson, who was then the proud owner of the telescope in question, kindly read the paper on my behalf.

These papers focussed on what I saw to be specific 'critical events' that had a profound effect upon Tebbutt's subsequent astronomical career, and I identified three others. First, there was his discovery of the Great Comet of 1861; then, in 1862 he was offered (but declined) the vacant Sydney Observatory directorship; and in 1863 he established the Windsor Observatory. Papers on these last two topics were published in 1988 (see Orchiston 1988a, b; see, also, Orchiston 2001).

Tebbutt's appearance on the new Australian $100 note in 1984 generated for the first time in nearly 70 years a heightened public awareness of who John Tebbutt was and a demand for information about his achievements. After submitting a letter to the editor of *Sky and Telescope*, I presented a review paper on Tebbutt as my Presidential Address to the Astronomical Society of Victoria on 20 February 1986, and followed this up with a feature article in *The Age* newspaper (Melbourne) on 8 February 1988 and a review paper in the English astronomical magazine, *Astronomy Now* (Orchiston 1988c). Tebbutt also featured prominently in 'The role of the amateur in early Australian astronomy' (Orchiston 1989), one of three commissioned review papers on the history of Australian astronomy that I prepared for *Search* in connection with the Australian Bicentennial.

An important milestone occurred in 1987 when Tebbutt's 1908 *Astronomical Memoirs* was reprinted (see Tebbutt 1986), complete with an excellent 'Introduction' by Graeme White (1986). White also took advantage of the return of Comet 1P/Halley in 1985 to examine Tebbutt's observations of this comet in 1910 (see Roser and White 1986). At about this time, I also published a number of short communications on different aspects of Tebbutt's work (Orchiston 1987, 1990, 1991).

In 1993, the Australian astronomical historian, Dr. Ragbir Bhathal, produced a short, popular paperback book titled *Australian Astronomer John Tebbutt: the Life and World of the Man on the $100 Note*, and in 1996 Raymond and Ros Haynes, David Malin and Dick McGee produced the masterly *Explorers of the Southern Kky* ..., which included numerous references to Tebbutt and his work.

In more recent times, I have prepared papers on Tebbutt's Great Comets of 1861 and 1881 (Orchiston 1998b, 1999a) and on the history of the Windsor Observatory (Orchiston 2001), an overview of his wide-ranging observational programmes (Orchiston 2004b) and a detailed account of his variable star work (Orchiston 2000a). A case study relating to comets and amateur-professional tension in Australasian astronomy (Orchiston 1999b), and review papers on nineteenth century astronomy education in Australia and on the formation and development of Australia's earliest astronomical societies (Orchiston 1997, 1998a), all contain substantial material about Tebbutt, and his long-running feud with Sydney Observatory Director, H.C. Russell, is examined in Orchiston (2000b, 2002). At the General Assembly of the International Astronomical Union, which was held in Sydney in July 2003, I promoted Tebbutt's achievements as Australia's foremost nineteenth century observational astronomer by preparing a poster paper about his observational work (a précis of the Orchiston 2004b paper mentioned above) and presenting an oral paper on the Tebbutt Collection of manuscript material in the Mitchell Library, Sydney (see Orchiston 2004a).

After working on a range of other topics, Ragbir Bhathal eventually returned to write about John Tebbutt in 2009, and the following January a short popular article appeared in the Royal Astronomical Society's *Astronomy & Geophysics* (Bhathal 2010). Meanwhile, I presented a paper titled 'The amateur-turned-professional syndrome: two Australian case studies' at the StephensonFest Conference at the University of Durham in April 2011, and while this dealt mainly with R.T.A. Innes and C.J. Merfield, I did provide a theoretical framework within which to examine the 'ATP Syndrome' and included John Tebbutt in this discussion (see Orchiston 2015, pp. 330–335).

The most recent presentation in which I specifically focussed on Tebbutt was titled 'John Tebbutt and the formation of Sydney's earliest astronomical societies'. This was the Keynote Paper presented at the 14 March 2015 seminar organised by the Sydney City Skywatchers and Sydney Observatory to celebrate the 120th anniversary of the founding of the New South Wales Branch of the British Astronomical Association. John Tebbutt was the inaugural President in 1895–1896, and it was an honour for me to occupy this same office a little over a century later.

The foregoing narrative indicates that, one way or another, John Tebbutt has been a conspicuous part of my astronomical consciousness for more than half a century, so it is only fitting that I should end up writing a book about him. And the timing is impeccable, for this year (2016) marks the centenary of his death.

I should mention that as an unexpected by-product of my Tebbutt research, the 'Tebbutt Collection'—but particularly the numerous volumes of inward correspondence—in the Mitchell Library led me to study a number of other nineteenth century or early twentieth century astronomers (e.g. Beattie, Biggs, Bone, Butterfield, Gale, Grigg, Hoskins, Howarth, Swindlehurst, Innes, Macdonnell, Merfield, Roseby, Ross, Scott, Ward, Wooster and Hugh Wright), various cometary discoveries, notable historic telescopes, and the astronomical work of New South Wales Lands Department. What began as a simple narrow-focus Tebbutt study became the catalyst for a much broader trans-Tasman canvas which eventually extended to Innes' work in South Africa at the Cape and Union Observatories!

This book attempts to bring some of these research themes together by drawing on my published works for many of the chapters. In the first Section of the book, I begin by overviewing John Tebbutt's life and achievements, and then discuss the remarkable 'Tebbutt Collection' of records in the Mitchell Library that made this study possible. Then follow two chapters which outline key developments in Australian astronomy during the nineteenth century. These two chapters provide a chronological context within which to place John Tebbutt's astronomical contributions. Then follow ten different 'Case Studies', each of which deals with a different specialised aspect of Tebbutt's wide-ranging 'astronomical portfolio'. Most of these are based on published material, but parts of Chaps. 8, 11, 14 and 15 are new.

Despite the broad brush-strokes adopted here, I have not attempted to paint the definitive picture of John Tebbutt, Australian astronomer extraordinaire. Rather, this book is an introduction to what I feel are some of the more interesting aspects of his illustrious astronomical career. In the course of the following chapters, I hope that some of Tebbutt's immense passion for astronomy comes through, together with the interesting interplay of power, patronage and politics which served to make him a key figure in late nineteenth century Australian astronomy.

Chiang Mai, Thailand Wayne Orchiston

## References

Ashbrook, J. (1972). John Tebbutt, his observatory, and a possible nova. *Sky and Telescope, 44,* 236, 240.

Bhathal, R. (1993). *Australian Astronomer John Tebbutt. The Life and World of the Man on the $100 Note*. Sydney: Kangaroo Press.

Bhathal, R. (2010). Australia's great comet hunter. *Astronomy & Geophysics, 51,* 1.23–1.25.

Haynes, R., Haynes, R., Malin, D., & McGee, R. (1996). *Explorers of the Southern Sky. A History of Australian Astronomy*. Cambridge: Cambridge University Press.

Kimpton, C. (1980). John Tebbutt – amateur astronomer. In *9th National Australian Convention of Amateur Astronomers* (pp. 89–94). Geelong: Geelong Astronomical Society.

McDonagh, E., & Orchiston, W. (1961). Out Windsor way (life of the astronomer John Tebbutt). *Journal of the North Shore Astronomical Society, 2* (1), 9–11.

Moyal, A. (1976). *Scientists in Nineteenth Century Australia: A Documentary History*. Sydney: Cassell Australia.

Orchiston, W. (1968). Biography of a gentleman astronomer: John Tebbutt, F.R.A.S. *British Astronomical Association New South Wales Branch Bulletin, 481,* 1–2.

Orchiston, W. (1981). Illuminating incidents in Antipodean astronomy: John Tebbutt and Great Comet of 1881. *Journal of the Astronomical Society of Victoria, 34,* 78–92.

Orchiston, W. (1982a). Illuminating incidents in Antipodean astronomy: John Tebbutt and the abortive Australian association of comet observers. *Journal of the Astronomical Society of Victoria, 35*, 70–83.

Orchiston, W. (1982b). John Tebbutt and the Whakatane Eight Inch Refractor: a review of "The Australian Connection". *Southern Stars, 29,* 215–244.

Orchiston, W. (1987). John Tebbutt: Australian astronomer extraordinaire. *Australian Journal of Astronomy, 2,* 44–46.

Orchiston, W. (1988a). Illuminating incidents in Antipodean astronomy: John Tebbutt and the founding of the Windsor Observatory. *Journal of the British Astronomical Association, 98,* 287–293.

Orchiston, W. (1988b). Illuminating incidents in Antipodean astronomy: John Tebbutt and the Sydney Observatory directorship of 1862. *Australian Journal of Astronomy, 2,* 149–158.

Orchiston, W. (1988c). John Tebbutt: an Australian astronomer of note. *Astronomy Now, 2* (5), 11–16.

Orchiston, W. (1989). The role of the amateur in early Australian astronomy. *Search, 20,* 16–23.

Orchiston, W. (1990). The *Astronomical Register*, John Tebbutt, and astronomy in the Antipodes. *Journal of the British Astronomical Association, 100,* 212.

Orchiston, W. (1991). Comet Tebbutt 1881III. *The Observatory, 111,* 313–314.

Orchiston, W. (1997). The role of the amateur in popularising astronomy: an Australian case study. *Australian Journal of Astronomy, 7,* 33–66.

Orchiston, W. (1998a). Amateur-professional collaboration in Australian science: the earliest astronomical groups and societies. *Historical Records of Australian Science, 12,* 163–182.

Orchiston, W. (1998b). Illuminating incidents in Antipodean astronomy: John Tebbutt and the Great Comet of 1861. *Irish Astronomical Journal, 25,* 167–178.

Orchiston, W. (1999a). C/1881 K1: a forgotten "Great Comet" of the nineteenth century. *Irish Astronomical Journal, 26*, 33–44.

Orchiston, W. (1999b). Comets and communication: amateur-professional tension in Australian astronomy. *Publications of the Astronomical Society of Australia, 16,* 212–221.

Orchiston, W. (2000a). John Tebbutt of Windsor, New South Wales: a pioneer southern hemisphere variable star observer. *Irish Astronomical Journal, 27,* 47–54.

Orchiston, W. (2000b). Politics and personalities in Australian astronomy: the Russell-Tebbutt feud. *AAO Newsletter, 95,* 8–11.

Orchiston, W. (2001) 'Sentinel of the Southern Heavens': the Windsor Observatory of John Tebbutt. *Journal of the Antique Telescope Society, 21,* 11–23.

Orchiston, W. (2002). Tebbutt vs Russell: passion, power and politics in nineteenth century Australian astronomy. In S.M.R. Ansari (Ed.), *History of Oriental Astronomy* (pp. 169–201). Dordrecht: Kluwer (IAU Joint Discussion 17).

Orchiston, W. (2004a). Highlighting the history of nineteenth century Australian astronomy: the Tebbutt Collection in the Mitchell Library, Sydney. *Journal of Astronomical Data, 10,* 77–103.

Orchiston, W. (2004b). John Tebbutt and observational astronomy at Windsor Observatory. *Journal of the British Astronomical Association, 114,* 141–154.

Orchiston, W. (2015). The amateur-turned-professional syndrome: two Australian case studies. In W. Orchiston, D.A. Green, & R. Strom (Eds.), *New Insights from Recent Studies in Historical Astronomy: Following in the Footsteps of F. Richard Stephenson* (pp. 259–350). Cham: Springer.

Orchiston, W., & Bembrick, C. (1995). The role of the large reflecting telescope in amateur astronomy: an Australian case study. *Australian Journal of Astronomy, 6,* 53–74.

Roser, S., & White, G.L. (1986). John Tebbutt's observations of Halley's Comet in 1910. *Proceedings of the Astronomical Society of Australia, 6*, 354 358.

Tebbutt, J. (1986). *Astronomical Memoirs*. Windsor: Hawkesbury Shire Council (Reprinted edition). (Although the reprint date is given as 1986, in fact this book was in fact printed in 1987 and officially was launched on 9 April 1987).

White, G. (1979). John Tebbutt and the astronomy at the Windsor Observatory. *Proceedings of the Astronomical Society of Australia, 3*, 408–411.

White, G. (1986). Introduction. In J. Tebbutt, *Astronomical Memoirs* (pp. vi–xix). Windsor: Hawkesbury Shire Council (Reprint).

Wood, H. (1976). Tebbutt, John (1834–1916). In G. Serle, & R. Ward (Eds.), *Australian Dictionary of Biography Volume 6: 1851–1890 R-Z* (pp. 251–252). Melbourne: Melbourne University Press.

# Acknowledgements

Completion of this book would have been almost impossible without the co-operation of John Halley Tebbutt of Windsor (the astronomer's great-grandson), and bibliographical assistance from Michelle Malley and Philippa Matla (both formerly Librarians at the Carter Observatory, Wellington, New Zealand), the late Peter Hingley and Mary Chibnall (formerly, Royal Astronomical Society Library, London), Professor Nick Lomb (formerly, Sydney Observatory, and now at the University of Southern Queensland, Toowoomba) and the staff of the Mitchell Library (Sydney).

I also wish to thank the following for their assistance in various ways: staff of the Alexander Turnbull Library (Wellington, New Zealand), staff of the State Records New South Wales (Sydney), the late Dr. Frank Bateson (Tauranga, New Zealand), Colin Bembrick (Bathurst), the late Reginald A. Biggs (Launceston), Adrian Brewer (Melbourne), the late Pat Burke (Newcastle), Dr. Aedeen Cremin (University of Sydney), P. de Villiers (South African Astronomical Observatory), the late Ignace Debono (Sydney), Dr. Richard Dodd (former Director of the Carter Observatory, Wellington, New Zealand), Janet Dudley (former Royal Greenwich Observatory Librarian and Archivist), the late Jim Duthie (Whakatane Astronomical Society, New Zealand), David Emanuel (Reserve Bank of Australia, Sydney), Peter Ferrett and Ian Symonds (Port Macquarie Museum), Alan Flint (Reserve Bank of Australia, Melbourne), Dr. David Frew (University of Hong Kong), Alan Gilmore (formerly, Mount John University Observatory, New Zealand), the late Dr. J. Hers (formerly Union Observatory, South Africa), Julian Holland (formerly Macleay Museum, Sydney), Professor Rod Home (University of Melbourne), Associate Professor Ian Jack (University of Sydney), Professor Ian Inkster (The Nottingham Trent University, England), Andrew James (Sydney), Dr. Bill Kitson (Department of Natural Resources, Brisbane), Ethleen Lastovica (formerly South African Astronomical Observatory Librarian), the Librarian at the Royal Observatory Edinburgh, Frank Moss (Gisborne, New Zealand), the late Sir Patrick Moore (Selsey, England), Mr. Nuttapon (Chongruk Studio, Chiang Mai, Thailand), Professor Jay Pasachoff (Williams College, Massachusetts), Ian Pearce (Archives Office of Tasmania, Hobart), the late John Perdrix (Perth), staff of the Queen Victoria Museum

(Launceston), Jeanette Regan (former Mount Stromlo and Siding Springs Observatory Librarian), Professor Robert Stebbins (University of Calgary, Canada), Ian Sullivan (Melbourne), staff of the Tasmanian Museum (Hobart), the late Dr. Con Tenukest (Sydney), Rosanne Walker (Australian Science Archives Project, Canberra), Professor Brian Warner (University of Cape Town, South Africa), Dr. Graeme White (Wagga Wagga), Steve Williams (formerly, John Tebbutt Observatories, Windsor) and the late Dr. Harley Wood (former Government Astronomer of New South Wales).

I am particularly grateful to Dr. Richard Dodd and Professor Nick Lomb for reading and commenting on drafts of this book, and to Sir Patrick Moore for kindly agreeing to write the Foreword many years ago, on the basis that eventually this book would be published.

A long-term research project like this is an expensive venture, and I wish to thank the Donovan Astronomical Trust (Sydney) for valued financial support. Without this funding, research trips to Sydney, first from Melbourne and later from New Zealand, would not have been possible.

Final thanks go to Maury Solomon at Springer for her help and encouragement and seeing this book through the press, and to my wife, Darunee Lingling, who provided me with the inspiration to finally complete the write-up of this all-too-prolonged 'Tebbutt Project'—and the environment in which to complete it.

# Contents

# List of Figures

# List of Tables

# List of Bio Boxes

# About the Author

**Wayne Orchiston** is a Senior Researcher at the National Astronomical Research Institute in Chiang Mai, Thailand, and an Adjunct Professor of Astronomy at the University of Southern Queensland in Australia. Formerly he worked in optical and radio astronomy in Australia and New Zealand. He has published on the history of Australian, English, French, Georgian, Indian, Indonesian, Iraqi, Japanese, New Zealand, Thai and US astronomy, and has supervised a large pool of graduate students. Wayne has played a leading role in developing history of astronomy through International Astronomical Union Commission 41 (and its successor C3), and was responsible for the formation of the Historic Radio Astronomy and the Transits of Venus Working Groups. Currently he is Vice- President of Commission C3. In 1998 Wayne co-founded the *Journal of Astronomical History and Heritage* and is the current Editor. He is on the Editorial Board of Springer's Historical and Cultural Astronomy series. In 2013 the IAU recognised his contribution to international astronomy by naming minor planet 48471 'Orchiston' after him.

# Part I
# Overview

# Chapter 1
# John Tebbutt: Australian Astronomer Extraordinaire

## 1.1 Introduction

John Tebbutt (Fig. 1.1) was one of the world's most remarkable nineteenth century astronomers (see Williams 1988). Entirely self-taught, from the geographical and intellectual isolation of his Windsor Observatory near Sydney, Australia, he carried out a wide-ranging observing programme, publishing prolifically in local and overseas journals, actively popularising astronomy, maintaining a local time-service and a meteorological station, and monitoring local floods and freshes. In 1862, he discovered a nova, and '... two of the grandest comets of the nineteenth century, namely, those of 1861 and 1881, both of which were objects of great interest when at their maximum splendour' (Steele 1916, Chapter XXII). In 1882, he formed Australia's first national astronomical group of any kind, the short-lived Australian Corps of Comet Observers, and in 1895 he served as the inaugural President of the New South Wales Branch of the British Astronomical Association when this group was founded.

Although offered the post of Government Astronomer of New South Wales and Director of Sydney Observatory in 1862, Tebbutt chose to retain his independence, and quickly went on to build an international reputation. By 1869, Windsor Observatory was deemed important enough to be listed in the *Nautical Almanac*, and later in the century it was reliably reported by a scientific colleague in Germany—arguably the leading nation for positional astronomy—that '... there are two men in Australia who were of wide fame in Europe: one was Baron von Mueller [the Melbourne botanist], and the other John Tebbutt' (Macdonnell 1904, p. 85). As a further indication of his international eminence, in 1905 Tebbutt was only the third recipient of the Royal Astronomical Society's Jackson Gwilt Medal and Gift.

Despite a remarkable record of achievement extending over more than half a century, in more recent times Tebbutt's name was largely unknown to Australian astronomers until a crater on the Moon was named after him in 1973 and his face and Windsor Observatory were chosen to feature on a new $100 Australian bank


W. Orchiston, *John Tebbutt*, Historical & Cultural Astronomy,
DOI 10.1007/978-3-319-44521-2_1

**Fig. 1.1** This photograph shows John Tebbutt and his distinctive handwriting (after Tebbutt 1908a, Frontispiece)

note issued in 1984. Two years later, his great-grandson began restoring the two surviving Windsor Observatory buildings, and these were opened in 1989 as a unique museum of astronomy.

Tebbutt now has a secure place in Australian scientific and astronomical history and has been described as a jewel in the crown of Australian astronomy (Haynes, et al. 1996, p. 115) and as '… an example of what may be achieved by an … observer with modest instrumental equipment applied in a methodical manner to the fundamental problems of astronomy' (Reports of the Branches 1923, p. 45).

This chapter presents an astronomical overview of Tebbutt's life and times, and 'sets the scene' for the remainder of the book by providing a framework within which the individual studies in Part III can be placed. Parts of this chapter have already appeared in print (see Orchiston 1982b, b).

## 1.2 Tebbutt: The Man

John Tebbutt was born in Windsor on 25 May 1834 and effectively was an only child for his younger sister, Ann, born in 1838, died at 8 months. His parents conducted a successful business in the main street of Windsor, so young John grew up in comparatively affluent surroundings. He attended local church schools where he excelled, and his enquiring mind soon turned to scientific and mechanical pursuits and ultimately to astronomy. This was mainly due to the influence of his first teacher, Mr. Edward Quaife (ca. 1802–1871), who '… although he did not possess an acquaintance with the refinements of the science [i.e. astronomy], had a very good general knowledge of it' (Astronomy 1875) and enjoyed observing. Quaife owned a number of books on astronomy and a small telescope (Quaife 1861), possibly a marine telescope (see Quaife 1869b).

In 1843, John Tebbutt's father sold his business in Windsor and bought a 250 acre (100 ha) tract of beautiful, fertile, alluvial land on the eastern outskirts of the town. Their farm was known as 'The Peninsula Estate' as it was almost surrounded by the Hawkesbury River and its tributary, South Creek (see Fig. 1.2). On the Estate, there was a small hill just 15 m above the local mean tide level, and in 1845 a comfortable two-storey homestead was constructed there (Tebbutt 1887, 1908a). This has survived to the present day (Fig. 1.3). Little could an 11-year-old John

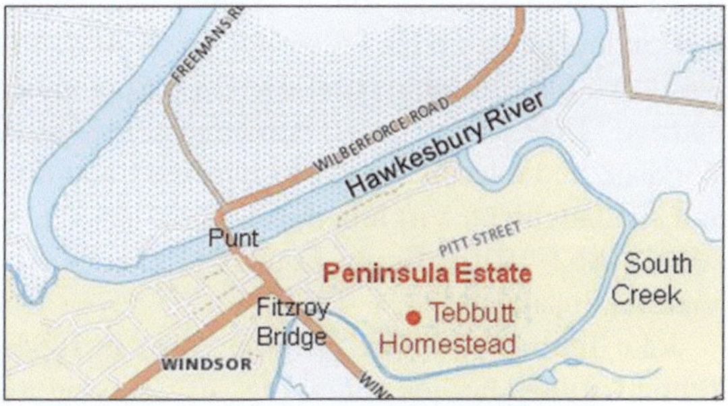


**Fig. 1.2** A map showing the location of the Peninsula Estate relative to Windsor, the Hawkesbury River and South Creek. When the Tebbutt Homestead was erected, there was no bridge over the Hawkesbury River (*Map modifications* Wayne Orchiston)

**Fig. 1.3** The historic Tebbutt Homestead in July 1973 (*Photograph* Wayne Orchiston)

Tebbutt have imagined that this would be his home for the rest of his life, or that it would witness the rise of Windsor Observatory and his own emergence as a world figure in astronomy.

Windsor originally was known as Green Hills, and when proclaimed a town in 1810 was one of the earliest settlements established in the colony of New South Wales. Later, it was renamed 'Windor' because of the physical resemblance to its English namesake on the banks of the Thames near London. By the 1840s, Windsor was the fourth-largest town in New South Wales, growing in population from 1440 in 1841 to 1679 in 1848. It also was the leading agricultural centre of the young colony, and because it was sited on the banks of the majestic Hawkesbury River, near the head of navigation, was home to small-scale ship-building and a vibrant river trade. In those days, there was no railway access (it came in 1864), so a visit to Sydney involved a lengthy boat trip (to the mouth of the Hawkesbury River and then down the coast to Port Jackson), or a far from enjoyable coach trip along a deplorable road (Steele 1916).

At 15 years of age young Tebbutt completed his formal schooling and began working full time on his father's farm (N.S.W. 1893). Farming the Peninsula Estate was to be his life-long vocation. On 8 September 1857, when 23 years of age, he married 19-year-old Jane Pendergast (1838–1903) at St. Matthew's Church, Windsor, and over the next 14 years they had one son, John Thomas (1868–1936), and six daughters: Virginia Sarah (1858–1931), Ann Elizabeth (1860–1891), Mary Alice (1862–1910), Emily Francis (1864–1933), Jane Margaret (1866–1943) and Victoria Susannah (1871–1889).

John Tebbutt's mother, Virginia, died in 1865 (aged 64) and his uncle, Thomas Tebbutt, joined her the following year, leaving a significant portion of his estate to his talented young nephew (Will of Thomas Tebbutt). When his own father (also named John) died in December 1870 (aged 76), 35-year-old John Tebbutt assumed ownership of the Peninsula Estate. Initially, he had avoided confusion by using the name 'John Tebbutt Jn', but soon after his father's death he abandoned the distinguishing affix (Tebbutt 1881).

As a man, Tebbutt was a hard task-master. Haynes et al. (1996, p. 115) found that 'Time and precision became, for him, almost an obsession, and friends who arrived late for an appointment risked his permanent displeasure'. He was also a stickler for accuracy and honesty. When Captain O.H. Stokes (a former Director of the Royal Engineers Observatory on Mauritius) settled in Sydney and claimed to be a Fellow of the Royal Astronomical Society, Tebbutt looked unsuccessfully for his name in the membership lists before writing to the Society to formally enquire (Tebbutt 1891a). Furthermore, he had strong views on who should be entitled to membership of this distinguished body and refused to support the nomination of a number of Australian astronomers who really were worthy of the honour, including Joseph Brooks (1847–1918) and Walter Gale (1865–1945).

These strong views went far beyond astronomy and were a feature of his personality. In 1891, he wrote: 'Whilst I am a lover of perspicuity in scientific descriptions,

I may say that I am also a cordial hater of anything in the shape of ambiguity' (Tebbutt 1891b). He also despised drunkenness and gambling, did not look favourably on horse-racing, and as other 'sins' identified '… those of the self-righteous and the haughty man; the proud and the unforgiving man …' (Tebbutt 1875b). He also believed that the 'little sins' of the thoughts, the tongue, the temper and of omission although appearing small, '… are of awful magnitude in the eyes of an infinitely pure and holy God' (ibid.). Some would say that with the passage of the years Tebbutt would himself become that very same 'proud and unforgiving man' that he viewed with contempt back in 1875.

Because of his own internationally-recognised achievements in astronomy, Tebbutt ended up with unrealistic expectations of others, and when they could not live up to the ideals that he aspired to he would sometimes display a degree of intolerance. This certainly occurred in connection with the ill-fated Australian Comet Corps (see Chap. 10), and the mounting support that Henry Chamberlain Russell (1836–1907) showed for meteorology at Sydney Observatory to the detriment of astronomy (see Chap. 12). It was also manifest in his attitude towards Dr. William Bone (1836–1885) of Castlemaine (Orchiston 1987) and must have motivated M.L. Allen, a school teacher from near Goulburn, to pen the following comments after he was chided by Tebbutt for getting his 'practice calculations' very wrong (he was hoping to become Tebbutt's computer):

> … it seemed to me as if it was rather harsh, considering that it is some fourteen years since I opened trigonometry and that I told you in the former letter that I was rather rusty … I am certain you did not mean to be harsh … I am truly sorry that I was so far out and I must acknowledge was a little bit hurt at your letter … (Allen 1886).

Despite Tebbutt's biting response, Allen maintained an amiable correspondence with him for many years. This intolerance, and even an element of arrogance, comes through in the following letter to the Royal Astronomical Society:

> It is very annoying and disheartening to find that my pamphlets presented and actually received are not now forthcoming, and unfortunately the R. Ast. Society's library is not the only one which is defective in this respect. I had trusted that my publications would have been bound and preserved and thus made available to future fellows of the Society. (Tebbutt 1903a).

Tebbutt rarely reveals his innermost feelings in his published writings, but his observing journals do contain one remarkable non-astronomical diversion which is worth reproducing here:

> Poor Jacko, my pet magpie and a remarkably intelligent bird, was unfortunately killed by one of the dogs … He was an exceedingly clever imitator, and my constant companion at the Observatory day and night for the past seven years. I buried him yesterday under the ladder which leads up to the maximum shade thermometer. On this ladder he was accustomed frequently to roost [Fig. 1.4] and when not asleep invariably saluted me on my passage to and fro between the main observatory and the round equatorial room. I was wonderfully attached to the poor bird and I feel his loss very keenly. Hence this little tribute to his memory. (Tebbutt 1894–96, 7 August 1895 entry).

**Fig. 1.4** A 1906 photograph of the two observatory buildings and between them the ladder (leading to the maximum shade thermometer) where Jacko liked to roost (Orchiston collection)

This paints a rather poignant picture of an ageing astronomer sharing his nightly vigils with a friendly old magpie. For a serious astronomer working in isolation astronomy could be a lonely preoccupation, and Tebbutt was not like others in Windsor. From his youth he had been driven by this passion for astronomy. The following poem by Kathleen O'Byrne Grece (1947) could just as easily have been written by Tebbutt about himself:

I march to hidden music, song of spheres
And roll of constellations in my ears.
The clouds surrender first, and in retreat,
Fling wide the starry carpets for my feet.
I Capture far horizons; flaming skies
Grow pale before the knowledge of my eyes.
I march to hidden music. Only they
Who seek the absolute may walk this way.

Tebbutt's long and productive life came to an end on 29 November 1916 when he was 82 years of age. He was predeceased by his wife (in 1903) and three of his six daughters. According to Wood (1976), his estate was valued for probate at £69,364. In this same year, 1916, another local identity, the historian James Steele (1861–1918), published the book *The Early Days of Windsor* in which he features Tebbutt on the Frontispiece and devotes a short chapter to the astronomer and his exploits. Steele starts this chapter with the following salute: 'Of all the names, both past and present, connected with Windsor, none is so widely known outside New South Wales as that of John Tebbutt, F.R.A.S. …' (Steele 1916, p. 186).

## 1.3 The Formative Years

Tebbutt was entirely self-taught as an astronomer (Death of Mr. John Tebbutt 1916). During his school days, he became acquainted with the works of Ovid, and particularly with *Hominis Creatio* and its pointed references to astronomy. 'What occupation could be more sublime and elevating?' Tebbutt was later to write:

> It is an occupation which, in a preeminent degree, distinguishes man from the beings who, the poet tells us, look downwards to the earth. The highest forms of music and poetry, as Proctor has shown us, are to be found among the stars. Is there not, he would ask, a music of the spheres, whose tones were too subtle and too sublime for the physical ear, and could there be grander poetry than that which had for its rhythm the mathematical march of countless worlds. (Reports of the Branches 1897, p. 378).

Tebbutt gives an interesting account of his introduction to astronomy. In his *Astronomical Memoirs*, he states that it was partly through the education he received from the Reverend Henry Tarlton Stiles, M.A. (1808–1867; Fig. 1.5; see Cable 1967; Steele 1916) and partly through his frequent contact with his old tutor, Mr. Edward Quaife, that his tastes were directed towards astronomy (Tebbutt 1908a, p. 9). British-born Quaife continued to teach in Windsor until 1865 (Steele 1916) and remained in close contact with his 'star pupil' throughout, and even after he moved away from Windsor. Their correspondence was only brought to a close with Quaife's death in 1871 (N.S.W. 1893). It is notable that Tebbutt makes no mention of viewing the Great Comet of 1843, which ranks as one of the most spectacular comets of the century (see Chambers 1909). He would have been 9 years of age at the time, and already Quaife had introduced him to some elementary astronomy. We do know that Quaife observed and recorded this comet (Quaife n.d.).

Young Tebbutt's interest in astronomy escalated towards the end of the 1840s, partly as a result of continuing contact with Quaife, and through reading some popular books on astronomy by the London astronomer J.R. Hind in his spare time (N.S.W. 1893). From 1840 to 1844, Nottingham-born John Russell Hind

**Fig. 1.5** The Reverend H.T. Stiles, Rector of St. Matthew's, Windsor, from 1833 to 1867, was one of John Tebbutt's teachers (after Steele 1916)

**Fig. 1.6** An undated photograph of John Russell Hind (https://en.wikipedia.org)

(1823–1895; Fig. 1.6) worked at Royal Greenwich Observatory, before taking over as observer at George Bishop's private observatory in Regent's Park, London. Between 1844 and December 1852, Hind discovered two comets (C/1846 O1 and C/1847 C1), new variable stars, a nova (Nova Ophiuchus 1848) and eight minor planets (Obituaries 1896), and books he wrote included *The Solar System: or the Sun, Moon, and Planets* (1844), *Recent Comets and the Elements of their Orbits* (1844), *An Astronomical Vocabulary: Being an Explanation of All Terms in Use Amongst Astronomers at the Present Day* (1852), *The Comets. A Descriptive Treatise upon these Bodies* (1852) and *The Solar System: A Descriptive Treatise upon the Sun, Moon, and Planets, Including an Account of all the Recent Discoveries* (1852). It is not known which two or more of these books Tebbutt mentally consumed, but it soon dawned on him that

> … the universe was really mechanism of the highest order, and being, as I have already said, mechanically inclined, I began to turn my attention to celestial mechanism. Thus began my love for astronomy, and I became fired with an ambition to do real work in its behalf. (Tebbutt 1908a, p. 10).

In May 1853, this 'armchair' interest underwent a sudden metamorphosis with the appearance of a conspicuous naked eye comet near Orion (see Fig. 1.7), and Tebbutt's 'career' as an observational astronomer was launched. He observed this celestial visitor with a small marine telescope provided by his father and recorded his observations and accompanying drawings in what was to be the first of his observing journals (Tebbutt 1853–59).

Later in the year, he added a second-hand sextant to his tiny stable of astronomical equipment, which included the family's grandfather clock. At about the same time, Tebbutt improvised a crude mounting for the little marine telescope and began sunspot observations using the projection method. In this way, he was able to confirm the rotational period of the Sun on the basis of his own observations.

**Fig. 1.7** A drawing made by John Tebbutt on 2 May 1853 showing Comet C/1853 G1 (Schweizer) and part of the constellation of Orion (after Tebbutt 1853-59; *Courtesy* Mitchell Library)

An unusually large group of spots was visible on the Sun in May 1854, and after observing this Tebbutt wrote an account of it which was published in the *Sydney Morning Herald* (Tebbutt 1854). In so doing, he commenced a practice that he was to pursue throughout his life: that of using the media to offer the general public accounts of his own observations and discoveries, and information on upcoming astronomical events or objects which might be of interest. Much later, a newspaper reporter was to write: 'Many were impressed with the discoveries he made known through the columns of local newspapers' (N.S.W. 1893).

So by the comparatively young age of 21, Tebbutt was committed to a life of astronomy, and

> Henceforth he lived in an atmosphere of moving planets. The science of astronomy became his ruling passion; all else to him went by the board, and he had perpetually in his mind, while pursuing his study of the starry heavens, a working model of the universe, whose mysteries it was the purpose of his life to unfathom. (Houghton 1917, p. 7).

The occurrence of a total solar eclipse potentially visible from New South Wales on 26 March 1857 provided Tebbutt with another opportunity to bring astronomy before the public via the newspapers. Unfortunately, the day turned out cloudy and no observations were obtained, but Tebbutt's efforts placed him squarely in the public eye (The Windsor (N.S.W.) Observatory 1889). Much later, he was to write that this eclipse '… gave a great impetus to my studies' (Tebbutt 1908a, p. 12).

In 1858, Comet C/1858 L1 (Donati) attracted Tebbutt's attention (see Fig. 1.8), and two years later he observed Comet C/1860 M1 (the Great Comet of that year). Both were naked eye objects, and Tebbutt determined their positions with his sextant and then went on to compute their orbital elements and publish these in the local newspapers. For good measure, he then calculated the orbit of the May 1853 comet, using his own crude observations as positional data. In those distant times before computers, such calculations were time-consuming and intellectually

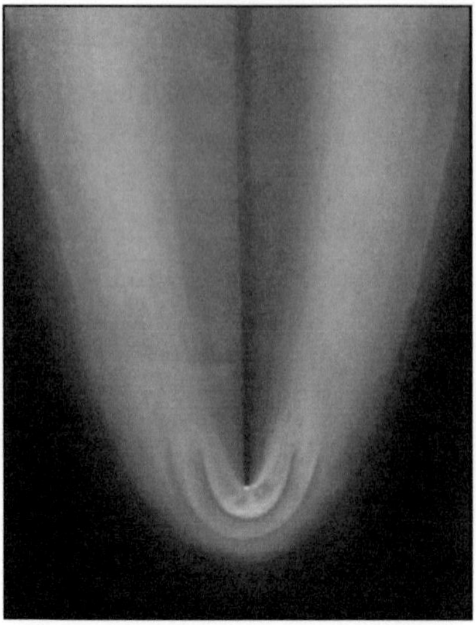

**Fig. 1.8** Comet C/1858 L1 (Donati) was famous not only for its impressive tail, but also for the changing nature of its head (after Bond et al. 1858)

demanding. Many years later, Tebbutt was to recount how he came by the requisite mathematical knowledge. While still a teenager, he read some popular articles on comets, and they

> … brought home to him how sadly deficient he was in mathematical knowledge, and how essential mathematics were to anyone seeking to make progress in astronomical studies.
>
> He accordingly resolved to master what mathematics he could. He began with algebra. He followed on with geometry and trigonometry. Interest increased as he advanced. When he reached the field of the higher problems he felt he would succeed. The stars looked down on him with different meanings in their faces. Day and night the heavens conversed with him in the newly-discovered speech. (N.S.W. 1893).

This really was a remarkable accomplishment for a young man working in virtual academic isolation.

Other objects and events to attract Tebbutt's attention during the 1850s were fireballs and meteors, Jupiter's satellites, lunar eclipses, lunar occultations of stars, the variable star Algol, aurorae and the zodiacal light (Tebbutt 1853–59). This catholic approach to observational astronomy was to be a notable feature of Tebbutt's later astronomical career, once he was properly equipped to make a useful contribution to the science.

## 1.4 The Great Comet of 1861

In 1861 came a discovery which was to propel Tebbutt's name into the annals of international astronomy. Let us quote from his *Astronomical Memoirs*, which reflects some of the excitement of the moment even though it was written more than 40 years after the great event:

> On the evening of May 13, 1861, while searching the western sky for comets, I detected a faint nebulous object near the star Lacaille 1316 in the constellation Eridanus. In my marine telescope the object appeared much diffused, and it was with the greatest difficulty that I measured its distance from three well known fixed stars. The object was hardly distinguishable in the small telescope attached to the sextant … Every comet hunter knows how necessary it is to the carrying out of his work to have at hand a copious catalogue of nebulae, but this valuable adjunct I unfortunately did not possess. I could not, however, find the object in the limited catalogues at my command. I accordingly made up my mind to watch it … I rose on the following morning to examine it, but failed to detect any change in its position … The evening of the 15th was cloudy, but on the following evening the nebula was found to occupy sensibly the same position as on the night of discovery. I had now almost given up hope of its proving to be a comet. Cloudy weather prevailed till the evening of the 21st, when … I now noted such a change in position as caused me to remark in my journal that "I was almost persuaded of the cometary nature of the nebula".... On the following evening, the 22nd, I again … found that it had very sensibly changed its position. There could, therefore, now be no doubt as to the cometary nature of the object of my solicitude … (Tebbutt 1908a, pp. 14–16).

The announcement of the discovery of this comet appeared in the local newspapers on 25 May, thus serving as a novel 27th birthday present for the talented young astronomer from Windsor.

After moving into northern skies—beyond the range of Australian telescopes—Comet C/1861 J1 (Tebbutt) became one of the most magnificent celestial visitors of the nineteenth century (Fig. 1.9), and at its prime featured an ever-changing head

**Fig 1.9** Richard Proctor's drawing of the Great Comet of 1861, which was discovered by John Tebbutt (after Weiss 1888)

and a prominent tail that extended for more than 100°. For a time, the comet was even visible in broad daylight. Back in Windsor, Tebbutt had forecast that the Earth would pass through the comet's tail on 30 June, which created considerable public excitement at the time (Macdonnell 1904), and throughout the evening in question he observed '... a peculiar whitish light throughout the sky, but more particularly along the eastern horizon' (Tebbutt 1908a, p. 23). Similar aurora-like phenomena were also reported by observers in the Northern Hemisphere.

In August 1861, Tebbutt thought he had found another comet, but was surprised to discover that it was simply his old comet returning to southern skies after rounding the Sun.

A detailed account of Comet C/1861 J1 (Tebbutt), also known as the Great Comet of 1861, is presented in Chap. 6.

## 1.5 The First Observatory

While the Great Comet of 1861 eventually brought Tebbutt international recognition, it also provided the impetus for his next notable achievements, the acquisition of an astronomical telescope and the establishment of an observatory. The telescope was purchased in Sydney in November 1861 and was a 3.25-in. (8.3-cm) refractor made by Jones of Liverpool. It had a finder and came complete with six eyepieces and an altazimuth mounting. Tebbutt (1908a, p. 25) saw this acquisition as another important stage in his early career.

The first serious observations Tebbutt carried out with this instrument were positional measurements of Comet 109P/Swift-Tuttle in 1862, and he used these to calculate the orbital elements of the comet and reported his results in *Monthly Notices of the Royal Astronomical Society*, *Astronomische Nachrichten* and the *Transactions of the Philosophical Society of New South Wales*. These were the first research papers that he published in professional astronomical journals, and as such they brought him a great deal of satisfaction (Tebbutt 1908a). They were certainly not to be his last!

Tebbutt then pressed on with the construction of an observatory, and this was completed by the end of 1863 (Orchiston 1988c, 2001b). It comprised a simple rectangular wooden building with a central tower for the Jones refractor (see Fig. 1.10). The main observatory room had two transit piers, one at each end of the building. Associated with each pier were wall and ceiling shutters which opened to expose a narrow band of sky. A small prime vertical room was appended to the south side of the building in 1867 and provided with a pier for a portable transit telescope.

In order to determine the latitude of his observatory and establish a local time-service, Tebbutt ordered a 2.1-in. (5.3-cm) transit telescope from the Sydney scientific instrument-maker, Angelo Tornaghi (1831–1906; Maguire 1985), and this arrived in September 1864. He also purchased a box chronometer, and continued to make use of the family's grandfather clock, which kept reliable time. The new

**Fig. 1.10** A woodcut showing Tebbutt's first and second observatories, erected in 1863 and 1874, respectively (Orchiston collection)

transit telescope, which was normally kept in the main observatory building under one of the north-south aligned roof shutters, was used to regulate this clock and the chronometer. The procedure was to record the precise times that selected stars transited the meridian on successive nights. Since the time elapsed from one night to the next was exactly 23 h 56 min 4.09 s, the clock and chronometer could be adjusted accordingly.

When he wanted to make observations for latitude, the transit telescope was temporarily transferred to the pier in the prime vertical room. By recording the declinations of stars which were observed to transit near the zenith, the latitude could be accurately established.

Tornaghi also provided a ring micrometer for the Jones refractor, so that Tebbutt could measure precise distances between adjacent celestial objects. This was most useful in establishing the distance between a comet and neighbouring reference stars (from which the right ascension and declination of the comet could be calculated), or measuring the separation and position angle between the components of a double star system.

Tebbutt completed the fitting out of his observatory by purchasing a full set of meteorological instruments (some of these are shown in Fig. 1.4), so that Windsor Observatory could function as a meteorological station. In this regard, he was simply mimicking the practice in vogue at the time at the government observatories in Melbourne and Sydney, which saw meteorology as part of the 'normal' charter of a professional astronomical observatory (see Orchiston 1988a).

In this book, Tebbutt's purchase of the Jones telescope and the early history of Windsor Observatory are dealt with in Chap. 8.

## 1.6 An Emerging Reputation

Soon after acquiring the 3.25-in. telescope, Tebbutt was offered but refused the post of Government Astronomer of New South Wales (Orchiston 1988d) which fell vacant when the founding Director, British-born and Cambridge-educated William Scott (1825–1917) took early retirement in 1862. This post included the Directorship of Sydney Observatory, one of Australia's leading professional observatories of the day. Tebbutt's credentials for the position are discussed in Chap. 7, along with possible reasons for his refusal.

After installing the Jones refractor and Tornaghi's transit telescope in his brand new Windsor Observatory, Tebbutt at last felt he was ready to do

> "… something respectable for astronomy. At the outset I resolved to devote myself to systematic observations of occultations of stars by the moon, of eclipses of Jupiter's satellites, and of positions of comets, with attention to phenomena of an occasional character. (Tebbutt 1908a, p. 30).

The phenomena of an occasional character which particularly interested him were transits of Mercury and Venus, and solar and lunar eclipses. The only other celestial object that he decided to keep under scrutiny was the unusual variable star, η Argus (now known as η Carinae).

The actual carrying out of the observations was simply part of the story. It was necessary to compute the local occurrence times of certain events (e.g. lunar occultations and eclipses) beforehand, and after most telescopic work came the task of reducing the observations and preparing them for publication. In the case of cometary positions and the subsequent orbital computations, the amount of labour involved was prodigious. Yet until 1881, Tebbutt carried out all of these calculations himself, and thereafter it was only with difficulty that he found computers competent to assist (see Eminent men of New South Wales 1889), until the surveyor, Joseph Brooks, and fellow astronomers, Robert Innes (1861–1933) and C.J. Merfield (1866–1931), came to his aid late in the century. When the well-known New South Wales astronomer William Macdonnell (1842–1910) summarised Tebbutt's lifelong observational achievements, he specifically referred to the onerous nature of these calculations: 'The reductions connected with the work which has been so systematically carried out for the last 40 years involve an immense amount of labour …' (Macdonnell 1904). Just how much Tebbutt managed to accomplish is brought out when we consider Sheepshank's warning: 'It is not enough that a splendid building is filled with magnificent instruments directed by a consummate astronomer; he must have at his disposal numerous subordinate assistants for observation and compilation …' (cited in Dewhirst 1985, p. 147). Far from having numerous subordinate assistants, Tebbutt could not boast a solitary one!

Despite turning down the Sydney Observatory post and retaining his independence, Tebbutt quickly acquired an international reputation as an astronomer and by 1869 had made sufficient impact for his Windsor Observatory to be included in the *Nautical Almanac's* listing of major world observatories. Just four years later, he was elected a Fellow of the Royal Astronomical Society. Upon hearing the good news, he immediately wrote to the Society: 'I cannot conclude without expressing the deep sense of gratitude which I entertain for the honour conferred upon me by my election as a member of so illustrious a body …' (Tebbutt 1873). Tebbutt took his membership very seriously and cherished the opportunity to use 'F.R.A.S.' after his name.

## 1.7 Still More Telescopes and Observatories

Further telescopes were added to the fold over the years and with them new observatory buildings (Orchiston 2001b). In 1872, Tebbutt purchased a fine equatorially-mounted 4.5-in. (11.4-cm) refractor by the leading British manufacturer, Thomas Cooke (1807–1868). Initially, this was housed in the 1863 'tower observatory', but in 1874 he erected a 12-ft (3.7-m) wooden observatory near the original observatory and it was transferred there. This little observatory building is shown on the right in Fig. 1.10.

Five years later (in 1879), Tebbutt constructed a substantial brick building, complete with dome room, transit room, prime vertical room, computing room and a library to the south of the original wooden observatory (see Fig. 1.11). A new 3-in. (7.6-cm) Cooke transit telescope was acquired for the transit room, and the 4.5-in. Cooke refractor was relocated to the new dome.

Finally, in 1886, Tebbutt took possession of an even larger telescope, an 8-in. (20.3-cm) Grubb refractor (Fig. 1.12) that was formerly the property of Dr. William Bone of Castlemaine, Victoria (Orchiston 1987). He now had what many professional astronomers by international standards would regard as minimal equipment (Turner 1918) although it was comparable to the largest refractors found in the Australian colonial observatories of the day. Tebbutt installed this telescope in the vacant small wooden observatory, but this proved rather cramped quarters for so large an instrument and eventually the ravages of the white ants forced the issue and demanded its demolition. In 1894, it was replaced by a more commodious brick observatory. The development of Windsor Observatory and acquisition of the Grubb telescope are discussed in Chaps. 8 and 11, respectively.

Armed with the Grubb refractor, the Cooke transit telescope, and an outstanding research and reference library (see Tebbutt n.d. (a)), Tebbutt was now in a position to compare his Windsor Observatory favourably with the government observatories in Sydney, Melbourne and Adelaide.

From the time the 4.5-in. Cooke refractor was installed at Windsor Observatory Tebbutt was able to carry out a truly formidable variety of observational programmes (see Orchiston 2004b), and these were expanded even further with the

**Fig. 1.11** A woodcut showing the substantial brick observatory, erected in 1879 (Orchiston collection)

**Fig. 1.12** The 8-in. Grubb telescope (after Tebbutt 1908a)

acquisition of the larger Grubb refractor (see Orchiston 1982b). In addition to the old observational stalwarts mentioned earlier, double stars, planets and minor planets also featured prominently. Unlike its smaller Cooke predecessor, the Grubb telescope was furnished with a drive, which made all micrometric observing much easier. In addition, the increased light grasp of the larger refractor allowed Tebbutt to follow comets for longer and thus obtain observations over an extended arc of their orbits. Chapter 11 documents the range of observations that Tebbutt undertook with the Grubb telescope, and how they related to his earlier observational aspirations.

## 1.8 A Succession of Societies

The first scientific group that Tebbutt joined was the Philosophical Society of New South Wales, when the Government Astronomer, William Scott, nominated him for membership in 1861. This elite scientific body was subsequently renamed the Royal Society of New South Wales, and when a *de facto* Astronomy Section was formed in 1876 Tebbutt was one of its strongest supporters. Despite this, the Section lasted a mere 6 years. A detailed account of Section A, as it was known, is presented in Orchiston and Bhathal (1991).

As we have seen, Tebbutt was elected a Fellow of the Royal Astronomical Society in 1873, and he valued his association with the Society. In 1915, he wrote to the Secretaries: ‘I have now been a fellow of the Society for 40 years and I think I have fulfilled the promise which I made when I was elected, namely, to further the interests of the Society to the best of my ability’ (Tebbutt 1915a). For much of his astronomical career, he saw the Royal Astronomical Society as the world’s premier body for professional astronomers.

Tebbutt’s fondness for comets led him to form Australia’s first national astronomical group in 1882. The ‘Australian Comet Corps’, as I prefer to call it, was modelled on a similar group established in the United States (Orchiston 1982a), and the idea was to assign selected areas of the southern sky to different observers who would then be responsible for searching their ‘zones’ for new comets. In principle the idea was a good one, but in reality Australia was not ready for such a group and it survived for just 1 year. The circumstances surrounding its formation and demise make interesting reading and are the subject of Chap. 10.

The Australian Comet Corps proved an anachronism for it was only in the 1890s that Australia gained its first successful astronomical groups, in Adelaide, Melbourne and Sydney (Orchiston 1998). The largest and most dynamic of these was the New South Wales Branch of the British Astronomical Association, founded in 1895 (Orchiston 1988b; Orchiston and Perdrix 1990, 2002), and Tebbutt was elected its first President. This was a clear indication of the esteem in which he was held by most members of the powerful Sydney astronomical community at the time. The early history of the New South Wales Branch and Tebbutt’s role in its development are portrayed in Chap. 13.

In addition to the aforementioned scientific groups and societies, at one time or another Tebbutt was a member of the Astronomical Society of the Pacific, and a Corresponding Member of Alliance Scientifique Universelle (Paris), the Sociedad Cientifica 'Antonio Alzate' (Mexico), and of the Queensland Branch of the Royal Geographical Society of Australasia.

## 1.9 A Commitment to Observational Astronomy

As we shall see in Chaps. 3 and 11, during the 1830s Australia's fine international reputation in astronomy was all but destroyed thanks to Dunlop's exploits at Parramatta Observatory and problems associated with the 'Parramatta Catalogue'. It then fell largely to Tebbutt and those at Williamstown, Melbourne and Sydney Observatories to address this situation, and during the 1850, 1860s and 1870s they gradually rebuilt and strengthened the foundations of Australian astronomy (though, at the time, they certainly did not view their observations in this light). What of Tebbutt's observations in this regard?

From the time he purchased the Jones refractor, Tebbutt carried out astronomical observations every single year until his formal 'retirement' at the end of 1903 at the age of 70, and intermittently thereafter through to 1915, just 1 year before his death (as outlined in Chap. 14). During his busiest years, he observed on a large percentage of clear nights.

Between 1863 and 1903, there was only one conspicuous hiatus when Tebbutt almost abandoned observational astronomy and that was between June 1871 and March 1873 when he was living in Parramatta (near Sydney) and only had access to his observatory on weekends—but not every weekend (Tebbutt 1870–73). This temporary change of domicile came at a most inopportune time being so soon after the death of his father, but it was probably prompted by illness. On 17 March 1871, Henry Russell wrote: 'I hope you will take the Doctors [*sic*] advice and get change of air for a time I am sure it will do you more good than anything he can give you' (Russell 1871a). Less than 3 months latter, Tebbutt was living in Parramatta. Later in the year, he was invited to join the Melbourne-Sydney Observatories' Queensland solar eclipse expedition (Lomb 2016) but declined because he and the family were ill (Russell 1871b; Scott 1871; cf. Raymond 1871). As Tebbutt's health improved so too did his appetite for astronomy and for a larger telescope, and it was while he was still living in Parramatta that he purchased the Cooke refractor.

One is almost overwhelmed by the variety of observational programmes that Tebbutt undertook during his long career as an astronomer (see Table 1.1). Although he had catholic tastes, his main emphases were on cometary, planetary and minor planet positions; Jovian satellite phenomena timings; double star position angle and distance measurements; lunar occultation timings; variable star magnitude estimates; and what he called 'phenomena of an occasional character' (such as eclipses, transits, lunar occultations of planets, planetary conjunctions and occultations of stars by planets). He also participated in international searches for the postulated intra-Mercurial planet, Vulcan (see Baum and Sheehan 1997).

**Table 1.1** Tebbutt's observational astronomy, 1853–1915

| Year | C* | D | LE | LP | LS | J | M | P | S | TM | TV | V |
|---|---|---|---|---|---|---|---|---|---|---|---|---|
| 1853 | × | | | | | | | | | | | |
| 1854 | | | | | | | | | | | | × |
| 1855 | | | | | | | | | | | | |
| 1856 | | | | | | | | | | | | |
| 1857 | | | | | | | | | × | | | |
| 1858 | × | | × | | | | | | | | | |
| 1859 | × | | | | | | | | | | | |
| 1860 | × | | | | | | | | | | | × |
| 1861 | × | | | | | | | | × | | | |
| 1862 | × | | | | | | | | | | | × |
| 1863 | | | | | × | | | | | | | × |
| 1864 | × | | | | × | × | | | | | | × |
| 1865 | × | | | | | | | | | | | × |
| 1866 | | | × | | × | × | | | | | | × |
| 1867 | | | × | | × | × | | | | | | × |
| 1868 | | | | | × | × | | | × | | | × |
| 1869 | | | × | | × | × | | × | | | | × |
| 1870 | | | × | | × | × | | | × | | | × |
| 1871 | | | | | | | | | × | | | × |
| 1872 | | | | | | | | | | | | × |
| 1873 | | | × | | × | × | | | | | | × |
| 1874 | × | | × | | × | × | | | | | × | × |
| 1875 | × | | | × | × | × | | | | | | × |
| 1876 | | | | | × | × | | | × | | | × |
| 1877 | | | × | | × | × | × | | | | | × |
| 1878 | × | | | | × | × | | × | × | × | | × |
| 1879 | × | | × | | | × | × | × | | | | × |
| 1880 | × | × | × | | × | × | × | × | | | | × |
| 1881 | × | × | × | | × | × | | × | | × | | × |
| 1882 | × | × | | | × | × | × | × | × | | | × |
| 1883 | × | | | | | | | | | | | × |
| 1884 | × | × | | | | | | | | | | × |
| 1885 | × | × | × | | × | × | | | | | | × |
| 1886 | × | × | | × | × | × | | × | | | | × |
| 1887 | × | × | × | | × | × | × | × | × | | | × |
| 1888 | × | × | | × | × | × | × | × | | | | × |
| 1889 | × | × | | | × | × | × | | | | | × |
| 1890 | × | × | | | × | × | × | | × | | | × |
| 1891 | × | × | × | | × | × | × | × | | × | | × |
| 1892 | × | × | | | × | × | | × | | | | × |
| 1893 | × | × | | × | × | | × | × | | | | × |
| 1894 | × | × | | × | × | | | × | | × | | × |

(continued)

**Table 1.1** (continued)

| Year | C* | D | LE | LP | LS | J | M | P | S | TM | TV | V |
|---|---|---|---|---|---|---|---|---|---|---|---|---|
| 1895 | | × | | | × | × | × | × | | | | × |
| 1896 | | × | | | × | × | × | | | | | × |
| 1897 | × | | | | × | × | × | × | | | | × |
| 1898 | × | | | | × | × | × | × | | | | × |
| 1899 | × | × | | | × | × | × | × | | | | |
| 1900 | | × | | × | × | × | × | × | | | | |
| 1901 | × | × | | | × | | × | × | | | | |
| 1902 | × | × | | | × | × | × | | | | | |
| 1903 | × | | | | | | × | | | | | |
| 1904 | × | | | | | | × | × | | | | |
| 1905 | | × | | | | | | × | | | | |
| 1906 | | × | | × | | | | | | | | |
| 1907 | | × | | | | | | | | | | |
| 1908 | | | | | × | × | | | | | | |
| 1909 | × | × | | | | | | | | | | |
| 1910 | × | | | | | | | | | | | |
| 1911 | | × | | | | | | | | | | |
| 1912 | × | | | | | | | | | | | |
| 1913 | | × | | | | | | | | | | |
| 1914 | | | | | | | | | | | | |
| 1915 | | × | | | | | | | | | | |

**Key*: *C* comets, *D* double stars, *LE* lunar eclipses, *LP* lunar occultations of planets, *LS* lunar occultations of stars, *J* Jovian satellite phenomena, *M* minor planets, *P* planetary positions, *S* solar eclipses, *TM* transits of Mercury, *TV* transits of Venus, *V* variable stars

These were precisely the types of programmes that astronomers tended to focus on during the second half of the nineteenth century although some were beginning to take seriously the research opportunities offered by astronomical photography and spectroscopy (see Clerke 1893). Tebbutt was aware of these international developments, for over the years he acquired a number of books and received many reprints about the so-called new astronomy, astrophysics (e.g. see the library listings in Tebbutt 1887 and his successive *Annual Reports*—Tebbutt 1889, 1890, 1891c, 1892, 1893, 1894, 1895, 1896, 1897, 1898a, 1899, 1900, 1901, 1902, 1903b, 1904), but it did not tempt him. In fact, he was somewhat dismissive when the whole question of the research potential of astronomical photography was raised (e.g. see Elkington 1901). Instead, Tebbutt was seen by those who outlived him as '… one of the few remaining links that connect the astronomy of to-day with the older form that Airy and his school recognized and practised. The late Mr.. Tebbutt, as a loyal member of that school, worked hard to record positions, [and] to deduce orbits …' (John Tebbutt 1917).

As this obituary suggests, of all celestial objects, comets were undoubtedly Tebbutt's favourite (see Proctor 1926, 1937), and there were two separate facets to

**Table 1.2** Numbers of comets observed by Tebbutt 1853–1912

| Year | No. | Year | No. | Year | No. | Year | No. | Year | No. |
|---|---|---|---|---|---|---|---|---|---|
| 1853 | 2 | 1874 | 1 | 1884 | 3 | 1892 | 4> | 1903 | 2 |
| 1858 | 1> | 1875 | 1 | 1885 | 1 | 1893 | 2 | 1904 | 1 |
| 1859 | 1 | 1878 | 1 | 1886 | 5 | 1894 | 3 | 1909 | 1> |
| 1860 | 1 | 1879 | 1 | 1887 | 1 | 1897 | 1 | 1910 | 1 |
| 1861 | 1 | 1880 | 1 | 1888 | 4> | 1898 | 2> | 1912 | 1 |
| 1862 | 2 | 1881 | 2 | 1889 | 3 | 1899 | 3 | | |
| 1864 | 1 | 1882 | 3> | 1890 | 2 | 1901 | 1 | | |
| 1865 | 2 | 1883 | 1 | 1891 | 2 | 1902 | 1> | | |

*Note*: Annual totals are based on data in Tebbutt (1889, 1890, 1891c, 1892, 1893, 1894, 1895, 1896, 1897, 1898a, 1899, 1900, 1901, 1902, 1903, 1904, 1908a, 1911, 1912a, 1913). Where a '>' occurs this indicates that a comet was observed through into the following year

this work that we must consider. Firstly, he knew that by using his micrometers to precisely determine the positions of known comets, he and others could use these measurements to calculate the orbital elements and thereby contribute significantly to cometary astronomy. Secondly, he quickly discovered that systematic comet-searching could be rewarding, for it not only produced new targets for those committed to cometary science but it also brought the discoverer considerable kudos—and kudos *was* important to Tebbutt.

In 1914, Melbourne Observatory Director Pietro Baracchi (1851–1926) described Tebbutt's cometary work as 'remarkable', and this is an apt evaluation. Between 1853 and 1912 he followed 59 different periodic and non-periodic comets (see Table 1.2), and in addition he searched unsuccessfully for known comets in 1889, 1890, 1892, 1894, 1896, 1898, 1900 and 1901. In some instances, he had been provided with incorrect positional data; other comets were simply too faint to be detected, or were inconveniently located in the sky.

Once he was furnished with proper instruments, Tebbutt's usual procedure was to obtain micrometric positions of each comet from the time it first became visible at Windsor or news of its discovery was received, up until it disappeared from southern skies or became too faint to detect any longer. Many comets were observed on every possible clear night and followed for extended intervals. Tebbutt became so proficient in cometary observation that this sphere of astronomy was sometimes left entirely to him by the Australian government observatories. For example, on 22 November 1892 Melbourne Observatory Director Robert Ellery (1827–1908) wrote Tebbutt: 'The Andromeda Comet [17P/Holmes] is just within our reach but we are not observing it as we know you are looking after it' (Ellery 1892b). This was by no means an isolated example (e.g. see Ellery 1892a).

The first comet that Tebbutt observed micrometrically was periodic comet 109P/Swift-Tuttle in 1862, and the longest series of micrometric observations he obtained for any one comet was of Comet C/1898 L1 (Coddington-Pauly) on 103 nights in 1898–1899 (Tebbutt 1900). But the comet which he observed most frequently was 2P/Encke, which is distinguished by having the shortest orbital period of any known comet (cometary designations used here are after Marsden and Williams 1996). As a result, with the passing of the years this became an all too familiar visitor, and Tebbutt observed it on eight different occasions: in 1862, 1865, 1875, 1878, 1888, 1894, 1898 and 1904. Tebbutt was the first to observe the comet during its 1878 and 1888 apparitions and is credited with the recovery on each occasion (Vsekhsvyatskii 1964). In 1898, he shared this honour with New Zealand's leading independent astronomer, John Grigg (1838–1920; see Orchiston 1993).

In the case of 12 of the comets that he observed, Tebbutt used his own positional data to compute the orbital elements. This was a considerable accomplishment, and in 1902 one of his Sydney colleagues, the Reverend Thomas Roseby (1844–1918) wrote:

> … the place of highest honour in Cometary Computation in Australia belongs to Mr. Tebbutt. For nearly forty years he must have had that field entirely to himself. From intimate acquaintance with Mr. Tebbutt and his work for the past fourteen years & more, I can say I hardly know whether to admire more his universally recognised accuracy & refinement of his observational work, or the mathematical skill … of his work—during all these years—in the field of Cometary calculation. (Roseby 1902).

In addition to following known comets, Tebbutt searched for new ones. In 1881, he detected his second comet almost 20 years to the day after his 1861 discovery, and this also turned out to be a 'Great Comet' (Orchiston 1981, 1999). Because of rapid developments in astronomical photography and spectroscopy at the time, Comet C/1881 K1 (Tebbutt) made a major contribution to cometary astronomy, as outlined in Chap. 9. Later in 1881, Tebbutt 'discovered' a second comet (C/1881 N1), only to learn that it had been found earlier by a northern hemisphere observer.

Tebbutt also played a leading role in securing observations of comets discovered by other Australian astronomers. This first occurred with the Great Comet of 1865 (C/1865 B1), which was discovered by Francis Abbott (1799–1883) of Hobart although it is not generally credited to him (see Orchiston 1997c). After this, Tebbutt provided 'support observations' of Comet C/1884 A1 (Ross), Comet C/1889 O1 (Davidson), Comet C/1894 G1 (Gale) and finally Comet C/1912 R1 (Gale). In the case of this last comet, Gale (1912) specially asked Tebbutt to confirm the discovery—which he did, at the grand old age of 78—and this event and his subsequent observations of this comet were to feature in his last cometary publications (Tebbutt 1912a, 1913).

Further details of Tebbutt's achievements in cometary astronomy are provided in Chaps. 6–11 and 13–14.

The second most important area of observational astronomy to which Tebbutt contributed was variable stars, even though he only made extensive observations of two different stars. One of these was η Argus (or η Carinae as it is now known) and the other R Carinae. Because a larger aperture was never called for, Tebbutt made all the telescopic observations with his Jones refractor.

Eta Carinae is an extremely unusual variable star and is of great interest to astrophysicists (see Morse et al. 1999). It is one of the most massive and luminous stars in our Galaxy and is the stand-out example of a small class of variable stars known as 'luminous blue variables' (LBVs). The precise nature of LBVs is still being debated (e.g. see Humphreys and Davidson 1994; van Genderen et al. 1995). Tebbutt observed η Carinae from 1854 until 1898 and contributed significantly to our knowledge of its unique and unpredictable light curve (but at that time there was little interest in understanding the dynamics of the star itself).

Tebbutt also observed the variable star R Carinae, which is a common Mira-type variable, and on the basis of his own observations between 1880 and 1898 and magnitude estimates obtained by others he was able to derive a period for the light variability.

While Tebbutt's observations of these two variable stars were valuable (White 1986), perhaps his most amazing achievement was the discovery of Nova V728 Scorpii 1862 (Tebbutt 1877a). What is remarkable is that he made the discovery quite by accident in 1877 while re-analysing some of his own 1862 observations of Comet 109P/Swift-Tuttle (see Ashbrook 1984; White 1979). Despite the strange circumstances, there seemed little doubt that Tebbutt did unwittingly observe a nova, and after considering the evidence Oxford University's Professor H.H. Turner (1861–1930) concluded that Tebbutt was too good an observer to have made a mistake (Turner 1918).

Tebbutt also carried out occasional observations of a few other variable stars, and during the 1870s investigated a number of stars whose variability was being questioned at the time. All of these are discussed in Chap. 11, along with his observations of η Argus and R Carinae, while further details of Nova V728 Scorpii 1862 appear in Chap. 8.

Double stars constitute a third class of astronomical objects for which Tebbutt made a useful contribution. Various astronomers combined his numerous micrometric measures of position angle and separation for selected double stars with comparable data obtained at earlier times by others to derive the orbital parameters of some of these stars.

Tebbutt carried out his first double star observations in 1880 and continued to observe double stars through to 1915, just 1 year before his death (see Table 1.3), making a great many observations in the years 1900, 1901 and 1902 in particular. In

**Table 1.3** Numbers of different double stars observed by Tebbutt, 1880–1915

| Year | No. | Year | No. | Year | No. | Year | No. | Year | No. |
|---|---|---|---|---|---|---|---|---|---|
| 1880 | 6 | 1888 | 28 | 1896 | 8 | 1904 | 0 | 1912 | 0 |
| 1881 | 12 | 1889 | 3 | 1897 | 0 | 1905 | 5 | 1913 | 2 |
| 1882 | 16 | 1890 | 7 | 1898 | 0 | 1906 | 1 | 1914 | 0 |
| 1883 | 0 | 1891 | 8 | 1899 | 2 | 1907 | 13 | 1915 | 1 |
| 1884 | 1 | 1892 | 14 | 1900 | 26 | 1908 | 0 | | |
| 1885 | 18 | 1893 | 1 | 1901 | 28 | 1909 | 7 | | |
| 1886 | 1 | 1894 | 11 | 1902 | 54 | 1910 | 0 | | |
| 1887 | 29 | 1895 | 17 | 1903 | 0 | 1911 | 5 | | |

*Note*: Annual totals are based on data in Tebbutt (1885, 1889, 1890, 1891c, 1892, 1893, 1894, 1895, 1896, 1897, 1898a, 1899, 1900, 1901, 1902, 1903, 1904, 1906a, 1908b, 1910, 1912b, 1915b)

**Table 1.4** Numbers of minor planets observed by Tebbutt, 1879–1904

| Year | No. | Year | No. | Year | No. | Year | No. | Year | No. |
|---|---|---|---|---|---|---|---|---|---|
| 1879 | 1 | 1885 | 2 | 1891 | 1 | 1897 | 4 | 1903 | 3 |
| 1880 | 1 | 1886 | 0 | 1892 | 0 | 1898 | 3 | 1904 | 1 |
| 1881 | 0 | 1887 | 2 | 1893 | 2 | 1899 | 3 | | |
| 1882 | 1 | 1888 | 1 | 1894 | 0 | 1900 | 3 | | |
| 1883 | 0 | 1889 | 3 | 1895 | 2 | 1901 | 4 | | |
| 1884 | 0 | 1890 | 4 | 1896 | 8 | 1902 | 5 | | |

*Note*: Annual totals are based on data in Tebbutt (1889, 1890, 1891c, 1892, 1893, 1894, 1895, 1896, 1897, 1898a, 1899, 1900, 1901, 1902, 1903b, 1904, 1908a)

all, he observed 133 different double stars, his most popular targets being α Centauri, p Eridani, γ Coronae Australis, γ Centauri, β Muscae and Lacaille 2145.

All of the stars he observed were known doubles. Unlike his Sydney Observatory colleagues, Henry Russell, James Arthur Pollock (1865–1922) and Richard Pickering Sellors, he did not carry out systematic searches for new stars (see Haynes et al. 1996). Further details of Tebbutt's double star work are provided in Chap. 11.

In a similar way, Tebbutt was able to contribute to minor planet astronomy, for his published micrometric positions were useful to those astronomers interested in refining the orbital parameters of known minor planets.

His first minor planet observations were of 4 Vesta in 1877 using the 4.5-in. Cooke telescope, but these observations became much more numerous after he acquired the larger telescope in 1886. Between 1877 and 1904, he at one time or another tracked 23 different minor planets (Table 1.4) and used his micrometers to record their positions relative to selected reference stars. His favourite targets were 1 Ceres, 2 Pallas, 3 Juno, 4 Vesta, 6 Hebe and 11 Parthenope.

Tebbutt never engaged in any systematic searches for new minor planets, as some astronomers were prone to do. His minor planet work is discussed further in Chap. 11.

A fifth area in which Tebbutt contributed was planetary positional astronomy. Between 1880 and 1905, he carried out micrometric observations of three different types of phenomena: mutual conjunctions of planets; conjunctions of planets and accurately-catalogued stars; and occultations of stars by planets. By far the most commonly observed were conjunctions of planets and stars, and one or more of these was observed almost every year from 1886 to 1901, inclusive.

These conjunctions and occultations had the potential to reveal planetary positions with considerable accuracy, and this was particularly important in the case of the outer planets, Uranus and Neptune, whose orbital characteristics still posed problems. However, Tebbutt did not discriminate: he observed phenomena associated with all of the known planets, but most common were stellar conjunctions involving Jupiter and Uranus. Further details of Tebbutt's planetary astronomy programme are provided in Chap. 11, while his participation in international Vulcan searches is discussed in Chap. 8.

Of the various 'phenomena of an occasional character' which interested Tebbutt, his observations of solar and lunar eclipses and transits of Mercury and Venus were of greatest significance.

Transits of Venus are rare events. Two were scheduled for the nineteenth century, in 1874 and 1882, and both were potentially visible from Australia (see Proctor 1874). Tebbutt realised that his observations, when combined with others, could be used to investigate that fundamental Solar System 'yardstick', the Astronomical Unit (AU), the distance from the Earth to the Sun. During the nineteenth century, accurate determination of the AU was one of the challenges of world astronomy (see Clerke 1893; Dick et al. 1998). As it turned out, Tebbutt successfully observed the 1874 event (see Tebbutt 1875a, 1883), and his contact times formed part of the data set used by Captain George Lyon Tupman (1838–1922) to derive a solar parallax of 8.8455 seconds. (Tupman 1878). In marked contrast, on the vital day in 1882 the sky was '… completely overcast with dense dark cloud, through which there was not the slightest prospect of obtaining a glimpse of the Sun' (Tebbutt 1882a).

Somewhat less spectacular in appearance and of lesser importance were transits of Mercury, and Tebbutt observed four of these between 1878 and 1894, paying particular attention to possible evidence of a planetary atmosphere or Mercurian satellite. Details of his transit of Venus and transit of Mercury programmes are presented in Chaps. 8 and 11, respectively.

Solar and lunar eclipses were visually-alluring events, but they also allowed checks to be made on the Moon's orbit. Probably both factors explain Tebbutt's interest in these events, and as we have seen, he first attempted to observe a solar eclipse—in this case a total eclipse of the Sun—in 1857, but was thwarted by clouds. Although this was the only total solar eclipse potentially visible from Windsor during his lifetime, he did succeed in observing 9 different partial solar eclipses between 1857 and 1891, and 15 different lunar eclipses. All of his eclipse observations are discussed in Chap. 11.

Yet another area of astronomy in which Tebbutt contributed was in the observation of Jovian satellite phenomena. Four different types of events were involved: eclipses, transits, shadow transits and occultations. An eclipse occurs when the satellite passes into the shadow of Jupiter cast by the Sun; a transit occurs when the satellite is seen crossing the Jovian disk; a shadow transit occurs when the shadow of the satellite is seen crossing the Jovian disk; and an occultation occurs when the satellite is occulted by Jupiter, as viewed from the Earth.

Tebbutt began observing Jovian satellite phenomena in earnest in 1866, and continued to contribute observations through to 1902 (see Table 1.5). Before the acquisition of the Grubb telescope he tended to confine his observations to eclipses, but thereafter he added occultation and transits to his repertoire. He knew that all three types of observations could be used by astronomers to investigate the orbital dynamics of the different satellites.

While Tebbutt generally confined his Jovian studies to the satellites, from time to time he was seduced into observing and recording some of the distinctive markings

**Table 1.5** Numbers of Jovian satellite phases observed by Tebbutt, 1866–1902

| Year | No. | Year | No. | Year | No. | Year | No. | Year | No. |
|---|---|---|---|---|---|---|---|---|---|
| 1866 | 18 | 1874 | 11 | 1880 | 14 | 1888 | 75 | 1896 | 43 |
| 1867 | 21 | 1875 | 14 | 1881 | 3 | 1889 | 26 | 1897 | 32 |
| 1868 | 5 | 1876 | 18 | 1882 | 4 | 1890 | 41 | 1898 | 11 |
| 1869 | 2 | 1877 | 8 | 1885 | 15 | 1891 | 31 | 1899 | 23 |
| 1870 | 3 | 1878 | 27 | 1886 | 27 | 1892 | 14 | 1900 | 8 |
| 1873 | 7 | 1879 | 19 | 1887 | 46 | 1895 | 24 | 1902 | 23 |

*Note*: Annual totals are based on data in Tebbutt (1889, 1890, 1891c, 1892, 1893, 1894, 1895, 1896, 1897, 1898a, 1899, 1900, 1901, 1902, 1903b, 1904, 1908a)

on the planet's disk. More information on these observations and on his Jovian satellite work is given in Chap. 11.

One other area of observational astronomy to which Tebbutt was particularly committed was the timing of lunar occultations of stars. Such observations were carried out in 32 different years during the 40-year period, 1864–1903 (inclusive). Tebbutt was aware that his observations could be used to provide a valuable check on anomalies in the Moon's orbit, and that his copious published occultation data would be utilised to establish the longitude of his observatory. In 1901, Elkington pointed out that through these lunar occultation observations '… the position of Windsor [Observatory] has been more accurately determined than that of any other place in New South Wales' (Elkington 1901, p. 42). As a result of his painstaking lunar occultation work over many years, Tebbutt succeeded in making his Observatory one of Australia's fundamental geodetic reference points.

In addition to lunar occultations of stars, on eight different occasions between 1875 and 1906 Tebbutt was fortunate enough to witness a much more tantalising phenomenon, the occultation of a planet by the Moon. All of Tebbutt's lunar occultation observations are discussed in Chap. 11.

To Tebbutt, observing was a means to an end: to provide colleagues with useful data. Publication therefore was the obvious climax of any viable observational project. In this regard, Tebbutt was particularly productive, as Table 1.6 illustrates.

In this table, the dominance of Tebbutt's cometary work stands out. It is also significant that those programmes ranking 2 and 4 (Jovian satellite phenomena and lunar occultations of stars) were less important in terms of internationally-significant research research results than many of the entries below them in the table. On this score, transits of Venus have the lowest ranking of all, yet Tebbutt made a particularly valuable contribution in 1874.

In conclusion, Tebbutt's overall observational output can best be summed up by quoting an anonymous writer from 1889: 'Mr. Tebbutt shows us an example of unflagging zeal, untiring patience and wonderful industry linked with the highest abilities …' (Eminent men of New South Wales 1889).

**Table 1.6** Number of published papers by Tebbutt dealing with different areas of astronomy

| Area of astronomy | Papers[a] |
|---|---|
| Comets | 162 |
| Jovian satellite phenomena | 33 |
| Planets | 33 |
| Occultations of stars | 32 |
| Variable stars | 30 |
| Double stars | 29 |
| Minor planets | 27 |
| Solar eclipses | 9 |
| Lunar eclipses | 9 |
| Occultations of planets | 8 |
| Latitude/longitude | 8 |
| Transits Mercury | 7 |
| Transits Venus | 5 |
| 'Other' | 14 |
| Total | 406 |

Note that the 'Papers' column does not give the total number of *individual* papers that Tebbutt published, as some papers dealt with more than one topic.
[a]Based mainly on entries in Tebbutt (1908a, pp. 118–132) and papers published subsequently

## 1.10 Astronomy and Beyond

The previous section deals with what Tebbutt referred to as 'Extra-meridian Work', but he also conducted regular transit telescope observations for time-keeping purposes. By observing the meridian transits of selected stars—known as 'clock stars'—he was able to regulate his clock and chronometers and thereby maintain a local time-service for his own use and for the citizens of and visitors to Windsor.

While Tebbutt channelled most of his astronomical efforts into observational programmes, he also showed an interest in history and was quick to realise that New South Wales astronomy in the second half of the nineteenth century rested on firm historical foundations.

In 1866, he was asked to prepare a paper on the history and current status of New South Wales astronomy for the Paris International Exhibition of 1867 (Murray 1866a, b), and in it he summarised the pioneering work carried out at Parramatta Observatory during the 1820s and 1830s, discussed the founding of Sydney Observatory in 1858 and its early history, and reviewed his own contribution to Australian astronomy (Tebbutt 1867). This important 10-page paper constituted the first review ever written on the history of New South Wales astronomy and météorology (see Tebbutt 1890), and remains a valuable record of early astronomical achievement in Australia. Because of its local relevance, Tebbutt was asked to revise and expand this paper, and it was published by the New South Wales Government in 1871 (Tebbutt 1871). This paper had special meaning for the colourful British astronomer, Charles Piazzi Smyth (1819–1900; Bruck and Bruck 1988), who became familiar with the southern skies during the 10 years that he spent at the Royal Observatory, Cape of Good Hope (Warner 1983):

> Your history of Astronomy in New South Wales, was quaint, curious & original; & full of new & to-the-purpose remarks to me who had known something of either the men, or the appointments, of Brisbane, Rumker, Dunlop, Scott & Smalley—when living and working in the same Southern latitude, but in a very different Longitude. (Smyth 1880).

In his twilight years, Tebbutt returned to this historical theme and prepared short papers on Sir Thomas Brisbane (Tebbutt 1915c), and William Scott, the founding Director of Sydney Observatory (Tebbutt 1905).

Tebbutt also was interested in the relationship between astronomy and religion (see Chalmers n.d., which was in his library), and even published pamphlets on this topic (e.g. Tebbutt 1878) based upon lectures that he presented. One of these is shown in Fig. 1.13. Tebbutt also managed to introduce some basic astronomy in pamphlets he penned on religion:

> Man, by the aid of science, is at least in a fair way of determining with accuracy the physical constitution of bodies so remote as the sun and stars. He has long since been able to trace out, by the most subtle mathematical methods, the influence and motions of bodies far

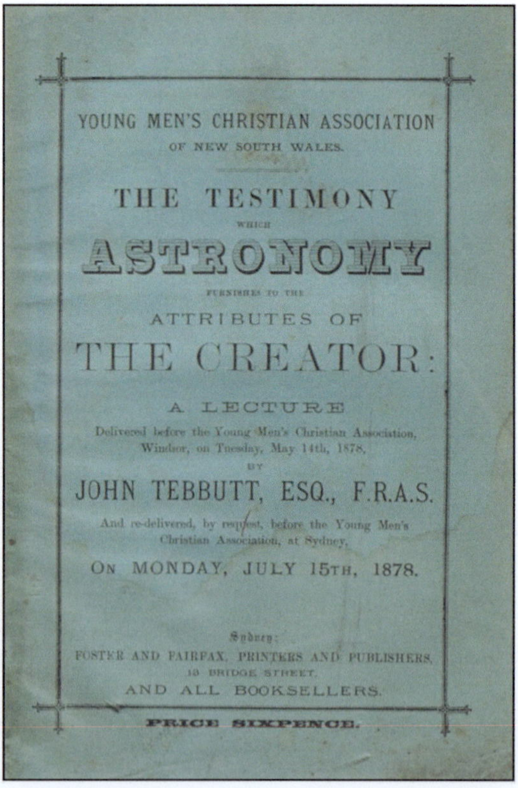


**Fig. 1.13** One of the booklets on the religion–astronomy interface published by Tebbutt (Orchiston collection)

> beyond the reach not only of the unassisted eye, but also of the most powerful telescope. (Tebbutt 1877c, p. 7).

Initially, Tebbutt was closely associated with St. Matthew's Church in Windsor (Steele 1916), but with the passing of the years he and formal religion grew apart (see Chap. 3 in Bhathal 1993). Despite this, St. Matthew's Church (Fig. 1.14) was to play a key role at the very end of his life as the venue for his funeral service and burial (and this is outlined in Chap. 14).

In addition to time-keeping and extra-meridian astronomy, Tebbutt's scientific interests embraced meteorology and tidal studies, which were also legitimate domains of the Australian government observatories during the nineteenth century (see Orchiston 1988a). This catholic perspective tends to reinforce Ashbrook's (1984, p. 66) perception that Tebbutt was trying to run a one-man Greenwich Observatory in the Southern Hemisphere.

As we have seen, in 1862–1863 Tebbutt set up a full meteorological station as part of his new Windsor Observatory. He issued his first monthly 'weather report' at the end of January 1863, and from that date he continued to provide reports to Sydney Observatory and to Windsor district and Sydney newspapers through until 1898, when a curtailed meteorological programme was adopted (and continued up to the time of his death in 1916). In addition to his numerous newspaper articles on meteorology, and responses to public inquirers, Tebbutt took his meteorological data to a pan-Australian and overseas audience by publishing his results in eight monographs (Tebbutt 1868, 1874, 1877b, 1882b, 1886, 1891d, 1898b, 1916) and distributing these widely. Tebbutt's meteorological station and his instruments are described in Chap. 8, and his important contribution to Australian meteorology is summarised in Chap. 14.

**Fig. 1.14** St. Matthew's Church was built in 1822 (*Photograph* Wayne Orchiston)

**Fig 1.15** A flood on South Creek, looking from Windsor down the main road to Sydney. The fully-submerged Fitzroy Bridge over South Creek (near Tebbutt's Peninsula Estate) is located between the two stands of willow trees (after Steele 1916)

As part of his 'weather-watch' and as a concerned farmer occupying low-lying land partially surrounded by a flood-prone water course, Tebbutt kept a healthy watch over South Creek and included information on floods (see Fig. 1.15) and freshes (minor floods) in his meteorological monographs. Extending from 1857 to 1915, this is an invaluable database on Windsor flooding.

As a natural spin-off from his study of floods and freshes, Tebbutt constructed a tide gauge on South Creek in 1889 and for a year monitored its level. Further details of Tebbutt's tide gauge, his papers on tides, and his study of floods and freshes on South Creek are presented in Chap. 8.

Finally, Tebbutt's interests went far beyond the physical sciences if we may judge from the wide range of topics traversed in his 'Letters to the Editor' of Sydney's daily newspapers. They included boxing (which he was vehemently opposed to), colonial history, daylight saving, religion, the economy and even prayers for rain. He also was vitally interested in politics (Elkington 1901).

## 1.11 Academic Publications Aplenty

During the course of his astronomical career, Tebbutt published 388 different research notes and research papers (Table 1.7). Most of these were short and descriptive and simply recorded raw data, but in the case of his cometary work, Tebbutt sometimes included orbital elements deduced from his own observations. In all, he

**Table 1.7** Tebbutt's publications, 1862–1915

| Journal | AJo | ANa | ARe | ASP | BAA | NSW | RMN | RMe | Obs | SSR | Total |
|---|---|---|---|---|---|---|---|---|---|---|---|
| Totals | 1 | 149 | 9 | 6 | 20 | 27 | 125 | 1 | 47 | 3 | 388 |

*Notes*: The various journal totals are based on the main entries in Tebbutt (1908a, pp. 118–132) and papers published subsequently

*Journal Key*: AJo *Astronomical Journal,* ANa *Astronomische Nachrichten,* ARe *Astronomical Register,* ASP *Publications of the Astronomical Society of the Pacific,* BAA *Journal of the British Astronomical Association,* NSW *Transactions of the Philosophical Society of New South Wales* plus its successor the *Journal and Proceedings of the Royal Society of New South Wales,* RMN *Monthly Notices of the Royal Astronomical Society,* RMe *Memoirs of the Royal Astronomical Society,* Obs *The Observatory,* SSR *Southern Science Record* [published in Melbourne]

published elements for 12 different comets, but some of these were reported in the local newspapers rather than in scientific journals.

As Table 1.7 indicates, these papers appeared in three local and eight international journals. All of the overseas journals were reputable ones, *Astronomische Nachrichten* and *Monthly Notices* being the leading publication outlets at that time for professional astronomers (though some would argue that this was to change at the end of the century when *Astronomische Nachrichten* was eclipsed by the *Astronomical Journal*, the *Astrophysical Journal* and *Publications of the Astronomical Society of the Pacific*).

In Australia, Tebbutt had to resort to the publications of the local Royal Society and a general scientific journal, the *Southern Science Record*, given the absence of any specialist astronomical journal at that time. In 1883, Dr. William Bone of Castlemaine suggested to Tebbutt that he should rectify this by launching an 'Amateur Astro. Journal' specifically for Australian astronomers, but after carefully considering the amount of time and effort such a venture would entail and the limited audience such a journal would service he decided not to take up the idea (see Orchiston 1990).

While the figures in Table 1.7 are certainly impressive, they do not tell the whole story, for as was the custom in the nineteenth century science, Tebbutt published some of his papers in duplicate (or occasionally even in triplicate) in two (or three) different journals. For example, of the 125 papers which appeared in *Monthly Notices of the Royal Astronomical Society*, 39 of them (31 %) also appeared in *Astronomische Nachrichten*. Likewise, about 25 % of all papers published in *The Observatory* also appeared in *Astronomische Nachrichten*, and some papers published locally by the Philosophical Society (later the Royal Society) of New South Wales also suffered this fate. However, Tebbutt rarely sought duplicate publication venues after 1885.

After identifying all such duplications and triplications, we can determine the total number of *individual papers and research notes* that Tebbutt wrote. His lifetime tally was 323, which is still a prodigious output and one that many a present-day astronomer would be proud of.

But even this does not signify his total output, for over the years Tebbutt also published a number of astronomy-related booklets at his own expense, and these

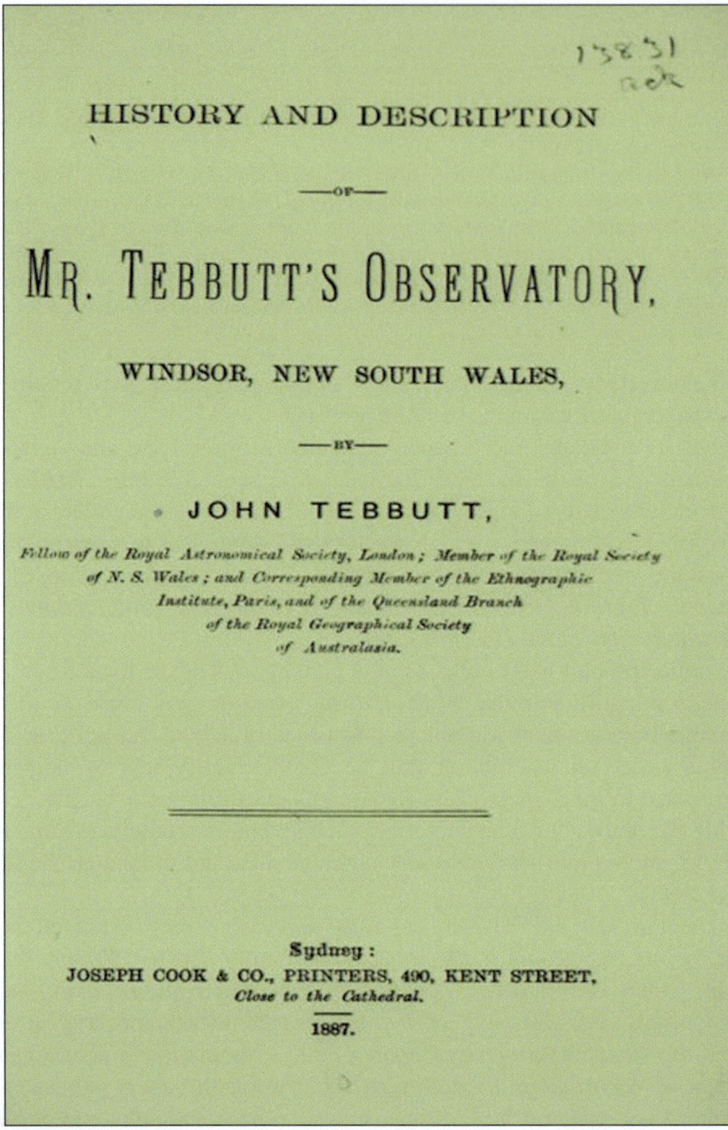

HISTORY AND DESCRIPTION

—OF—

MR. TEBBUTT'S OBSERVATORY,

WINDSOR, NEW SOUTH WALES,

—BY—

JOHN TEBBUTT,

*Fellow of the Royal Astronomical Society, London; Member of the Royal Society of N. S. Wales; and Corresponding Member of the Ethnographic Institute, Paris, and of the Queensland Branch of the Royal Geographical Society of Australasia.*

Sydney:
JOSEPH COOK & CO., PRINTERS, 490, KENT STREET,
*Close to the Cathedral.*

1887.

**Fig. 1.16** The front cover of *History and Description. Mr. Tebbutt's Observatory, Windsor, New South Wales,* which John Tebbutt published at his own expense in 1887 (Orchiston collection)

were widely distributed to Australian and overseas institutions, scientific colleagues and others. The first of these private publications to appear was a 74-page soft-cover booklet titled *History and Description of Mr. Tebbutt's Observatory*, *Windsor*, *New South Wales* (Tebbutt 1887; see Fig. 1.16), and this was followed from 1888 to 1903 (inclusive) by soft-cover *Annual Reports* summarising the instruments at and the astronomical activities of Windsor Observatory (see Tebbutt 1889, 1890, 1891c, 1892, 1893, 1894, 1895, 1896, 1897, 1898a, 1899, 1900, 1901, 1902, 1903b, 1904). This was an ambitious undertaking, but reflected Tebbutt's professional approach to astronomy and his realisation that the wide circulation of such publications was surely an effective way of promoting his work and his Observatory. To some extent, this was also the motivation behind the preparation of his *Astronomical Memoirs*, which appeared in 1908 as a 132-page soft-cover book (Tebbutt 1908a). With the passage of time, this became a collectors' item, and it was reprinted, with additional supporting material, in 1987 (although the publication date is listed as 1986—see Tebbutt 1986). The cover (Fig. 1.17) features a colourful image by Dr. David Malin of a well-known spiral galaxy, and is a far cry from the plain nondescript cover of the original edition.

In addition to these 18 astronomical books and booklets, Tebbutt also wrote a number of booklets on the astronomy–religion interface (e.g. Fig. 1.13), published the 8 meteorological monographs referred to earlier and produced astronomy chapters for 2 different books (in 1867 and 1871). When all of these published works are considered alongside his very numerous newspaper articles, we are looking at a phenomenal output—especially for a solitary self-taught astronomer—and '... a monument to Mr. Tebbutt's untiring energy and unremitted diligence' (Eminent men of New South Wales 1889).

## 1.12 Promoting Popular Interest in Astronomy

As well as research, Tebbutt also was committed to the popularisation of astronomy (see Orchiston 1991, 1997a), and through his efforts much was done to stimulate the growth of the science in New South Wales (John Tebbutt 1917).

One of the ways he achieved this was to submit 'Letters to the Editor', reports and short articles to the Sydney and Hawkesbury district newspapers (e.g. see Table 1.8) in the hope of creating a heightened public interest in astronomy (The Windsor (N.S.W.) Observatory 1889). He wrote his first newspaper report, on sunspots, in 1854 (Tebbutt 1854), and by 1889 they exceeded 250 (Eminent men of New South Wales 1889). These contributions continued to appear almost up to the time of his death (see 'Many Happy Returns' ... 1915), and Tebbutt was able to bring the joy and excitement of astronomy to a great many people (e.g. see his accounts of the Great Comet of 1861 in Chap. 6). But these newspaper contributions were not restricted to astronomy, for he also wrote about the weather and local flooding (e.g. see Fig. 1.18).

In this way, Tebbutt had a profound impact on the general public and enhanced their familiarity with and tolerance of astronomy. At times, he exploited this ability to influence public opinion when he had political points to make (e.g. see Chap. 12), and in this regard he was in an unusual position for a non-government scientist in Australia.

**Fig. 1.17** The colourful cover of the 1987 reprint of Tebbutt's *Astronomical Memoirs* (*Photograph* Wayne Orchiston)

**Table 1.8** Newspaper articles and reports by Tebbutt, 1851–1885 (after Tebbutt 1887)

| Discipline | 1851–1855 | 1856–1860 | 1861–1865 | 1866–1870 | 1871–1875 | 1876–1880 | 1881–1885 |
|---|---|---|---|---|---|---|---|
| Astronomy | 1 | 16 | 38 | 23 | 23 | 56 | 85 |
| Meteorology | 0 | 0 | 8 | 5 | 6 | 4 | 8 |
| 'Other' | 0 | 0 | 0 | 1 | 1 | 1 | 1 |
| Totals | 1 | 16 | 46 | 29 | 30 | 61 | 94 |

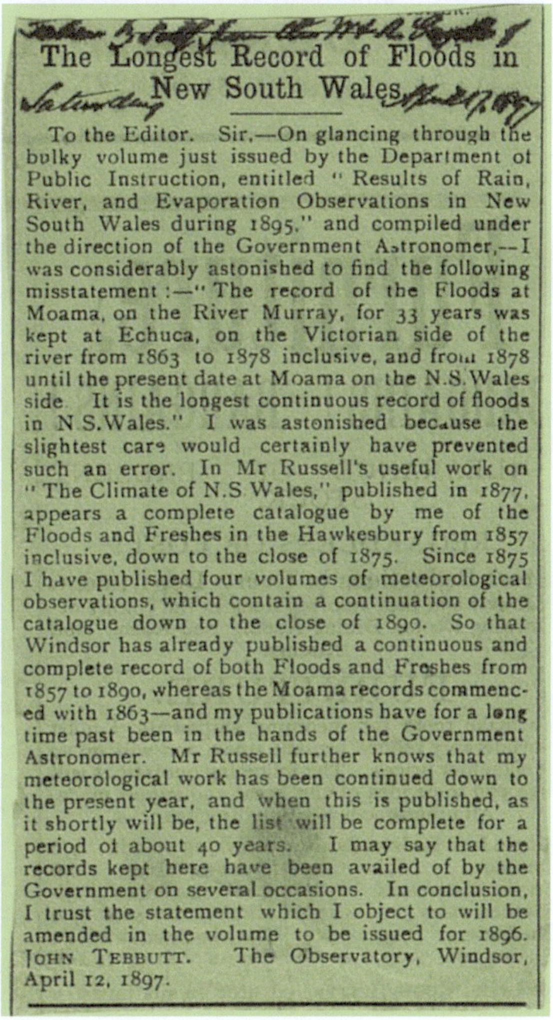

The Longest Record of Floods in New South Wales.

To the Editor. Sir,—On glancing through the bulky volume just issued by the Department of Public Instruction, entitled "Results of Rain, River, and Evaporation Observations in New South Wales during 1895," and compiled under the direction of the Government Astronomer,—I was considerably astonished to find the following misstatement :—"The record of the Floods at Moama, on the River Murray, for 33 years was kept at Echuca, on the Victorian side of the river from 1863 to 1878 inclusive, and from 1878 until the present date at Moama on the N.S.Wales side. It is the longest continuous record of floods in N.S.Wales." I was astonished because the slightest care would certainly have prevented such an error. In Mr Russell's useful work on "The Climate of N.S.Wales," published in 1877, appears a complete catalogue by me of the Floods and Freshes in the Hawkesbury from 1857 inclusive, down to the close of 1875. Since 1875 I have published four volumes of meteorological observations, which contain a continuation of the catalogue down to the close of 1890. So that Windsor has already published a continuous and complete record of both Floods and Freshes from 1857 to 1890, whereas the Moama records commenced with 1863—and my publications have for a long time past been in the hands of the Government Astronomer. Mr Russell further knows that my meteorological work has been continued down to the present year, and when this is published, as it shortly will be, the list will be complete for a period of about 40 years. I may say that the records kept here have been availed of by the Government on several occasions. In conclusion, I trust the statement which I object to will be amended in the volume to be issued for 1896. JOHN TEBBUTT. The Observatory, Windsor, April 12, 1897.

**Fig. 1.18** One of John Tebbutt's newspaper articles about flooding in Windsor (Orchiston Collection)

Tebbutt also firmly believed that astronomy should be taught in schools, and even through Schools of Art (Thorne 1911). He expounded on this view in a newspaper article about 'nature studies' written in 1906:

> In this country, unfortunately, the oldest and grandest of all nature studies is completely in the shade, and it is on this account that ignorance with regard to the mechanism of the universe is so strongly and painfully prevalent amongst us. Surely children can be as much interested and intellectually improved by a study of the heavens as they can be by an examination of the life history of the frog or an insect. (Tebbutt 1906b).

Another way in which Tebbutt promoted astronomy was to run occasional private viewing nights at Windsor Observatory for friends and acquaintances, but after a number of unfortunate incidents he was careful to regulate these and by 1892 his attitude had become rather negative:

> I endeavour to avoid visitors as much as possible, for I find there is nothing but ill luck in the reception of them. They come without five minutes preparation for what they wish to see, and go away just as edified as before their arrival. (Tebbutt 1890–92, 15 July 1891 entry).

Tebbutt was also an inspiring lecturer, but because of the relative difficulty of getting to and from Sydney (Turner 1918) had to turn down many of the invitations he received. Those lectures that he did give were not solely on astronomy. For example, in 1878 he spoke to the Young Men's Christian Association in Windsor on 'The Testimony Which Astronomy Furnishes to the Attributes of the Creator' and repeated this in Sydney 2 months later.

Usually, Tebbutt was happy to supply genuine inquirers with information about astronomy and meteorology, and he was known for his quick responses (generally within 1–2 days) and eagerness to assist those with a sincere interest in science. This message comes through time and again when one reads the volumes of inward letters in the 'Tebbutt Collection' in the Mitchell Library (e.g. see Biggs 1881; Eedy 1893; Foley 1911; Furber 1899; Merfield 1900, 1904; Morris 1883; Smith 1901; Streatfeild 1878; Wright 1896, 1915). When Swindlehurst (1877a) wrote Tebbutt on 18 January 1877 requesting information on sextants and astronomical books, he received an almost instant reply to which he responded on 3 February:

> I am sure that I do not know what to say sufficiently to thank you for your exceedingly kind reply to my first letter. I certainly hoped to receive <u>a few lines</u> from you—but such a full friendly epistle, so full of good matured advice and useful hints I really had not dreamed of receiving. (Swindlehurst 1877b; his underlining).

This response is typical of those forwarded by satisfied inquirers.

## 1.13 A Magnificent Research and Reference Library

One of the reasons that Tebbutt was so successful in research and in popularising astronomy was because he possessed '… a library such as few private establishments can boast of' (Tebbutt 1908a, p. 112). This contained a goodly selection of astronomical books and monographs, runs of journals and an extensive collection of reprints, some of which were unavailable elsewhere in New South Wales—including at Sydney Observatory (see Knibbs 1900). The Windsor Observatory library holdings, as at 1887, are listed between pages 47 and 74 in *History and Description of Mr. Tebbutt's Observatory* ... (Tebbutt 1887), and each year thereafter, new acquisitions are listed in successive *Annual Reports* (Tebbutt 1889, 1890, 1891c, 1892, 1893, 1894, 1895, 1896, 1897, 1898a, 1899, 1900, 1901, 1902, 1903b, 1904). We are talking about a major private library.

It was partly through an extensive international publications exchange network (Table 1.9) involving his *Annual Reports*, other booklets and reprints of his papers that Tebbutt was able to build up this magnificent library. In addition to private and professional observatories, he received exchanges and gifts from

**Table 1.9** Numbers of institutions which exchanged publications with Windsor Observatory (after Tebbutt 1908a)

| Country | Number of Institutions/Bodies |
|---|---|
| England | 19 |
| Scotland | 3 |
| Ireland | 3 |
| Portugal | 1 |
| France | 6 |
| Italy | 9 |
| Switzerland | 1 |
| Austria | 11 |
| Germany | 16 |
| Belgium | 1 |
| Holland | 9 |
| Denmark | 2 |
| Sweden | 1 |
| Norway | 2 |
| Russia | 11 |
| South Africa | 3 |
| India | 3 |
| Australia | 14 |
| New Zealand | 1 |
| Hong Kong | 1 |
| Japan | 2 |
| Canada | 3 |
| The United States | 35 |
| Mexico | 4 |
| San Salvador | 1 |
| Brazil | 1 |
| Uruguay | 2 |
| Argentine | 2 |
| Chile | 1 |
| Total: | 168 |

various government departments, universities, scientific societies and of course from individuals. By the end of the nineteenth century, most of the world's major observatories provided Windsor Observatory with gratis copies of their publications, which says a great deal about Tebbutt's international reputation.

Table 1.9 may seem impressive, but it does not indicate the full geographical range of countries represented, as we now understand them. For instance, some of the observatories in the 'Austria' listing in Tebbutt (1908a) are now located in Czechoslovakia and Hungary, while the 'Russia' entries include Finnish and Polish observatories.

The Windsor Observatory library contained works in many languages other than English, and in order to keep up with astronomical developments published in French and German, Tebbutt taught himself both these languages (Elkington 1901).

## 1.14 Status and Recognition

Status and recognition were important to Tebbutt, and he was keen to succeed on his own merits and not through the vehicle of a public service post. As a result of his newspaper articles of the 1850s and his discovery of the Great Comet of 1861, 'Tebbutt' became a household name in New South Wales, and after the founding of Windsor Observatory it did not take long for him to achieve international recognition. This particularly delighted his old mentor, Edward Quaife, who wrote: '… it gives me much pleasure to hear of your great success in scientific matters and of the value set on your labour by the learned men of Europe and elsewhere' (Quaife 1869a).

By the late 1880s, Windsor Observatory was being openly touted as the only private one of importance in the Southern Hemisphere (Clerke 1891; Tischner 1888), and Tebbutt's wide-ranging astronomical and meteorological work was much appreciated by northern hemisphere colleagues (e.g. see Harkness 1887; Peters 1888; See 1893; Smyth 1887; Turner 1894). Overseas astronomers solicited observations from him (e.g. Burnham 1898; Marth 1893; Perrine 1898; Pickering 1894), as seen in the first page of the letter shown in Fig. 1.19; took him into their confidence on delicate matters of staffing appointments (Gill 1895); asked his advice on possible Australian sites for southern stations (Boss 1903; Campbell 1900); and one even asked him for a reference (See 1898)! For his part, Tebbutt greatly relished these contacts and the marks of appreciation that he received from some of the leading overseas astronomers (Swindlehurst 1877c).

Both George Frederick Chambers (1841–1915; 1908) and Charles Piazzi Smyth (1887) pointed out that England had no independent astronomers of Tebbutt's status, and the latter went one step further: 'I wonder whether it would be within the range of possibilities that you should come home & do for Great Britain and her Cometic Astronomy what you have been doing so long for that of Australia …'. This discussion arose because Tebbutt felt that he was not appreciated in his own country, and let Smyth know this. Tebbutt believed that credit should be given where

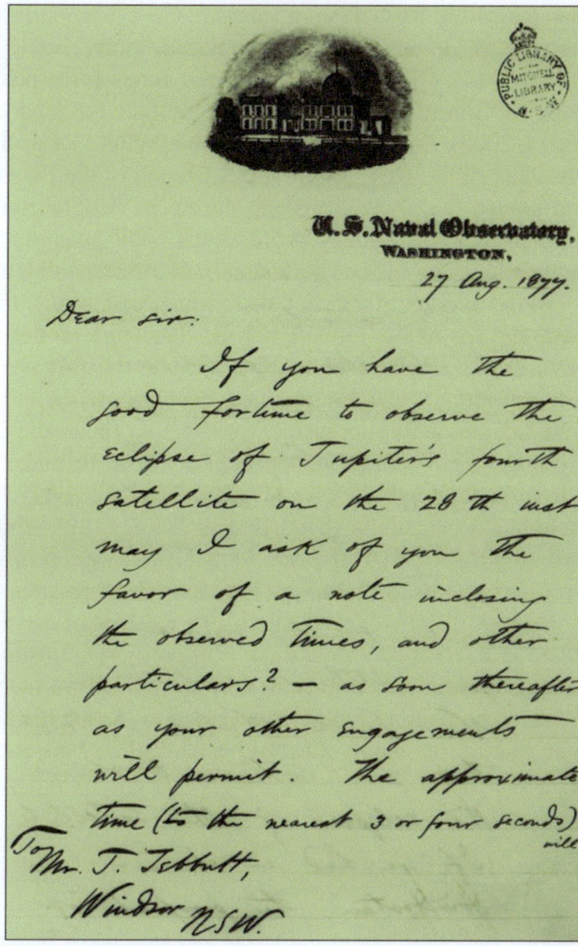


U.S. Naval Observatory,
Washington,
27 Aug. 1877.

Dear sir:

If you have the good fortune to observe the eclipse of Jupiter's fourth satellite on the 28th inst may I ask of you the favor of a note inclosing the observed times, and other particulars? — as soon thereafter as your other engagements will permit. The approximate time (to the nearest 3 or four seconds) will

To Mr. T. Tebbutt,
Windsor NSW.

**Fig. 1.19** The first page of a letter from the U.S. Naval Observatory requesting observations of Jovian satellite phenomena (*Courtesy* Mitchell Library)

it is due, and one of his principal complaints in later years was that this did not always occur. In his *Astronomical Memoirs*, he wrote: 'Since 1861 I have scarcely ever been directly attacked on astronomical matters; the main opposition locally to my efforts has been by the method of ignoring, a mode of opposition which is difficult to meet' (Tebbutt 1908a, p. 28). This was an oblique reference to Henry Chamberlain Russell, the one-time Government Astronomer of New South Wales. Russell was just 2 years Tebbutt's junior and initially the two men had been friends, but with the passage of time they grew increasingly estranged and by 1891 were openly feuding. The deteriorating Russell–Tebbutt relationship, which also embroiled other leading Sydney astronomers, is detailed in Chap. 12.

Part of the problem was that Tebbutt in some instances usurped the rightful role of the Sydney professional astronomers. For example, Sir Terence Aubrey Murray (1810–1873), the Executive Commissioner for the New South Wales exhibit at the Paris International Exhibition of 1867, invited Tebbutt (Murray 1866a, b) rather than Russell or Russell's predecessor, George Robarts Smalley (1822–1870) to prepare the astronomy essay for the Exhibition (Tebbutt 1867), and when David Gill (1843–1914) was considering Robert Innes for a post at the Royal Observatory, Cape of Good Hope, he turned to Tebbutt for advice and action (see Orchiston 2001a, 2003). Similarly, it was Tebbutt and not Russell who served as the founding President of the New South Wales Branch of the British Astronomical Association and guided that dynamic group during its fledgling years (see Chap. 13). Tebbutt also upstaged his Sydney professional colleagues with his over-abundant research publications and his constant stream of articles and letters to the local press.

To the average Australian astronomer of the day this output was much appreciated, and for many Tebbutt was an ideal role model (Swindlehurst 1890; Wright 1896, 1915). His international renown was the ultimate achievement for an independent astronomer, and though many may have wished to emulate him, few tried. Two who did, and were successful, were R.T.A. Innes and C.J. Merfield, and after coming under Tebbutt's influence both went on to make their marks in professional astronomy in South Africa and Australia, respectively (see Orchiston 2015). Others, including Tasmania's Alfred Barrett Biggs (1825–1900; see Orchiston 1985) and Sydney's Reverend Thomas Roseby (1909) were profoundly influenced by Tebbutt over many years, even though they had no professional aspirations.

By 1890, Tebbutt was the undisputed doyen of Australian astronomy and a world figure in international positional astronomy (Orchiston 1989), and although strictly an independent astronomer in that he derived no income from astronomy (to the contrary, he paid a high personal price to pursue his hobby), he was regarded as a pseudo-professional by most of his Australian astronomical contemporaries.

## 1.15 Just Rewards

Despite this, tangible rewards rarely came his way while he was alive, partly one suspects because of the influence that Russell was able to exert as Government Astronomer of New South Wales and one-time Vice-Chancellor of Sydney University. Thus, Tebbutt did not receive any honorary degrees or civil honours, and although it could be argued that his overall contribution to science exceeded that of Ellery (Melbourne Observatory Director; Gascoigne 1992), Russell (from Sydney Observatory; Bhathal 1991) or Sir Charles Todd (1826–1910; Director of Adelaide Observatory; Edwards 1993), he was not invited to join these three professional astronomers as a Fellow of the Royal Society. Such was the price that he paid for pursuing an independent career in science, but perhaps he derived some satisfaction from an 1889 magazine article which referred to '… the man of Science, who needs all his time which he can employ far more usefully in the quiet of his own study than in doing the dreary round of meetings and banquets, which are the engines for the display of political talent' (Eminent men of New South Wales 1889).

Tebbutt received just two notable honours during his lifetime. Firstly, in 1867 he was awarded a Silver Medal for his Paris International Exhibition paper (Fig. 1.20),

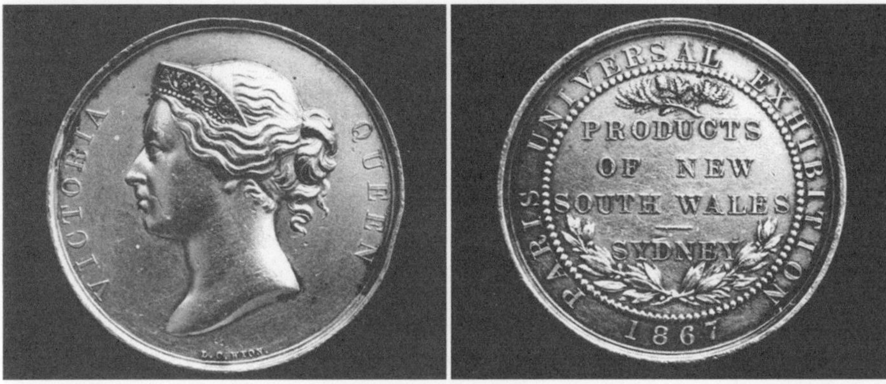


**Fig. 1.20** Obverse and reverse views of the Silver Medal that Tebbutt was awarded for his Paris Exhibition paper on Astronomy (after Tebbutt 1986, p. 132)

and then in 1905 he was only the third recipient of the Jackson-Gwilt Medal and Gift which was presented to him by the Royal Astronomical Society for his important observations of comets and double stars, and his long services to Australian astronomy extending over 40 years (Tebbutt 1908a, p. 109). In putting a case to the Society, R.T.A. Innes (1895) had earlier mentioned how 'Single handed, without any kindred spirit near him for over 40 years he has made valuable observations. His results command a very high value, higher than many State Observatories'. Tebbutt greatly valued this award. Further details of its presentation and of other honours of lesser moment that he received are given in Chap. 14.

## 1.16 An Enduring Legacy

Tebbutt passed away on 29 November 1916 (Death of Mr. John Tebbutt 1916; Houghton 1917; Turner 1918), and his funeral was a major event in Windsor (see Chap. 14). Soon afterwards, his son John Thomas Tebbutt donated the library and most of the records of Windsor Observatory to the Mitchell Library in Sydney. As outlined in the following chapter of this book, the 'Tebbutt Collection' comprises more than 100 volumes of manuscript notebooks, observing journals, calculations, meteorological and other data, and private correspondence (Orchiston 2004a) and is an incomparable resource for those studying the history of the nineteenth century and early twentieth century Australian astronomy and meteorology.

Since Tebbutt's death, there has been increasing public recognition of his remarkable contribution to Australian science. Streets have been named after him, and in 1973 Picard G, a crater on the Moon, was renamed 'Tebbutt'. Finally, as a pioneer explorer of the southern skies, Tebbutt was one of two 'explorers' chosen to grace Australia's new $100 bank note which was released in 1984 (Fig. 1.21). He shared this honour with the much better-known Sir Douglas Mawson (1882–1958) of Antarctic fame.

**Fig. 1.21** A black and white mock-up of the side of the 1984 Australian $100 bank note featuring John Tebbutt (Orchiston collection)

During the 1920s, the majority of the Windsor Observatory instruments were dispersed and the current whereabouts of most of the meteorological instruments and some of the telescopes remains a mystery. However, the chain of ownership of the 8-in. Grubb refractor is well known (see Orchiston 1997b; Orchiston and Bembrick 1997), and in 1986 this historic instrument was returned to Windsor to eventually form a centrepiece of the refurbished Windsor Observatory.

The costly and time-consuming task of turning the two surviving observatory buildings (see Fig. 1.22) into a museum of astronomy was undertaken by the astronomer's great-grandson, John Halley Tebbutt, and was completed in 1989 (Fig. 1.23). More information on this challenging project and on how Tebbutt's face came to appear on the $100 note are presented in Chap. 15.

## 1.17 Concluding Remarks

For the last three decades of the nineteenth century, Tebbutt was Australia's leading astronomer, and he enjoyed a considerable local, national and international reputation.

As an independent astronomer, he had few international peers. From humble beginnings and minimal instrumentation, he improved his situation as his increasing means allowed until his Windsor Observatory was in a position to compete effectively with larger government-funded northern hemisphere observatories. Through exchanges with many of these, he built up an outstanding research and reference library which he made extensive use of.

Characterised by unbounded energy and total dedication, Tebbutt made optimal use of his modest astronomical equipment and was able to contribute significantly

**Fig. 1.22** A photograph of the two surviving Windsor Observatory buildings in July 1973 (*Photograph* Wayne Orchiston)

**Fig. 1.23** The refurbished John Tebbutt Observatories in November 1996 (*Photograph* Wayne Orchiston)

to cometary, double star, minor planet, variable star and Jovian satellite astronomy and the international transit of Venus programme in 1874. He discovered two of the 'Great Comets' of the century, was responsible for three recoveries of Comet 2P/Encke, discovered a nova and provided observations which were used in establishing the geodetic framework for New South Wales.

Although an independent astronomer, he was regarded as a professional astronomer by most Australian astronomers and as the doyen of Australian astronomy. His observatory rivalled those maintained by the colonial governments in New South Wales, Victoria, South Australia and Western Australia; he observed the same types of astronomical objects and celestial events as they did; and published in the same international journals—but far more frequently as his lifetime total of more than 300 individual papers and research notes testifies. To these must be added his books, booklets and chapters of books, in all a phenomenal output for a single individual.

Over the years, he was a strong supporter of the Royal Astronomical Society and the British Astronomical Association, and in recognition of his pre-eminent status in Australian astronomy was elected founding President of the New South Wales Branch of the latter society when this was formed in 1895. Back in 1882, he had formed Australia's first national astronomical group, the Australian Comet Corps, but this was an anachronism and after struggling to survive folded 1 year later.

Along with the likes of Walter Gale and Thomas Roseby in Sydney and Alfred Barrett Biggs of Launceston, John Tebbutt was an Australian pioneer in the popularisation of astronomy. Through his innumerable contributions to Sydney newspapers, his popular lectures and his viewing nights at Windsor Observatory, the general population gradually became more astronomically literate, and a number of individuals who later made names for themselves were wooed onto astronomy as a hobby. Although a busy man, Tebbutt somehow managed to maintain a voluminous correspondence. He always had time to reply to those with serious inquiries and was renowned for the encouragement he gave to beginners in astronomy.

During the last three decades of the nineteenth century, Tebbutt was a power-broker in New South Wales astronomy, and his advice and support were sought by astronomers from within Australia and overseas. As such, he was the ideal role model for those independent astronomers aspiring to greatness, but his high profile came with a price for he became increasingly estranged from Russell, the Government Astronomer of New South Wales, and by 1891 an open feud existed between the two men.

Finally, in any assessment of Tebbutt's contribution to Australian science it is important to remember that for more than half a century he maintained a time-service for the citizens of Windsor; ran a meteorological station and published his data in local newspapers and in a series of eight monographs; recorded local floods and freshes; and for a short time carried out local tidal studies. In addition to his astronomical achievements, Tebbutt should be seen as an Australian pioneer in the field of meteorology.

But while we may applaud his achievements, Tebbutt was frustrated by what might have been, and in 1875 he published the following perceptive comments to the pages of the *Sydney Morning Herald*:

> I am gratified to find that what little I have, under God's blessing, been enabled to do, is of some service to science, but I feel that what has been accomplished is only a fraction of what should be done to represent Windsor in the astronomical world. Sublumary matters and a somewhat delicate state of health prevent me from bestowing that attention on the subject which it deserves. I have long lived in the hope of meeting someone who, with a

> certain amount of leisure at his command, could assist me in my work, partly out of love for the science, and partly for a remuneration. If such an one can be found, I shall feel myself in a position to add to my instrument appliances, and so make my work more effective. It is almost needless to remark that there is a wide and fruitful field open to the enthusiast in our southern skies. (Tebbutt 1875c).

It is illuminating to imagine just how much more Tebbutt might have achieved had such an assistant materialised. Despite this apparent setback, Houghton (1917, pp. 6–7) characterised Tebbutt as the first great Australian-born astronomer, and for more than half a century he was able to make a major contribution to astronomy. As Ellery wrote in 1901: 'To Mr. Tebbutt's enterprise, persevering and well-directed personal work belongs no small share of Australia's contribution to astronomy during the last 30 years' (Ellery 1901, p. 9).

Yet few tangible rewards came his way during life, although since his decease a crater on the Moon has been named after him and he has featured on an Australian $100 banknote. The refurbished John Tebbutt Observatories at Windsor now operate as a commemorative museum to one of Australia's most remarkable astronomers.

In closing this introductory chapter, I can do no better than to quote the distinguished British astronomer, Oxford University's Professor H.H. Turner, who wrote: 'John Tebbutt was one of those enthusiastic non-professional astronomers who, with instrumental appliances of the humblest order [by world standards] and no outside help, made an indelible mark on our records' (Turner 1918). Turner corresponded with Tebbutt for many years, but the two only met in 1914 when the British Association for the Advancement of Science held its 84th meeting in Australia (see Knibbs 1914) and a carload of astronomers—Turner included—made a pilgrimage to Windsor to meet a frail John Tebbutt. Just 2 years later, he was dead.

**Acknowledgements** I wish to thank Professor Nick Lomb (University of Southern Queensland, Toowoomba), the staff at the Mitchell Library (Sydney) and the Royal Astronomical Society Archives (London), and John Halley Tebbutt, for their assistance, and the Mitchell Library for kindly supplying Figs. 1.7 and 1.19.

## References[1]

Allen, M.L. (1886). Letter to John Tebbutt, dated 4 November. In TL.

Ashbrook, J. (1984). John Tebbutt, his observatory, and a probable nova. In J. Ashbrook, *The Astronomical Scrapbook. Skywatchers, Pioneers and Seekers in Astronomy* (pp. 66–71). Cambridge (Mass.): Sky Publishing Corporation.

Astronomy. *Sydney Morning Herald*, 22 March 1875.

Baracchi, P. (1914). Astronomy and geodesy in Australia. In Knibbs, 326–390.

Baum, R., & Sheehan, W. (1997). *In Search of Planet Vulcan. The Ghost in Newton's Clockwork Universe*. New York: Plenum.

[1] The following abbreviation is used: TL = Bound manuscript tetters to John Tebbutt, in the Mitchell Library, Sydney.

Bhathal, R. (1991). Henry Chamberlain Russell – astronomer, meteorologist and scientific entrepreneur. *Journal and Proceedings of the Royal Society of New South Wales, 124,* 1–21.
Bhathal, R. (1993). *Australian Astronomer John Tebbutt. The Life and World of the Man on the $100 Note*. Sydney: Kangaroo Press.
Biggs, A.B. (1881). Letter to John Tebbutt, dated 17 November. In TL.
Bond, G.P., Fette, H.G. & Watts, J.W. (1858). *An Account of Donati's Comet of 1858.* Cambridge (Mass.): John Bartlett.
Bone, W. (1883). Letter to John Tebbutt, dated 14 November. In TL.
Boss, L. (1903). Letter to John Tebbutt, dated July [no day given]. In TL.
Bruck, H.A., & Bruck, M.T. (1988). *The Peripatetic Astronomer. The Life of Charles Piazzi Smyth.* Bristol: Adam Hilger.
Burnham, S.W. (1898). Letter to John Tebbutt, dated 12 March. In TL.
Cable, K.J. (1967). Stiles, Henry Tarlton (1808–1867). In A.G.L. Shaw, & C.M.H. Clark, (Eds.), *Australian Dictionary of Biography. Volume 2: 1788–1850* (pp. 483–484). Melbourne: Melbourne University Press.
Campbell, W.W. (1900). Letter to John Tebbutt, dated 17 December. In TL.
Chalmers, T. (n.d.). *Discourses on the Christian Revelation, Viewed in Connection with the Modern Astronomy*. Glasgow: Collins.
Chamber, G.F. (1908). Letter to John Tebbutt, dated 15 December. In TL.
Chamber, G.F. (1909). *The Story of Comets Simply Told for General Readers*. Oxford: Clarendon Press.
Clerke, A. (1891). Letter to John Tebbutt, dated 8 June. In TL.
Clerke, A. (1893). *A Popular History of Astronomy during the Nineteenth Century*. London: Adam and Charles Black.
Death of Mr. John Tebbutt. *The Surveyor, 29,* 156–157 (1916).
Dewhirst, D.W. (1985). Meridian astronomy in the private and university observatories of the United Kingdom: rise and fall. *Vistas in Astronomy, 28,* 147–158.
Dick, S., Orchiston, W., & Love, T. (1998). Simon Newcomb, William Harkness and the nineteenth century American transit of Venus expeditions. *Journal for the History of Astronomy, 29,* 221–255.
Edwards, P.G. (1993). Charles Todd and the Adelaide Observatory. *Proceedings of the Astronomical Society of Australia, 10,* 349–354.
Eedy, A.M. (1893). Letter to John Tebbutt, dated 28 August. In TL.
Elkington, H. (1901). Mr. John Tebbutt, F.R.A.S., interviewed. *The Speaker, 2* (2), 40–42.
Ellery, R.J. (1892a). Letter to John Tebbutt, dated 8 April. In TL.
Ellery, R.J. (1892b). Letter to John Tebbutt, dated 22 November. In TL.
Ellery, R.J. (1901). A brief history of the beginnings and growth of astronomy in Australasia. In *Report of the Eighth Meeting of the Australasian Association for the Advancement of Science* (pp. 1–17). Melbourne: AAAS.
Eminent men of New South Wales. Mr John Tebbutt, F.R.A.S. *The Sydney Quarterly Magazine, 6,* 164–170 (1889).
Foley, M. (1911). Letter to John Tebbutt, dated 18 December. In TL.
Furber, T.F. (1899). Letter to John Tebbutt, dated 29 March. In TL.
Gale, W. (1912). Letter to John Tebbutt, dated 9 September. In TL.
Gascoigne, S.C.B. (1992). Robert L.J. Ellery, his life and times. *Proceedings of the Astronomical Society of Australia, 10,* 170–176.
Gill, D. (1895). Letter to John Tebbutt, dated 18 October. In TL.
Grece, K.O. (1947). Ad Astra. *Sky and Telescope, 6* (6), 14.
Harkness, W. (1887). Letter to John Tebbutt, dated 13 September. In TL.
Haynes, R., Haynes, R., Malin, D., & McGee, R.X. (1996). *Explorers of the Southern Sky. A History of Australian Astronomy*. Cambridge: Cambridge University Press.
Houghton, T.H., 1917. Presidential address. *Journal and Proceedings of the Royal Society of New South Wales, 51,* 1–70.

Humphreys, R.M. & Davidson, K., 1994. The Luminous Blue Variables: astrophysical geysers. *Publications of the Astronomical Society of the Pacific, 106,* 1025–1051.

Innes, R.T.A. (1895). Letter to W.H. Wesley, dated 30 September. In RAS Letter Archives.

John Tebbutt. *The Observatory, 40,* 141 (1917).

Knibbs, G.H. (1900). Letter to John Tebbutt, dated 1 December. In TL.

Knibbs, G.H. (ed.) (1914). *Federal Handbook Prepared in Connection with the Eighty-fourth Meeting of the British Association for the Advancement of Science, Held in Australia, August, 1914.* Melbourne: Government Printer.

Lomb, N. (2016). Australian solar eclipse expeditions: the voyage to Cape York in 1871. Journal of Astronomical History and Heritage, 19, 79–95.

Macdonnell, W.J. (1904). Presidential address. *Journal of the British Astronomical Association, 15,* 80–86.

Maguire, R. (1985). Angelo Tornaghi. An inventive Italian of the nineteenth century. *The Australian Antique Collector, 29,* 42–46.

"Many Happy Returns." Mr. John Tebbutt, F.R.A.S., who celebrates the eighty-first anniversary of his birthday to-day. *Daily Telegraph,* 25 May 1915.

Marsden, B.G. & Williams, G.V. (1996). *Catalogue of Cometary Orbits 1996.* Cambridge (Mass.): Minor Planet Center.

Marth, A. (1893). Letter to John Tebbutt, dated 24 January. In TL.

Merfield, C.J. (1900). Letter to John Tebbutt, dated 9 November. In TL.

Merfield, C.J. (1904). Letter to John Tebbutt, dated 30 April. In TL.

Morris, E.R. (1883). Letter to John Tebbutt, dated 3 October. In TL.

Morse, J.A., Humphreys, R.M., & Daminelli, A. (Eds.) (1999). *Eta Carinae at the Millennium.* San Francisco: Astronomical Society of the Pacific (ASP Conference Series, Volume 179).

Murray, Sir T.A. (1866a). Letter to John Tebbutt, dated 30 April. In TL.

Murray, Sir T.A. (1866b). Letter to John Tebbutt, dated 22 May. In TL.

N.S.W. (1893). A self-taught astronomer. *The Australasian,* 26 August.

Obituaries: John Russell Hind. *Monthly Notices of the Royal Astronomical Society, 56,* 200–205 (1896).

Orchiston, W. (1981). 1981. Illuminating incidents in Antipodean astronomy: John Tebbutt and Great Comet of 1881. *Journal of the Astronomical Society of Victoria, 34,* 78–92.

Orchiston, W. (1982a). Illuminating incidents in Antipodean astronomy: John Tebbutt and the abortive Australian association of comet observers. *Journal of the Astronomical Society of Victoria, 35,* 70–83.

Orchiston, W. (1982b). John Tebbutt and the Whakatane Eight Inch Refractor: a review of "The Australian Connection". *Southern Stars, 29,* 215–244.

Orchiston, W. (1985). The contribution of Alfred Barrett Biggs to Australian and Tasmanian astronomy. *Records of the Queen Victoria Museum, 89,* 1–28.

Orchiston, W. (1987). Dr William Bone, and the role of the amateur observatory in Australian astronomy. *Southern Stars, 32,* 111–128.

Orchiston, W. (1988a). The emergence of professional astronomy in nineteenth century Australia. *Search, 19,* 76–81.

Orchiston, W. (1988b). Illuminating incidents in Antipodean astronomy: formation of the New South Wales Branch of the British Astronomical Association. *Journal of the British Astronomical Association, 98,* 75–84.

Orchiston, W. (1988c). Illuminating incidents in Antipodean astronomy: John Tebbutt and the founding of the Windsor Observatory. *Journal of the British Astronomical Association, 98,* 287–293.

Orchiston, W. (1988d). Illuminating incidents in Antipodean astronomy: John Tebbutt and the Sydney Observatory directorship of 1862. *Australian Journal of Astronomy, 2,* 149–158.

Orchiston, W. (1989). The role of the amateur in early Australian astronomy. *Search, 20,* 16–23.

Orchiston, W. (1990). The Astronomical Register, John Tebbutt, and astronomy in the Antipodes. *Journal of the British Astronomical Association, 100,* 212.

Orchiston, W. (1991). Amateur-professional co-operation in astronomy education: a nineteenth century Australian model. *Proceedings of the Astronomical Society of Australia, 9,* 186–188.
Orchiston, W. (1993). John Grigg, and the genesis of cometary astronomy in New Zealand. *Journal of the British Astronomical Associationi, 103,* 67–76.
Orchiston, W. (1997a). The role of the amateur in popularising astronomy: an Australian case study. *Australian Journal of Astronomy, 7,* 33–66.
Orchiston, W. (1997b). The role of the large refracting telescope in Australian amateur astronomy: an historical perspective. *Australian Journal of Astronomy, 7,* 89–114.
Orchiston, W. (1997c). The "tyranny of distance" and Antipodean cometary astronomy. *Australian Journal of Astronomy, 7,* 115–126.
Orchiston, W. (1998). Amateur-professional collaboration in Australian science: the earliest astronomical groups and societies. *Historical Records of Australian Science, 12,* 163–182.
Orchiston, W. (1999). C/1881 K1: A forgotten "Great Comet" of the nineteenth century. *Irish Astronomical Journal, 26,* 33–44.
Orchiston, W. (2001a). From amateur astronomer to observatory director: the curious case of R.T.A. Innes. *Publications of the Astronomical Society of Australia, 18,* 317–327.
Orchiston, W. (2001b). 'Sentinel of the Southern Heavens': The Windsor Observatory of John Tebbutt. *Journal of the Antique Telescope Society, 21,* 11–23.
Orchiston, W. (2003). Passion and patronage: the Afro-Australian astronomy of Robert Thorburn Ayton Innes. *Monthly Notices of the Astronomical Society of Southern Africa, 62,* 4–13.
Orchiston, W. (2004a). Highlighting the history of nineteenth century Australian astronomy: the Tebbutt Collection in the Mitchell Library, Sydney. *Journal of Astronomical Data, 10,* 77–103.
Orchiston, W. (2004b). John Tebbutt and observational astronomy at Windsor Observatory. *Journal of the British Astronomical Association, 114,* 141–154.
Orchiston, W. (2015). The amateur-turned professional syndrome: two Australian case studies. In W. Orchiston, D.A. Green & R. Strom (Eds.), *New Insights from Recent Studies in Historical Astronomy: Following in the Footsteps of F. Richard Stephenson* (pp. 259–350). Cham: Springer.
Orchiston, W. & Bembrick, C. (1997). The potency of patronage: George Hoskins and the New South Wales Branch of the British Astronomical Association. *Australian Journal of Astronomy,* **7**, 1–15.
Orchiston, W. & Bhathal, R. (1991). Illuminating incidents in Antipodean astronomy: Section A of the Royal Society of New South Wales. *Australian Journal of Astronomy, 4,* 49–67.
Orchiston, W. & Perdrix, J. (1990). The British Astronomical Association in Australia: a centennial perspective. *Australian Journal of Astronomy, 3,* 163–179.
Orchiston, W. & Perdrix, J. (2002). A history of the British Astronomical Association in Australia: the fate of the Branches. *Journal of the British Astronomical Association, 112,* 68–77
Perrine, C.D. (1898). Letter to John Tebbutt, dated 12 July. In TL.
Peters, E.A. (1888). Letter to John Tebbutt, dated 29 January. In TL
Pickering, W.H. (1894). Letter to John Tebbutt, dated 24 March. In TL.
Proctor, M. (1926). *The Romance of Comets*. New York: Harper & Brothers.
Proctor, M. (1937). *Comets. Their Nature, Origin, and Place in the Science of Astronomy*. London: Technical Press.
Proctor, R. (1874). *Transits of Venus. A Popular Account of Past and Coming Transits*. London: Longmans.
Quaife, E. (n.d.). Undated note to John Tebbutt. In TL.
Quaife, E. (1861). Letter to John Tebbutt, dated 6 November. In TL.
Quaife, E. (1869a). Letter to John Tebbutt, dated 27 July. In TL.
Quaife, E. (1869b). Letter to John Tebbutt, dated 27 August. In TL.
Raymond, F. R. (1871). Letter to John Tebbutt, dated 20 October. In TL.
Reports of the Branches. New South Wales Branch. *Journal of the British Astronomical Association, 7,* 376–378 (1897).
Reports of the Branches. New South Wales Branch. *Journal of the British Astronomical Association, 34,* 44–46 (1923).

Roseby, T. (1902). Letter to the Secretary, dated 3 December. In RAS Letter Archives.
Roseby, T. (1909). Letter to John Tebbutt, dated 22 February. In TL.
Russell, H.C. (1871a). Letter to John Tebbutt, dated 17 March. In TL
Russell, H.C. (1871b). Letter to John Tebbutt, dated 6 November. In TL.
Scott, W. (1871). Letter to John Tebbutt, dated 4 November. In TL.
See, T.J.J. (1893). Letter to John Tebbutt, dated 15 July. In TL.
See, T.J.J. (1898). Letter to John Tebbutt, dated 20 September. In TL.
Smith, J.D. (1901). Letter to John Tebbutt, dated 21 September. In TL.
Smyth, P. (1880). Letter to John Tebbutt, dated 8 September. In TL.
Smyth, P. (1887). Letter to John Tebbutt, dated 3 May. In TL.
Steele, J. (1916). *The Early Days of Windsor*. Sydney: Tyrrells.
Streatfeild, G. (1878). Letter to John Tebbutt, dated 16 August. In TL.
Swindlehurst, W. (1877a). Letter to John Tebbutt, dated 18 January. In TL.
Swindlehurst, W. (1877b). Letter to John Tebbutt, dated 3 February. In TL.
Swindlehurst, W. (1877c). Letter to John Tebbutt, dated 24 February. In TL.
Swindlehurst, W. (1890). Letter to John Tebbutt, dated 15 April. In TL.
Tebbutt, J. (n.d. (a)). Catalogue of the Observatory Library of John Tebbutt. MSS, Mitchell Library (A3781).
Tebbutt, J. (1853–59). Observations made … at the Peninsula, near Windsor, in the Colony of New South Wales. MS, Mitchell Library, Sydney (A3745$^{-1}$).
Tebbutt, J. (1854). Solar phenomena. *Sydney Morning Herald,* 13 May.
Tebbutt, J. (1867). On the progress and present state of astronomical science in New South Wales. In *Catalogue of the Natural and Industrial Products of New South Wales. Forwarded to the Paris Universal Exhibition of 1867, by the New South Wales Exhibition Commissioners* (pp. 55–64). Sydney: Government Printer [Appendix].
Tebbutt, J. (1868). *Meteorological Observation Made at the Private Observatory of John Tebbutt, Junr., The Peninsula, Windsor, New South Wales, in the Years 1863, 1864, 1865, and 1866.* Sydney: printed for the author.
Tebbutt, J. (1870–73). Journal of the Observatory. The Peninsula, Windsor, New South Wales. MS, Mitchell Library, Sydney (A3748).
Tebbutt, J. (1871). On the progress and present state of astronomical science in New South Wales. In *The Industrial Progress of New South Wales: Being a Report of the Intercolonial Exhibition of 1870, at Sydney; Together with a Variety of Papers Illustrative of the Industrial Resources of the Colony* (pp. 617–632). Sydney: Government Printer.
Tebbutt, J. (1873). Letter to E. Dunkin, dated 18 April. In RAS Letter Archives.
Tebbutt, J. (1874). *Meteorological Observations Made at the Private Observatory of John Tebbutt, The Peninsula, Windsor, New South Wales, in the Years 1867, 1868, 1869, and 1870.* Sydney: printed for the author.
Tebbutt, J. (1875a). Beobachtung des Venusdurchganges in Windsor (N.-S.-Wales). *Astronomische Nachrichten, 85,* 173–176.
Tebbutt, J. (1875b). *The Broad and Narrow Way. An Address, Delivered before the United Religious Mission …* Windsor: printed for the author.
Tebbutt, J. (1875c). The Windsor Observatory. *Sydney Morning Herald,* 27 March.
Tebbutt, J. (1877a). On a new and remarkable variable star in the constellation Ara. *Journal of the Royal Society of New South Wales, 11,* 197–202.
Tebbutt, J. (1877b). *Results of Meteorological Observations Made at the Private Observatory of John Tebbutt, The Peninsula, Windsor, New South Wales, in the Years 1871, 1872, 1873, 1874, 1875 & 1875*. Sydney: printed for the author.
Tebbutt, J. (1877c). *Thoughts on the Fifth Chapter of St. Paul's First Epistle to the Thessalonians.* Windsor: printed for the author.
Tebbutt, J. (1878). *The Testimony which Astronomy Furnishes to the Attributes of The Creator*. Sydney: Young Men's Christian Association.
Tebbutt, J. (1881). Letter to the Secretaries, dated 23 December. In RAS Letter Archives.
Tebbutt, J. (1882a). Letter to the Secretaries, dated 9 December. In RAS Letter Archives.

Tebbutt, J. (1882b). *Results of Meteorological Observations Made at the Private Observatory of John Tebbutt, The Peninsula, Windsor, New South Wales, in the Years 1877, 1878, 1879, 1880, & 1881*. Parramatta & Sydney: printed for the author.
Tebbutt, J. (1883). Observations of the transit of Venus, 1874, December 8–9, made at Windsor, N.S.W. *Memoirs of the Royal Astronomical Society, 47*, 89–92.
Tebbutt, J. (1885). Observations of double stars at Windsor, New South Wales. *Monthly Notices of the Royal Astronomical Society, 46*, 50–55.
Tebbutt, J. (1886). *Results of Meteorological Observations Made at the Private Observatory of John Tebbutt, The Peninsula, Windsor, New South Wales, in the Years 1882, 1883, 1884, & 1885*. Parramatta & Sydney: printed for the author.
Tebbutt, J. (1887). *History and Description of Mr. Tebbutt's Observatory, Windsor, New South Wales*. Sydney: printed for the author.
Tebbutt, J. (1889). *Report of Mr. Tebbutt's Observatory, The Peninsula, Windsor, New South Wales, for the Year 1888*. Sydney: printed for the author.
Tebbutt, J. (1890). *Report of Mr. Tebbutt's Observatory, The Peninsula, Windsor, New South Wales, for the Year 1889*. Sydney: printed for the author.
Tebbutt, J. (1890–92). Journal for Astronomical Observations. The Peninsula, Windsor, New South Wales. MS, Mitchell Library, Sydney (A3754).
Tebbutt, J. (1891a). Letter to the Secretaries, dated 31 March. In RAS Letter Archives.
Tebbutt, J. (1891b). Observing occultations – misuse of scientific symbols. *The Observatory, 14*, 310–311.
Tebbutt, J. (1891c). *Report of Mr. Tebbutt's Observatory, The Peninsula, Windsor, New South Wales, for the Year 1890*. Sydney: printed for the author.
Tebbutt, J. (1891d). *Results of Meteorological Observations Made at the Private Observatory of John Tebbutt, The Peninsula, Windsor, N.S. Wales, in the Years 1886, 1887, 1888, 1889, & 1890*. Sydney: printed for the author.
Tebbutt, J. (1892). *Report of Mr. Tebbutt's Observatory, The Peninsula, Windsor, New South Wales, for the Year 1891*. Sydney: printed for the author.
Tebbutt, J. (1893). *Report of Mr. Tebbutt's Observatory, The Peninsula, Windsor, New South Wales, for the Year 1892*. Sydney: printed for the author.
Tebbutt, J. (1894). *Report of Mr. Tebbutt's Observatory, The Peninsula, Windsor, New South Wales, for the Year 1893*. Sydney: printed for the author.
Tebbutt, J. (1894–96). Untitled journal. MS, Mitchell Library, Sydney (A3756).
Tebbutt, J. (1895). *Report of Mr. Tebbutt's Observatory, The Peninsula, Windsor, New South Wales, for the Year 1894*. Sydney: printed for the author.
Tebbutt, J. (1896). *Report of Mr. Tebbutt's Observatory, The Peninsula, Windsor, New South Wales, for the Year 1895*. Sydney: printed for the author.
Tebbutt, J. (1897). *Report of Mr. Tebbutt's Observatory, The Peninsula, Windsor, New South Wales, for the Year 1896*. Sydney: printed for the author.
Tebbutt, J. (1898a). *Report of Mr. Tebbutt's Observatory, The Peninsula, Windsor, New South Wales, for the Year 1897*. Sydney: printed for the author.
Tebbutt, J. (1898b). *Results of Meteorological Observations Made at the Private Observatory of John Tebbutt, The Peninsula, New South Wales, in the Years 1891, 1892, 1893, 1894, 1895, 1896, and 1897*. Sydney: printed for the author.
Tebbutt, J. (1899). *Report of Mr. Tebbutt's Observatory, The Peninsula, Windsor, New South Wales, for the Year 1898*. Sydney: printed for the author.
Tebbutt, J. (1900). *Report of Mr Tebbutt's Observatory, The Peninsula, Windsor, New South Wales, for the Year 1899*. Sydney: printed for the author.
Tebbutt, J. (1901). *Report of Mr. Tebbutt's Observatory, The Peninsula, Windsor, New South Wales, for the Year 1900*. Sydney: printed for the author.
Tebbutt, J. (1902). *Report of Mr. Tebbutt's Observatory, The Peninsula, Windsor, New South Wales, for the Year 1901*. Sydney: printed for the author.
Tebbutt, J. (1903a). Letter to W.H. Wesley, dated 25 March. In RAS Letter Archives.
Tebbutt, J. (1903b). *Report of Mr. Tebbutt's Observatory, The Peninsula, Windsor, New South Wales, for the Year 1902*. Sydney: printed for the author.

Tebbutt, J. (1904). *Report of Mr. Tebbutt's Observatory, The Peninsula, Windsor, New South Wales, for the Year 1903*. Sydney: printed for the author.

Tebbutt, J. (1905). The first director of Sydney Observatory. *The Observatory, 28,* 104–105.

Tebbutt, J. (1906a). Measures of southern binary stars. *Monthly Notices of the Royal Astronomical Society, 66,* 547.

Tebbutt, J. (1906b). The proposed nature study exhibition at Richmond. *The Observatory, 30,* 112.

Tebbutt, J. (1908a). *Astronomical Memoirs*. Sydney: printed for the author

Tebbutt, J. (1908b). Measures of southern binary stars in 1907. *Monthly Notices of the Royal Astronomical Society, 68,* 396.

Tebbutt, J. (1910). Measures of southern binary stars in 1909. *Monthly Notices of the Royal Astronomical Society, 70,* 655.

Tebbutt, J. (1911). Observations of Halley's Comet at Windsor, N.S. Wales. *Monthly Notices of the Royal Astronomical Society, 71,* 224–226.

Tebbutt, J. (1912a). Gale's Comet. *Journal of the British Astronomical Association, 23,* 139–140.

Tebbutt, J. (1912b). Measures of southern binary stars in 1911. *Monthly Notices of the Royal Astronomical Society, 72,* 563.

Tebbutt, J. (1913). Komet 1912a (Gale). Observations at Windsor, N.S. Wales. *Astronomische Nachrichten, 193,* 313–314.

Tebbutt, J. (1915a). Letter to the Secretaries, dated 13 February. In RAS Letter Archives.

Tebbutt, J. (1915b). Measures of southern binary stars. *Monthly Notices of the Royal Astronomical Society, 76,* 36.

Tebbutt, J. (1915c). Sir Thomas Makdougall Brisbane. *The Observatory, 38,* 221.

Tebbutt, J. (1916). *Results of Meteorological Observations at Mr. Tebbutt's Observatory The Peninsula, Windsor, New South Wales, During the Period 1898–1915*. Windsor: printed for the author.

Tebbutt, J. (1986). *Astronomical Memoirs*. Windsor: Hawkesbury Shire Council (Reprinted edition). (Although the reprint date is given as 1986, in fact this book was in fact printed in 1987 and was officially launched on 9 April 1987).

The Windsor (N.S.W.) Observatory. *The Town and Country Journal,* 2 November 1889.

Thorne, T.W. (1911). Letter to John Tebbutt, dated 17 July. In TL.

Tischner, A. (1888). Letter to John Tebbutt, dated 20 September. In TL.

Tupman, G.L. (1878). On the mean solar parallax as derived from the observations of the transit of Venus, 1874. *Monthly Notices of the Royal Astronomical Society, 38,* 429–457.

Turner, H.H. (1894). Letter to John Tebbutt, dated 19 January. In TL.

Turner, H.H. (1918). Obituary: John Tebbutt. *Monthly Notices of the Royal Astronomical Society, 78,* 252–255.

van Genderen, et al., 1995. A pulsating star inside η Carinae. I. Light variations, 1992–1994. *Astronomy and Astrophysics, 304,* 415–430.

Vsekhsvyatskii, S.K., 1964. *Physical Characteristics of Comets.* Jerusalem: Israel Program for Scientific Translations.

Warner, B. (1983). *Charles Piazzi Smyth Astronomer-Artist. His Cape Years 1835–1845*. Cape Town: Balkema.

Weiss, E. (1888). Bilderatlas der Sternewelt: Eine Astronomie *fur* Jedermann. Stuttgart: J.F. Schreiber.

White, G. (1979). John Tebbutt and the astronomy at the Windsor Observatory. *Proceedings of the Astronomical Society of Australia, 3,* 408–411.

White, G. (1986). Introduction. In Tebbutt, J. *Astronomical Memoirs* (pp. vi–xix). Windsor: Hawkesbury Shire Council (Reprint).

Will of Thomas Tebbutt. In Tebbutt Papers, Mitchell Library, Sydney (MSS.5588).

Williams, T.R. (1988). A galaxy of amateur astronomers. *Sky and Telescope, 76*, 484–486.

Wood, H.W. (1976). Tebbutt, John (1834–1916). In G. Serle & R. Ward, R. (Eds.), *Australian Dictionary of Biography. Volume 6: 1851–1890. R-Z* (pp. 251–252). Melbourne: Melbourne University Press.

Wright, H. (1896). Letter to John Tebbutt, dated 6 February. In TL.

Wright, H (1915). Letter to John Tebbutt, 28 May. In TL.

# Chapter 2
# An Invaluable Resource: The 'Tebbutt Collection' in the Mitchell Library, Sydney

## 2.1 Introduction: The C41/ICHA Archives Working Group

The Archives Working Group of the International Astronomical Union (IAU) was formed in 1991 with a view to (1) establishing a register of the whereabouts of surviving historically-significant astronomical archives, and (2) encouraging observatories and other institutions to preserve, catalogue and conserve such archives (Débarbat 2002).

In response to the first of these goals, at the Manchester IAU General Assembly in 2000 I presented a paper on the historically-significant astronomical archives in Australian and New Zealand repositories (see Orchiston 2004a), with a view to providing basic national inventories for these two Oceanic nations.

Given these two national lists, the task now confronting astronomical historians is to describe and document the contents of individual collections or repositories in these two countries. This chapter focusses on the 'Tebbutt Collection' in the Mitchell Library, Sydney (mentioned, in passing, in Mourot 1973), and is based on a paper that I presented on this topic at the 2003 IAU General Assembly in Sydney, which subsequently was published (see Orchiston 2004b). Background material about John Tebbutt has already been presented in the previous chapter of this book, and further details appear in Chaps. 6–15.

## 2.2 The Tebbutt Collection

### *2.2.1 Background and Listing*

When John Tebbutt died in 1916, none of his surviving children was interested in astronomy, so dispersal of his astronomical records, his library and his telescopes was inevitable. In July 1917, just 7 months after Tebbutt's death, his son, John T. Tebbutt, arranged for most of the Windsor Observatory astronomical library, volumes of inward



W. Orchiston, *John Tebbutt*, Historical & Cultural Astronomy,
DOI 10.1007/978-3-319-44521-2_2

letters, and original observing books, journals, notebooks, diaries and other astronomical and meteorological records to be donated to the Mitchell Library in Sydney, in memory of his father. The Mitchell Library is a repository of books and manuscripts relating to the history of the Australia-Pacific region. A major research facility by international standards, the Mitchell Library is part of the State Library of New South Wales in Sydney and was founded when David Scott Mitchell (1836–1907)

> … bequeathed his unrivalled collection, with an endowment, to the Trustees of the Public Library of New South Wales. A condition of this bequest is that the collection and additions to it must be permanently kept as a separate library. Since its foundation the Mitchell Library has been further enriched by many important gifts as well as purchases. (The Mitchell Library … 1967).

The Tebbutt Collection was one such important gift, and its donation was entirely in accordance with John Tebbutt's wishes. Indeed, it was his close British Astronomical Association New South Wales Branch colleague, Hugh Wright (Bio Box 2.1), who first raised this delicate matter in writing with him back in 1908:

> There is a matter I wish you would consider—that is the preservation, for posterity, of your MS. journals, letter books, letters received, &c. Will descendants value this material? You cannot be sure. It would be a pity for your records to be dispersed … (Wright 1908; his underlining).

**Bio Box 2.1: Hugh Wright**

Hugh Wright (1868–1957; Fig. 2.1) was born in Sheffield, England (Mr. Hugh Wright … 1924), in 1868 and in 1882 emigrated to Sydney with his parents. While attending school during the day, he studied science and languages in the evenings at Sydney Technical College and the Sydney School of Arts. He also made regular use of the Public Library, and in 1884 was appointed to a position there '… on the strict understanding that he should remain 6 months. It lasted 48 years' (Mr. Hugh Wright … 1924)! His abilities and diligence were carefully noticed, and he was given more and more responsibilities, eventually being offered the post of 'Mitchell Librarian' in 1909 when a decision was made to found the Mitchell Library. Wright only retired as Mitchell Librarian in 1932 when he reached the age of 65.

**Fig. 2.1** Hugh Wright (Orchiston collection)

(continued)

Given his Mitchell Library position, it is no surprise that Hugh Wright was a Fellow and a Past President of the Royal Australian Historical Society, and he edited the Society's journal for many years (ibid.).

Wright's childhood interest in science translated into a teenage passion for astronomy, and later he owned a fine equatorially-mounted 8.5-in. (21.6-cm) Browning-With reflector that he used mainly to observe lunar occultations and double stars. Wright was a prominent member of the New South Wales Branch of the British Astronomical Association. He was one of those who petitioned the parent body in 1894 requesting the formation of a New South Wales Branch and was a member of the inaugural Council. Soon after he served as Secretary, and later was elected President. He published a number of short papers in the *Journal of the British Astronomical Association* (see Orchiston 1988).

Hugh Wright, O.B.E., died in 1957 after a long and eventful life.

Like Tebbutt, Wright was an observational astronomer (see Fig. 2.2), and as a librarian he was keen to make sure that Tebbutt's astronomical records were stored safely. Thus, in his 1908 letter he also mentioned that the new Mitchell Library would be a fire-proof building, where the material could be kept together as the 'Tebbutt Collection'.

When Wright raised this matter again in 1912, he was the inaugural Mitchell Librarian, and Tebbutt was in his twilight years at 78 years of age. In this same year, he accepted on behalf of the Library bound copies of Tebbutt's papers published by the Royal Astronomical Society and the Royal Society of New South Wales and was moved to write: 'In collected form the papers will be readily available to any student who desires either to work up your scientific life-work or to study the achievements in Astronomy in Australia' (Wright 1912).

These sentiments reflect equally on the Mitchell Library's entire Tebbutt Collection, which is an incomparable resource for historians of nineteenth century Australian astronomy. It comprises 118 different record lots, which include inward letters; journals, diaries and observing notebooks; reductions of observations, and manuscripts of some completed papers; instrumentation records; latitude and longitude determinations; catalogues of the Windsor Observatory library; and meteorological observations (see Table 2.1).

Some of the items in this table are self-explanatory and of little moment to most historians of astronomy, and therefore do not warrant further comment here. In contrast, other items contain invaluable research data. These are discussed below.

**Fig. 2.2** Hugh Wright's 8.5-in. (21.6-cm) Browning-With reflector, with H.V. Barclay C.E. (who designed and made the mounting) on the right (Orchiston collection)

## *2.2.2 Inward letters*

These 47 weighty, large-format bound volumes of letters, telegrams, postcards and other documents received by Tebbutt between 1860 and 1915 contain a wealth of information about Tebbutt and Windsor Observatory, and about Australian and overseas astronomers and observatories.

**Table 2.1** Summary of the Tebbutt Collection

| Item | Volumes[a] | Catalogue number(s) |
|---|---|---|
| Inward letters, 1860–1915 | 47 | A3682–A3728 |
| Astronomical journals, 1853–1902 | 17 | A3745$^{-1}$, A3747–A3758, A3783–A3786 |
| Observational notebooks, 1862–1865, 1879–1882, 1884–1905 | 15 | A3760–A3774 |
| Calculations; reduction of observations, 1853–1910 | 19 | A3729–A3744, A3745$^{-2}$, A3746, A3775, A3780 |
| Instrumental determinations, 1882–1901 | | A3759 |
| Values of the wire of the transit instrument, 1885–1903 | | A3776 |
| Determination of latitude, 1882 | | A3778 |
| Determination of longitude, 1885–1887 | | A3777 |
| Catalogue of the Observatory Library | | A3781 |
| Extracts from periodicals, etc. | | A3782 |
| Observations during hot winds, 1862–1863 | 1 | A3779 |
| Meteorological observations, 1863–1913 | 13 | A3787–A3799 |

[a]Not all items are in notebooks or bound volumes and can be easily quantified. Sometimes items are simply loose, or in folders, in tied labelled bundles

Tebbutt maintained a voluminous correspondence with a great many astronomers, and Australian and foreign names of interest are given in Table 2.2. Tebbutt's principal Australian astronomical correspondents were concentrated in towns and cities along the coast from Brisbane to Adelaide, and in Tasmania and inland Victoria (see Fig. 2.3).

In addition to Australian astronomy, and Tebbutt's own achievements, the inward letters cast useful light on aspects of overseas astronomy during the second half of the nineteenth century and early in the twentieth century, at a time when astronomy worldwide was witnessing the emergence of astrophysics, and the gradual decline of positional astronomy.

Obviously, there is no space here to discuss the many thousands of inward letters in detail, so we will focus on New Zealand astronomy and just the Grigg and Stock letters to illustrate the research potential of the Tebbutt Collection.

By the end of the nineteenth century, John Grigg (Fig. 2.4) was New Zealand's foremost independent astronomer (see Orchiston 1998b, 2016, Chaps. 10, 17, 22), and he maintained a small private observatory with a 3.5-in. (8.9-cm) equatorial Wray refractor and a small transit telescope in the wealthy North Island gold-mining city of Thames. Although there are only 24 letters, dating between 1902 and 1907, from Grigg to Tebbutt in the Tebbutt Collection, and a number of letters from other astronomers that document aspects of Grigg's astronomy, these collectively tell us a great deal about Grigg's instrumentation and observatories (see Orchiston 2001) and his various cometary discoveries (see Orchiston 1993)—including the evolving friction, at this time, between Australian and New Zealand independent astronomers

**Table 2.2** List of Tebbutt's key astronomical correspondents

| Names | Affiliation |
|---|---|
| Cooke, Dodwell, Todd | Adelaide Observatory |
| Kingsmill, Shortt | Hobart Observatory |
| Baracchi, Ellery, Merfield, White | Melbourne Observatory |
| Brooks | NSW Lands Department |
| Cooke | Perth Observatory |
| Cooke, Egeson, Lenehan, Masters, Merfield, Pollock, Russell, Scott, Sellors, Smalley | Sydney Observatory |
| Tornaghi | Scientific instrument-maker |
| Beattie, Biggs, Bone, Brindley, Brownrigg, Butterfield, Close, Cobham, Colyer, Craven, Davidson, Fitchett, Furber, Gale, Heath, Hirst, Hoskins, Innes, Knibbs, Macdonnell, Martin, Megginson, Merfield, Morris, Nangle, Roseby, Ross, Swindlehurst, Thomson, Wooster, Hugh Wright | Independent Australian astronomers |
| Chambers, Clerke, Crommelin, Denning, Gore, Johnson, Lynn, Maunder, Orr, Mary Proctor, Piazzi Smyth, Turner, Tupman, Wesley | the United Kingdom |
| Auwers, Clemens, Kreutz, Rümker, Wolf | Germany |
| Veenstra | the Netherlands |
| Gill, Innes, Roberts, Voute | South Africa |
| Bickerton, Grigg, Stock, Ward | New Zealand |
| Mary Evershed | India |
| Doberth | Hong Kong |
| Burnham, W.W. Campbell, Coddington, Harkness, Holden, Newcomb, Perrine, Peters, E.C. Pickering, W.H. Pickering, See, Todd | the United States |
| Hussey, Perrine | Argentina and the United States |

and Melbourne Observatory's Pietro Baracchi (Orchiston 1999b). Grigg was also a New Zealand astrophotography pioneer (Orchiston 1995), and one of the letters even includes photographs of Comet C/1901 G1 (for an example of one, see Fig. 2.5).

The talented Reverend (later Archdeacon) Arthur Stock (1823–1901; Fig. 2.6; Orchiston 2016) was the first Astronomical Observer at the government's Provincial Observatory when this was set up in the nation's capital, Wellington, but he also maintained a private observatory with a 3.1-in. (8-cm) and later a 4-in. (10.4-cm) refractor at his home (see Orchiston 1998b, pp. 99–101). Twenty letters to Tebbutt dating between 1869 and 1886 document his independent discovery of Comet C/1881 K1 just one day after Tebbutt and discuss his various observations of other comets and of the Moon, planets, lunar occultations, transits of Mercury and Venus, Jovian satellite phenomena, double stars and that enigmatic southern variable star, η Carinae. They also record something of his efforts to actively popularise astronomy, through the publication of two booklets, one on the upcoming 1874 transit of Venus (see Orchiston 1986). In addition, one of his 1869 letters outlines the princi-

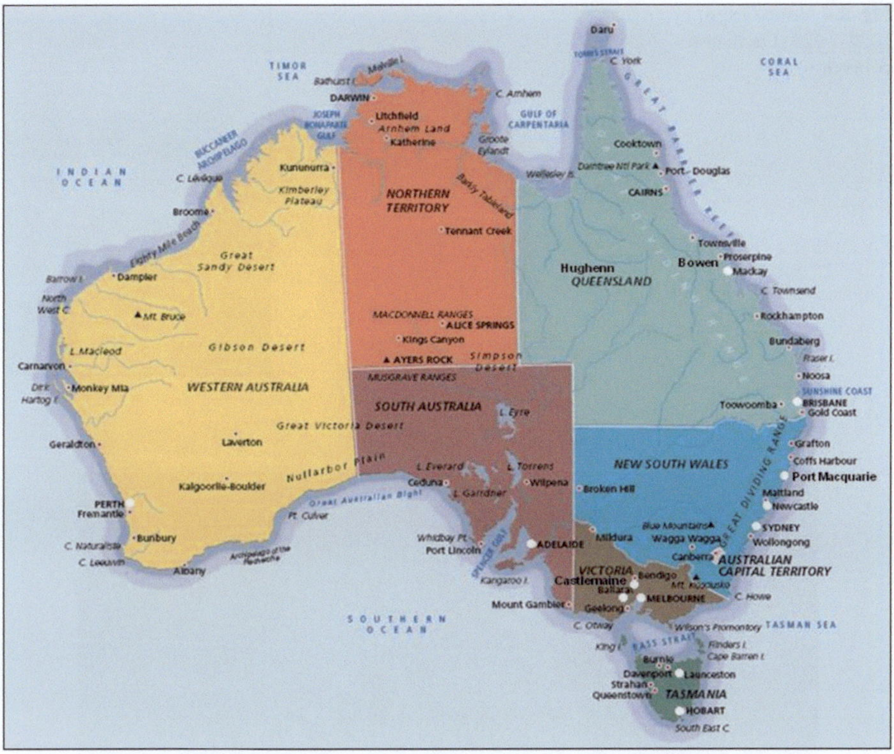


**Fig. 2.3** A map showing the geographical distribution of Tebbutt's *principal* Australian astronomical correspondents. Other astronomical correspondents and non-astronomical correspondents resided mainly in towns throughout country New South Wales and along the Queensland coast (*Map* Wayne Orchiston)

ples of the coronagraph, fully 60 years before Lyot perfected the design and 11 years before William Huggins (1824–1910) came up with a similar idea.

One other New Zealand-related letter is of special interest and deserves a mention here. Lieutenant-Colonel G.L. Tupman was in New Zealand in 1882 to co-ordinate the British transit of Venus programme (see Orchiston 2004c), and during a visit to the South Island city of Dunedin he met and was so impressed with a local astronomer named Arthur Beverly that he was moved to write Tebbutt the following account (*inter alia*):

> Since I have been in this wonderful country I have discovered "a truly bright light under a bushel" in the person of Mr. Arthur Beverley [sic.] of Dunedin. Like many distinguished Astronomers he is a mechanic of a high order, making his own microscope objectives on his own formulae; inventing ingenious apparatus and possessing high mathematical attainments. In Europe he would be in the first rank among physicists & Astronomers. In New Zealand scarcely anyone knows his name—not one soul knows his merits. A watchmaker, I think, by trade, from an obscure part of Scotland, with consumption and another about as deadly disease, he made achromatic doublets of surpassing excellence for Sir David Brewster, Prof Thwaites & other distinguished men until he was able to pay his passage to New Zealand. There he soon saved as much as he required, or acquired enough somehow,

**Fig. 2.4** John Grigg, 1838–1920 (Orchiston collection)

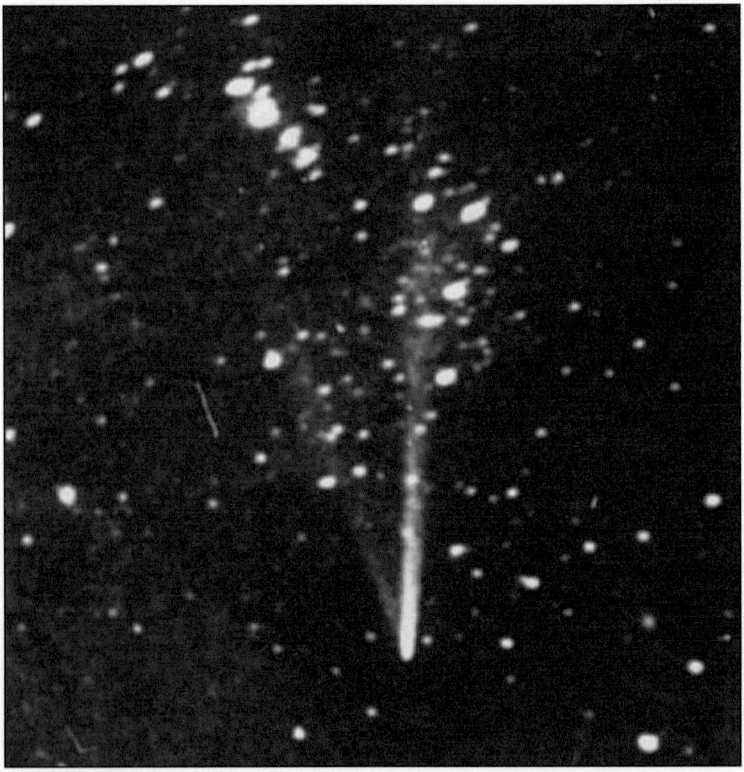

**Fig. 2.5** One of Grigg's photographs of Comet C/1901 G1 (*Courtesy* Mitchell Library)

**Fig. 2.6** Arthur Stock (Orchiston collection)

**Fig. 2.7** Arthur Beverly (Orchiston collection)

> chiefly by plying his trade, and at once gave up business to live & revel in botany, physics & astronomy. He quickly gets out the orbit of any comet that he sees and probably was the first to find that of the present comet …
>
> It is right that you should know what a neighbour you have. He only possesses a three inch achromatic, but it is very good and very well mounted equatorially, with circles graduated to 3 0 … [and] an excellent chronometer … (Tupman 1883).

Beverly (1822–1907; Fig. 2.7) came to New Zealand in 1856 and quickly activated his astronomical interests. In addition to comets, he observed the 1882 transit of Venus and a total solar eclipse that was visible from New Zealand in 1885 (Campbell 2001; Orchiston 1998b, 2016).

The foregoing excerpts give some indication of the wealth of documentation available through the Tebbutt Letters, but they represent only a small sample, even for New Zealand. In all, the Tebbutt Letters contain 88 different items, from 17 different authors, that relate in one way or another to New Zealand astronomy. These are discussed in more detail in Orchiston (1985b).

### 2.2.3 Astronomical Journals, 1853–1902

Variously referred to in the Mitchell Library shelf lists as 'Astronomical Observations', 'Diaries', 'Journals' and a 'Minute Book', these 17 volumes mainly contain day-by-day journal entries from 1853 to 1902 of all things astronomical. They therefore provide a detailed account of Tebbutt's astronomical observations from the date of his very first glimpse of the Great Comet of 1853 (see Fig. 1.7) through almost to his official retirement from observing at the end of 1903. But they do more than this, for they also include data on Windsor Observatory, Tebbutt's telescopes and his other astronomical instruments; sky conditions; visitors to the Observatory; and its latitude and longitude (but see, also, Sect. 2.2.7 below). They also record his astronomically-motivated visits to Sydney.

Of particular interest are the journals dating between 1853 and 1861 since most of the observations recorded in them never made it into print. These were Tebbutt's 'apprenticeship years', when anything and everything celestial captivated him, and we see this reflected in these manuscripts. In addition to a search for Vulcan and observations of sunspots, meteors and fireballs, a transit of Mercury, solar and lunar eclipses, lunar occultations, Saturn, Jovian satellites, comets, the zodiacal light, double stars and variable stars, we find records of atmospheric phenomena (solar and lunar haloes, double rainbows and aurorae), and observations of the Sun and stars made in order to determine the latitude and longitude of the Peninsula Estate Homestead (Fig. 1.2—at this time Windsor Observatory was still a dream).

Many drawings, particularly of sunspots, accompany the journal entries, and these and other renditions (particularly of comets) reveal that Tebbutt was no natural artist. But he was an avid observer, keen to learn proper observing techniques, and trial and error played a key role. For example, on 17 September 1855 he went to observe his very first lunar occultation, and 'The star [8 Scorpii] was gradually approaching the limb, when my breath accidently entered the eyepiece of the telescope, and this of course rendered the star indistinct' (Tebbutt 1853–1859). By the time, he managed to carefully wipe the lens the star was gone. As he perfected his observing and mathematical skills, Tebbutt also reassessed his potential targets and decided to abandon some of the objects (such as sunspots) that particularly took his fancy in these early formative years. The days of experimentation were over!

Once furnished with a proper astronomical telescope, and observatory (Fig. 1.10), Tebbutt embarked on what can only be described as a prolific observing programme, and although many of the observations recorded in his post-1863 journals found their way into print, lots of the detailed documentation, most of his field sketches and some of his earliest experimental observations did not. For instance, Tebbutt (1873–1879) tells us that after looking unsuccessfully for Comet 6P/d'Arrest on 13 April 1877, he located the minor planet Vesta, and used the ring micrometer to obtain nine positional measurements (which were recorded in his journal). Minor planets only became an important part of Tebbutt's research repertoire 9 years later, once he had access to the much larger clock-driven Grubb telescope (Fig. 1.12).

Generally, Tebbutt's observations are presented neatly and clinically, but every so often an entry appears that reveals his deep love of astronomy and his fascination of the night sky. Thus, 11 May 1864 was 'A brilliant night, the milky-way beautifully white against the dark background of the sky' (Tebbutt 1864–1869), while on the evening of 22 September 1879 he was enthralled by Jupiter's appearance:

> In addition to the two long dusky red equatorial belts, there was another large oblong and bright pink patch below or apparently north of the equator. It appeared to be edged with white, of a whiter colour than the general ground of the disc. The general ground I regard as a light yellow. *At this time the planet's disc presented one of the most beautiful views I have ever seen of it.*" (Tebbutt 1879–1881; my italics).

The following year, the lunar eclipse of 22–23 June 1880 was reportedly '… very interesting owing to her [the Moon] being at the time in the eastern branch of the Milky Way' (ibid.), and on the morning of 26 May 1881,

> While looking for the comet I was surprised *by one of the most charming sights I have ever beheld.* This was the group just above the eastern horizon comprising the moon, now a crescent, and Venus, Jupiter and Saturn. All three of the planets were very bright in the clear morning sky, more particularly Venus which shone with extraordinary brilliancy. A white fog creeping along the low points, but not rising to the horizon, set off the spectacle to great advantage. (ibid.; my italics).

Accompanying this account is a field sketch. In another example, on 2 August 1882, Tebbutt (1882–1884) noted that 'The contrast between the colours of the planets as seen in the comet eyepiece was very marked and beautiful, Mars was fiery red while Venus was beautifully white'.

Tebbutt's first astronomical acquisitions were a sextant and a copy of *A Complete Epitome of Practical Navigation* by William John Norie (1772–1843), and his 1853–1859 journal describes the associated coach trip to Sydney on 23 September 1853. The sextant was a valued supplement to the marine telescope made available by his father, while Norie's volume marked the start of what would later grow into a major research and reference library. From this date, there are accounts as each new instrument is acquired, but particularly invaluable is a 13-page overview at the start of the 1879–1881 journal, with details of all three Windsor Observatory buildings (see Fig. 2.8), the sextant and artificial horizon, the marine telescope, the Jones and Cooke refractors, both transit telescopes, the chronometer, a microscope and the various meteorological instruments (see Fig. 1.3).

Only occasionally do the journal entries reflect the obvious affection that Tebbutt held for the instruments that brought him such international renown, but this certainly comes through when he describes the Jones refractor in the Introduction to his 1890–1892 journal:

> Since the acquisition of the 4½ equatorial it [i.e. the 3.25-in Jones refractor] has been restored to its tripod stand and frequently employed in observations of variable stars and in examination of the moon's dark limb for probable occultations. It is to me a very interesting instrument as it is associated with my earliest efforts at astronomical observation.

By this time, the Jones telescope had been with him for nearly 30 years, and although surpassed in aperture by the Cooke and Grubb refractors and inappropriate

**Fig. 2.8** The Tebbutt Homestead in 1880 and three Windsor Observatory buildings (after Tebbutt 1908)

for most of the observing programmes that then held sway at Windsor Observatory, Tebbutt still found a way for it to make a meaningful research contribution.

From time to time evening visitors came to Windsor Observatory (Orchiston 1997), and in his early years Tebbutt welcomed them, but with the passage of the years he became increasingly less tolerant of the requisite loss of valuable observing time and the danger that strangers could pose to his delicate scientific instruments. Compare, for example, the following journal entries, which date to 19 September 1864 and 15 July 1891, respectively:

> I did not observe the comet last night [i.e. 18 September], owing to the Rev John Mosley being at the Observatory. I showed him the comet, which was faint, Jupiter, the double star α Centauri, and other objects. He was much gratified with the visit. (Tebbutt 1864–1869).
>
> Last evening a Windsor gentleman came to the Observatory according to appointment to see the moon, but instead of bringing only another visitor with him as stipulated he brought three others, making five visitors in all. The consequence was that my small round equatorial chamber was well filled. The accident which happened to me on the 8th instant and is recorded above [the spider hairs on the micrometer eyepiece were broken] was due to the agitation produced by a visitor at unseasonable hours, and now again last evening I had the misfortune to break a small tube which holds the square-bar micrometer … and the micrometer fell to the floor … I endeavour to avoid visitors as much as possible [now], for I find there is nothing but ill luck in the reception of them. They come without five minutes preparation for what they wish to see, and go away just as edified as before their arrival. (Tebbutt 1890–1892).

In addition to visitors to Windsor Observatory, Tebbutt's observing journals also mention his visits to Sydney Observatory, and his other astronomical journeys to Sydney by coach or train, mostly to attend meetings or to purchase instruments. One interesting exception has a particularly historical ring to it. In his entry for 18 January 1868, Tebbutt reveals that he

> … visited for the first time the ruins of the old Parramatta Observatory, erected by Sir Thomas Brisbane in the Domain. The old stone transit piers are in a tolerable state of preservation … I felt much pleasure in visiting a place so intimately associated with astronomical science. (Tebbutt 1864–1869).

Only occasionally do Tebbutt's rather clinical journal entries reveal something of his personality. His first published observations were of Comet 109P/Swift-Tuttle in 1862 and appeared in two different issues of *Astronomische Nachrichten*. On 23 June 1863, he received four issues of this journal and recorded in his Diary: 'No 1404 contains my second letter to the Editor on the subject of Comet II. He has styled me "Director der Sternwarte zu Windsor in Neu-Süd-Wales" which is very annoying, considering that my astronomical observatory is merely a portable one' (Tebbutt 1863; his underlining). Others might view this as an unintended compliment, but Tebbutt was a stickler for 'the truth', so it was an indiscretion! Another all-too-brief glimpse of the 'inner man' is revealed in the fascinating journal entry for 7 August 1895 (see Tebbutt 1894–1896) when we learn for the first time about Tebbutt's beloved night-time companion, the magpie 'Jacko' (as documented in the preceding chapter).

### 2.2.4 Observational Notebooks, 1862–1865, 1879–1882, 1884–1905

These little notebooks contain event timings, micrometric measurements, notes, descriptions and field sketches, all recorded 'live', while at the telescope. Most of these details were subsequently transferred to the astronomical journals although some of the sketches and notes were not. The notebooks therefore contain some useful supplementary data for those researching specific aspects of Tebbutt's observing programmes, particularly his cometary work.

### 2.2.5 Determination of Latitude, 1882, and Determination of Longitude, 1885–1887

Precise knowledge of the latitude and longitude were prerequisites for any nineteenth century observatory engaged in positional astronomy, and Windsor Observatory was no exception. Tebbutt was keen to determine both parameters right from the founding of the Observatory, but it was only with the construction of the more substantial brick building in 1879 (see Fig. 1.10) that he derived a figure for the latitude that was used throughout the remainder of his observing 'career'. This was 33°36′30.8″ S, which compared favourably with the value obtained for the original Observatory building 33°36′28.9″ S (Tebbutt 1884). In the 'Definitive Determination of Latitude, 1882' record lot, there are two crudely bound bundles of papers that relate to this new latitude determination.

While establishing latitude was a relatively straightforward matter, there was considerable debate over the best way of determining the longitude of an observatory (e.g. see Challis 1879). Initially, Tebbutt used lunar occultations to derive a figure of 10 h 03 m 21.81 s E (see Tebbutt 1880), but between 1885 and 1887 he used two other methods, which gave slightly different results. One was the telegraphic exchange of time signals between his Observatory and Sydney and Melbourne Observatories, and the other involved physically transporting a Windsor Observatory chronometer to Sydney Observatory and comparing it with the time shown there. These two methods produced values of 10 h 03 m 20.11 s E and 10 h 03 m 20.21 s E, respectively, and relevant here are the six crudely bound bundles of papers in the 'Papers re Telegraphic Determination of Longitude…' record lot. As a matter of interest, a fourth longitude determination method, linking Windsor Observatory to the local trigonometric survey, gave a very similar result (10 h 03 m 20.88 s) when this was later attempted (see Tebbutt 1888).

By the end of the nineteenth century, the latitude and longitude of Windsor Observatory were known with considerable precision, and this private establishment was able to join Melbourne and Sydney Observatories in providing the fundamental reference framework for all Australian astronomical observatories.

### *2.2.6 Extracts from Periodicals etc.*

This record lot includes extracts of various research papers (mainly about specific comets) that appeared in *Astronomische Nachrichten*, *Comptes Rendues des Sciences…*, *English Mechanic* and *Nature*, as well as two items about the computation of cometary orbits (something in which Tebbutt became particularly proficient), and one on the reduction of lunar occultations.

As indicated in Chap. 1, Tebbutt was responsible for forming a Corps of Comet Observers, Australia's first national astronomical group of any kind (see Orchiston 1982). This particular Tebbutt Collection record lot includes a 4-page document about the proposed *modus operandi* of the group and is almost certainly a draft that Tebbutt sent Launceston's Alfred Barrett Biggs for his evaluation and comment. Biggs was a noted Tebbutt protégé (see Orchiston 1985a), and apart from Tebbutt was the only other active member of this short-lived group.

Still on the topic of astronomical groups and societies, two items of special interest in this record lot are the full texts of Tebbutt's Presidential addresses prepared for the Inaugural Meeting and the second Annual General Meeting of the New South Wales Branch of the British Astronomical Association. These manuscripts run to 14.5 and 13 manuscript pages, respectively, and contain a lot of useful information in that only 1.5–2 page summaries of them were published in the Association's *Journal* (see Reports of the Branches… 1895, 1897). As its founding President, Tebbutt added international credibility and visibility to the Branch and was one of those who was instrumental in seeing it develop rapidly during the early years (Orchiston 1988).

Finally, this record lot includes a fascinating 47.5-page manuscript about Christianity and astronomy. On 3 June 1909, Tebbutt wrote and initialled the following description on the first page of this document: 'Rough Draft of a Lecture made some years ago, but not since utilised'. For much of his life, Tebbutt was interested in the interplay between astronomy and religion and sought to reconcile their sometimes disparate views. This was a theme upon which he occasionally lectured, and he even published a booklet on the topic (see Tebbutt 1878 and Fig. 1.12).

### 2.2.7 Hot Winds, 1862–1863; and Meteorological Observations, 1863–1913

The first of these record lots consists of an exercise book with detailed meteorological observations on successive days between 24 December 1862 and 14 January 1863, a period when Windsor experienced prolonged hot winds. Extreme maximum shade temperatures were recorded on three different days: 107.0° Fahrenheit (41.7 °C) on 24 December, 113.4° (45.2 °C) on 5 January and 106.7° (41.5 °C) on 9 January. These observations came straight after Tebbutt acquired a range of instruments and set up a fully-equipped meteorological station. At the end of January 1863, he submitted his first monthly meteorological report (Russell 1863), and from this date he continued to provide regular reports to Sydney Observatory and to Windsor district and Sydney newspapers through until 1898, when a curtailed meteorological programme was commenced and continued up until his death in 1916.

The 13 bound volumes making up the record lot, ‘Meteorological Observations, 1863–1913’, contain the various observations made by Tebbutt at Windsor Observatory. Not only were these summarised in his monthly reports, but at his own expense he also published eight different monographs reporting on this work (Tebbutt 1868, 1874, 1877, 1882, 1886, 1891, 1898b, 1916). These monographs (e.g. see Fig. 2.9), which were widely distributed within Australia and overseas, collectively total 389 pages and provide a document of the weather and flooding at Windsor for more than half a century. In addition, Tebbutt published research papers on Australian storms and Saxby’s weather-prediction model, and a series of newspaper articles on periodicity in local rainfall.

It is apparent that in addition to his outstanding astronomical record, John Tebbutt deserves to be recognised as a pioneer of Australian meteorology (Todd 1893) and that a separate detailed study of his overall contribution in this important scientific field is long overdue.

### 2.2.8 Catalogue of the Windsor Observatory Library

One of the reasons for Tebbutt’s success as a researcher and populariser was that he could rely on an outstanding private research and reference library, and the ‘Catalogue of the Observatory Library’ record lot documents this. A perusal of the entries reveals a good selection of books and monographs, runs of journals and an extensive collection of reprints. In order to publicise the Windsor Observatory

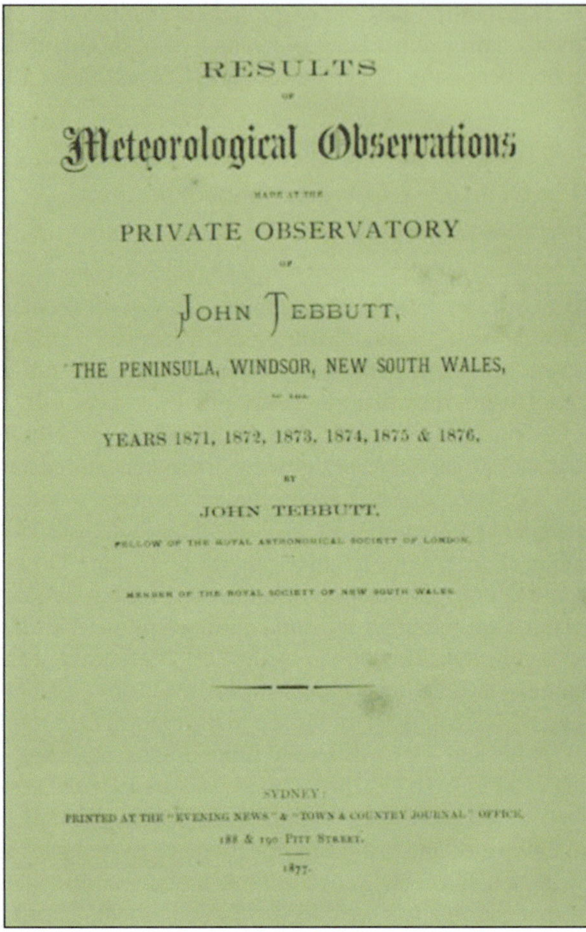
RESULTS

OF

Meteorological Observations

MADE AT THE

PRIVATE OBSERVATORY

OF

JOHN TEBBUTT,

THE PENINSULA, WINDSOR, NEW SOUTH WALES,

FOR THE

YEARS 1871, 1872, 1873, 1874, 1875 & 1876,

BY

JOHN TEBBUTT,

FELLOW OF THE ROYAL ASTRONOMICAL SOCIETY OF LONDON,

MEMBER OF THE ROYAL SOCIETY OF NEW SOUTH WALES.

SYDNEY:

PRINTED AT THE "EVENING NEWS" & "TOWN & COUNTRY JOURNAL" OFFICE,

188 & 190 PITT STREET.

1877.

**Fig. 2.9** An example of the title page of one of Tebbutt's meteorological monographs (Orchiston collection)

Library and at the same time provide appropriate acknowledgement for donors, Tebbutt (1887, pp. 47–74) outlined the Library holdings in his *History and Description of Mr. Tebbutt's Observatory …*, and each year thereafter all new acquisitions were listed in successive *Annual Reports* of the Windsor Observatory.

It was partly these *Annual Reports* that allowed Tebbutt to successfully grow his library, as these little 19–33 page booklets, other books and booklets that he from time to time published, and reprints of his papers, formed the basis of an elaborate international publications exchange network. By the end of the nineteenth century this involved 168 different donors in 33 countries, and most of the world's foremost observatories were supplying Windsor Observatory with gratis publications. In addition to private and professional observatories, he received exchanges and gifts from various government departments, universities, scientific societies and individuals.

### 2.2.9 Other Records

The three remaining record lots, 'Calculations; Reduction of Observations, 1853–1910', 'Instrumental Determinations, 1882–1901' and 'Values of Wire of Transit Instrument, 1885–1903' are self-explanatory and hardly deserve comment here. Many of the reduced times or positions contained in the first record lot appeared in Tebbutt's publications, including newspaper articles that appeared before he began publishing in local and international journals. Meanwhile, some of the data in the second and third record lots were reproduced in his *History and Description of Mr. Tebbutt's Observatory* ... (Tebbutt 1887) and in the *Annual Reports* that he published as little booklets between 1888 and 1903 (inclusive).

## 2.3 The Missing Records

Even a cursory examination of the surviving Tebbutt manuscripts in the Mitchell Library reveals that there are a number of curious gaps in the collection. One of the glaring omissions relates to copies of Tebbutt's outward letters (to which many of the inward letters relate), and their whereabouts has long remained a mystery. Given his professional approach to record-keeping and his methodical nature, it is reasonable to suppose that Tebbutt maintained an outward letter file, and the following 'Memorandum' in the Mitchell Library puts the matter beyond dispute:

> The M.S. found herewith contains the draft of Results prepared at the request of Professor Auwers of Berlin, as per his letter of May 31st 1886. A clean copy of this draft was forwarded by post yesterday to his address. *See Letter Book, page 44*. (Tebbutt n.d.; my italics).

This memorandum is dated 3 November 1886, and at that time Auwers was investigating the longitude of Windsor Observatory and had asked Tebbutt to send him some lunar occultation observations from the 1860s.

If Tebbutt's outward Letter Books had survived through to the year of his death they would have run to many volumes, and like the inward letters would have occupied several linear metres of bookshelf space. Checks with Mitchell Library staff indicated that all of the donated Tebbutt manuscript material has definitely been catalogued, and that the 'missing' volumes are not languishing somewhere in the basement, unrecognised and long forgotten (Irvine 1990).

Given their importance in the overall scheme of things, it is unlikely that the outward letter books would have been retained by the family in 1917 had they survived. This suggests that most if not all of these books did not survive, and a number of possible explanations come to mind. Floods were a comparatively common occurrence at Windsor during the nineteenth and early twentieth centuries (e.g. see Fig. 1.15) and impacted directly on Windsor Observatory (see Table 2.3), but it is unlikely that the missing volumes were destroyed during one or more of these since all of the other records apparently survived these natural disasters. Besides, Tebbutt

**Table 2.3** Major floods recorded on South Creek, 1857–1915[a]

| Year | Date of maximum | Height (m) | Comments |
|---|---|---|---|
| 1857 | 29 July | 9.8 | |
| | 22 August | 11.3 | Very destructive flood |
| 1860 | 29–30 April | 11.2 | Calamitous flood |
| | 26 July | 10.5 | Disastrous flood |
| | 19 November | 10.8 | Calamitous flood |
| 1864 | 13 June | 14.4 | Highest flood for many years; observatory transit room floor 54.8 cm above the flood level |
| | 16 July | 10.9 | |
| 1867 | 23 June | 19.1 | Highest flood in recorded history; water was 4.1 m deep on the floor of the observatory |
| 1868 | 18 February | 8.9 | |
| 1869 | 09 May | 11.1 | |
| 1870 | 28 April | 13.5 | |
| | 13–14 May | 10.6 | |
| 1871 | 30 April | 9.5 | |
| | 02 May | 11.1 | |
| 1873 | 26–27 February | 12.5 | |
| 1875 | 07 June | 11.7 | |
| 1877 | 02–03 May | 9.1 | |
| 1879 | 11 September | 13.0 | |
| | 17 September | 10.4 | |
| 1889 | 29 May | 11.6 | |
| 1890 | 13 March | 11.7 | |
| | 26 March | 10.3 | |
| 1891 | 26 June | 10.8 | |
| 1894 | 23 March | 9.7 | |
| 1895 | 24 January | 9.3 | |
| 1898 | 15–16 February | 9.7 | |
| 1900 | 07 July | 14.1 | Third highest flood of past 60 years; threatened the Observatory buildings |
| 1904 | 12 July | 12.2 | |

[a]Based on data published in Tebbutt (1877, 1882, 1886, 1891, 1898b, 1916)

(1908) makes no specific mention of the loss of any such records in his *Astronomical Memoirs*, which was written in 1907. Nor is there any reference to such a loss in his *Annual Reports*, in any of the inward letters, or in newspaper articles and reports.

However, several letters do refer to another catastrophe: a fire that occurred in December 1897. A news item in the 11 December issue of the *Daily Telegraph* reports:

> A fire broke out in the storeroom at Mr. John Tebbutt's Windsor Observatory early this morning, the building and contents being consumed. But for the exertions of many hands the main buildings would have caught. The walls were considerably charred as it was. (Fire at Windsor 1897).

The specific wording of this account, and a similar one in the *Sydney Morning Herald*, give the impression that Windsor Observatory itself and its entire contents were destroyed and, amongst others, Joseph Brooks (1897), Walter Gale (1897), George Handley Knibbs (1858–1929; 1897) and C.J. Merfield (1897), four well-known Sydney astronomers, immediately wrote expressing their concerns and dismay. However, a letter from J.A. Dick (1897) indicates that this fire was actually located in one of the free-standing buildings adjacent to the Observatory. Thanks to a map of the Peninsula Estate buildings produced on page 120 in the 1986 reprinting of *Astronomical Memoirs* (Tebbutt 1986) we can identify this storeroom, for this shows a rectangular building, "H" near the kitchen wing of the homestead and described as 'Site of granary and storeroom destroyed by fire Dec 1897 (now rebuilt)'.

I believe that all of the outward Letter Books, except the one current at that time, were housed in this store room in 1897 and were destroyed or damaged beyond repair in the course of the fire. Regrettably, there is no mention of the fire in Tebbutt's *Annual Reports* for 1897 or 1898 (Tebbutt 1898a, 1899) or in his *Astronomical Memoirs* (Tebbutt 1908), and there are no entries at all in his Journal (Tebbutt 1896–1898) between the dates of 5 and 23 December 1897, but some of the Tebbutt Collection items in the Mitchell Library do have water damage (e.g. A3779), while others display fire damage or have a distinctly smoky smell.

This scenario would explain the disappearance of the Letter Books predating 1898, but what of those letters written by Tebbutt between that date and his death in 1916? In 1992, the present owner of the property, John Halley Tebbutt (a great-grandson of the astronomer), informed me (pers. comm.) that the search of the various Peninsula Estate buildings at one time produced what appears to have been a single Letter Book (from his recollection it contained pages of carbon copies of letters in John Tebbutt's distinctive writing). Unfortunately, he was unable to locate this at the time of my visit, or on subsequent ones. I believe that when this document eventually reappears it will prove to be the, or one of the, missing 'post-fire' Letter Books.

There are also a number of other obvious gaps in the Mitchell Library's Tebbutt Collection. We know that Tebbutt continued to observe almost up to the time of his death and that he processed these observations (e.g. Tebbutt 1905–1915) and published them, but there is no manuscript journal summarising his observations after 1903, even though it is highly likely that one was maintained. Nor is there an 'Observational Notebook' for the period 1906–1915 although Tebbutt's post-retirement publications would indicate that one was definitely kept. The other glaring inconsistency is the absence of any observational notebooks for the interval 1866–1878. We know that notebooks were kept at this time since observations were made, reduced and published, and a journal was maintained.

I believe that these 'missing' manuscripts were retained by the Tebbutt family when the bulk of the Tebbutt Collection went to the Mitchell Library (perhaps because the family planned to write about their illustrious relative, and most of these documents built on the autobiography that Tebbutt had published in 1908), but were subsequently destroyed. Two possible scenarios come to mind. When I first visited Windsor Observatory, in 1959, the Peninsula Estate was owned by Bruce Tebbutt

(1904–1963), a grandson of the astronomer, and he kept various historic memorabilia in a large safe in the cottage in which he lived (at that time the 1845 Tebbutt homestead and the 1879 observatory building were rented out). Soon after his death in 1963, this safe was opened for the first time in a great many years and was found to be full of water from the 1961 flood—and possibly from an earlier flood or floods. Apart from some coins and rings, and the 1905 Jackson-Gwilt Medal and a silver medal that Tebbutt received in connection with the 1867 Paris Universal Exhibition, this safe contained nothing else but 'black slime' (J.H. Tebbutt, pers. comm., 1996) which could represent the decayed remains of the missing manuscripts. Alternatively, in 1959 I discovered a large assemblage of books (but mainly on topics other than astronomy), assorted papers, and various astronomical items (e.g. copies of *Annual Reports*, *Astronomical Memoirs* and meteorological monographs, and reprints of research papers) that had been dumped in the 1894 circular brick observatory building. They were not systematically grouped or stored, and many items were in very poor condition, with abundant evidence of silverfish and water damage. In addition, much paper-based material had disintegrated to the point where it was totally unrecognisable and literally fell to pieces if touched. It is quite possible that some or all of the 'missing manuscripts' were amongst these items.

## 2.4 Concluding Remarks

We must count ourselves fortunate that the Tebbutt Collection is located in the Mitchell Library, one of Australia's foremost archival repositories. As such, it is cared for by well-trained staff, and stored in facilities that reflect the best internationally-accepted standards of preventative conservation. Meanwhile, if remedial action is called for through accidents or excessive use of the Collection by researchers, the Mitchell Librarian can call on the services of the State Library's Conservation Department. It is reassuring to know that the future of the Collection is secure.

On the other hand, the research potential of the Collection has yet to be optimised. The inward letters will only be effectively utilised as data sources on Australian and non-Australian astronomical history when master lists of the different letters—and more importantly their contents—are available to scholars. Similarly, Tebbutt's observational records contain invaluable data and diagrams that were not included in his numerous publications, and it is only when detailed descriptions of the contents of all of the logs, journals, diaries and notebooks have been published that the full research value of this corpus of archives will be realised (even though some useful case studies, incorporating such data, have already been published—e.g. see Orchiston 1998a, 1999a, 2000). The Tebbutt Collection also contains a rich history of Windsor meteorology, extending over more than half a century, and this invaluable data set will only be properly probed and utilised when the story of Tebbutt's role as a pioneering Australian meteorologist is written. But in order to do this effectively, and for many other Tebbutt case studies, it would help immeasurably if we could access Tebbutt's outward letters. Since his file copies cannot be found, this will involve an extensive and expensive survey of many overseas repositories in a quest for the originals. Some of these (such as those in the RAS Archives in London) have already been examined, but much remains to be done.

John Tebbutt was Australia's foremost late nineteenth century astronomer, and the Mitchell Library's Tebbutt Collection therefore will long remain a vital resource for those researching nineteenth century Australian astronomy.

**Acknowledgements** I am grateful to Professor Nick Lomb (University of Southern Queensland, Toowoomba) and the staff at the Mitchell Library for their assistance and for kindly supplying Fig. 2.5.

## References[1]

Brooks, J. (1897). Letter to John Tebbutt, dated 13 December. In TL.

Campbell, R. (2001). Arthur Beverly 1822–1907. *Southern Stars, 40* (3), 23–24.

Challis, J. (1879). *Lectures on Practical Astronomy and Astronomical Instruments*. London: Deighton, Bell and Co.

Débarbat, S. (2002). The IAU, astronomical archives and Commissions 41 and the ICHA. *Journal of Astronomical History and Heritage, 5,* 181–183.

Dick, J.A. (1897). Letter to John Tebbutt, dated 11 December. In TL.

Fire at Windsor. *Daily Telegraph,* 11 December 1897.

Gale, W. (1897). Letter to John Tebbutt, dated 13 December. In TL.

Irvine, K. (1990). Letter to Wayne Orchiston, dated 13 November. In Orchiston Collection.

Knibbs, G.H. (1897). Letter to John Tebbutt, dated 11 December. In TL.

Merfield, C.J. (1897). Letter to John Tebbutt, dated 16 December. In TL.

Mourot, S. (1973). Astronomical archives in Australia (1). *Journal for the History of Astronomy, 4,* 66–68.

Mr. Hugh Wright. Distinguished Librarian to retire. *The Sydney Morning Herald,* 27 September 1924, page 8.

Orchiston, W. (1982). Illuminating incidents in Antipodean astronomy: John Tebbutt and the abortive Australian association of comet observers. *Astronomical Society of Victoria, 35,* 70–83.

Orchiston, W. (1985a). The contribution of Alfred Barrett Biggs to Australian and Tasmanian astronomy. *Records of the Queen Victoria Museum, 89,* 1–28.

Orchiston, W. (1985b). A survey of manuscript material in the Mitchell Library (Sydney) relating to the history of astronomy in New Zealand. *Southern Stars, 31,* 225–238.

Orchiston, W. (1986). Taking Stock of early astronomy books. *Southern Stars, 31,* 326–328.

Orchiston, W. (1988). Illuminating incidents in Antipodean astronomy: formation of the New South Wales Branch of the British Astronomical Association. *Journal of the British Astronomical Association, 98,* 75–84.

Orchiston, W. (1993). John Grigg, and the genesis of cometary astronomy in New Zealand. *Journal of the British Astronomical Association, 103,* 67–76.

Orchiston, W. (1995). John Grigg, and the development of astrophotography in New Zealand. *Australian Journal of Astronomy, 6,* 1–14.

Orchiston, w. (1997). The role of the amateur in popularising astronomy: an Australian case study. *Australian Journal of Astronomy, 7,* 33–66.

[1]The following abbreviation is used: TL = Bound manuscript tetters to John Tebbutt, in the Mitchell Library, Sydney.

Orchiston, W. (1998a). Illuminating incidents in Antipodean astronomy: John Tebbutt and the Great Comet of 1861. *Irish Astronomical Journal, 25,* 167–178.

Orchiston, W. (1998b). *Nautical Astronomy in New Zealand. The Voyages of James Cook.* Wellington: Carter Observatory (Occasional Papers No. 1).

Orchiston, W. (1999a). C/1881 K1: a forgotten "Great Comet" of the nineteenth century. *Irish Astronomical Journal, 26,* 33–44.

Orchiston, W. (1999b). Comets and communication:amateur-professional tension in Australian astronomy. *Publications of the Astronomical Society of Australia, 16,* 212–221.

Orchiston, W. (2000). John Tebbutt of Windsor, New South Wales: a pioneer southern hemisphere vartiable star observer. *Irish Astronomical Journal, 27,* 47–54.

Orchiston, W. (2001). The Thames observatories of John Grigg. *Southern Stars, 40* (1), 14–22.

Orchiston, W. (2004a). An introduction to the astronomical archives of Australia and New Zealand. *Journal of Astronomical Data, 10,* 67–76.

Orchiston, W. (2004b). Highlighting the history of nineteenth century Australian astronomy: the Tebbutt Collection in the Mitchell Library, Sydney. *Journal of Astronomical Data, 10,* 77–103.

Orchiston, W. (2004c). The nineteenth century transits of Venus: an Australian and New Zealand overview. *Journal of Astronomical Data, 10,* 219–308.

Orchiston, W. (2016). *Exploring the History of New Zealand Astronomy: Trials, Tribulations, Telescopes and Transits.* Cham: Springer.

Reports of the Branches. New South Wales Branch. *Journal of the British Astronomical Association, 5,* 243–245 (1895).

Reports of the Branches. New South Wales Branch (Sydney). *Journal of the British Astronomical Association, 7,* 376–378 (1897).

Russell, H.C. (1863). Letter to John Tebbutt, dated 13 February. In TL.

Tebbutt, J. (n.d.).Tebbutt Papers. Miscellaneous MSS, Mitchell Library (A3775).

Tebbutt, J. (1853–1859). *Observations made by John Tebbutt Jun$^r$. at the Peninsula, near Windsor, in the Colony of New South Wales, during the years 1853, 1854, 1855, 1856, 1857, 1858, and 1859.* MS, Mitchell Library (A3745$^{-1}$).

Tebbutt, J. (1863). Diary. MS, Mitchell Library (A3786).

Tebbutt, J. (1864–1869). *Journal. Observatory, Peninsula, Windsor, New South Wales.* MS, Mitchell Library (A3747).

Tebbutt, J. (1868). *Meteorological Observations Made at the Private Observatory of John Tebbutt, Junr., The Peninsula, Windsor, New South Wales, in the Years 1863, 1864, 1865, and 1866.* Sydney: printed for the author.

Tebbutt, J. (1873–1879). Untitled journal. MS, Mitchell Library (A3783).

Tebbutt, J. (1874). *Meteorological Observations Made at the Private Observatory of John Tebbutt, The Peninsula, Windsor, New South Wales, in the Years 1867, 1868, 1869, and 1870.* Sydney: printed for the author.

Tebbutt, J. (1877). *Results of Meteorological Observations Made at the Private Observatory of John Tebbutt, The Peninsula, Windsor, New South Wales in the Years 1871, 1872, 1873, 1874, 1875 & 1876.* Sydney: printed for the author.

Tebbutt, J. (1878). *The Testimony Which Astronomy Furnishes to the Attributes of the Creator.* Sydney: Young Men's Christian Association.

Tebbutt, J. (1879–1881). *Journal for Astronomical Observations.* MS, Mitchell Library (A3749).

Tebbutt, J. (1880). On the longitude of the Observatory, Windsor, N.S. Wales. *Monthly Notices of the Royal Astronomical Society, 40,* 440.

Tebbutt, J. (1882). *Results of Meteorological Observations Made at the Private Observatory of John Tebbutt, The Peninsula, Windsor, New South Wales, in the Years 1877, 1878, 1879, 1880, & 1881.* Sydney: printed for the author.

Tebbutt, J. (1882–1884). *Journal for Astronomical Observations. The Peninsula, Windsor, N.S. Wales. MS,* Mitchell Library (A3750).

Tebbutt, J. (1884). On the latitude of the Observatory, Windsor, N.S. Wales. *Monthly Notices of the Royal Astronomical Society, 44,* 101–103.

Tebbutt, J. (1886). *Results of Meteorological Observations Made at the Private Observatory of John Tebbutt, The Peninsula, Windsor, New South Wales, in the Years 1882, 1883, 1884, & 1885*. Sydney: printed for the author.

Tebbutt, J. (1887). *History and Description of Mr. Tebbutt's Observatory, Windsor, New South Wales*. Sydney: printed for the author.

Tebbutt, J. (1888). On the difference of longitude between Mr Tebbutt's Observatory, Windor, New South Wales, and the Government Observatories at Sydney and Melbourne. *Monthly Notices of the Royal Astronomical Society, 48,* 313–316.

Tebbutt, J. (1890–1892). *Journal for Astronomical Observations. The Peninsula, Windsor, New South Wales*. MS, Mitchell Library (A3754).

Tebbutt, J. (1891). *Results of Meteorological Observations Made at the Private Observatory of John Tebbutt, The Peninsula, Windsor, N.S, Wales, in the Years 1886, 1887, 1888, 1889, & 1890.* Sydney: printed for the author.

Tebbutt, J. (1894–1896). Untitled journal. MS, Mitchell Library (A3756).

Tebbutt, J. (1896–1898). Untitled journal. MS, Mitchell Library (A3757).

Tebbutt, J. (1898a). *Report of Mr. Tebbutt's Observatory, The Peninsula, Windsor, New South Wales, For the Year 1897*. Sydney: printed for the author.

Tebbutt, J. (1898b). *Results of Meteorological Observations Made at the Private Observatory of John Tebbutt, The Peninsula, Windsor, New South Wales, in the Years 1891, 1892, 1893, 1894, 1895, 1896, and 1897*. Sydney: printed for the author.

Tebbutt, J. (1899). *Report of Mr. Tebbutt's Observatory, The Peninsula, Windsor, New South Wales, For the Year 1898*. Sydney: printed for the author.

Tebbutt, J. (1905–1915). *Reductions of Observations of Binary Stars 1905 to 1915*. MS, Mitchell Library (A3745[2]).

Tebbutt, J. (1908). *Astronomical Memoirs*. Sydney: printed for the author.

Tebbutt, J. (1916). *Results of Meteorological Observations Made at Mr. Tebbutt's Observatory The Peninsula, Windsor, New South Wales, During the Period 1898–1915*. Sydney: printed for the author.

Tebbutt, J. (1986). *Astronomical Memoirs*. Windsor: Hawkesbury Shire Council (Reprinted Edition, with additional material added).

'The Mitchell Library Sydney N.S.W.Guide for Readers.' Cyclostyled notes, Mitchell Library (1967).

Todd, Sir Charles (1893). Meteorological work in Australia: a review. In R. Tate, E.H. Rennie & W.H. Bragg (Eds.), *Report of the Fifth Meeting of the Australasian Association for the Advancement of Science, Held in Adelaide, South Australia, September, 1893* (pp. 246–270). Melbourne: AAAS.

Tupman, G.L. (1883). Letter to John Tebbutt, dated 16 January. In TL.

Wright, H. (1908). Letter to John Tebbutt, dated 25 October. In TL.

Wright, H. (1912). Letter to John Tebbutt, dated 21 August. In TL.

# Part II
# Background Context

# Chapter 3
# The Development of Professional Astronomy in the Nineteenth Century Australia

## 3.1 Introduction

Australia has a long and interesting history of astronomical achievement (see Haynes et al. 1996). The first astronomical observations made from this island continent took place more than 50,000 years ago when the progenitors of the Aboriginal Australian people began exploring this strange new world after venturing forth from their tropical ancestral homeland in what is now South-East Asia. However, from the very start Aboriginal astronomy was bound up with mythology, religion and the natural landscape (e.g. see Clarke 2014; Fuller et al. 2014; Hamacher 2012, Haynes 2000; Johnston 1998; Norris and Harney 2014; Norris and Norris 2009), and differed markedly from the scientific astronomy that blossomed in Europe and Britain during the last 500 years or so.

Scientific astronomy came to Australia with the 'First Fleet' in 1788, thus initiating an era of Antipodean scientific inquiry that has continued through to the present day. When we review the development of Australian astronomy during the past two centuries or so, it is convenient to identify four discrete chronological phases which, by international analogy, can be best termed the Formative, Classical, Transitional and Modern Eras.

The Formative Era (1788–1847) began with the arrival of the First Fleet, witnessed the construction of the ill-fated Dawes Observatory, and ended with the closing of Parramatta Observatory. Australian astronomy gained international visibility during this era, thanks to the achievements of the staff at Parramatta Observatory, but the final years of this Observatory also brought notoriety as the British Astronomer Royal George Biddell Airy (1801–1892; Fig. 3.1; Airy 1896) painted a picture of disaster.

Australia's initial 'golden age' of astronomy occurred during the Classical Era, from 1848 to 1923, and the first task was to repair the damaged reputation inherited from Parramatta Observatory. Between 1855 and 1880, three professional observatories and one distinguished non-professional observatory largely were responsible


W. Orchiston, *John Tebbutt*, Historical & Cultural Astronomy,
DOI 10.1007/978-3-319-44521-2_3

**Fig. 3.1** An undated photograph of Astronomer Royal, George Biddell Airy (https://en.wikipedia.org)

for repairing Australia's international reputation in astronomy and rebuilding and strengthening the foundations of Australian astronomy. They were Williamstown Observatory and its successor Melbourne Observatory in the colony of Victoria and Sydney Observatory in New South Wales, while the fourth observatory was at Windsor, near Sydney, the home base of John Tebbutt. As we shall see, the Classical Era also was an expansionist era when the different colonial governments decided to fund their own observatories, as well as eclipse and transit expeditions. Many of the staff employed in the government observatories were amongst the power-brokers in emergent Australian science. As a result of the substantial publications output of the nation's astronomers (professional and non-professional), Australia had a very high level of astronomical visibility and Australia and South Africa were viewed as the world's leading southern hemisphere astronomical nations.

The Transition Era (1924–1945) started with the founding of the Commonwealth Solar Observatory and closed with the termination of World War II. This era was characterised by a marked shift in research direction from positional astronomy to astrophysics (and solar physics), and by 1945 the two surviving state observatories had drifted far from the international mainstream of astronomical research and into relative international obscurity.

The Modern Era began immediately after World War II and continues to the present day. It is Australian astronomy's second 'golden age', one characterised by significant developments in cosmic ray, gamma ray, X-ray, ultraviolet, optical, infrared, and radio astronomy, and notable Australian breakthroughs in both instrumentation and theory (Haynes et al. 1996). Australia is once more at the cutting edge of contemporary astronomy (e.g. see Bhathal et al. 2013).

While the above model is a useful one for the purposes of discussion and analysis, it is important to recognise that the gradualist pattern of astronomical development was punctuated throughout by intervals of stagnation (e.g. during the depressions of the 1890s and 1930s) and periods of accelerated growth. There was also a wholesale change in theoretical emphasis, as the pre-eminent place of the nineteenth century positional astronomy was slowly replaced during the 1930s and 1940s by astrophysics (Orchiston 1989a).

After a brief diversion into the records of Parramatta Observatory, this chapter reviews the emergence of professional astronomy in Australia during the Classical Era when John Tebbutt was active and provides a national context for his achievements. This introductory chapter has been developed from a review paper that was published back in 1988 (Orchiston 1988a) and also follows closely the introductory section of Orchiston (2015).

## 3.2 Sir Thomas Brisbane: Pioneering Days at Parramatta Observatory

Apart from some valuable meteorological records (McAfee 1981) nothing of lasting scientific merit was achieved at the short-lived Dawes Observatory erected in Sydney in 1788 (see Laurie 1988), but this was not true of its successor, Parramatta Observatory (Tebbutt 1867). Prior to his appointment as Governor of New South Wales in 1821, Thomas Brisbane (1773–1860; Saunders 1990) foresaw the enormous potential that a properly-equipped observatory offered under the neglected skies of the Southern Hemisphere and proposed that he be provided with appropriate instruments.

When the British authorities chose not to accede to his request Brisbane acted alone, bringing his own equipment (Lomb 2004), and securing the services of two astronomers, the German Charles Rumker and Scotland's James Dunlop. All were quickly installed at Parramatta Observatory (shown in Fig. 3.2), adjacent to Brisbane's residence. In the course of the next 15 years, pioneering astronomical work was carried out at this institution (see Saunders 2004), the main achievement being compilation of the so-called Parramatta Catalogue of star positions (Richardson 1835), even though astronomers later became aware that this was not without serious problems (see Bhathal 2012). However, in addition, observations of the planets and comets were made, and Dunlop and Rumker discovered new comets. Parramatta Observatory also served as the fledgling colony's meteorological and time-keeping centre (Russell 1888).

During the 1840s, the Observatory fell into disrepair, and it was finally closed in 1847 (Orchiston 1988d), much to the relief of Astronomer Royal Airy (cf. 1846). Later, when the idea of a new observatory in or near Sydney was voiced, Airy (1849; my italics) was vehemently opposed to this:

> … I have no design of recommending the renewed establishment of an Astronomical Observatory in that character; as my experience of the conduct of observatories in general, and *my strong feeling of the discreditable mismanagement of that at Parramatta*, will effectively prevent me from making such a recommendation.

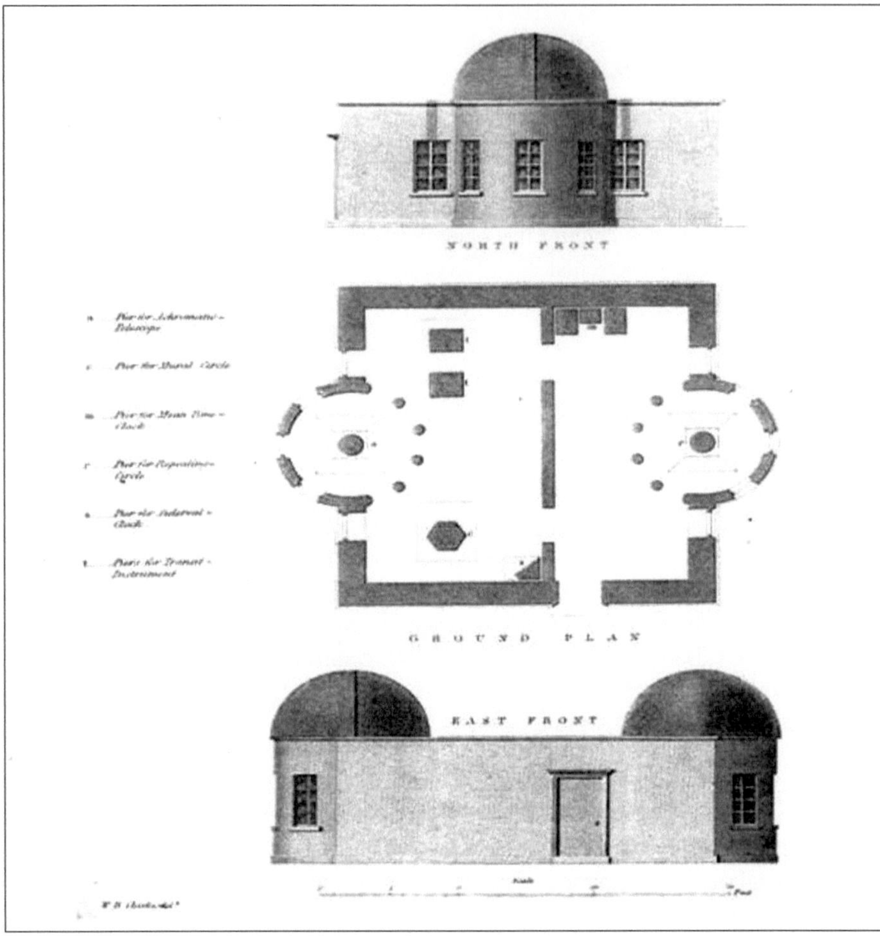


**Fig. 3.2** Plan, end and side elevation drawings of Parramatta Observatory made by the W.B. Clarke in 1825 (after Robertson et al. 2010, p. 404)

Notwithstanding these comments, in his review of Australian astronomy, Ellery (1901) has this to say about Parramatta Observatory:

> I regard [this observatory] as the earliest indication of the growth of astronomy in Australasia. For the first time we have a permanent and well-founded observatory established, with an accomplished staff of observers, who did splendid astronomical work, and left behind them valuable contributions to the science …

It was through Parramatta Observatory that Australia first gained international credibility as an astronomical nation, but as we have seen, Airy's attitude and comments also guaranteed that this Observatory would end up with a bad reputation.

## 3.3 From Imperial to Independent Astronomy: The Development of the Colonial Observatories

Apart from Phillip Parker King (Orchiston 1988d), no one was particularly active in astronomy in Australia following the demise of Parramatta Observatory, and it was only in the 1850s that an urgent need for observatories emerged, in both Sydney and Melbourne, as the lure of the goldfields brought an avalanche of immigrants (for localities mentioned in the text, see Fig. P.1). The primary demands of an accurate time-service and up-to-date meteorological data were best satisfied by the establishment of an observatory, and only Tasmania was adequately supplied in this regard, although its Hobart-based Rossbank Observatory (which was set up primarily as a magnetic facility) was destined to close in 1854 (Savours and McConnell 1982).

Thus, it was that professional observatories proper were established in the various Australian colonies during the second half of the nineteenth century. The first of these was at Williamstown, Victoria (Fig. 3.3), in 1853 (Andropoulos 2014) and the last at Perth in 1896 (Baracchi 1914). Initially, only Adelaide (Fig. 3.4; Dodwell 1910), Melbourne (Fig. 3.5; Perdrix 1961) and Sydney Observatories (Fig. 3.6; Wood 1958) contained sufficient staff and equipment to warrant ranking on an international scale, but Perth (Fig. 3.7) was able to join this trio at the end of the century (Utting 1989, 1992). Information about the various government observatories is given in Table 3.1. Melbourne Observatory only came into existence when the Victorian Government decided to close down the Melbourne-based geomagnetic-meteorological Flagstaff Observatory (Home 1991; Home and Kretzer 1991); Neumayer (1858) *and* Williamstown Observatory and combine their functions in a single institution (Andropoulos 2014; Ross 1918).

**Fig. 3.3** Williamstown Observatory in 1862 (Orchiston collection)

**Fig. 3.4** Adelaide Observatory (*Courtesy* setterfield.org)

**Fig. 3.5** A woodcut showing Melbourne Observatory in 1863 (after New Observatory 1863)

**Fig. 3.6** A sketch showing Sydney Observatory in the 1870s (Orchiston collection)

**Fig. 3.7** The old Perth Observatory main building, which is now the headquarters of the National Trust of WA (*Courtesy* westaussieweddding.typad.com)

**Table 3.1** Australian government observatories, 1850–1916

| Observatory | Founding year | Main telescope(s)[a] | References |
|---|---|---|---|
| Williamstown | 1853 | 4.5-in. OG | Andropoulos (2014), Ellery (1869) |
| Flagstaff | 1858 | Small OG | Neumayer (1858), Perdrix (1990), Weiderkehr (1988) |
| Melbourne | 1863 | 48-in. spec | Andropoulos (2014), Gascoigne (1992), Gillespie (2011), Perdrix (1961, 1970) |
| | | 8-in. OG | |
| | | 13-in. astr | |
| Sydney | 1858 | 11.5-in. OG | Lomb (2011), Orchiston (1988b), Wood (1958) |
| | | 7.25-in. OG | |
| | | 13-in. astr | |
| Adelaide | 1874 | 8-in. OG | Edwards (1993, 1994) |
| Brisbane | 1879 | Small OG | Haynes et al. (1993) |
| | | Small tran | |
| Hobart | 1882 | Small tran | Meteorological Department (1900) |
| Perth | 1896 | 12-in. spec | Hutchison (1980, 1981), Utting (1989, 1992) |
| | | 13-in. astr | |

[a]*Key*: astr = astrograph; OG = refractor; spec = reflector; tran = transit telescope

In addition to the above eight institutions, an observatory was founded by the Jesuits at Riverview College, Sydney, in 1909 (Orchiston 1985), but since the primary initial concern was seismology (see O'Connell 1952) it has been omitted from Table 3.1. The only other Australian institution with a substantial nineteenth century involvement in observational astronomy was the New South Wales Lands Department, which even created the official position of 'Field Astronomer' in 1886 (Orchiston 1987).

When considering the 'astronomical observatories' listed in Table 3.1, it is important to realise that astronomy was in fact only one of their charters. To varying degrees, these colonial observatories also were involved in geomagnetism, meteorology, seismology, tidal studies, trigonometrical survey work and providing a local time service (see Haynes et al. 1996), and the percentage of time that an observatory's staff devoted to each of these fields normally depended upon government priorities (which were usually reflected in the range of scientific instruments available), staffing levels and the specific interests of the directors. While Adelaide, Melbourne, Perth, Sydney and Williamstown Observatories adopted more wide-ranging policies, the short-lived Flagstaff Observatory was concerned with geomagnetism and meteorology, while the small rather modestly-equipped observatories in Brisbane (Fig. 4.7) and Hobart focussed almost exclusively on meteorology and time-keeping (Baracchi 1914).

The discipline-mix of the various observatories changed rather dramatically in 1908, after Australian federation, when meteorology was centralised under the banner of the newly-formed Commonwealth Bureau of Meteorology (Home and Livingston 1994; Wood 1968). This was a 'two-edge sword', for although it left more time free for astronomy it also was accompanied in most instances by the almost inevitable funding cuts.

A commonly-held misconception is that those observatories which were engaged in astronomy, after taking observations for time-keeping purposes, spent the remainder of their effort on research. This was not so, for each institution also had important educational and informational roles to fulfil (see Orchiston 1991). Because of chronic understaffing, and the emphasis that governments placed on public-viewing nights some colonial observatories found their intended research programmes severely disrupted (e.g. see Fig. 3.8).

To some extent the types of astronomical research which could be carried out by the colonial observatories were influenced by geography. Since all were founded in the first place to provide local time services, they tended to be situated within or on the boundaries of their respective towns—generally within direct visual range of the port areas—and such locations were rarely suited to non-meridian astronomical work. This situation was exacerbated with the passage of time as towns grew into cities and levels of light and other pollution increased.

Despite these shortcomings, Adelaide, Melbourne, Perth, Sydney and Williamstown Observatories all succeeded in carrying out worthwhile astronomical research and placing their names before the international astronomical community. Their research interests were in positional astronomy, which involved establishing the precise positions of celestial objects, either through micrometric observations or

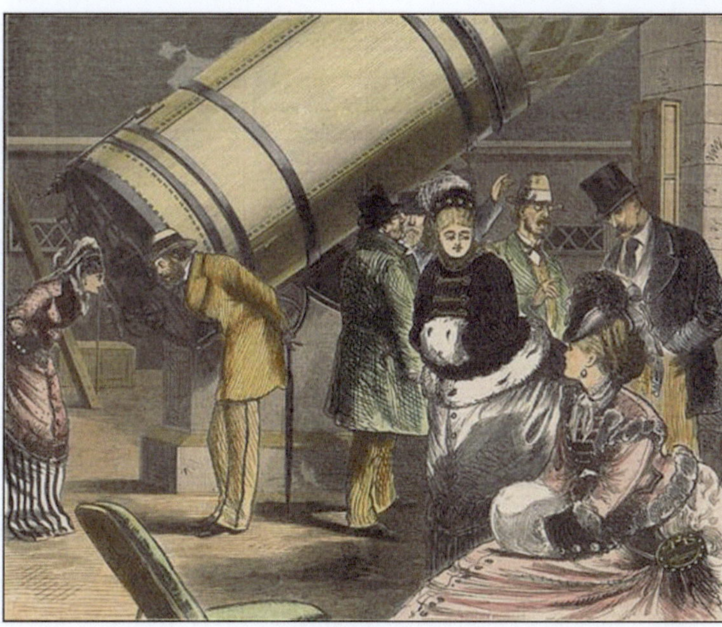

**Fig. 3.8** An engraving of 'A party of amateur astronomers' at the Great Melbourne Telescope, in the *Australasian Sketcher* of 13 June 1874 (*Courtesy* Museum Victoria)

**Table 3.2** Types of astronomical research programmes most commonly undertaken by the leading Australian government observatories, 1850–1916

| |
|---|
| Transitory events |
| Eclipses of the Moon |
| Eclipses of the Sun |
| Lunar occultations of planets |
| Lunar occultations of stars |
| Phenomena of Jupiter's satellites |
| Transits of Mercury |
| Transits of Venus |
| Short-term Monitoring Projects |
| Comets (positions and appearance) |
| Long-term Monitoring Projects |
| Double stars (separation and position angle) |
| Planets (appearance) |
| Sun (surface details) |
| Search Programs |
| New Double stars |
| Sky Survey Work |
| Star positions |

by accurately timing the occurrence of particular events. The range of objects and events that drew their attention is indicated in Table 3.2. In addition, all of these Australian observatories were interested in accurately mapping the positions of stars in specific regions of the sky.

Research astronomy required the right combination of instruments and staff (see Tables 3.1 and 3.3), and in this regard the four major Australian observatories were well served, by international standards. Sydney Observatory established an excellent reputation for quality staff, the first two directors, Scott (Fig. 6.1) and Smalley, being Cambridge mathematicians appointed in England by Airy, the Astronomer Royal (Orchiston 1998b; Wood 1958). All new senior subordinate staff attached to the Observatory came with excellent newly-acquired academic qualifications, or else with substantial experience in observational and mathematical astronomy. The most prominent of these was Henry Russell (Fig. 12.1), who joined the staff in 1859 and was Director and Government Astronomer of New South Wales from 1870 to 1905 (Bhathal 1991). After Dr. Harley Wood (Director from 1943 to 1974; see Fig. 12.8), Russell would rank as Sydney Observatory's second greatest director.

A properly-trained professional also was appointed to direct Adelaide Observatory, in the person of Charles (later Sir Charles) Todd (Fig. 12.28). As a 15-year-old, Todd went straight from school into a professional astronomy post in England, and by the time he came to Australia had worked at both Royal Observatory, Greenwich and Cambridge Observatory (Obituaries 1911). Todd was also the South Australian Superintendent of Telegraphs, and in 1870 added the post of Postmaster General to the list!

**Table 3.3** Australian observatory directors, 1850–1916

| Observatory | Officer-in-charge | Years in service | References |
|---|---|---|---|
| Williamstown | Ellery | 1853–1863 | Gascoigne (1992) |
| Flagstaff | Neumayer | 1858–1864 | Swan (1974) |
| Melbourne | Ellery | 1863–1895 | Gascoigne (1992) |
| | Baracchi | 1895–1915 | Perdrix (1979) |
| | Baldwin | 1915– | Simon (1979) |
| Sydney | Scott | 1856–1862 | Orchiston (1998b) |
| | Smalley | 1862–1870 | Wood (1976) |
| | Russell | 1870–1905 | Bhathal (1991) |
| | Lenehan | 1905–1908 | Wood (1986) |
| | Raymond | 1908–1912 | Wood (1958) |
| | Cooke | 1912– | Wood (1958) |
| Adelaide | Todd | 1855–1906 | Edwards (1993), Symes (1976) |
| | Griffiths | 1907–1909 | Edwards (1994) |
| | Dodwell | 1909– | Edwards (1994) |
| Brisbane | MacDonnell | 1879–1885 | Haynes et al. (1993) |
| | Wragge | 1887–1891 | Haynes et al. (1993) |
| | McDowall | 1891–1902 | Wilson (1986) |
| | Spowers | 1902– | Haynes et al. (1993) |
| Hobart | Shortt | 1882–1892 | Obituary (1892) |
| | Kingsmill | 1893–1909 | Meteorological Department (1900), The late Mr. Kingsmill (1909) |
| Perth | Cooke | 1896–1912 | Hutchison (1980, 1981) |
| | Curlewis | 1912– | Utting (1993) |

Todd's assistant and understudy at Adelaide Observatory was an enthusiastic young local graduate named William Cooke (Fig. 3.9), who became founding Government Astronomer of Western Australia when Perth Observatory was opened in 1896 (see Utting 1989, 1992) and eventually took charge of Sydney Observatory.

For a long time, Melbourne was the only major Australian observatory without professionally-trained senior staff. Robert Ellery (Fig. 7.16), second Director at Williamstown Observatory and subsequently the founding Director at Melbourne Observatory, had trained as a medical practitioner (Gascoigne 1992) and acquired his initial post by being a concerned citizen who happened to speak up at the right time. He used a local newspaper to champion the need for a local observatory (in order to establish an accurate local time system); the government liked what he said, and they promptly responded by installing him as Director! Ellery's First Assistant, Edward John White (Fig. 6.8; Obituaries 1914), was a former independent astronomer, and a professionally-trained individual (a surveyor no less) only joined the senior staff in 1882 when Pietro Baracchi (Fig. 11.10) was appointed (Perdrix 1979).

**Fig. 3.9** William Ernest Cooke (after Hutchison 1981)

**Fig. 3.10** The 8-in. Cooke refractor at Adelaide Observatory (*Courtesy* Museum Victoria)

While the Directors at Adelaide, Melbourne, Perth and Sydney also served as the Government Astronomers of their respective colonies, the Brisbane and Hobart Observatories were not so well blessed. From the start, these small observatories were placed in the care of their colonies' respective Meteorological Observers, but at some point in time late in the century the Surveyor-General of Queensland took over responsibility for the Brisbane Observatory (see Haynes et al. 1993).

The emphasis here has been placed on the observatory directors, but this should not blind us to the importance of their subordinates. After all, these men were

> The forgotten heroes of nineteenth century astronomy … who did most of the work; their salaries small, their duty 'to observe all night and to calculate all day' (Airy 1828), and their names barely more than mentioned in the volumes of published results. (Dewhirst 1985, p. 153).

When it came to astronomical instruments, the major Australian observatories were well equipped, partly due to the high public profile of the 1874 and 1882 transits of Venus (see Ellery 1901) and the pressure that the colonial observatory directors therefore were able to apply to their respective governments for special 'equipment funding'. The principal refractors at Adelaide (Fig. 3.10), Melbourne (Fig. 3.11) and Sydney Observatories were all acquired in this way and were finely-finished instruments made by highly-respected English or European manufacturers. They were provided with equatorial mountings, setting circles, drives and micrometres—all prerequisites for useful positional work. The largest of these telescopes had an objective 11.5 in. in diameter (Fig. 8.8), while the largest refractor in the world at the time (1874) was the 26-in. instrument at the U.S. Naval Observatory in Washington (Welther 1984). So, by world standards the Australian refractors were modest, but if assigned to suitable research projects were capable of good work nonetheless.

Each observatory also was supplied with a high-quality transit telescope (Fig. 3.12) and one or more astronomical clocks, the mainstays of a valid time-keeping service, and with the introduction of the international Astrographic

**Fig. 3.11** A sketch of Ellery using the 8-in. Troughton and Simms refractor at Melbourne Observatory during the observation of the 1874 transit of Venus (*The Australasian Sketcher* 1874)

**Fig. 3.12** The 8-in. Troughton and Simms transit telescope at Melbourne Observatory (*Courtesy* Museum Victoria)

Catalogue and Carte du Ciel project in 1890, Melbourne and Sydney Observatories, and later Perth, acquired astrographs with 13-in. Grubb objectives (Russell 1892a). All were carefully built to exacting specifications for this challenging project (see Fig. 3.13).

Of all the colonial observatory instruments, by far the most impressive was the Great Melbourne Telescope (Fig. 3.14), which after some debate (Warner 1982) was eventually commissioned and paid for by the Victorian Government and became operational in 1869 (Gillespie 2011). Its 48-in. (1.22-m) speculum metal mirror by Grubb made this the largest operational telescope in the world at the time (Robinson and Grubb 1869), and with such a gain in light grasp and a precision equatorial mounting, it was expected to revolutionise Southern Hemisphere nebular astronomy.

Despite its promise, the Great Melbourne Telescope did not live up to expectations (Gascoigne 1995). During its first couple of years of operation, Le Sueur observed a wide range of objects (e.g. various nebulae, comets, the Moon, Jupiter, Uranus, Neptune and double stars), but thereafter the focus was on 'the southern nebulae', with a view to revising the catalogue that Sir John Herschel produced during his time at the Cape of Good Hope in 1834–38 (Herschel 1847). Although a large number of nebulae and star clusters were observed, and many of these featured in *Observations of the Southern Nebulae Made With the Great Melbourne Telescope* (Ellery 1885), drawings based on naked eye observations of these tenuous yet complex objects were rapidly replaced during the last two decades of the nineteenth century by innovations in astronomical photography (Hughes 2013).

By a cruel quirk of fate, W. de la Rue, one of the pioneers of astrophotography (see Arnold 1992), was involved in planning the Great Melbourne Telescope and

**Fig. 3.13** The Melbourne Observatory 13-in. (33-cm) Grubb astrograph and 10.25-in. (26-cm) guidescope. After Melbourne Observatory closed, this telescope was purchased by Sydney Observatory (Orchiston collection)

made certain that it was provided with the latest equipment in this regard, but because of its lengthy open-lattice tube the telescope suffered from vibrations in all but the stillest of conditions. This limited photography to very short exposures and although some outstanding images of the Moon were obtained during the early 1870s other experiments in astrophotography were not a success. Likewise, attempted spectroscopic observations of stars were doomed to failure.

Had the Great Melbourne Telescope been housed in a conventional dome rather than its roll-off roof observatory and a glass mirror of considerably shorter focal length been used then it is probable that this instrument would have made a major contribution to astrophysics. Instead, it proved to be a 'white elephant'. Furthermore, Ritchie (1904) believed that

> ... the failure of the Melbourne reflector [was] ... one of the greatest calamities in the history of instrumental astronomy; for by destroying confidence in the usefulness of great reflecting telescopes, it has hindered the development of this type of instrument, so wonderfully efficient in photographic and spectroscopic work, for nearly a third of a century.

Apart from the design issues and chronic under-staffing at Melbourne Observatory, the research potential of the Great Melbourne Telescope also was

**Fig. 3.14** The Great Melbourne Telescope (Orchiston collection)

squandered because there were other calls placed on the telescope. In his *Annual Report* for 1876, Ellery (1877) explains:

> During the year there were 150 nights more or less fit for observing with the great telescope, but of these forty were solely occupied with visitors, and quite lost for astronomical work. Perhaps as an act of policy, this great interference in the observing duties of the staff may be necessary in a Colonial Observatory, but it seems rather too great a sacrifice to give up more than a quarter of the stated work of the observers to satisfy the curiosity of unprofessional star-gazers.

Figure 3.9 is particularly apposite in this regard.

With severe staff cuts during the depression of the 1890s and increased demands from the international Astrographic Catalogue and Carte du Ciel project, the Great Melbourne Telescope was all but decommissioned, and thereafter was rarely used. The other Australian professional observatories also suffered during this depression (see Wood 1958).

If the Great Melbourne Telescope offered little for Australia to boast about, the research output of the refracting telescopes at the major colonial observatories stood in a much better light. These instruments were used on most of the programmes listed in Table 3.2, and results were published in the leading overseas astronomical journals of the day and in the respective publication series of the individual observatories. When research papers by the nation's leading independent astronomers (Orchiston 1989b) are added to those produced by the colonial observatories, Australia's overall publications record in astronomy during the second half of the nineteenth century is rather impressive (Andropoulos 2014).

With the benefit of historical perspective, some of the projects undertaken by the colonial observatories particularly stand out. For instance, Henry Russell and his astronomical assistants at Sydney Observatory established fine reputations for themselves through their southern double star work (see Innes 1899), and Russell also was one of these who pioneered astronomical photography in Australia (e.g. see Russell 1890a, b). One of his remarkable southern Milky Way images is shown in Fig. 3.15.

In Adelaide, Todd did not always see his Observatory as a research institution (Haynes et al. 1996), but Sells and then Cooke were able to assemble a long series of excellent detailed drawings showing Jupiter's changing atmospheric features (Dodwell 1913).

However, the premier accolades belonged to Melbourne, for during the 1870s this Observatory produced what Astronomer Royal, George Airy, described as the '... best catalogue of stars of the Southern Hemisphere ever published' (*Eleventh Report* ... 1876, p. 8), and for 20 years, starting in 1874 and weather permitting,

**Fig. 3.15** A photograph of the η Carinae region taken by Russell on 9 April 1891 (*Courtesy* Dr. Harley Wood)

**Fig. 3.16** The 4-in. Dallmeyer photoheliograph at Melbourne Observatory (*Courtesy* Dr. Barry Clark)

daily solar photographs were taken with a 4-in. (10.2-cm) Dallmeyer photoheliograph (Fig. 3.16) and forwarded to the Royal Observatory, Greenwich (Haynes et al. 1996). By about 1880, Melbourne was ranked up with the best of the world's observatories, the only such Australian institution to achieve this distinction, but by the end of the first decade of the twentieth century Perth had emerged as the nation's premier regional observatory (ibid.).

One of the key international projects that most of the major Australian government observatories participated in was the quest to improve the value for the solar parallax, and hence the Earth–Sun distance or Astronomical Unit (Clerke 1893). A favoured method was to observe transits of Venus from far northern and southern stations, and the United States and Britain dispatched observing parties to Australia (amongst other countries) for the 1874 and 1882 transits, respectively (e.g. Airy 1881; Dick et al. 1998; Orchiston et al. 2015; Stone 1887). In addition, Adelaide, Melbourne and Sydney Observatories all set up observing stations (Edwards 2004; Lomb 2011; Russell 1892b), so Australia was able to contribute significantly to this challenging international attempt to improve our knowledge of a fundamental astronomical yardstick (Orchiston 2004; see Fig. 3.17).

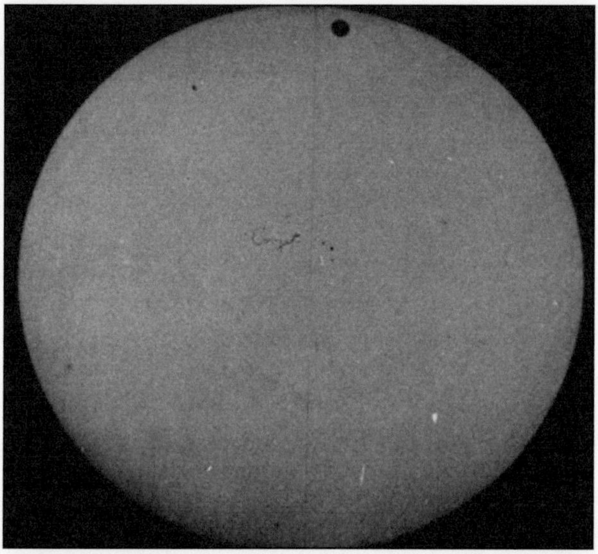

**Fig. 3.17** A photograph of the 1874 transit of Venus taken by the American party at Campbell Town, Tasmania (*Courtesy* Queen Victoria Museum and Art Gallery)

Despite this, transits of Venus did present some serious observational problems, and in order to keep their options open astronomers also looked to other methods. In this context, international programmes were arranged to observe Mars during its 1862 and 1877 oppositions, and Ellery and Todd (respectively) participated in these (Edwards 1993; Gascoigne 1992).

Research was not the only astronomical contribution made by the staff at the Australian colonial observatories, for both Ellery and Russell were active in grinding, polishing and figuring telescope mirrors. Ellery produced speculum metal mirrors and had an ulterior motive for it was his task to refurbish the Great Melbourne Telescope mirrors when they tarnished; from all accounts he became rather proficient in this (see Gascoigne 1992). On the other hand, during the 1870s Russell experimented with glass rather than speculum metal, and in 1880 completed his *piece de resistance*, a mirror with a diameter of 15 in. (38.1 cm). The telescope employed a novel equatorial mounting of Russell's own design (and somewhat reminiscent of that used later for the Great Palomar Telescope—see Fig. 12.14), and this was installed in a 26-ft (7.9-m) diameter observatory of equally unusual design (Orchiston 2000, 2001). At the time it was constructed, this was probably the second largest reflecting telescope in Australia after the Great Melbourne Telescope. As well as making his own telescopes, Russell provided tuition for a number of amateur astronomers who also completed their own mirrors (see Orchiston and Bembrick 1995).

Another area where staff from the colonial observatories were active was in the formation and development of local astronomical groups (see Orchiston 1998a). During the 1870s, Russell and Ellery were responsible for founding what quickly evolved into astronomical sections of the Royal Societies of New South Wales and Victoria, respectively, and they were the key power-brokers during the short-lived histories of these groups (see Orchiston and Bhathal 1991 regarding the former group). Similarly, Todd and his Adelaide Observatory colleagues were amongst the leading members of the Astronomical Section of the Royal Society of South Australia, which was formed in 1891 and soon evolved into the independent Astronomical Society of South Australia (Waters 1980, 1981).

During the 1890s, Melbourne Observatory professional astronomers were very much to the fore when the Victoria Branch of the British Astronomical Association was formed in Melbourne (see Orchiston and Perdrix 1990, 2002). In stark contrast, senior staff at Sydney Observatory only became heavily involved in the affairs of the New South Branch of the British Astronomical Association following Russell's retirement in 1905 (see Orchiston 1988c).

As well as their involvement in astronomical sections, Baracchi, Ellery, Russell and Todd were extremely active in their respective Royal Societies over many years, and each served at one time or another as President (e.g. see Maiden 1918; Pescott 1961). Russell also was a founder of the Australasian Association for the Advancement of Science and was its inaugural President (Hoare 1976; MacLeod 1988). Ellery, Russell and Todd were all elected Fellows of the Royal Society, while Todd was knighted in 1893 (Edwards 1993) and Baracchi received an equivalent honour from the Italian monarch in 1897 (Perdrix 1979).

By the end of the nineteenth century, Australia was viewed—along with South Africa—as a major force in Southern Hemisphere astronomy, and Ellery, Russell and Todd were names well known to Northern Hemisphere colleagues, particularly those immersed in positional astronomy. But following the deaths of these three scientists, the situation in which the Australian state observatories found themselves was soon to change. Part of the problem lay with the international Astrographic Catalogue and Carte du Ciel project, an ambitious scheme devised initially by Sir David Gill (Fig. 3.18) from the Cape Observatory (Glass 1991; Turner 1912). The aim was to produce a photographic map of the entire sky down to magnitude 14, accompanied by a precision catalogue of stars to the 11th magnitude. This mammoth

**Fig. 3.18** Sir David Gill in about 1905 (adapted from Glass 1991, p. 122)

undertaking eventually involved twenty different observatories from around the world, all using astrographs with virtually identical 13-in. (33-cm) objectives (Lankford 1984). Australian participants were Melbourne, Perth and Sydney Observatories (White 1988)

Melbourne and Sydney Observatories joined the scheme when it was first mooted in 1887, and Perth Observatory added its name to the list in 1900. The Melbourne and Sydney astrographs are shown in Figs. 3.14 and 12.15, respectively. While it took many decades, all three zones of the sky undertaken by Sydney, Melbourne and Perth observatories were observed, measured and published. Progress was slow, and major difficulties abounded. Ultimately, this project became a millstone around the necks of these three observatories, relegating them to a relatively insignificant backwater in world astronomy as they watched overseas observatories prosper that had embraced astrophysics.

Australia was not the only nation that suffered in this way. As Ashbrook (1984) has noted, 'The whole outlook of French astronomy was confined by the Carte du Ciel project. Perhaps astrophysical research in France would have begun to flower earlier if the brothers Henry had not made such good lenses'. However, Professor Nick Lomb (pers. comm., June 2016) has pointed out that this

> … negative judgement by science historians needs to be revised in view of recent developments. The USNO has published a star catalogue based on the digitisation of the entire Astrographic Catalogue and with the advent of space-based astrometry, especially the spacecraft HIPPARCOS and GAIA, the Astrographic Catalogue is seen as a unique and invaluable resource to astronomy.

## 3.4 Concluding Remarks

Astrophysics had a long and at times painful gestation during the second half of the nineteenth century, but with its successful delivery during the 1880s and 1890s came a theoretical and philosophical revolution in world astronomy (Clerke 1903). Also appropriately termed the 'New Astronomy' (Langley 1884), astrophysics involved a marriage of astronomy, spectroscopy, photography and photometry (Hearnshaw 1986, 1996; Hughes 2013). Gone was that preoccupation with celestial mechanics, for astrophysics was concerned with

> … the physical characteristics of celestial bodies, that is, their luminosity and spectroscopic peculiarities, their temperature and radiation, the nature and condition of their atmosphere, surface and interior, their qualitative and quantitative composition, and finally all the phenomena arising from these physical conditions. (Abetti 1954, p. 18).

Astrophysics was about patterns, processes and mechanisms; changes through time; and interrelationships. It was largely non-descriptive, and emphasised the dynamic rather than the static. It was very different from the positional astronomy which had been the mainstay of the government observatories in Australia.

We may wonder why Australia's leading professional astronomers failed to take up the exciting challenge of astrophysics at this time. After all, they had access to the international journals which reported this new work with increasing frequency, and their observatories were equipped with suitable instruments.

In fact, Melbourne and Sydney Observatories did take up the challenge, but only briefly. During the late 1880s, Baracchi (1889) and Ellery (1889) carried out extensive stellar spectroscopic observations with the 8-in. refractor, while even earlier Le Sueur had used the Great Melbourne Telescope to obtain the spectrum of η Argus (now η Carinae) and to carry out the first-ever spectroscopic observation of an extragalactic object, 30 Doradus in the Large Magellanic Cloud (Hearnshaw 1986). Meanwhile, in Sydney, Russell (1881) conducted a spectroscopic examination of the Great Comet of 1881 (C/1881 K1). The problem was that spectroscopy was still in its infancy at the time, and all four astronomers faced problems in trying to interpret their observations.

A further issue may have been the almost inane suspicion with which many professional astronomers viewed astrophysics during these formative years (Meadows 1984a, b), partly because of the important role that was played by some non-professional astronomers. While most professional astronomers saw serious independent astronomers as colleagues, their perceived duty was to perpetuate the craft of 'traditional' astronomy, not to pioneer new fields (Lankford 1981).

By the 1890s, the leading interpretive problems in astronomical spectroscopy had been resolved, and there was a growing and highly-visible pool of reputable professional astronomers worldwide espousing astrophysics, but by now the Australian observatories were totally committed to the international Astrographic Catalogue and Carte du Ciel project and the economic depression of the day meant that financial constraints prevented them from expanding their research portfolios.

Only with the founding of the Commonwealth Solar Observatory in 1924 did astrophysics gain a firm foothold on Australian shores (Allen 1978; Love 1985; Orchiston 1989a), but by this time the state observatories were rapidly heading for extinction.

**Acknowledgements** I wish to thank Dr Richard Gillespie (Museum Victoria), Professor Nick Lomb (University of Southern Queensland, Toowoomba) and the late Dr. Harley Wood (formerly Sydney Observatory) for their assistance, and Dr. Barry Clark (Melbourne), Museum Victoria (Melbourne) and the Queen Victoria Museum and Art Gallery (Launceston, Tasmania) for kindly supplying Figs. 3.8, 3.10, 3.12, 3.16 and 3.17.

## References

Abetti, G. (1954). *The History of Astronomy*. London: Sidgwick and Jackson.

Airy, G.B. (1828). Memorandum to the Trustees of the Plumian Professorship (cited in Stratton 1949: 5).

Airy, G. (1846). Letter to Sir Robert Peel, dated 16 April. In Department of Palaeography and Diplomatic, University of Durham.

Airy, G. (1849). Letter to J. Parker, dated 9 November. In Observatory and Time Ball. *New South Wales. Votes and Proceedings of the Legislative Council, During the Session of the Year 1852. With Various Documents Connected Therewith* (pp. 1349–1369). Sydney: Government Printer.

Airy, Sir George. (1881). *Transit of Venus. Account of Observations of the Transit of Venus, 1874, December 8, Made Under the Authority of the British Government: and the Reduction of the Observations*. London: HMSO.

Airy, W. (Ed.) (1896). *Autobiography of Sir George Biddell Airy, K.C.B.* Cambridge: Cambridge University Press.

Allen, C.W. (1978). The beginnings of the Commonwealth Solar Observatory. *Records of the Australian Academy of Science, 4,* 27–49.

Andropoulos, J. (2014). *The Astronomical Publications of Melbourne Observatory*. Ph.D. Thesis, James Cook University, Townsville, Australia.

Arnold, H.J.P. (1992). Eclipse of the Sun: the first images. In S.J. Edberg (Ed.), *Research Amateur Astronomy* (pp. 189–204). San Francisco: Astronomical Society of the Pacific (Conference Series, Volume 33).

Ashbrook, J. (1984). The brothers Henry and the Carte de Ciel. In *The Astronomical Scrapbook. Skywatchers, Pioneers, and Seekers in Astronomy* (pp. 436–440). Cambridge (Mass.): Sky Publishing Corporation.

Baracchi, P. (1889). Spectra of southern stars observed at the Melbourne Observatory with the McClean direct-vision spectroscope attached to the south equatorial. *Monthly Notices of the Royal Astronomical Society, 50,* 66–71.

Baracchi, P. (1914). Astronomy and geodesy in Australia. In G.H. Knibbs (Ed.), *Federal Handbook Prepared in Connection with the Eighty-fourth Meeting of the British Association for the Advancement of Science, Held in Australia, August, 1914* (pp. 326–390). Melbourne: Government Printer.

Bhathal, R. (1991). Henry Chamberlain Russell – astronomer, meteorologist and scientific entrepreneur. *Journal and Proceedings of the Royal Society of New South Wales, 124,* 1–21.

Bhathal, R. (2012). Some scientific aspects of Parramatta Observatory. *Journal and Proceedigs of the Royal Society of New South Wales, 145,* 111–127.

Bhathal, R., Sutherland, R. & Butcher, H. (2013). *Mt Stromlo Observatory: From Bush Observatory to the Nobel Prize*. Melbourne: CSIRO Publishing.

Clarke, P.A. (2014). The Aboriginal Australian cosmic landscape, Part 1: the ethnobotany of the sky world. *Journal of Astronomical History and Heritage, 17,* 307–325.

Clerke, A. (1893). *A Popular History of Astronomy during the Nineteenth Century*. London: Adam and Charles Black.

Clerke, A. (1903). *Problems in Astrophysics*. London: Adam and Charles Black.

Dewhirst, D.W. (1985). Meridian astronomy in the private and university observatories of the United Kingdom: rise and fall. *Vistas in Astronomy, 28,* 147–158.

Dick, S., Orchiston, W. & Love, T. (1998). Simon Newcomb, William Harkness and the nineteenth century American transit of Venus expeditions. *Journal for the History of Astronomy, 29,* 221–255.

Dodwell, G.F. (1910). *Brief Account of the Adelaide Observatory*. Adelaide: Government Printer (South Australian Intelligence Bulletin, Part II).

Dodwell, G.F. (1913). *Physical Observations of Jupiter Made at the Adelaide Observatory ... During the Years 1884–1893*. Adelaide: Government Printer.

Edwards, P.G. (1993). Charles Todd and the Adelaide Observatory. *Proceedings of the Astronomical Society of Australia, 10,* 349–354.

Edwards, P.G. (1994). The Adelaide Observatory after Todd. *Proceedings of the Astronomical Society of Australia, 11,* 206–210.

Edwards, P.G. (2004). Charles Todd's observations of the transits of Venus. *Journal of Astronomical History and Heritage, 7,* 1–7.

*Eleventh Report of the Board of Visitors to the Observatory, With the Annual Report of the Government Astronomer*. Melbourne, Government Printer (1876).

Ellery, R.J. (1869). *Astronomical Observations Made at the Williamstown Observatory in the Years 1861, 1862, and 1863*. Melbourne: Government Printer.

Ellery, R.J. (1877). *Report of the Government Astronomer*. Melbourne: Government Printer.

Ellery, R.J. (1885). *Observations of the Southern Nebulae Made With the Great Melbourne Telescope from 1869 to 1885. Part 1*. Melbourne: Government Printer.

Ellery, R.J. (1889). Preliminary spectroscopic survey of southern stars, made at Melbourne Observatory with a McClean direct-vision spectroscope on the 8-inch equatorial. *Monthly Notices of the Royal Astronomical Society, 49,* 439–445.

Ellery, R.J. (1901). A brief history of the beginnings and growth of astronomy in Australasia. In *Report of the Eighth Meeting of the Australasian Association for the Advancement of Science* (pp. 1–17). Melbourne: AAAS.

Fuller, R.S., Norris, R.P. & Trudgett, M. (2014).The astronomy of the Kamilaroi People and their neighbours. *Australian Aboriginal Studies, 2014* (2), 3–27.

Gascoigne, S.C.B. (1992). Robert L.J. Ellery, his life and times. *Proceedings of the Astronomical Society of Australia, 10,* 170–176.

Gascoigne, S.C.B. (1995). The Great Melbourne Telescope and other 19th century reflectors. *Historical Records of Australian Science, 10,* 223–245.

Gillespie, R. (2011). *The Great Melbourne Telescope*. Melbourne: MuseumVictoria.

Gingerich, O. (Ed.) (1984). *The General History of Astronomy. Volume 4. Astrophysics and Twentieth Century Astronomy to 1950: Part A*. Cambridge: Cambridge University Press.

Glass, I.S. (1991). Gill, Grubb and the astrographic telescopes – a correspondence. *Monthly Notices of the Astronomical Society of Southern Africa, 50,* 121–136.

Hamacher, D.W. (2012). *On the Astronomical Knowledge and Traditions of Aboriginal Australians*. Ph.D. Thesis (by publication), Department of Indigenous Studies, Macquarie University, Sydney, Australia.

Haynes, R.D. (2000). Astronomy and the dreaming: the astronomy of the Aboriginal Australians. In H. Selin (Ed.), *Astronomy Across Cultures: The History of Non-Western Astronomy* (pp. 53–90). Dordrecht: Kluwer.

Haynes, R.D, Haynes, R.A. & Kitson, W.S. (1993). The history of astronomy in Queensland. *Vistas in Astronomy, 36,* 231–252.

Haynes, R., Haynes, R., Malin, D. & McGee, R. (1996). *Explorers of the Southern Sky. A History of Australian Astronomy*. Cambridge: Cambridge University Press.

Hearnshaw, J.B. (1986). *The Analysis of Starlight. One Hundred and Fifty Years of Astronomical Spectroscopy*. Cambridge: Cambridge University Press.

Hearnshaw, J.B. (1996). *The Measurement of Starlight. Two Centuries of Astronomical Photometry*. Cambridge: Cambridge University Press.

Herschel, Sir John F.W. (1847). *Results of Astronomical Observations Made During the Years 1834, 5, 6, 7, 8 at the Cape of Good Hope*. London: Smith, Elder.

Hoare, M. (1976). The intercolonial science movement in Australasia, 1870–1890. *Records of the Australian Academy of Science, 3,* 7–28.

Home, R.W. (1991). George Neumayer and the Flagstaff Observatory. In J. Tampke & D. Walker (Eds.), *From Berlin to the Burdekin* (pp. 40–53). Sydney: New South Wales University Press.

Home, R.W. & Kretzer, H.-J. (1991). The Flagstaff Observatory, Melbourne: new documents relating to its foundation. *Historical Records of Australian Science, 8,* 213–243.

Home, R. & Livingston, K.T. (1994). Science and technology in the story of Australian federation: the case of meteorology, 1876–1908. *Historical Records of Australian Science, 10,* 109–127.

Hughes, S. (2013). *Catchers of the Light. The Forgotten Lives of the Men and Women Who First Photographed the Heavens. Two Volumes*. Paphos:ArtDeCiel Publishing.

Hutchison, D. (1980). William Ernest Cooke: first Government Astronomer. *Early Days, 8,* 93–114.

Hutchison, D. (1981). William Ernest Cooke, astronomer 1863–1947. *Historical Records of Australian Science, 5,* 58–77.

Innes, R.T.A. (1899). Reference catalogue of southern double stars. *Annals of the Cape Observatory, 2* (2).

Johnston, D.D. (1998). *The Night Skies of Aboriginal Australia – a Noctury*. Sydney: University of Sydney Press (Oceania Monograph 47).

Langley, S.P. (1884). *The New Astronomy*. New York: Houghton Mifflin.

Lankford, J. (1981). Amateurs and astrophysics: a neglected aspect in the development of a scientific speciality. *Social Studies in Science, 11,* 275–303.

Lankford, J. (1984). The impact of photography on astronomy. In Gingerich, 16–39.
Laurie, P.S. (1988). William Dawes, and Australia's first observatory. *Quarterly Journal of the Royal Astronomical Society, 29,* 469–482.
Lomb, N. (2004). The instruments from the Parramatta Observatory. *Historical Records of Australian Science, 15,* 211–222.
Lomb, N. (2011). *Transit of Venus: 1639 to the Present.* Sydney: NewSouth and Powerhouse Museum.
Love, R. (1985). Science and government in Australia 1905–14: Geoffrey Duffield and the foundation of the Commonwealth Solar Observatory. *Historical Records of Australian Science, 6,* 171–188.
McAfee, R.J. (1981). *Dawes Meteorological Journal.* Canberra: Australian Government Publishing Service (Bureau of Meteorology Historical Note No. 2).
MacLeod, R. (Ed.), (1988). *The Commonwealth of Science: ANZAAS and the Scientific Enterprise in Australasia, 1888–1988.* Melbourne: Oxford University Press.
Maiden, J.H. (1918). A contribution to the history of the Royal Society of New South Wales, with information in regard to other New South Wales societies. *Journal and Proceedings of the Royal Society of New South Wales, 52,* 215–361.
Meadows, J. (1984a). The new astronomy. In Gingerich, 59–72.
Meadows, J. (1984b). The origins of astrophysics. In Gingerich, 3–15.
Meteorological Department. The Observatory. In *Cyclopedia of Tasmania* … Hobart: Maitland and Krone (1900).
Neumayer, G. (1858). Description and system of working of the Flagstaff Observatory. *Transactions of the Philosophical Society of Victoria, 3,* 94–103.
New Observatory, St Kilda Road. *The Illustrated Melbourne Post*, Volume 2, No. 46, p. 1 (1863).
Norris, RP. & Harney, B.Y. (2014). Songlines and navigation in Wardaman and other Australian Aboriginal cultures. *Journal of Astronomical History and Heritage, 17,* 141–148.
Norris, R.P. & Norris, C.M. (2009). *Emu Dreaming: An Introduction to Australian Aboriginal Astronomy*. Sydney: Emu Dreaming Press.
O'Connell, D.K.J. (1952.) Father Edward Francis Pigot, S.J. *Studies, 1,* 189–196; *2,* 323–332.
Obituaries. Sir Charles Todd. *Monthly Notices of the Royal Astronomical Society, 71,* 272–273 (1911).
Obituaries. Edward John White. *Monthly Notices of the Royal Astronomical Society, 74,* 280–281 (1914).
Obituary [Captain John Shortt]. *The Mercury* 27 June 1892.
Orchiston, W. (1985). Life-history of a telescope: the 17.8-cm refractor at the Riverview Observatory in Sydney. *Australian Journal of Astronomy, 1,* 57–80.
Orchiston, W. (1987). Contribution of the Lands Department to the development of astronomy in New South Wales during the nineteenth century. *Australian Journal of Astronomy, 2,* 65–74.
Orchiston, W. (1988a). The emergence of professional astronomy in nineteenth century Australia. *Search, 19,* 76–81.
Orchiston, W. (1988b). From research to recreation: the rise and fall of Sydney Observatory. *Vistas in Astronomy, 32,* 49–63.
Orchiston, W. (1988c). Illuminating incidents in Antipodean astronomy: formation of the New South Wales Branch of the British Astronomical Association. *Journal of the British Astronomical Association, 98,* 75–84.
Orchiston, W. (1988d). Illuminating incidents in Antipodean astronomy: Phillip Parker King and the founding of Sydney Observatory. *Vistas in Astronomy, 32,* 285–301.
Orchiston, W. (1989a). Astrophysics and the evolution of twentieth century Australian astronomy. *Search, 20,* 198–202.
Orchiston, W. (1989b). The role of the amateur in early Australian astronomy. *Search, 20,* 16–23.
Orchiston, W. (1991). Amateur-professional co-operation in astronomy education: a nineteenth century Australian model. *Proceedings of the Astronomical Society of Australia, 9,* 186–188.

Orchiston, W. (1998a). Amateur-professional collaboration in Australian science: the earliest astronomical groups and societies. *Historical Records of Australian Science, 12,* 163–182.

Orchiston, W. (1998b). Mission impossible: William Scott and the first Sydney Observatory directorship. *Journal of Astronomical History and Heritage, 1,* 21–43.

Orchiston, W. (2000). Illuminating incidents in Antipodean astronomy: H.C. Russell and the origin of the Palomar-type mounting. *Journal of the Antique Telescope, 19,* 13–15.

Orchiston, W. (2001). Australian innovation in telescope design: origin of the horseshoe mounting. *AAO Newsletter, 97,* 12–13.

Orchiston, W. (2004). The nineteenth century transits of Venus: an Australian and New Zealand overview. *Journal of Astronomical Data, 10,* 219–308.

Orchiston, W. (2015).The amateur-turned-professional syndrome: two Australian case studies. In W. Orchiston, D.A. Green & R. Strom (Eds.), *New Insights from Recent Studies in Historical Astronomy: Following in the Footsteps of F. Richard Stephenson* (pp. 259–350). Cham: Springer.

Orchiston, W. & Bembrick, C. (1995). The role of the large reflecting telescope in amateur astronomy: an Australian case study. *Australian Journal of Astronomy, 6,* 53–74.

Orchiston, W. & Bhathal, R. (1991). Illuminating incidents in Antipodean astronomy: Section A of the Royal Society of New South Wales. *Australian Journal of Astronomy, 4,* 49–67.

Orchiston, W. & Perdrix, J. (1990). The British Astronomical Association in Australia: a centennial perspective. *Australian Journal of Astronomy, 3,* 163–179.

Orchiston, W. & Perdrix, J. (2002). A history of the British Astronomical Association in Australia: the fate of the Branches. *Journal of the British Astronomical Association, 112,* 68–77.

Orchiston, W., Buchanan, A., Price, G. & Sprent, T. (2015). Highlighting our international astronomical heritage: Tasmanian relics of the 1874 transit of Venus. *Journal of Astronomical History and Heritage, 18,* 149–160.

Perdrix, J. (1961). The Melbourne Observatory. *Journal of the Astronomical Society of Victoria, 14,* 19–25, 45–48, 60–64.

Perdrix, J. (1970). The Great Melbourne Telescope. *Journal of the Astronomical Society of Victoria, 23,* 54–67.

Perdrix, J. (1979). Baracchi, Pietro Paulo Giovanni Ernesto (1851–1926). In B. Nairn & G. Serle (Eds.), *Australian Dictionary of Biography. Volume 7: 1891–1939. A-Ch* (pp. 166–167). Melbourne: Melbourne University Press.

Perdrix, J.L. (1990). Flagstaff Hill and its observatory. *Australian Journal of Astronomy, 3,* 113–124.

Pescott, R.T.M. (1961). The Royal Society of Victoria from then, 1854 to now, 1959. *Proceedings of the Royal Society of Victoria, 73,* 1–40.

Richardson, W. (1835). *Catalogue of 7385 Stars, Chiefly in the Southern Hemisphere, Prepared from Observations Made in the Years 1822, 1823, 1824, 1825, and 1826, at the Observatory at Paramatta [sic], New South Wales …* London: His Majesty's Stationery Office.

Ritchie, G.W. (1904). On the modern reflecting telescope. *Smithsonian Contributions to Knowledge, 34,* 1–149.

Robertson, P., Cozens, G., Orchiston, W. & Slee, B. (2010). Early optical and radio observations of Centaurus A. *Publications of the Astronomical Society of Australia, 27,* 402–430.

Robinson, T.R. & Grubb, T. (1869). Description of the Great Melbourne Telescope. *Philosophical Transactions of the Royal Society* (London), *159,* 127–161.

Ross, C.S. (1918). Our Observatory: the story of its establishment. *Victorian Historical Magazine, 6,* 134–144.

Russell, H.C (1881). Spectrum and appearance of the recent comet. *Journal and Proceedings of the Royal Society of New South Wales, 15,* 81–86.

Russell, H.C (1888). Astronomical and meteorological workers in New South Wales, 1778 to 1860. In A. Liversidge & R. Etheridge (Eds.), *Report of the First Meeting of the Australasian Association for the Advancement of Science …* (pp. 45–94). Sydney: AAAS.

Russell, H.C (1890a). On some photographs of the Milky-Way, recently taken at Sydney Observatory. *Journal and Proceedings of the Royal Society of New South Wales, 24,* 106–111.

Russell, H.C (1890b). *Photographs of the Milky-Way and Nebeculae at Sydney Observatory in 1890*. Sydney: Government Printer.

Russell, H.C (1892a). *Description of the Star Camera, at Sydney Observatory*. Sydney: Government Printer.

Russell, H.C (1892b). *Observations of the Transit of Venus, 9 December, 1874; Made at Stations in New South Wales*. Sydney: Government Printer.

Saunders, S.D. (1990). *Astronomy in Colonial New South Wales: 1788 to 1858*. Ph.D. Thesis, Department of General Philosophy, University of Sydney, Sydney, Australia.

Saunders, S. (2004). Sir Thomas Brisbane's legacy to colonial science: colonial astronomy at Parramatta Observatory, 1822–1848. *Historical Records of Australian Science, 15,* 177–209.

Savours, A., and McConnell, A. (1982). The history of the Rossbank Observatory, Tasmania. *Annals of Science, 39,* 527–564.

Simon, P. (1979). Baldwin, Joseph Mason (1878–1945). In B. Nairn & G. Serle (Eds.), *Australian Dictionary of Biography. Volume 7: 1891–1939. A-Ch* (pp. 157–158). Melbourne: Melbourne University Press.

Stone, E.J (1887). *Transit of Venus, 1882*. London: HMSO.

Stratton, F.J.M. (1949). The history of the Cambridge Observatories. *Annals of the Solar Physics Observatory, 1*.

Swan, R.A. (1974). Neumayer, Georg Balthasar (1826–1909). In N.B. Nairn, A.G. Serle & R.B. Ward (Eds.), *Australian Dictionary of Biography. Volume 5: 1851–1890. K-Q* (pp. 55–64). Melbourne: Melbourne University Press.

Symes, G.W. (1976). Todd, Sir Charles (1826–1910). In G. Serle & R. Ward (Eds.), Australian Dictionary of Biography. Vol. 6: 1851–1890 (pp. 280–282). Melbourne: Melbourne University Press.

Tebbutt, J. (1867). On the progress and present state of astronomical science in New South Wales. In *Catalogue of the Natural and Industrial Products of New South Wales. Forwarded to the Paris Universal Exhibition of 1867, by the New South Wales Exhibition Commissioners* (pp. 55–64). Sydney: Government Printer (Appendix).

*The Australasian Sketcher*, 26 December 1874.

The late Mr. Kingsmill. *The Mercury*, 19 July 1909.

Turner, H.H. (1912). *The Great Star Map. Being a General Account of the International Project Known as the Astrographic Chart*. London: Murray.

Utting, M. (1989). *Cooke's Perth Observatory*. Perth, Perth Observatory.

Utting, M. (1992). *Astronomy in Western Australia. Volume 1: 1896 to 1912*. Perth: Perth Observatory.

Utting, M. (1993). *Astronomy in Western Australia. Volume 2: 1912 to 1940*. Perth: Perth Observatory.

Warner, B. (1982). The large Southern Telescope: Cape or Melbourne. *Quarterly Journal of the Royal Astronomical Society, 23,* 505–514.

Waters, B. (1980). *A Reference History of the Astronomical Society of South Australia Inc. Volume 1. The First Decade 1891–1901*. Adelaide: printed for the author.

Waters, B. (1981). *A Reference History of the Astronomical Society of South Australia Inc. Volume 2. The Second Decade 1902–1911*. Adelaide: printed for the author.

Weiderkehr, V.H. (1988). Georg Neumayer und die Grundung eines geophysikalisch-nautischen Observatoriums in Melbourne 1857. In W. Schroder (Ed.), *Past, Present and Future Trends in Geophysical Research* (pp. 111–125). Bremen-Roennebeck: International Commission on History of AIGA.

Welther, B.L. (1984). The world's largest telescopes, 1850–1950. In Gingerich, Ai–Avi.

White, G. (1988). The Carte du Ciel – the Australian connection. In S. Débarbat, J.A. Eddy, H.K. Eichhorn, & A.R. Upgren (Eds.), *Mapping the Sky. Past Heritage and Future Directions* (pp. 45–51). Dordrecht: Kluwer.

Wilson, P.D. (1986). McDowall, Archibald (1841–1918). In B. Nairn, B & G. Serle (Eds.), *Australian Dictionary of Biography. Volume 10: 1891–1939. Lat-Ner.* (pp. 261–262). Melbourne: Melbourne University Press.

Wood, H. (1958). *Sydney Observatory 1858 to 1958*. Sydney: Government Printer (Sydney Observatory Papers No. 31).
Wood, H. (1968). The sky and the weather, Australia, 1866–1966. In *A Century of Scientific Progress* (pp. 379–430). Sydney: Royal Society of New South Wales.
Wood, H. (1976). Smalley, George Robarts (1822–1870). In G. Serle & R. Ward (Eds.), *Australian Dictionary of Biography. Volume 6: 1851–1890. R-Z* (pp. 136–137). Melbourne: Melbourne University Press.
Wood, H. (1986). Lenehan, Henry Alfred (1843–1908). In Nairn, B., and Serle, G. (Eds.), *Australian Dictionary of Biography. Volume 10: 1891–1939. Lat-Ner* (pp. 73–74). Melbourne: Melbourne University Press.

# Chapter 4
# The Role of the Independent Astronomer in Early Australian Astronomy

## 4.1 Introduction

As we saw in the previous chapter, the Classical Era (1848–1923) was a golden age as Australian astronomy rapidly rebuilt its fledgling international reputation following the fiasco of Parramatta Observatory. Five different 'fundamental elements' can be identified in this evolutionary process:

1. Establishment of the various colonial (i.e. professional) observatories
2. Establishment of a large number of significant non-government observatories
3. Emergence of the first notable Australian-based telescope-makers
4. Blossoming of popular interest in and awareness of astronomy
5. Formation of the earliest formal astronomical groups and societies

Independent astronomers (those who failed to receive a *primary income* from their astronomical activities) have a proud record of international achievement (e.g. see Ashbrook 1984; Chapman 1998; Clerke 1893; Dunlop and Gerbaldi 1988; Lankford 1981a; Meadows and Henbest 1981; Orchiston 2006), and Australia was no different in this regard (see Bhathal and White 1991; Haynes et al. 1996). Australian non-professional astronomers were involved in all of the above 'fundamental elements'.

This chapter has been developed from a review paper published in 1989 (Orchiston 1989b) and examines the contribution that independent astronomers made to Australian astronomy during the Classical Era, when John Tebbutt was active.



W. Orchiston, *John Tebbutt*, Historical & Cultural Astronomy,
DOI 10.1007/978-3-319-44521-2_4

## 4.2 A Cavalcade of Comets and Transits: The Growth of Popular Interest in Astronomy

During the second half of the nineteenth century, there was a remarkable blossoming of popular interest in astronomy (Ellery 1901), which left as its legacy a general public much more aware of matters celestial and for the most part sympathetic—or at least neutral—towards astronomy.

Part of the reason for this enlightened public view was the amazing succession of spectacular astronomical events and objects visible in Australian skies during the comparatively short period from 1874 to 1887 (Table 4.1).

The comets were all prominent naked eye objects (e.g. see Fig. 4.1), and that of 1881 was found by an Australian, none other than John Tebbutt (Orchiston 1981, 1999)! This discovery inspired other local successes, and between 1881 and 1889 Australians contributed more than 10 % of all new comets detected worldwide (see Orchiston 1983). The honour associated with this record of achievement did not escape the Australian public, particularly when viewed alongside well-publicised local observations of known comets.

The public interest generated by the 1874 transit of Venus subsequently generalised to transits of Mercury, even though these equally-fascinating events were far less important scientifically. But they were significant enough to inspire some of the colonial observatories to establish observing stations. During the period under review, two transits of Mercury were visible from Australia, in 1878 and 1881.

Astronomy also succeeded in gaining a strong public following in Australia during the 1880s because of the entrepreneurial efforts of a number of astronomers, both professional and non-professional (Orchiston 1991, 1997a, 2003b). They were committed to popularising astronomy, and achieved this by making planispheres; writing popular books (e.g. Bonwick 1866), and articles, observational reports and 'Letters to the Editor' for their local newspapers; by lecturing; and in some cases by offering courses in astronomy (mainly through church groups, Schools of Mines, Schools of Arts and scientific societies). What Australia lacked were the itinerant

**Table 4.1** Impressive astronomical events and objects visible from Australia, 1874–1887

| Year | Event/object |
|---|---|
| 1874 | Coggia's Comet (C/1874 H1) |
| 1874 | Transit of Venus |
| 1878 | Transit of Mercury |
| 1880 | Great Southern Comet (C/1880 C1) |
| 1881 | Transit of Mercury |
| 1881 | Great Comet (C/1881 K1 (Tebbutt)) |
| 1882 | Transit of Venus |
| 1882 | Great Comet (C/1882 R1) |
| 1887 | Great Southern Comet (C/1887 B1) |

**Fig. 4.1** The Great Comet of 1880 (C/1880 C1) was independently discovered by Castlemaine's Dr. William Bone and attracted enormous public interest (Orchiston collection)

lecturers and showmen who were so prominent earlier in the century in England and the United States (e.g. see Inkster 1978, 1980, 1982), although the noted British astronomy populariser, Richard Proctor (North 1975), did undertake a lecture tour of Australia in 1880.

While comets were undoubtedly the most exciting astronomical curiosities to whet the public appetite during the 1870s and 1880s, the two transits of Venus also attracted very considerable attention. Observations of the 1761 and 1769 transits had already been used to determine the Astronomical Unit (the distance from the Earth to the Sun), but the nineteenth century transits offered opportunities to refine this value (Dick et al. 1998; Woolf 1959). As we saw in the previous chapter,

**Fig. 4.2** Members of the various Sydney Observatory 1874 transit of Venus parties. Fairfax, Hirst, Macdonnell and Wright all were prominent independent astronomers (after Russell 1892, Frontispiece)

observing stations were set up by Adelaide, Melbourne and Sydney Observatories and by visiting British and American astronomers. Most stations reported success in 1874 (e.g. see Lomb 2011; Orchiston 2004b; Russell 1892) and independent astronomers played a key role (e.g. see Fig. 4.2), but clouds covered much of the southeastern sector of the continent in 1882 (Baracchi 1914).

Popularisers had a profound effect upon public attitudes towards and knowledge of astronomy, and they gained many converts to astronomy. If we define an 'active independent astronomer' as a non-professional astronomer who engages on a regular basis in *any* form of astronomical activity, we can document the numbers of such individuals in the Sydney-Windsor region alone on the basis of data provided by Orchiston (1982, 1987a, 1988a), Orchiston and Bhathal (1984, 1991) and Russell (1892). As Fig. 4.3 shows, the resulting plot is an exponential growth curve (cf. Herrmann 1973). In 1860, there were only 3 active independent astronomers in the region, but by 1896 the number had grown to about 60. Although it is reasonable to assume that similarly-shaped curves characterise other major Australian population centres, this has yet to be established empirically.

While most independent astronomers found in the nineteenth century Australia were local converts, the substantial population influx during the period 1860–1890 (initially associated with a succession of gold rushes) also brought its share of astronomers, and a number of these went on to achieve considerable prominence.

Amongst the growing Australian population of independent astronomers during the nineteenth century there were some who were happy to remain 'armchair astronomers', but for many the ultimate goal—though oft unrealised—was to establish their own observatory.

**Fig. 4.3** A plot of the number of active independent astronomers in the Sydney-Windsor region of New South Wales between 1850 and 1899 (after Orchiston 1991)

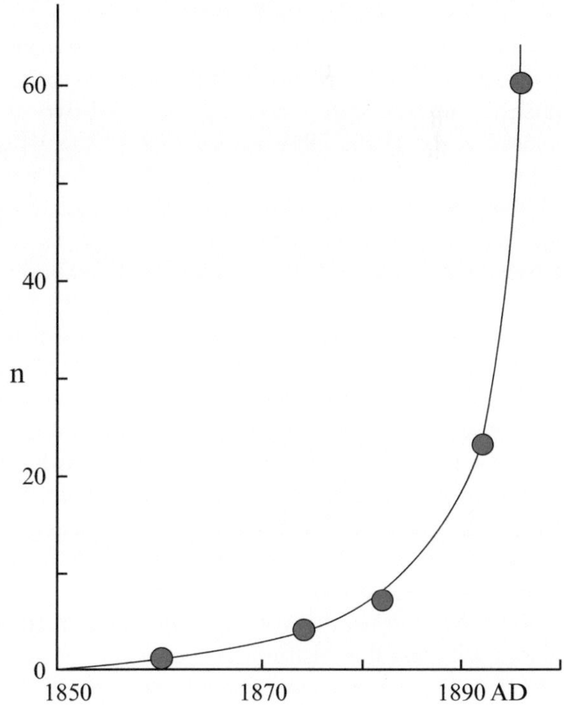

## 4.3 Independent Astronomers to the Fore: The Rise of the Private Observatories

A particularly important feature of the nineteenth century and early twentieth century Australian astronomy was the emergence of a network of private observatories across the nation, established by independent astronomers. Table 4.2 lists the more significant observatories of this kind, where their pan-continental coastal or near-coastal distribution and their mushrooming in the Sydney region between 1890 and 1910 are immediately apparent.

Three different types of observatory are included in Table 4.2. A number of institutions were devoted solely or primarily to public astronomy in that their proprietors were committed to offering the range of services normally provided on a regular basis for the general public by professional observatories (see Orchiston 1997a). Thus, all conducted their observatories as astronomical information and education centres, and most ran public viewing nights at the telescope; some also provided their local townsfolk with a time service and with up-to-date meteorological data.

The earliest example of an observatory committed solely to public astronomy is that erected by Manning in Fremantle during the 1850s (see Fig. 3.3 for localities mentioned in the text). Other observatories of this type were operated by Wooster in Ballarat (Bio Box 13.2), Jones in Bendigo, Dobbie in Adelaide, O'Reilly and Eglinton in Brisbane and McFarlane in Irvinebank (Fig. 4.4), but by far the most impressive was the private observatory set up in 1886 by the wealthy Ballarat businessman, James Oddie (Burk 1986).

Oddie appointed a well-known Victorian astronomer, Captain Henry Evans Baker (Davis 1990), as Director of this observatory. Baker (Fig. 4.5) was an accomplished telescope-maker and engineer, and two of his reflecting telescopes were housed at the Oddie Observatory. One of them featured a 26-in. (66-cm) mirror (Fig. 4.5), and was the second largest telescope in Australia, after the Great Melbourne Telescope. Following Oddie's death, the Observatory was taken over by the local Council, renovated, and reopened in 1918 as the 'Ballarat Municipal Observatory', a true public observatory. In addition to the two Baker reflectors and a transit instrument, the new complex included a 5-in. (12.7-cm) refractor and dome donated by a local engineering firm, a large lecture room (with attractive ornamental stained-glass windows) and a valuable reference library. The Observatory was once again able to take up its role as a leader in astronomical education.

The second kind of observatory listed in Table 4.2 is that constructed solely, or almost solely, for research. Research observatories were scattered along the eastern seaboard of the nation, but with a marked concentration in the Sydney region. Colyer's observatory has been included in this category because extensive use was made of it by G.D. Hirst (see Bio Box 10.3). In contrast, Colyer rarely found the time to carry out astronomical observations.

Given that research work was intended at these observatories, all were furnished with sizeable instruments, similar to or only marginally smaller than those found in the major Australian government observatories (the Great Melbourne

**Table 4.2** Significant Australian private observatories, 1850–1920

| Found'g decade | Astronomer | Location | Main instruments[a] | Function | References |
|---|---|---|---|---|---|
| 1850s | Abbott | Hobart | 4.5-in. OG<br>~5.5-in. OG<br>trans | MPRT | Orchiston (1992, 1997c) |
| | Manning | Fremantle | ? | MP | Utting (1991) |
| 1860s | O'Reilly | Brisbane | Small OG<br>trans | T | Haynes et al. (1993), Darlington and Orchiston (2017) |
| | Tebbutt | Windsor | 4.5-in. OG<br>8-in. OG<br>trans | MPRT | Orchiston (2004a), Tebbutt (1908) |
| 1870s | Biggs | Launceston | 8.5-in. sp<br>trans | MPRT | Giordano (1995), Orchiston (1985a) |
| | Bone | Castlemaine | 8-in. OG<br>trans? | MPRT | Orchiston (1986, 1987b) |
| | Colyer | Sydney | 10.25-in. sp | R | Baracchi (1914) |
| | Jones | Bendigo | ~4.75-in. sp<br>trans | PT | Martin and Orchiston (1987) |
| | Morris | Melbourne | 8.5-in. sp | R | Orchiston (1987a) |
| | Dr. Wright | Sydney | 8.5-in. sp | R | Baracchi (1914), Orchiston and Bhathal (1991) |
| 1880s | Davidson | Mackay | 6-in. OG | PR | Darlington and Orchiston (2017) |
| | Dobbie | Adelaide | 12-in. sp and 18-in. sp | P | Orchiston and Bembrick (1995), Waters (1980, 1981a, b) |
| | Macdonnell | Port Macquarie | 6-in. OG<br>trans | RT | Orchiston (2001b) |
| | Morris | Sydney | 8.5-in. sp | R | Orchiston (1987a) |
| | Oddie (Baker) | Ballarat | 12-in. sp and 26-in. sp<br>trans | PT | Burk (1986), Orchiston (1997a) |
| 1890s | Beebe | Bendigo | ~4.75-in. sp<br>trans | PRT | Martin and Orchiston (1987) |
| | Eglinton | Brisbane | 12-in. sp | P | Page (1959) |
| | Gale | Sydney and Newcastle | 6-in. OG<br>18-in. sp | PR | Orchiston (1997a, b), Orchiston and Bembrick (1997) |
| | Innes | Sydney | 6.25-in. OG<br>16.5-in. sp | R | Orchiston (2001a, 2003c, 2015) |
| | Merfield | Sydney | 7-in. sp | R | Orchiston (2015) |
| | Macdonnell | Sydney | 6-in. OG | R | Orchiston (2001b) |
| | Ross | Melbourne | 12-in. sp | R | Orchiston and Brewer (1990) |
| | Wooster | Ballarat | 12.25-in. sp | PR | Orchiston (1997a) |
| | Mr Wright | Sydney | 8.5-in. sp | R | Baracchi (1914) |
| 1900s | Beattie | Sydney | 6-in. OG | RT | Orchiston (1997b) |
| | Nangle | Sydney | 6.25-in. OG | R | Orchiston (1997b) |
| 1910s | McFarlane | Irvinebank | 7-in. OG<br>trans | MPT | Orchiston (1985b) |

[a]*Key*: OG = refractor; sp = reflector; trans = transit telescope
M = meteorological station; P = popularising astronomy; R = astronomical research; T = local time service

**Fig. 4.4** Dr. McFarlane's residence and observatory in the north Queensland tin-mining town of Irvinebank; the observatory housed a fine 7-in. Cooke refractor (Orchiston collection)

**Fig. 4.5** Two local astronomers pose in 1895 with the equatorially-mounted 26-in. Baker reflector, which was housed in a wooden run-off roof observatory. The *inset* shows Henry Evans Baker (both images Orchiston collection)

**Fig. 4.6** William Macdonnell posing outside his Port Macquarie Observatory in about 1885. This housed a 6-in. Grubb refractor and a transit telescope. In the 1890s, Macdonnell transferred these instruments to Sydney (*Courtesy*: Port Macquarie Museum, William Macdonnell's Photo Album, A 57)

Telescope excepted, of course). Reflectors and refractors were favoured almost equally. Apart from the locally-manufactured reflectors owned by Ross (Bio Box 9.3), Innes (Bio Box 8.2) and Merfield (Bio Box 11.4), and one of Gale's 18-in. (45.7-cm) instruments (Bio Box 12.1), all of the reflectors referred to in Table 4.2 were imported. Of the observatories listed that were devoted solely to research, those operated by Gale, Innes, Macdonnell (Bio Box 9.2; see, also, Fig. 4.6), Merfield and Ross produced the greatest output in terms of published papers and research notes. As Table 4.3 indicates, the types of research undertaken also closely mirrored that carried out at the major government observatories and, not unexpectedly, positional astronomy was the predominant focus.

The third type of observatory listed in Table 4.2 is that in which both public astronomy and research astronomy were followed. Again, a wide geographical distribution of observatories is apparent, and large telescope apertures are involved; the bias towards refractors is marked. Of the astronomers listed in this category, all but Bone (Bio Box 8.3), Beebe and Davidson (Bio Box 9.4) carried out wide-ranging series of observations and published prolifically. Dr. Bone was denied this chance through his untimely death, while Beebe moved interstate to a post which involved considerable fieldwork, thus preventing him from carrying out systematic astronomical work. Tebbutt was undoubtedly the leading Australian astronomer in this third category (Orchiston 2004a).

**Table 4.3** Principal observational programmes undertaken by leading Australian independent astronomers, 1850–1916

| |
|---|
| Transitory Events |
| Eclipses of the Moon |
| Eclipses of the Sun |
| Lunar occultations of planets |
| Lunar occultations of stars |
| Phenomena of Jupiter's satellites |
| Transits of Mercury |
| Transits of Venus |
| Short-term Monitoring Projects |
| Comets (positions and appearance) |
| Meteors (fireballs and shower activity) |
| Minor planets (positions) |
| Planets (positions) |
| Long-term Monitoring Projects |
| Double stars (separation and position angle) |
| Planets (appearance) |
| Sun (surface details) |
| Variable stars (magnitude variations) |
| Search Programmes |
| Coloured stars |
| New double stars |
| New variable stars |

## 4.4 An Evolving Science: Impact of the Leading Non-professional Astronomers on Australian Astronomy

If Tebbutt was Australian leading independent astronomer, then Francis Abbott of Hobart (Bio Box 6.4) would have to be ranked second, based on the quality and quantity of his publications, with Merfield third, followed by Innes and then Gale. All five astronomers, together with Beattie (Bio Box 13.3) and Nangle (Bio Box 13.4), were of professional standard in both intellect and mathematical prowess.

Given the calibre of these and other leading Australian independent astronomers, we should not underestimate their collective impact on research astronomy in Australia. Their combined publications tally far surpassed that of their professional colleagues, and like their government counterparts they published in the same local and overseas journals. Most popular were *Astronomische Nachrichten*, *Journal of the British Astronomical Association*, *Monthly Notices of the Royal Astronomical Society* and *The Observatory*, all highly-respected international journals. Through the sheer weight and quality of their publications, these and other non-professional astronomers helped to strengthen Australia's international astronomical reputation.

In some instances, independent astronomers were responsible for the actual founding of government observatories. Sydney Observatory resulted from the untiring efforts of P.P. King (see Orchiston 1988c, 2003b; Bio Box 7.1), while R.L.J. Ellery

**Fig. 4.7** Brisbane Observatory was a modest facility that was founded in 1879 through the actions of local astronomers (after Haynes et al. 1993)

(Bio Box 7.3) agitated successfully for a colonial observatory in Melbourne. In addition, lobbying, by groups of independent astronomers in both Brisbane and Hobart subsequently led to the establishment of small government-funded observatories in these cities (e.g. see Fig. 4.7).

Independent astronomers then provided a recruiting ground, albeit a limited one, for professional astronomy. Ellery and White (Bio Box 6.3) of Melbourne Observatory made the transition early, before they had established international reputations as astronomers, but this was not so of Innes and Merfield (Orchiston 2015). Innes went to the Cape Observatory in 1896 and later was founding Director of what became known as the Union Observatory in Johannesburg (Orchiston 2003c, 2015), while Merfield started working at Sydney Observatory in 1905 but soon transferred to Melbourne Observatory, where he eventually advanced to the position of Deputy Director (Orchiston 2015).

By the time they became professional astronomers, Innes and Merfield had achieved international recognition through their publications in mathematical astronomy; Innes was also highly regarded for his pioneering work on southern double stars. Both men were Tebbutt protégés, and their letters to him make it clear that they were out to emulate the brilliant example set by their mentor. In this respect, we should remember that in 1862 Tebbutt was offered but declined the post of Government Astronomer of New South Wales, which included the directorship of Sydney Observatory (Orchiston 1988b).

Most independent Australian astronomers never graduated to professional ranks. Nonetheless, many were able to offer invaluable assistance to their government observatory colleagues by providing much-needed observations (in both astronomy and meteorology), and by serving as willing collaborators when manpower was required for transit of Mercury, transit of Venus and solar eclipse expeditions. For

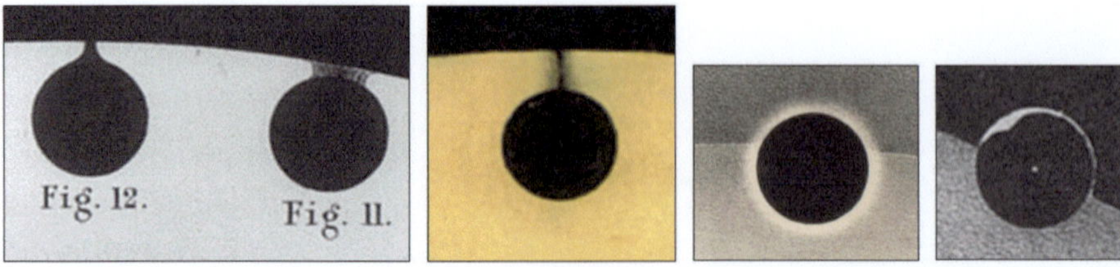


**Fig. 4.8** (*left* to *right*) Two ingress contact drawings (Allerding); the 'black drop effect' (Hirst); a halo around Venus (Macdonnell); Venus illuminated beyond the solar limb (Wright) (all after Orchiston 2004b)

example, most of those manning the three country stations established by Sydney Observatory for the 1874 transit of Venus were prominent independent astronomers (Russell 1892). Figure 4.8 shows examples of 1874 transit of Venus drawings made by four of these.

Independent astronomers also were able to occasionally contribute to professional astronomy by serving as sources of instruments. Thus, when O'Reilly died in Brisbane in 1879, the Queensland government purchased his observatory *in toto* and used it as a basis for a new colonial observatory (see Haynes et al. 1993). Similarly, in 1910 James Oddie gifted his 9-in. (22.9-cm) refractor (Fig. 4.9) to the Commonwealth, and after being used for site testing it became the founding instrument of the Commonwealth Solar Observatory (Allen 1978; Baracchi 1914; Morris-Kennedy 1989).

At times, professional astronomers also relied on instruments supplied by non-professional astronomers for the success of their research programmes. For instance, one of the telescopes used by the Woodford transit of Venus observing team in 1874 was a fine 5-in. (12.7-cm) Schroeder refractor provided by A. Fairfax, on whose property the party was stationed, while the team based at Sydney Observatory had the use of Colyer's 10.25-in. (26-cm) reflector (Russell 1892).

Certain independent astronomers also played a critical role in providing other astronomers with ready access to telescopes. Initially, astronomy was an expensive hobby for anyone wanting to carry out telescopic work, and especially if serious programmes were contemplated. Refracting telescopes by Cooke (Andrews 1992), Grubb (Andrews 1993) and Wray were preferred, but these precision instruments were highly priced and had to be imported from England. Quality British reflectors by Calver (Dall 1975) or With-Browning (Marriott 1996), although most desirable, were also beyond the price range of most Australian non-professional astronomers.

The only option for those without sufficient financial means was to manufacture their own reflecting telescopes or purchase a locally-made one, and this became a realistic proposition with the advent of the silver-on-glass reflector (see Tobin 1987). As we saw in the previous chapter, in Sydney Henry Russell experimented successfully with telescope-making during the 1870s and soon inspired some local non-professional astronomers to follow his lead (Orchiston and Bembrick 1995).

**Fig. 4.9** Oddie's 9-in. refractor at Melbourne Observatory, being prepared for the Mt Stromlo site-testing programme (Orchiston collection)

Many reflecting telescopes were produced in Australia during the last two decades of the nineteenth century (see Orchiston 2000, 2003a), and although most of these were small (in the 6–8-in. aperture range), a number of independent astronomers succeeded in grinding, polishing and figuring much larger mirrors. The leading pioneers were Edmonds, Gale and Madsen in Sydney (Orchiston and Bembrick 1995; Madsen 1886), Ross and Wigmore in Melbourne (Orchiston and Bembrick 1995; Orchiston and Brewer 1990; Orchiston and Perdrix 1990, 2002), and Dobbie in Adelaide (Orchiston and Bembrick 1995), all of whom completed mirrors in the 12–18-in. range (e.g. see Fig. 4.10), but the doyen of Australian telescope-makers at this time was surely Baker of Ballarat (Burk 1986). In 1988, 100 years after its manufacture, his 26-in. reflector (Fig. 4.5) was still the largest operational non-Dobsonian telescope in Victoria!

It is interesting to note that most of these early large instruments were committed to the popularisation of astronomy; only Ross and Gale employed their telescopes for serious observational programmes, and although Innes did discover some new double stars with the 16.5-in. (41.9-cm) reflector made for him by Edmonds, he preferred to use an old 6.25-in. (15.9-cm) Cooke refractor for his searches.

**Fig 4.10** Wigmore's 18-in. (45.7-cm) fork-mounted reflector later was donated to Swinburne Technical College (after The 18-in. … 1959)

The formation of organised groups and societies is a critical element in the evolution of any discipline, and this was no more apparent than in the nineteenth century Australian astronomy, with its vast geographical distances, physical isolation from the main international centres of astronomical activity and comparatively small number of astronomers. Australia's earliest astronomical groups and societies are listed in Table 4.4, and independent astronomers played a key role in their formation and operation.

Henry Russell from Sydney Observatory and Robert Ellery from Melbourne Observatory were the driving forces behind the short-lived Astronomy Sections of the Royal Societies of New South Wales and Victoria, respectively, but independent astronomers operating alone or in collaboration with professionals were intimately involved in the formation of all of the other groups listed (Ellery 1901; Baracchi 1914). In addition, Gale and Innes spent time planning the formation of the Australian Astronomical Society in 1892 (Chap. 13), but in the end decided not to proceed (see Orchiston and Bhathal 1984).

Although Tebbutt founded a national comet-seekers' group in 1882 (see Chap. 10), this was short-lived, and it was only in the 1890s that the first enduring astronomical groups were formed. These were generalist in that they dealt with all aspects of astronomy and were localised rather than national. By this time, Australian astronomy had evolved to the stage where a number of population

**Table 4.4** Australia's earliest formal astronomical groups and societies

| Duration | Founder(s)[a] | Name of group | References |
|---|---|---|---|
| 1876–1881 | Russell (p) | Royal Society of New South Wales, Section A | Orchiston and Bhathal (1991) |
| 1879–? | Ellery (p) | Royal Society of Victoria, Astronomy Section | Orchiston (1998) |
| 1882–1883 | Tebbutt (i) | Australian Comet Corps | Orchiston (1982) |
| 1892– | Farr (i) | Royal Society of South Australia, Astronomy Section | Waters (1980) |
| 1895– | Gale and Innes (i) | British Astronomical Association, NSW Branch (now=Sydney City Skywatchers) | Orchiston (1988a), Orchiston and Perdrix (2002) |
| 1895–1901 | 76 (i) | Brisbane Astronomical Society | Orchiston (1998) |
| 1897–1907 | Ross and Wigmore (i) | British Astronomical Association, Victoria Branch | Orchiston and Perdrix (1990, 2002) |
| 1912–1928 | Curlewis (p) and Hilton (i) | Astronomical Society of Western Australia | Western Australian Astronomical Society (1914) |

[a]i = independent astronomer; p = professional astronomer

centres could sustain their own formal groups: there were sufficient active astronomers present; support from both professional and non-professional astronomers was forthcoming; and astronomers of prominence existed who were willing to provide the requisite leadership. Material on Tebbutt's role in the early history of the New South Wales Branch of the British Astronomical Association is presented in Chap. 13.

All of the groups in Table 4.4, except perhaps the Brisbane Astronomical Society, played an important role in formalising astronomy in Australia, and cementing the close relations that existed between professional and non-professional astronomers (see Orchiston 1998). Formal groups gave newcomers to astronomy a structured way of getting started on their hobby, and provided all members with regular meetings, library facilities, and avenues for publication.

In any review of the achievement of Australia's independent astronomers, the outstanding contribution that they collectively made to public astronomy cannot be ignored (see Orchiston 1997a). Through their activities, an understanding or at least heightened public awareness of astronomy reached many more people than the government observatories alone could ever have serviced. In this respect, they supplemented the work of these official institutions, but all the while on a voluntary basis and at considerable personal sacrifice in terms of both time and money. In some towns and cities devoid of professional observatories, independent astronomers ran their own institutions as small-scale *de facto* city observatories (e.g. see Orchiston 1987b), and one of these is shown in Fig. 4.11. Without these individuals, and others who focussed on the popularisation of astronomy through their local news media, the Australian public would have been far less astronomically literate than was indeed the case.

**Fig. 4.11** John Beebe's East Bendigo Observatory served as a *de facto* public observatory for the citizens of Bendigo, Victoria (after Martin and Orchiston 1987: 5)

Finally, independent astronomers as a group had a political role to play. Late in the nineteenth century, they became a force to be reckoned with as watchdog bodies over government support for professional astronomy, and the activities of the professional observatories themselves. During the 1880s, Henry Russell increasingly turned the focus of Sydney Observatory from astronomy to meteorology, and in so doing estranged himself from many in the powerful Sydney astronomical fraternity, much to the consternation of some of his own staff members. Matters came to a head when Tebbutt intervened and made this a public issue (see Orchiston 2002). This illuminating case study in the deteriorating relationship between two of the nation's leading astronomers is discussed in some detail in Chap. 12.

## 4.5 Professionals and Non-professionals: Colleagues or Competitors?

Back in the late 1970s, the University of Calgary sociologist, Robert A. Stebbins, pioneered a whole new field of research, the sociology of 'amateurs' (e.g. Stebbins 1977, 1978, 1979, 1980a, 1992). Amateurs—or 'independent astronomers' in the context of this book—straddle the nebulous boundary between work and leisure. They are 'amators' in the strict sense in that they love their hobby and are happy to invest time, money and effort in it for the sake of the expected 'rewards' (see, also, Williams 1987). Five of Stebbins's later papers (Stebbins 1980b, 1981, 1982a, b, 1987) deal specifically with avocational astronomy, and these form an important contribution to our discipline.

Using 'dedication' as a criterion, Stebbins distinguishes 'devotees' from 'dabblers'. Australia's leading nineteenth century independent astronomers were all devotees, individuals who were happy to make a substantial commitment to the science in terms of both time and money. Using another dimension, 'knowledge and involvement', Stebbins differentiates between 'active' and 'armchair' astronomers. In the nineteenth century, leading active independent astronomers were engaged in observational and mathematical astronomy and in instrument-making. Stebbins also was able to categorise individual active astronomers within an 'apprentice–journeyman–master continuum'. Apprentices were beginning their astronomical 'careers', while masters were the acknowledged experts who were making a meaningful contribution to science whatever their area(s) of astronomical involvement. Most of the Australian independent astronomers discussed in this chapter were devotees who were active masters.

Because of the prominent status of many of Australia's independent astronomers and Tebbutt's anomalous position as the nation's foremost astronomer, the distinctions between professionals and serious non-professional astronomers were blurred in late nineteenth century Australian astronomy. Until Melbourne, Perth and Sydney Observatories became involved in the international Astrographic Catalogue and Carte du Ciel project, the nation's serious independent astronomers carried out much the same range of observations as their professional counterparts, with instruments of similar size and precision; published similar papers, in the same journals; joined the same scientific societies; served on the same committees; vied for the same prizes and awards; and used the very same tactics to promote a popular interest in their chosen discipline. In general, the intellectual and academic backgrounds of the two groups were not dissimilar and, as we have already noted, transfer from non-professional to professional ranks was sometimes possible.

Mobility along the professional–non-professional astronomy continuum only was possible in late nineteenth century Australia because the forces that emerged to separate independent astronomers and professional astronomers once and for all in other parts of the world between 1880 and 1920 simply by-passed Australia altogether at this time. The professional penchant for giant refractors (apertures >50-cm) was nowhere apparent, and the widely-held view that reflectors were simply 'the playthings of the non-professionals' never took hold. To the contrary, the nineteenth century Australian professional astronomers had a certain (uncharacteristic) fondness for mirrors, and at a time of giant refractor domination worldwide, Australia remained one of the few professional bastions of the beleaguered reflector.

The other prime reason for the rapid separation of independent and professional astronomers elsewhere was the emergence of astrophysics, which involved advanced training in mathematics and physics, access to elaborate instrumentation and research funding, and subject specialisation. No longer could a professional astronomer afford to be a generalist. As we saw in the previous chapter, Baracchi, Ellery, le Sueur and Russell all had fleeting spectroscopic escapades, but their exploits were matched in part by the Launceston independent astronomer, Alfred Barrett Biggs (see Biggs 1884a, b). Meanwhile, a few years earlier, Hobart's Francis Abbott had published three popular booklets on astronomy (Abbott 1878, 1879, 1880), and the first two contained a great deal of useful material on astronomical spectroscopy (see Orchiston 1992).

**Fig. 4.12** The 12-in. (30.5-cm) reflector made by David Ross, which he used very successfully for astrophotography (Orchiston collection)

The only major international trend to impact on Australian professional astronomy at the time was the international Astrographic Catalogue and Carte du Ciel project (Turner 1912), but this did not lead to immediate obvious differences between the nation's professional and non-professional astronomers for the government observatories also continued their non-astrographic work, while a number of independent astronomers experimented successfully with astrophotography. The most notable of these were David Ross in Melbourne (Fig. 4.12; Orchiston and Brewer 1990) and Sydney's Walter Gale (Reports of the Branches 1895). Thus, professional–non-professional communication was preserved, notwithstanding this innovation.

The professional–non-professional astronomy split of the 1880–1920 period so graphically described in the United Kingdom and the United States by Hetherington (1976), Lankford (1979, 1981a, b) and Rothenberg (1981) only surfaced in Australia after World War II (Orchiston 1989a), with the emergence of non-solar astronomy at Mt Stromlo Observatory (Davies 1982; Frame and Faulkner 2003; Bhathal et al. 2013) and the phenomenal growth of radio astronomy in Sydney (see Orchiston and Slee 2005; Robertson 1992; Sullivan 2005, 2009).

## 4.6 Concluding Remarks

Independent astronomers played a critical role in the evolution of early Australian astronomy. Largely because of the efforts of Abbott, Innes, Merfield and Tebbutt, their research contribution was out of all proportion to their numbers, and helped rebuild Australia's nineteenth century reputation in positional astronomy.

Through their instrument-making activities non-professional astronomers brought telescopes within the financial range of the average person, and by running public observing sessions at their observatories and publicising astronomy in local newspapers and through lectures and courses they were able to bring the joys of astronomy and the beauty of the night sky to the general public.

When formal astronomical groups and societies finally came to Australia, independent astronomers played a particularly important role in their formation and successful operation, usually alongside their professional colleagues. They also assisted astronomers from the government observatories by providing observations, joining them on special astronomical projects, and when necessary loaning their own instruments. Serious independent astronomers also provided a pool from which the government observatories could draw new staff, and on occasions they served as political lobby group for professional astronomy.

To the Australian professional astronomer, the serious independent astronomer was a colleague, someone to collaborate with rather than control, and someone to respect. Consequently, for the most part, professional–non-professional relations flourished in early Australian astronomy, and the overall contribution of the independent astronomers is one of the notable success stories in the history of Australian science.

**Acknowledgements** I wish to thank Professor Nick Lomb (University of Southern Queensland, Toowoomba), Adrian Brewer (Melbourne), Daryl Martin (Melbourne), the late John Perdrix (Perth), Clive Smith (Port Macquarie Historical Society) and the late Dr. Harley Wood (Sydney Observatory) for their assistance, and the Port Macquarie Museum for kindly supplying Fig. 4.6.

## References

Abbott, F. (1878) *Modern Astronomy* … Hobart: printed for the author.

Abbott, F. (1879) *Bibliographical Contribution to the Resume on Modern Astronomy*. Hobart: printed for the author.

Abbott, F. (1880). *Sidereal Systems of Modern Astronomy* … Hobart: printed for the author.

Allen, C.W. (1978). The beginnings of the Commonwealth Solar Observatory. *Records of the Australian Academy of Science, 4,* 27–49.

Andrews, A.D., (1992). Cyclopaedia of telescope makers. Part 1 (A-F). *The Irish Astronomical Journal, 20,* 102–183.

Andrews, A.D. (1993). Cyclopaedia of telescope makers. Part 2 (G-J). *The Irish Astronomical Journal, 21,* 1–82.

Ashbrook, J. (1984). *The Astronomical Scrapbook. Skywatchers, Pioneers and Seekers in Astronomy.* Cambridge (Mass.): Sky Publishing Corporation.

Baracchi, P. (1914). Astronomy and geodesy in Australia. In G.H. Knibbs (Ed.), *Federal Handbook Prepared in Connection with the Eighty-fourth Meeting of the British Association for the Advancement of Science, Held in Australia, August, 1914* (pp. 326–390). Melbourne: Government Printer.

Bhathal, R. & White, G. (1991). *Under the Southern Cross. A Brief History of Astronomy in Australia*. Sydney: Kangaroo Press.

Bhathal, R., Sutherland, R. & Butcher, H. (2013). *Mt Stromlo Observatory: From Bush Observatory to the Nobel Prize*. Melbourne: CSIRO Publishing.

Biggs, A.B. (1884a). Notes on spectroscopic observations of Comet "Pons"... *Papers and Proceedings of the Royal Society of Tasmania,* 200–201.

Biggs, A.B. (1984b). Report of spectroscopic observation of the twilight glows of February and March, 1884. *Papers and Proceedings of the Royal Society of Tasmania,* 202–203.

Bonwick, J. (1866). *Astronomy for Young Australians*. Melbourne: Mullen.

Burk, K. (1986). A history of James Oddie, Captain Henry Evans Baker and the Ballarat Municipal Observatory. *The Oddie-Baker Bulletin, 8,* 3–15.

Chapman, A. (1998). *The Victorian Amateur Astronomer. Independent Astronomical Research in Britain 1820–1920*. Chichester: Wiley.

Clerke, A. (1893). *A Popular History of Astronomy During the Nineteenth Century*. London: Adam and Charles Black.

Dall, H.E. (1975). George Calver – East Anglian telescope maker. *Journal of the British Astronomical Association, 86,* 49–52.

Darlington, V. & Orchiston, W. (2017). North Queensland and the 1882 transit of Venus: our final role in the quest for the Astronomical Unit. *Records of the Queensland Museum*, in press.

Davies, S. (1982). R.v.D.R. Woolley in Australia. *Historical Records of Australian Science, 6,* 59–69.

Davis, M.E. (1990). *Henry Evans Baker. A Brave, Able, Original Man*. Brisbane: printed for the author.

Dick, S., Orchiston, W. & Love, T. (1998). Simon Newcomb, William Harkness and the nineteenth century American transit of Venus expeditions. *Journal for the History of Astronomy, 29,* 221–255.

Dunlop, S. & Gerbaldi, M. (1988). *Stargazers: The Contribution of Amateurs to Astronomy*. Berlin: Springer-Verlag.

Ellery, R.J. (1901). A brief history of the beginnings and growth of astronomy in Australasia. In *Report of the Eighth Meeting of the Australasian Association for the Advancement of Science* (pp. 1–17). Melbourne: AAAS.

Frame, T. & Faulkner, D. (2003). *Stromlo: An Australian Observatory*. Sydney: Allen & Unwin.

Giordano, M. (1995). *Watcher of the Skies. The Story of Alfred Barrett Biggs. Pioneer in Astronomy, Seismology and Telecommunications*. Launceston: Regal Publications.

Haynes, R.D., Haynes, R.A. & Kitson, W.S. (1993). The history of astronomy in Queensland. *Vistas in Astronomy, 36,* 231–252.

Haynes, R., Haynes, R., Malin, D. & McGee, R. (1996). *Explorers of the Southern Sky. A History of Australian Astronomy*. Cambridge: Cambridge University Press.

Herrmann, D.B. (1973). An exponential law for the establishment of observatories in the nineteenth century. *Journal for the History of Astronomy, 4,* 57–58.

Hetherington, N.S. (1976). Amateur versus professional. The British Astronomical Association and the controversy over canals on Mars. *Journal of the British Astronomical Association, 86,* 303–308.

Inkster, I. (1978). Robert Goodacre's astronomy lectures (1823–25) and the structure of scientific culture in Philadelphia. *Annals of Science, 35,* 353–363.

Inkster, I. (1980). Robert Goodacre (1777–1835) and popular astronomy. *Journal of the British Astronomical Association, 90,* 245–252.

Inkster, I. (1982). Advocates and audiences – aspects of popular astronomy in England, 1750–1850. *Journal of the British Astronomical Association, 92,* 119–123.

Lankford, J. (1979). Amateur versus professional: the transatlantic debate over the measurement of Jovian longitude. *Journal of the British Astronomical Association, 89,* 574–582.

Lankford, J. (1981a). Amateurs and astrophysics: a neglected aspect in the development of a scientific speciality. *Social Studies in Science, 11,* 275–303.

Lankford, J. (1981b). Amateurs versus professionals: the controversy over telescope size in late Victorian science. *Isis,* 72, 11–28.

Lomb, N. (2011). *Transit of Venus: 1631 to the Present.* Sydney: NewSouth Publishing.

Madsen, H.F. (1886). Notes on the process of polishing and figuring 18 inch glass specula by hand. *Journal and Proceedings of the Royal Society of New South Wales, 20,* 79–92.

Marriott, R.A. (1996). The life and legacy of G.H. With, 1827–1904. *Journal of the British Astronomical Association, 106,* 257–264.

Martin, D. & Orchiston, W. (1987). *Observatories of Victoria. 2: The Historic Bendigo Observatories of J. Nelson Jones and John Beebe.* Melbourne: Astronomical Society of Victoria (Historical Notes No. 4).

Meadows, J. & Henbest, N. (1981). Astronomy; how amateurs got there first. *New Scientist, 90,* 170–171.

Morris-Kennedy, P. (1989). A celestial centenary. *Southern Astronomy, 2,* 30–34.

North, J.D. (1975). Proctor, Richard Anthony. In C.C. Gillespie (Ed.), *Dictionary of Scientific Biography. Volume XI: PITC-RUSH* (pp. 162–163). New York: Scribner.

Orchiston, W. (1981). Illuminating incidents in Antipodean astronomy: John Tebbutt and Great Comet of 1881. *Journal of the Astronomical Society of Victoria, 34,* 78–92.

Orchiston, W. (1982). Illuminating incidents in Antipodean astronomy: John Tebbutt and the abortive Australian association of comet observers. *Journal of the Astronomical Society of Victoria, 35,* 70–83.

Orchiston, W. (1983). Illuminating incidents in Antipodean astronomy: the mysterious Tasmanian comet of December 1883. *Proceedings of the Astronomical Society of Australia, 5,* 282–284.

Orchiston, W. (1985a). The contribution of Alfred Barrett Biggs to Australian and Tasmanian astronomy. *Records of the Queen Victoria Museum, 89,* 1–28.

Orchiston, W. (1985b). Life-history of a telescope: the 17.8-cm refractor at the Riverview Observatory in Sydney. *Australian Journal of Astronomy, 1,* 57–80.

Orchiston, W. (1986). *Observatories of Victoria. 1: Dr William Bone and the Castlemaine Observatory.* Melbourne: Astronomical Society of Victoria (Historical Notes No. 3).

Orchiston, W. (1987a). Contribution of the Lands Department to the development of astronomy in New South Wales during the nineteenth century. *Australian Journal of Astronomy, 2,* 65–74.

Orchiston, W. (1987b). Dr William Bone, and the role of the amateur observatory in Australian astronomy. *Southern Stars, 32,* 111–128.

Orchiston, W. (1988a). Illuminating incidents in Antipodean astronomy: formation of the New South Wales Branch of the British Astronomical Association. *Journal of the British Astronomical Association, 98,* 75–84.

Orchiston, W. (1988b). Illuminating incidents in Antipodean astronomy: John Tebbutt and the Sydney Observatory directorship of 1862. *Australian Journal of Astronomy, 2,* 149–158.

Orchiston, W. (1988c). Illuminating incidents in Antipodean astronomy: Phillip Parker King and the founding of Sydney Observatory. *Vistas in Astronomy, 32,* 285–301.

Orchiston, W. (1989a). Astrophysics and the evolution of twentieth century Australian astronomy. *Search, 20,* 198–202.

Orchiston, W. (1989b). The role of the amateur in early Australian astronomy. *Search, 20,* 16–23.

Orchiston, W. (1991). Amateur-professional co-operation in astronomy education: a nineteenth century Australian model. *Proceedings of the Astronomical Society of Australia, 9,* 186–188.

Orchiston, W. (1992). The contribution of Francis Abbott to Tasmanian and Australian astronomy. *Vistas in Astronomy, 35,* 315–344.

Orchiston, W. (1997a). The role of the amateur in popularising astronomy: an Australian case study. *Australian Journal of Astronomy, 7,* 33–66.

Orchiston, W. (1997b). The role of the large refracting telescope in Australian amateur astronomy: an historical perspective. *Australian Journal of Astronomy, 7,* 89–114.

Orchiston, W. (1997c). The "tyranny of distance" and Antipodean cometary astronomy. *Australian Journal of Astronomy, 7,* 115–126.

Orchiston, W. (1998). Amateur-professional collaboration in Australian science: the earliest astronomical groups and societies. *Historical Records of Australian Science, 12,* 163–182.

Orchiston, W. (1999). C/1881 K1: A forgotten "Great Comet" of the nineteenth century. *Irish Astronomical Journal, 26,* 33–44.

Orchiston, W. (2000). Amateur telescope-making in Australia: an historical perspective. *Amateur Telescope Making Journal, 15,* 10–26.

Orchiston, W. (2001a). From amateur astronomer to observatory director: the curious case of R.T.A. Innes. *Publications of the Astronomical Society of Australia, 18,* 317–327.

Orchiston, W. (2001b). William John Macdonnell and the development of astronomy in New South Wales. *Journal of the British Astronomical Association, 111,* 13–25.

Orchiston, W. (2002). Tebbutt vs Russell: passion, power and politics in nineteenth century Australian astronomy. In S.M.R. Ansari (Ed.), *History of Oriental Astronomy* (pp. 169–201). Dordrecht: Kluwer (IAU Joint Discussion 17).

Orchiston, W. (2003a). Amateur telescope making in Australia. An historical perspective. In W.J. Cook (Ed.), *The Best of Amateur Telescope Making Journal. Volume 2* (pp. 208–239). Richmond, Willmann-Bell.

Orchiston, W. (2003b). Australia's earliest planispheres. *Journal of the British Astronomical Association, 113,* 329–332.

Orchiston, W. (2003c). Passion and patronage: the Afro-Australian astronomy of Robert Thorburn Ayton Innes. *Monthly Notices of the Astronomical Society of Southern Africa, 62,* 4–13.

Orchiston, W. (2004a). John Tebbutt and observational astronomy at Windsor Observatory. *Journal of the British Astronomical Association, 114,* 141–154.

Orchiston, W. (2004b). The nineteenth century transits of Venus: an Australian and New Zealand overview. *Journal of Astronomical Data, 10,* 219–308.

Orchiston, W. (2006). Amateurs in the Antipodes: the common denominators of nineteenth and early twentieth century South African, Australian and New Zealand astronomy. *Monthly Notices of the Astronomical Society of Southern Africa, 65,* 148–159, 194–208.

Orchiston, W. (2015). The amateur-turned-professional syndrome: two Australian case studies. In W. Orchiston, D.A. Green & R.G. Strom (Eds.), *New Insights from Recent Studies in Historical Astronomy: Following in the Footsteps of F. Richard Stephenson* (pp. 259–350). Cham: Springer.

Orchiston, W. & Bembrick, C. (1995). The role of the large reflecting telescope in amateur astronomy: an Australian case study. *Australian Journal of Astronomy, 6,* 53–74.

Orchiston, W. & Bembrick, C. (1997). The potency of patronage: George Hoskins and the New South Wales Branch of the British Astronomical Association. *Australian Journal of Astronomy, 7,* 1–15.

Orchiston, W. & Bhathal, R. (1984). Illuminating incidents in Antipodean astronomy: attempted formation of the Australian Astronomical Society in 1892. *Proceedings of the Astronomical Society of Australia, 5,* 422–425.

Orchiston, W. & Bhathal, R. (1991). Illuminating incidents in Antipodean astronomy: Section A of the Royal Society of New South Wales. *Australian Journal of Astronomy, 4,* 49–67.

Orchiston, W. & Brewer, A. (1990). David Ross and the development of amateur astronomy in Victoria. *Journal of the British Astronomical Association, 100,* 173–181.

Orchiston, W. & Perdrix, J. (1990). The British Astronomical Association in Australia: a centennial perspective. *Australian Journal of Astronomy, 3,* 163–179.

Orchiston, W. & Perdrix, J. (2002). A history of the British Astronomical Association in Australia: the fate of the Branches. *Journal of the British Astronomical Association, 112,* 68–77.

Orchiston, W. & Slee, B. (2005). The Radiophysics field stations and the early development of radio astronomy. In W. Orchiston (Ed.), *The New Astronomy: Opening the Electromagnetic Window and Expanding our View of Planet Earth* (pp. 119–168). Dordrecht: Springer.

Page, A. (1959). Astronomy in Queensland 1849–1959. *Proceedings of the Astronomical Society of Queensland,* 19–24.

Reports of the Branches. New South Wales Branch. *Journal of the British Astronomical Association, 5,* 343–345 (1895).

Robertson, P. (1992). *Beyond Southern Skies: Radio Astronomy and the Parkes Radio Telescope.* Cambridge: Cambridge University Press.

Rothenberg, M. (1981). Organisation and control: professionals and amateurs in American astronomy, 1899–1918. *Social Studies in Science, 11,* 305–325.

Russell, H.C. (1892). *Observations of the Transit of Venus, 9 December, 1874; Made at Stations in New South Wales.* Sydney: Government Printer.

Stebbins, R.A. (1977). The amateur: two sociological definitions. *Pacific Sociological Review, 20,* 582–606.

Stebbins, R.A. (1978). Towards amateur sociology: a proposal for the profession. *The American Sociologist, 13,* 239–252.

Stebbins, R.A. (1979). *Amateurs: On the Margin Between Work and Leisure.* Beverley Hills: Sage Publications.

Stebbins, R.A. (1980a). "Amateur" and "hobbyist" as concepts for the study of leisure problems. *Social Problems, 27,* 413–417.

Stebbins, R.A. (1980b). Avocational science: the amateur routine in archaeology and astronomy. *International Journal of Comparative Sociology, 21,* 34–48.

Stebbins, R.A. (1981). Science amators? Rewards and cost in amateur astronomy and archaeology. *Journal of Leisure Research, 13,* 289–304.

Stebbins, R.A. (1982a). Amateur and professional astronomers: a study of their interrelationships. *Urban Life, 10,* 433–454.

Stebbins, R.A. (1982b). Looking downwards: sociological images of the vocation and avocation of astronomy. *Southern Stars, 29,* 177–188.

Stebbins, R.A. (1987). Amateurs and their place in professional science. In D.S. Hayes, R.M. Genet, & D.R. Genet (Eds.), *New Generation Small Telescopes* (pp. 217–225). Mesa, Fairborn Press.

Stebbins, R.A. (1992). *Amateurs, Professionals, and Serious Leisure.* Montreal: McGill-Queens University Press.

Sullivan, W.T. III (2005). The beginnings of Australian radio astronomy. *Journal of Astronomical History and Heritage, 8,* 11–342.

Sullivan, W.T. III (2009). *Cosmic Noise. An Early History of Radio Astronomy.* Cambridge: Cambridge University Press.

Tebbutt, J. (1908). *Astronomical Memoirs.* Windsor: printed for the author.

The 18-inch reflector at Swinburne Technical College. *Journal of the Astronomical Society of Victoria, 12,* 26–27 (1959).

Tobin, W. (1987). Foucault's invention of the silvered-glass reflecting telescope and the history of his 80-cm reflector at the Observatoire de Marseilles. *Vistas in Astronomy, 30,* 153–184.

Turner, H.H. (1912). *The Great Star Map. Being a General Account of the International Project Known as the Astrographic Chart.* London: Murray.

Utting, M. (1991). *Windows to the Southern Sky.* Perth: Murdoch University and Perth Observatory.

Waters, B., (1980). *A Reference History of the Astronomical Society of South Australia Inc. Volume 1. The First Decade 1891–1901.* Adelaide: printed for the author.

Waters, B., (1981a). *A Reference History of the Astronomical Society of South Australia Inc. Volume 2. The Second Decade 1902–1911.* Adelaide: printed for the author.

Waters, B., (1981b). *A Reference History of the Astronomical Society of South Australia Inc. Volume 3. The Third Decade 1912–1921.* Adelaide: printed for the author.

Western Australian Astronomical Society (1914). Organisation and membership. *Proceedings of the Western Australian Astronomical Society, 1,* 1–2.

Williams, T.R. (1987). Criteria for classifying an astronomer as an amateur. Paper presented at IAU Colloquium 98, Paris.

Woolf, H. (1959). *The Transits of Venus: A Study of Eighteenth Century Science.* Princeton: Princeton University Press.

# Part III
# Case Studies

# Chapter 5
# Introduction

John Tebbutt was an important figure in the nineteenth century Australian astronomy, and the aim of Part II is to provide a readable overview of his principal achievements in astronomy and meteorology and the *milieu* within which they occurred. As such, we must examine his discovery of two of the Great Comets of the nineteenth century and explore the instrumentation at his Windsor Observatory and the ways in which this was utilised to contribute to world astronomy. We also must trace Tebbutt's early associations with Sydney Observatory, and his involvement in the early history of the colony's first formal astronomical groups. We also must delve into the politics of New South Wales science and show how Tebbutt's abiding passion for astronomy brought him into open conflict with his one-time friend and confidant, Henry Russell—the Government Astronomer of New South Wales—and how their deteriorating relationship impacted directly or indirectly on most of Sydney's leading astronomers during the last two decades of the century and coloured the flavour of New South Wales astronomy.

This rich tapestry of science and politics, with its underlying elements of power and prestige, offered a number of different ways of organising the chapters in this section of the book, and after experimenting with various options I chose to pick out a succession of 'key events' and to present these in some sense of chronological order. The following 'key elements' were identified:

- Discovery of the Great Comet of 1861 (Chap. 6)
- Offer of the Sydney Observatory directorship in 1862 (Chap. 7)
- The founding and development of Windsor Observatory (Chap. 8)
- Discovery of the Great Comet of 1881 (Chap. 9)
- Formation of the Australian Comet Corps in 1882 (Chap. 10)
- Acquisition of the 8-in. Grubb telescope in 1886 (Chap. 11)
- The deteriorating relationship between Russell and Tebbutt (Chap. 12)
- Tebbutt's role in the formation and early development of the New South Wales Branch of the British Astronomical Association (Chap. 13)
- Tebbutt's 'official' retirement from astronomy (Chap. 14)
- The fate of the library, instruments and Windsor Observatory buildings (Chap. 15)



W. Orchiston, *John Tebbutt*, Historical & Cultural Astronomy,
DOI 10.1007/978-3-319-44521-2_5

Some of these 'key events' were literally that: they were short-term, one-off affairs (e.g. the discovery of the two Great Comets and the formation of the Australian Comet Corps). Others served as 'trigger events' to initiate ongoing developments (e.g. the founding of Windsor Observatory or the purchase of the Grubb Telescope). In this context, then, Chaps. 8, 11 and 12 all span considerable time intervals.

This approach allows for a logical chapter-by-chapter presentation of Tebbutt's significant contributions to Australian and international astronomy, and there are only two aspects of his work which are not concentrated within the confines of single chapters. One is his astronomical observations and the other is his involvement in new astronomical groups and societies, and these are spread over several different chapters. His astronomical observations are summarised and discussed in three separate chronologically-sequential instalments, in Chaps. 7 (for the period 1853–1863), 8 (1864–1885) and 11 (1886–1915), while his involvement in early astronomical groups and societies is presented, mainly, in Chaps. 10 and 13.

While this book is primarily about astronomy, we would be remiss to neglect Tebbutt's important contribution to Australian meteorology. His meteorological instruments are listed (but not described) in the chapter on Windsor Observatory (Chap. 8), and his meteorological observations and publications are discussed in Chap. 14. Meanwhile, his tide gauge is described in Chap. 8, along with his tidal work and the records he kept of local floods and freshes (minor floods). This same chapter also deals with his—some would say—obsessive attempts to determine successively more precise values for the latitude and longitude of his Observatory.

Although the focus throughout must, of necessity, be on Tebbutt, we do use him to introduce many other Australian independent astronomers along the way. Thus, short 'thumbnail biographies' of the following individuals are to be found: Hugh Wright (Chap. 2), Abbott and Hawkins (Chap. 6); King (Chap. 7); Bone and Innes (Chap. 8); Butterfield, Davidson, Macdonnell and Ross (Chap. 9); Biggs, Brindley, Hirst, Martin, Morris, Swindlehurst, Ward and Dr. Wright (Chap. 10); Merfield (Chap. 11); Gale, Roseby and Thomson (Chap. 12); Beattie, Nangle and Wooster (Chap. 13); Grigg, (Chap. 14); and Hoskins and Howarth (Chap. 15). Nor are professional astronomers, professional instrument-makers and university academics neglected, for similar treatment has been afforded Scott and White (Chap. 6); Ellery and Lenehan (Chap. 7); Brooks, Furber and Tornaghi (Chap. 8); Conder (Chap. 10); Baracchi, Dunlop and Rümker (Chap. 11); Pollock, Russell, Sellors and Todd (Chap. 12); and Knibbs (Chap. 13). For the most part, these mini-biographies are presented in clearly identified 'Bio Boxes', so that they do not disrupt the flow of the narrative.

All of the above individuals are men, reflecting an era when independent astronomy was largely unpopular with women (e.g. see Chapman 1998). While most of these astronomers were from New South Wales, Queensland (Davidson and Thomson), South Australia (Todd), Tasmania (Abbott and Biggs) and Victoria (Baracchi, Bone, Ellery, Ross, White and Wooster) also are represented.

Tebbutt's association with Sydney Observatory was central to the development of New South Wales astronomy during the second half of the nineteenth century, and by reading Chaps. 7 and 12 (in that order) one may gain an insight into the pattern of astronomical research at this institution and the interplay between astronomy and meteorology. Useful supplementary material is also to be found in Chaps. 6, 8 and 9. Meanwhile, some of White's early work at Melbourne Observatory is mentioned in Chap. 6.

While this book is basically about Australian astronomy, in fact its geographical net has been spread somewhat more widely to encompass New Zealand—albeit only briefly. Thus, there are accounts of work carried out by Stock (Chap. 2), Grigg (Chaps. 2, 13 and 14) and Bateson (Chap. 15).

## Reference

Chapman, A. (1998). *The Victorian Amateur Astronomer. Independent Astronomical Research in Britain 1820–1920*. Wiley: Chichester.

# Chapter 6
# The Great Comet of 1861: A Magnificent Career Catalyst

## 6.1 Introduction

In 1861, John Tebbutt discovered his first comet, when still 26 years of age. Fate decreed that this was to become one of the Great Comets of the nineteenth century and through it Tebbutt would become a household name in Australia. Yet this discovery was no accident. It was the culmination of 8 years of intensive self-education during which Tebbutt learnt about astronomy and taught himself mathematics, gained observational experience using what, even in those days, would have been regarded as rather primitive instruments, and honed his skills in effectively using the media both to publish his own observations and to alert the public to upcoming astronomical events of potential interest. In other words, the arrival of Tebbutt's Great Comet marked the end of his 'astronomical apprenticeship'.

While discovery of the 1861 comet brought Tebbutt an element of local fame, it also exposed him to ridicule—something for which he was ill prepared. With the exception of one minor incident, the discovery also served to strengthen the close working relations that had developed by this time between Tebbutt and the Reverend William Scott (see Bio Box 6.1), the Government Astronomer of New South Wales and founding Director of Sydney Observatory (see Orchiston 1988b; Wood 1958 for general histories of the Observatory).

This chapter draws closely on material in Orchiston (1998a) and documents Tebbutt's discovery and subsequent observation of Comet C/1861 J1 (Tebbutt), as it is now known. Along the way, we repeatedly meet up with Scott, and also encounter a number of little-known or anonymous individuals who took great delight in belittling both Tebbutt and Scott. We end the chapter by tracing the development of the comet in all its majesty after its migration into northern skies and examining Tebbutt's later attempt to claim priority as discoverer of what proved to be one of the most magnificent comets of the century.



W. Orchiston, *John Tebbutt*, Historical & Cultural Astronomy,
DOI 10.1007/978-3-319-44521-2_6

**Bio Box 6.1: William Scott**

The Reverend William Scott (Fig. 6.1) was born at Hartland, North Devon, in 1825 and obtained a B.A. degree from Cambridge University in 1848, taking honours as Third Wrangler (Houghton 1917). He was made a deacon in 1849 and was ordained a priest in 1850. In 1851, he married a widow with three sons and inherited family responsibilities which prompted him to become a mathematics coach (Cable 1976), and 2 years later he published a small textbook titled *Elementary Treatise on Plane Co-ordinate Geometry* (Lynn 1904). At the time he was offered the founding directorship of Sydney Observatory Scott was a Mathematics Lecturer at Sidney Sussex College, Cambridge.

**Fig. 6.1** The Reverend William Scott (adapted from Russell 1892, Frontispiece)

During his short time at the Observatory, in addition to advising on the design of the building and acquired new astronomical instruments and staff, Scott determined its latitude and longitude; conducted observations of comets, a partial solar eclipse and a transit of Mercury; carried out a survey of southern stars; published four volumes of *Astronomical [and Meteorological] Observations Made at the Sydney Observatory …*; initiated an astronomical education programme; set up a network of country meteorological stations; and carried out a magnetic survey of the colony. It was a frantically busy 6 years, which impacted negatively on his health and led to his premature resignation.

After leaving the Observatory in 1862, he turned to the less-demanding task of teaching and took over the Cook's River Collegiate School in Sydney. From 1865 to 1878 he was Warden of St. Paul's College at the University of Sydney and then committed the remainder of his life to the ministry (ibid.), serving at Bungendore, Gunning and Queanbeyan in country New South Wales.

Despite his short career as an astronomer, Scott retained an active interest in science and held several offices in the Royal Society of New South Wales (Maiden 1918). In 1874, he led the Sydney Observatory transit of Venus team that was based at Eden (see Scott 1874).

(continued)

Upon retiring, Scott and his wife settled at Kurrajong Heights near Sydney, where they owned an orchard. Scott lived to the ripe old age of 91, dying in Sydney on 29 March 1917. He was survived by a son and two daughters and had an estate valued at nearly £7,500 (Cable 1976).

Dr. Harley Weston Wood (1911–1984) and Henry Chamberlain Russell are typically paraded as the two most accomplished Government Astronomers of New South Wales (e.g. see Bhathal 1991; Haynes et al. 1996; Robertson 1985a, b), and Scott's remarkable contribution during his few short years at Sydney Observatory is usually overlooked. A detailed evaluation of his many achievements (see Doyle 2008; Orchiston 1998b and this Chapter) soon reveals that this is a short-sighted attitude.

## 6.2 A 'Nebula' Is Noted: Discovery of the Great Comet of 1861

In May 1861, Tebbutt made the discovery which was to propel his name before the professional astronomers of the world and assign to him forever a notable place in the history of cometary astronomy. He discovered what turned out to be one of the 'Great Comets' of the nineteenth century.

Although this is not specifically mentioned in his observing diary at the time, Tebbutt was later to recount that on 13 May he was searching the western sky for comets (Tebbutt 1908, p. 14) soon after sunset and in the process '… mistook a small nebula for a small comet …' (Tebbutt 1859–63). He used his sextant to take six different measurements of the distances between the 'nebula' and the first magnitude stars Procyon, Sirius, and Canopus, and from these derived a position for the mystery object.

Tebbutt's next task was to compare this position with nebulae in the only catalogue of any note then in his fledgling library, one first published by the French astronomer, Lacaille (in 1755), while based temporarily at the Cape of Good Hope between 1751 and 1753 (see Glass 2013 ). His conclusion was important: 'I could not find any nebula in Lacaille's Catalogue occupying this position; and this fact led me to think the object might be a comet' (Tebbutt 1859–63).

If this was indeed a comet, rather than a faint unrecorded nebula, galaxy or globular cluster, then Tebbutt could expect his 'nebula' to move relative to the background stars. Observations over the next few nights would prove critical. Filled with anticipation, Tebbutt rose before sunrise on the 14th, and '… again saw the nebula near a small star of the sixth or seventh magnitude, but could not on account of fog and the approach of twilight ascertain whether the star was the same which I observed near the nebula last night' (ibid.). That same evening was clear and he was able to observe the mystery object, but it appeared to be in exactly the same position as earlier, which convinced Tebbutt '… that the nebulous object is not a comet …' (ibid.).

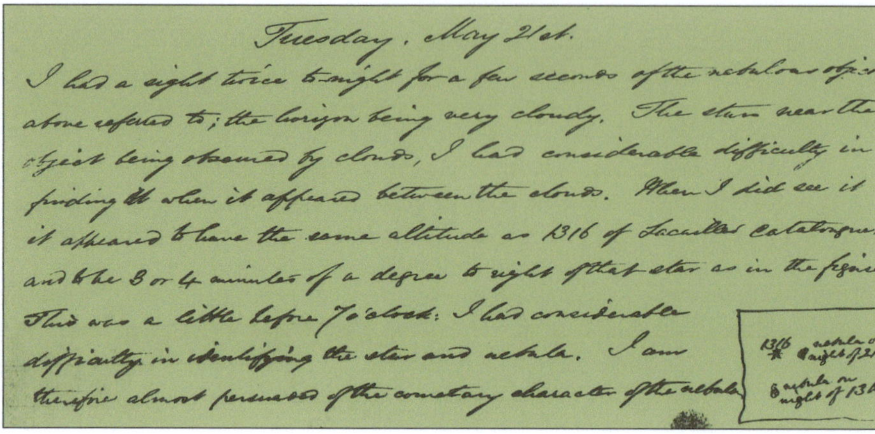

Tuesday. May 21st.

I had a sight twice to-night for a few seconds of the nebulous object above referred to; the horizon being very cloudy. The stars near the object being obscured by clouds, I had considerable difficulty in finding it when it appeared between the clouds. When I did see it it appeared to have the same altitude as 1316 of Lacaille's Catalogue, and to be 3 or 4 minutes of a degree to right of that star as in the figure. This was a little before 7 o'clock: I had considerable difficulty in identifying the star and nebula. I am therefore almost persuaded of the cometary character of the nebula

1316 nebula on night of 21st
nebula on night of 13th

**Fig. 6.2** Tebbutt's entry for 21 May 1861 in his diary, where he concluded that he had probably discovered a comet (after Tebbutt 1859–63; *Courtesy* Mitchell Library)

The next clear evening was on 21 May, and it was simply a matter of confirming the 14 May conclusion. Tebbutt's handwritten entry in his diary is reproduced here as Fig. 6.2 and indicates that although it was cloudy near the horizon, he twice caught sight of the 'nebula' for a few seconds, and with great difficulty made an observation. However, it is the last two lines of this diary entry that display the exciting outcome of his evaluation: 'I am therefore almost persuaded of the cometary character of the nebula' (ibid.)

The simple field sketch in the bottom right-hand corner of Tebbutt's diary entry, enlarged in Fig. 6.3 so that it is now legible, shows that between 13 and 21 May the 'nebula' appeared to have moved with respect to the reference star 1316 in Lacaille's catalogue. But Tebbutt rightfully decided that a confirmatory observation was required, under much better observing conditions than offered that evening. Nonetheless, merely in order 'to play it safe', he decided to send a letter off to Scott at Sydney Observatory reporting his suspicions (Tebbutt 1908).

May 22 must have been a day of great excitement and anticipation for Tebbutt as he waited for sundown. The hoped-for clear evening duly materialised and with it came the vital observation:

> … had a good view of the nebulous object again tonight; *there can now be no doubt as to its cometary character*. The following shows its position with reference to 1316 of Lacaille's Catalogue between 6h30m & 7h. The comet being excessively faint I could not ascertain its position with the sextant. I got one position from Procyon, namely 63°3½′ but this may be several minutes in error. The faintness of the comet arose from the moon's presence. I observed the passage of the star and comet across the field of the day and night telescope, but I could not observe the ingress & egress with anything like precision. The comet's right ascension and south polar distance were evidently a few minutes greater than those of the star, say 6 or 7 minutes. (Tebbutt 1859–63; my italics).

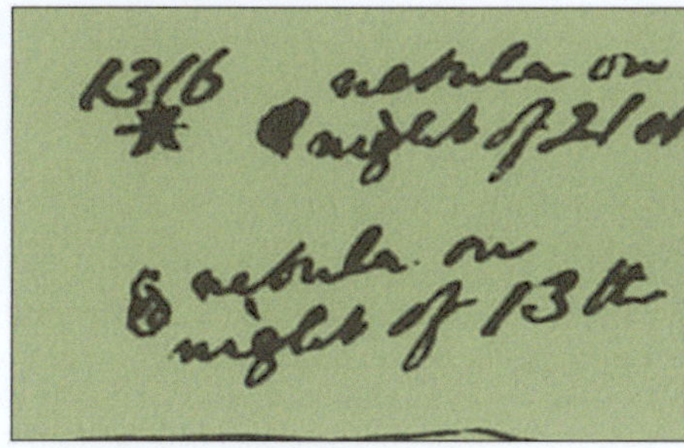


**Fig. 6.3** An enlargement of the field sketch accompanying Tebbutt's 21 May diary entry, showing the motion of the nebulous object (after Tebbutt 1859–63; *Courtesy* Mitchell Library)

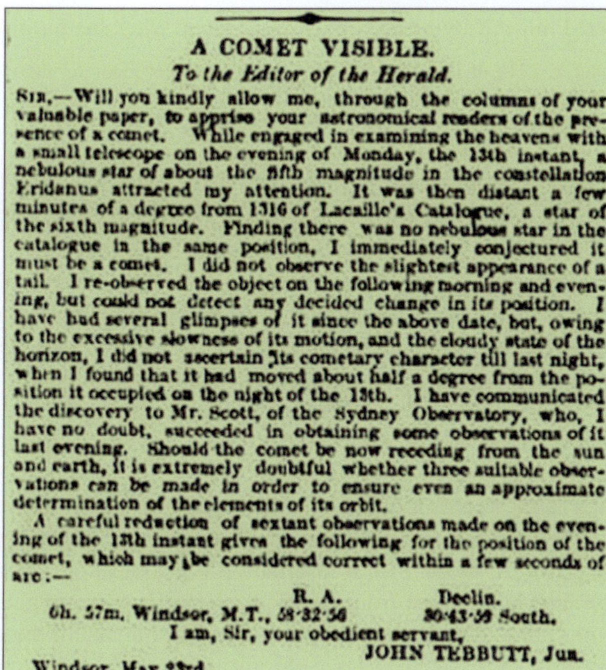

A COMET VISIBLE.

*To the Editor of the Herald.*

SIR,—Will you kindly allow me, through the columns of your valuable paper, to apprise your astronomical readers of the presence of a comet. While engaged in examining the heavens with a small telescope on the evening of Monday, the 13th instant, a nebulous star of about the fifth magnitude in the constellation Eridanus attracted my attention. It was then distant a few minutes of a degree from 1316 of Lacaille's Catalogue, a star of the sixth magnitude. Finding there was no nebulous star in the catalogue in the same position, I immediately conjectured it must be a comet. I did not observe the slightest appearance of a tail. I re-observed the object on the following morning and evening, but could not detect any decided change in its position. I have had several glimpses of it since the above date, but, owing to the excessive slowness of its motion, and the cloudy state of the horizon, I did not ascertain its cometary character till last night, when I found that it had moved about half a degree from the position it occupied on the night of the 13th. I have communicated the discovery to Mr. Scott, of the Sydney Observatory, who, I have no doubt, succeeded in obtaining some observations of it last evening. Should the comet be now receding from the sun and earth, it is extremely doubtful whether three suitable observations can be made in order to ensure even an approximate determination of the elements of its orbit.

A careful reduction of sextant observations made on the evening of the 13th instant gives the following for the position of the comet, which may be considered correct within a few seconds of arc:—

| | R. A. | Declin. |
|---|---|---|
| 6h. 57m. Windsor, M.T., | 58·32·56 | 30·43·39 South. |

I am, Sir, your obedient servant,
JOHN TEBBUTT, Jun.

Windsor, May 23rd.

**Fig. 6.4** John Tebbutt's (1861b) Letter to the Editor of the *Sydney Morning Herald*, reporting his discovery of the comet, which was published on 25 May, his 27th birthday (*Courtesy* TROVE)

By now Tebbutt was convinced that the 'nebula' was indeed a new comet, and on the following day, 23 May, he announced his discovery through a letter submitted to the Sydney daily newspaper, the *Sydney Morning Herald*. As fate would have it, this appeared in print on the 25 May, thereby presenting him with an ideal 27th birthday present! His letter is reproduced here as Fig. 6.4, and the critical phrase is: 'I did not ascertain its cometary character till last night, when I found that it had moved about half a degree from the position it occupied on the night of the 13th'.

## 6.3 Ridicule and Rancour: The Price of Publicising Further Observations

The day after sending his discovery announcement to the *Herald* Tebbutt observed the comet again:

> I saw the comet again to-night, but it was too faint for observation with a sextant owing to the moon's presence. I made the following [two] observations with Lynch's telescope (day or night glass) … The observations were not precise, as the star & comet were faint, being near the horizon … The comet had changed its position sensibly since Wednesday night. (Tebbutt 1859–63).

He also treated himself by observing it twice on 25 May, his birthday:

> I saw the comet & 1316 of Lacaille's Catalogue very distinct this morning in the day or night glass, and got … precise observations … I could get only one observation with the sextant, the comet being faint …
>
> The moon not having long risen and the atmosphere being very clear I could see the comet indistinctly with the naked eye to-night … [It] appears to be increasing in brilliancy. (ibid.).

Two days after Tebbutt reported his discovery in the *Sydney Morning Herald*, a letter by Scott appeared in the same newspaper. He mentioned that when he received Tebbutt's letter on 22 May he was busy trying to bring a new 7.25-in. (18.4-cm) Merz refractor into operation (Fig. 6.5) and had to use the old 3.25-in. (8.3-cm ex-Parramatta Observatory refractor to obtain a rough positional measurement. On 24 May, he adapted a micrometer that came with the Merz to the Parramatta telescope and obtained a pretty accurate determination of the comet's position (Scott 1861f).

While Scott's letter was important in that it gave official confirmation of the existence of the new comet and credited Tebbutt with the discovery, it served to open the 'flood gates' to those with a fascination for comets or a political 'axe to grind'. The first such letter came from a self-proclaimed cometary authority named Gledson, who not only contradicted Tebbutt's account but also threatened his burgeoning reputation:

> As my observations have, however, led me to different conclusions from that gentleman, they may possibly be of sufficient interest to be worthy of a place in your columns.
>
> Instead of "receding from the sun and earth" it is evident to me that this comet is rapidly approaching them …
>
> The data I have collected leads me to infer that this object will be most distinctly seen in the S.E. heavens, about from the 2nd to the 5th of June, from about 10.45 p.m. on the 2nd, allowing an hour later for each succeeding evening …
>
> My own impression is that the present is the comet known to astronomers as that of Amalfi, and to more recent enquirers as number three of La Bass, and if so, what a theme for reflection in the changes which "have shadowed o'er earth's surface" since its pale light served the brave but superstitious burghers of Ghent to resist the Moslem power, and brave its vengeance. (Gledson 1861).

One can imagine that Tebbutt and Scott were far from overjoyed by this letter, and the following evening the latter was moved to respond:

**Fig. 6.5** The beautiful new 7.25-in. (18.4-cm) Merz refractor that Scott was busy installing at Sydney Observatory at the time Tebbutt discovered the comet (*Courtesy* Dr. Harley Wood)

> I am not sure whether your correspondent Mr. Gledson was in earnest in what he wrote about the comet, or merely intending to make fun of the astronomers. At all events his results, which he calls upon me to confirm, are altogether wrong. The comet sets a few minutes earlier every night, and in a few days will only be visible in the morning before sunrise. (Scott 1861g).

In this same letter, Scott goes on to state that the comet is moving slowing eastwards and northwards, and that on the 27th '… I detected the existence of a tail, very faint and diffused, pointing nearly to the south pole'. Scott also published positions obtained on three different nights, and in the following day's *Herald* provided corrections to these and supplied a new position obtained on the evening of 30 May (Scott 1861h).

For his part, Tebbutt ignored Gledson's letter as he channelled his energies into further observations. On 27 May, he obtained good positional measurements and noted that the comet was '… very distinct in the night telescope: the tail was beautifully distinct for a short distance from the head; it appeared to be very broad' (Tebbutt 1859–63). On the morning of the 29th and the evening of 30 May, he secured further positions, and on both occasions he noted that the tail was distinct (ibid.).

After the initial flurry of press publicity, there was a hiatus in the correspondence until June 15, and during this interval both Tebbutt and Scott were busy obtaining positional measurements and computing the orbital elements of the comet. For his part, Tebbutt secured measurements on 3, 8, 11, 13, 14 and 15 June. On the 8th, the comet was described as '… a beautiful object, its tail having assumed considerable dimensions, and being plainly visible to the naked eye' (ibid.). When he observed it on the morning of the 11th,

> The head was very nearly, if not quite, as bright as a star of the second magnitude; its tail as near I could judge was about 9 to 10 degrees in length … The comet is evidently increasing in brilliancy, and is probably approaching the earth. (ibid.).

On 14 June, the seeing was not very good, but Tebbutt was still able to trace the tail for a distance of 11° (ibid.). The following morning was beautifully clear and in his diary Tebbutt (ibid.) described the comet as '… a magnificent object. I could trace the tail very readily as far as 20 degrees from the head, and with difficulty somewhat further'.

Following his initial confirmation of the comet on 27 May, Scott then observed it on the 30th. His next observations were conducted on the morning of 8 June, when it was very conspicuous to the naked eye and equal to a star of the fourth magnitude (Scott 1861i). These were his first observations with the newly-acquired Merz refractor, and the comet certainly proved to be an ideal test case, for as he was later to remind Astronomer Royal, George Airy:

> You are aware that I never made an Astronomical Observation before I left England [for the Sydney position] so will not be surprised at my losing a considerable time in training myself to the use of so formidable an Instrument. (Scott 1861l).

Meanwhile, Tebbutt was completing his initial calculation of the comet's orbital elements based on his own observations, and these were published in the 15 June issue of the *Sydney Morning Herald*. These are listed in Table 6.1, and Tebbutt (1861c) emphasised that they should be seen as preliminary, and that they may be subject to substantial correction when a longer data set of observations comes to hand. Tebbutt concluded his letter with a remarkable prediction: 'My calculations show that the comet will soon move rapidly towards the north, and that on the 29th instant the earth will be at no great distance from the extremity of its tail. There is some probability of the comet's becoming visible in full daylight about that date'. (ibid.). This was the first public intimation that the Earth would in fact pass through the tail of the comet although missing the main concentration, but its significance apparently was lost on most readers.

Two days later, on 17 June, further news of the comet appeared in the Sydney press, but this time in a rival newspaper, *The Empire*. And it came in the form of an article by Scott (1861a), which reviewed the discovery circumstances and subsequent observations made at Sydney Observatory. This article was written on 15 June in response to inquiries from the newspaper, and Scott noted that on that same morning the tail was 18° long. In describing Tebbutt's discovery of the comet, Scott erroneously characterised this as 'accidental', and Tebbutt was quick to chide him over this *faux pas*. In a long letter dated 20 June, but published in *The Empire* 2 days later, he explains that:

> I have been in the habit during the last three years of examining the heavens frequently in the evening, and occasionally in the morning, with a view to detect the existence of comets, and particularly the return of the great comet of 1556, which is still expected by astronomers. I was engaged in my customary examination on the night of the 13th of May … (Tebbutt 1861a).

**Table 6.1** Orbital elements of Comet C/1861 J1 (Tebbutt) according to different computers

| Computer | Tebbutt (1861c) | Scott (1861b) | Hawkins (1861b) | White (1861b) | Kreutz (1880) |
|---|---|---|---|---|---|
| Perihelion passage (June) | 13.7253 | 11.6892 | 11.75 | 11.2035 | 12.0068 |
| Perihelion distance (in AU) | 0.82033 | 0.82218 | 0.82215 | 0.82156 | 0.82238 |
| Longitude of ascending node | 280°00′44″ | 278°57′52″ | 279°01′41″ | 278°58′36″ | 280°12′39″ |
| Inclination of the orbit | 86°18′42″ | 85°38′11″ | 85°37′46″ | 85°25′42″ | 85°26′27″ |
| Distance of perihelion from node | | 29°42′58″ | 29°38′59″ | | 29°54′35″ |
| Heliocentric motion | Direct | Direct | Direct | Direct | Direct |

Scott obviously meant no offence and was a little surprised by Tebbutt's action in immediately running to the newspapers. He responded by writing Tebbutt a personal letter in which he emphasised his regard for the Windsor astronomer:

> I am quite grieved that you should have suspected me of a desire to detract from your merits as the discoverer of the Comet.
>
> I really thought it an accident, not being aware that you were in search of Comets at the time. I have mentioned your name as the discoverer, in my report to the Astronomical publications of Europe and have not used the objectionable word.
>
> I shall certainly write to the Empire expressing my regret for having unintentionally disparaged your discovery …
>
> I trust you will believe me when I assure you that not a particle of jealousy exists in my mind towards you and that in my almost isolated position the assistance of so valuable a computer as yourself is of too great importance to be lightly forfeited. (Scott 1861j).

This letter, and Tebbutt's letter in *The Empire* in which he 'set the record straight', both show for the first time what a touchy individual Tebbutt could be if he felt in any way belittled. A cometary discovery resulting from a systematic search programme showed a level of dedication, organisation and professionalism, whereas a fortuitous discovery did not. And it is clear that Tebbutt was a totally-committed and highly-organised astronomer who took a thoroughly-professional approach in his observational work. He had all the qualities of a professional astronomer, but he was only an independent astronomer! Later in life, this dilemma was to return time and again to haunt him.

After describing his post-discovery observations of the new comet in his 20 June letter to *The Empire*, Tebbutt (1861a) provided a glowing account of the tail and then concluded by expanding on his earlier pronouncement about its close proximity of the Earth on 29 June:

> On the last two or three mornings I have observed the tail to be divided into two branches, which emanate from the main part of the tail, at a distance of about six degrees from the head. The upper or western branch is the most distinct, and I could trace it to a distance of 42 degrees from the head. The tail, supposing it to point directly from the sun, will cross the earth's path about the 29th instant, at a point which will be occupied by the earth on the 2nd July; so that it appears that the earth will have a narrow escape … (Tebbutt 1861a).

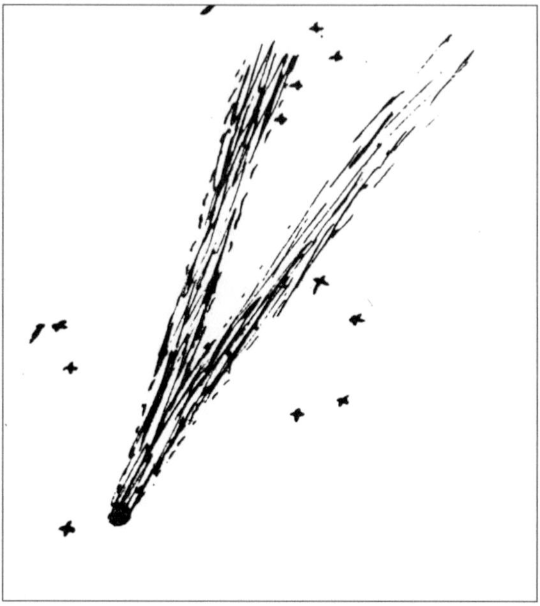

**Fig. 6.6** Tebbutt's 19 June field sketch in his diary showing the dust and ion tails of the comet, with the former extending about 35° across the sky (after Tebbutt 1859–63; *Courtesy* Mitchell Library)

This time his point was not lost on the local populace and as Macdonnell (1904) was later to reminisce, 'This prediction created considerable excitement at the time'.

While this 'media sparring' was underway, Tebbutt continued his observations from the geographical isolation of Windsor. By 18 June, the comet really was a 'magnificent object', and its tail could be traced for about 35° (Tebbutt 1859–63). The following night he described the tail in more detail in his diary:

> The comet was beautifully distinct; its tail was very bright to a distance of several degrees from the head, where it divided into two branches, the western being the more distinct of the two. They extended to a distance of 30 or 35 degrees from the head. (ibid.).

Accompanying the diary description is a crude field sketch (Fig. 6.6), which shows a long slightly curved dust tail and a shorter ion tail. Rahe et al. (n.d.) note that only ~30% of comets exhibit ion tails.

The morning of 20 June also was clear, and on this occasion Tebbutt was able to trace the western tail (as he called it) for 42° (ibid.), almost one-quarter of the way across the sky. The magnificent weather continued, and on the 21st the comet was very distinct and Tebbutt felt that the nucleus had increased in brilliancy (ibid.). The following morning, Scott described the comet as

> … brighter this morning than on any previous occasion being visible without a Telescope until 40 minutes before sunrise.
>
> The nucleus is bright round and well defined but I cannot detect any peculiarities or changes in appearance; the tail subtended an angle of about 18°… [and] its appearance was as represented in the rough sketch below, one narrow stream of light extending more than twice as far as the rest. (Scott 1861i).

This interesting account was included in a letter Scott wrote the German-born US astronomer from Hamilton College, Professor Christian Heinrich Friedrich Peters (1813–1890; Sheehan 1999), and never reached the general public.

Meanwhile, back in Sydney, the readers of *The Empire* and the *Sydney Morning Herald* were not overjoyed by the highly technical information (such as orbital elements) provided by their astronomers, which did little to satisfy their basic curiosity about this particular comet, and comets in general. On 19 June, one reader of the *Sydney Morning Herald* (using the pseudonym 'Orion') wrote to the Editor laying the blame for this deficiency squarely (but quite unfairly) at Scott's feet:

> Will you be kind enough to prevail upon the Astronomer of New South Wales to spare a little of his valuable time in giving the public some *popular* information respecting the new comet, whose appearance Mr. Tebbutt informed us of several weeks ago. If the gentleman cannot amuse us with any ingenious speculations as to when the luminous visitor last appeared, when he may appear again, the nature of the path he is travelling, and so forth, he can, at least, tell us in *plain terms* when and where the comet is to be seen by the naked eye above our horizon, the direction it is taking, and how long it may be visible to us. Some good might, perhaps, be effected if, in addition to the above, our Astronomer would give us a brief outline of the theories of Lardner, Arago, and others on "Cometary Influences", showing that no possible danger can arise to our earth from their presence, and thus allay or prevent the terrors of the alarmist. (Orion 1861; his italics).

Not content with this, 'Orion' continues his attack, focussing this time on the policy of the Observatory, something that had been the subject of earlier public criticism (Scott 1861k). Then, fearing that Scott would not find the time to furnish the information requested on 'the comet' and comets in general, 'Orion' proceeded to supply the details himself! He concluded with a description of the comet:

> It is a beautiful object, well worth turning out of bed a few hours earlier to have a look at. The "nucleus" is bright, and the "tail", which stretches nearly S.S.W., is said to be larger than that of any comet seen in this hemisphere since the memorable one of 1843. (ibid.)

To appreciate the majesty of Tebbutt's comet, we need only compare it to the drawing of the Great Comet of 1843 (C/1843 D1) shown here in Fig. 6.7.

Meanwhile, another anonymous correspondent, simply using the initials 'W.M.', penned a similar appeal for information to *The Empire* on 21 June, but on this occasion—no doubt, much to Scott's relief—without a single mention of Sydney Observatory. Instead, W.M. of Newtown lauds Tebbutt:

> … our neighbour, Mr Tebbutt, of Windsor, will very likely be acknowledged the discoverer; at all events, it was reported by him first, and long before it was visible to the naked eye. My notion is, that when a man who has his business to attend to during the day can, in those cold nights, put off his night-cap, and sit for hours carefully examining the "vasty depths" of the heavens, it is something more than accidental to discover a comet, and more so if he can describe its orbit. To such a man let all honour be awarded, and let it be known at home that we have such a man in the colony. (W.M. 1861).

By 'home', W.M. was of course referring to mother England.

On 22 June, Scott in fact gave up enough of his 'valuable time' to respond to Orion's misgivings in the *Herald*, and supplied the necessary information about Tebbutt's comet. He also described the remarkable tail: '… the usual diverging

**Fig 6.7** A drawing of Comet C/1843 D1 made by Mary Morton Allport (1806–1895) in March 1843 from Hobart, Tasmania (https://en.wikipedia.org)

brush-like appearance extends for about eight degrees … and beyond it extends a narrow stream of light in the same direction, and reaching about ten degrees further …' (Scott 1861b). It is interesting to note that on 20 June, under the darker skies of Windsor, Tebbutt was able to trace the tail for 42°, more than twice the distance recorded by Scott 2 days later.

In order to allay any public fears, Scott concluded his 22 June letter by discussing what in all likelihood will happen when the Earth passed through the tail, as forecast earlier by Tebbutt. He reassuringly concluded that this '… would probably produce no perceptible effect; or, if any, a very slight diminution in the light and heat which we receive from the sun' (ibid.). Following Scott's letter was another anonymous one, this time by 'F.C.B.' (1861a) defending Sydney Observatory and its Director.

Now, it was Tebbutt's turn for some literary vitriol! In the 25 June issue of *The Empire*, 'Diogenes' proceeded to defend Scott and query Tebbutt's competence:

> Now, whether Mr. Tebbutt has taken his observations correctly—whether the course and the position he assigns to the comet, be in reality, its true course, and real position, it must be left to those who have in their possession certain astronomical instruments, to determine.
>
> But, Sir, it does seem evident to me, that Mr. Tebbutt has made a series of egregious blunders ... (Diogenes 1861a).

Diogenes then proceeded to discuss these by taking, dissecting and contradicting Tebbutt's published statements about the length of the comet's tail, the distance of the comet from the Earth, when the comet would cross the ascending node and its distance at that time, the date when the Earth would arrive in this position, and the comet's perihelion date and distance. He concluded: 'The above are but a few of the inconsistencies which are deducible from the dictum of Mr. Tebbutt. I trust they are sufficiently illustrative of that gentleman's computative powers ...' (ibid.).

If Scott's unintended slight about his 'accidental' discovery of the comet riled Tebbutt, then this letter must have totally infuriated him (see Tebbutt 1908). But instead of responding immediately, he chose to concentrate on his observational work. He next saw the comet on 23 June, but because of the bright Moon it was washed out and he drew it with a single tail only in his diary (Tebbutt 1859–63). On the morning of the 24th, the nucleus was more brilliant, and he obtained a better view, but did not provide a description or sketch of the tail. On the 25th, he was only able to get a quick look before the sky clouded over, but he was far more successful on 26 June:

> The sky was very clear, and the nucleus much brighter than I had ever noticed it before; the tail was also very broad and well defined and extended several degrees from the head. The tail would have been magnificent, had it not been for the moon's presence. (ibid.).

With the Moon holding sway and the comet moving rapidly towards the Sun and into northern skies, the best time for observing the magnificent tail had all but passed for those in the southern hemisphere. As it turned out, Tebbutt was to get his last pre-perihelion view of the comet on 27 June, and as was his practice with almost every observation, he once more measured its position relative to selected bright stars. Because he lacked a telescope and micrometer and had to resort to the best instrument at his disposal, a simple marine sextant, Tebbutt realised that his positional data were crude (especially when compared with those obtained by Scott at Sydney Observatory), but they were the best that he could do under the circumstances.

Meanwhile, back in Sydney the locals were as enthralled by the media exchanges as they were by the comet itself, and it was a local citizen named Thomas Dawson who responded to Diogenes's attacks on Tebbutt. In a letter to *The Empire*, he pointed out that he only knew Tebbutt through his many interesting contributions to the local press over the years, but that when an attack is made on an individual who works in a voluntary capacity for the benefit of science a response is called for. In concluding, Dawson (1861) shows real literary flair in suggesting that

> The feeble light with which the modern "Diogenes" has sought to illuminate the world, would have been better employed by him, as by his ancient namesake, in seeking to find an honest man therein, or to assist him to find his own whereabouts, and not go longer wandering about in the dark.

Wishing not to be outdone, the *Sydney Morning Herald* joined the fray the same day (26 June), with another attack on Scott and Sydney Observatory, but written in a rather novel way. In a letter titled 'The Comet's Grievance', a correspondent using the name of 'Comet of 1861' writes:

> It is not often that I honour the people of New South Wales with a visit, but after the extremely shabby manner in which I have been treated by the local astronomers, I take good care that my future visitations shall be "far between". Through my influence, I have kept the elements clear, in order to allow your "scientific few" to have their eyes on me to make their observations of my onward progress, yet nothing has officially appeared in print about me …
>
> My object, in appearing in this hemisphere, was to put the ability of your local (paid and amateur) astronomers to the test, in order to give them an opportunity of furnishing an interesting dissertation to the common people, but my object and wishes were frustrated. I must confess that had it not been for the "amateur" at Windsor, in all probability my visit would have passed off unreported. Mr. Tebbutt has my thanks for informing Mr. Scott of my visit and whereabouts. At the same time it is very humiliating for a great man so often to play "second fiddle".

Given the contents, it would appear that this imaginative contribution was written by 'Orion', using a new *nom de plume*, or by one of his sympathisers.

As might be expected, the 'Comet's' complaints elicited a stern reply from Scott. His letter dated 26 June was published 1 day later and referred to

> The total ignorance respecting the work of an Observatory, and the duties of an Astronomer manifested by those of your correspondents who undertake to criticise my conduct … must not be allowed to pass unnoticed. (Scott 1861c).

Scott then proceeded to point out that he had prepared scientific accounts of the comet for both *Astronomische Nachrichten* (Scott 1861n) and *Monthly Notices of the Royal Astronomical Society*, (Scott 1861m) but that these were unsuitable for public consumption, there being only two or three people in New South Wales who would be interested (Scott 1861c). Scott concluded by commenting on his role and status vis-a-vis Tebbutt in the field of cometary astronomy:

> I do not feel in the least degree "humiliated" by being indebted to an amateur for the first discovery of the comet. Such will probably be the general result of my arrangements, as the knowledge that so able an ally as Mr. Tebbutt is engaged in comet hunting (so peculiarly the province of the amateurs), will enable me to give the time which might be so occupied to other equally important pursuits …
>
> In conclusion, allow me to assure the numerous critics who favour me with their remarks in the daily papers, that no amount of letter writing will induce me to deviate from the rule which I have laid down for myself, which is to devote myself to the promotion of astronomical science rather than the gratification of public curiosity. (ibid.)

As a postscript to his letter, Scott provided the (two or three) members of the interested public with the comet's orbital elements, which he had calculated. Despite Diogenes's concerns, they proved to be very similar to those published earlier by Tebbutt (see Table 6.1 above), which is a great credit to the Windsor astronomer given his limited instrumental means. In this same letter, Scott also reaffirmed Tebbutt's earlier claim that on 29 June, in just 2 days time, the Earth would pass near the tail of the comet.

The same issue of the *Sydney Morning Herald* also contained a letter by James H. Wilkinson of Sydney confirming Tebbutt's orbital elements and the 29 June event. With regard to the latter, Wilkinson (1861) forecast that '… its advent will be atmospherically perceptible, though not sufficiently so to warrant apprehension. The probability is, that there will be an unusual haziness in the firmament, and a sensible increase of temperature'.

The third comet-related contribution in this issue of the *Herald* was a second letter from 'F.C.B.' (1861b) supporting Scott against the attack by 'Comet of 1861'. Less than a week later, 'X.X.' (1861) was to add his support.

Meanwhile, Tebbutt was fuming over the published attack on his reputation and mathematical competence, and on 26 June he wrote to *The Empire* and made an identical claim for Diogenes. He also emphasised that the charges against Scott were unjust and that it was his desire to work with Scott not against him (Tebbutt 1861d). His final comments are pertinent as they reveal his astronomical priorities when faced with the very same options that Scott alluded to, namely the promotion of astronomical science versus the gratification of public curiosity:

> I am now quite exhausted by the laborious calculations which I have made respecting the comet … I have made observations on nearly every morning since the comet's first appearance, and have reduced a considerable proportion of them. Not having the advantage of an assistant, I have been obliged to check my own calculations. I have now everything in readiness for the computation of a more correct orbit, which, when obtained, will, I have no doubt, bear favourable comparison with Mr. Scott's accurate results. (ibid.)

But Tebbutt's letter did not end *The Empire's* cometary coverage on 28 June, for it was followed by two others, both of which also were written on the 26th. In the first, penned by one with a flair for original *nom de plumes*, 'G' (1861) lamented the fact that Tebbutt '… should be made the mark for the mud of every unwashed individual who chooses to live in a tub …'. In the next letter, Diogenes (1861b) emerges from his 'tub' once more to deny that he made any sort of personal attack on Tebbutt! After further protestations of innocence, he concluded by returning Dawson's earlier compliments: 'If Mr. Dawson will provide himself with a microscope of a high magnifying power, I have little doubt that I shall be enabled to afford him sufficient light to view his own magnitude on a scale of intellect …'.

The next round in this entertaining newspaper saga of science and stupidity commenced on 1 July when a letter written back on 26 June (a popular day for writing about comets!) appeared in the *Sydney Morning Herald*. In it, H.S. Hawkins of Goulburn (Bio Box 6.2) published his calculated orbital elements of the comet, based upon Scott's reported observations, and noted that they mirrored those of Tebbutt and Scott. 'This agreement,' Hawkins (1861a) noted, 'furnishes a sufficient answer to those who have published their doubts as to the correctness of Mr. Tebbutt's observations'. Apart from Tebbutt, Hawkins was one of the few non-professional scientists in the colony of New South Wales at that time with the mathematical ability to compute cometary orbital elements.

**Bio Box 6.2: Henry Spendlove Hawkins**

Henry Spendlove Hawkins was born in Kingston-upon-Hull in England on 22 April 1824. In 1847, he graduated B.A. (seventh Wrangler) from St. John's College Cambridge, following with an M.A. 3 years later. In 1850, he moved to Australia and the position of Mathematical Master at Sydney Grammar School and later moved to Kings School in nearby Paramatta. By 1858, he was the Headmaster of Goulburn Grammar School (Venn and Venn 2011, p. 294). In 1874, while still teaching, he was contracted by the New South Wales Government to check on calculations made to establish the trigonometrical survey base line at Lake George (Trigonometrical Survey … 1876), and in 1877 was appointed a 'Computer' in the Trigonometrical Survey Branch of the New South Wales Lands Department in Sydney; in 1882, he was promoted to 'Chief Computer'.

Hawkins had a long-standing passion for astronomy, which was first publicised in New South Wales when he published an account of Comet C/1858 L1 (Donati) in the *Sydney Morning Herald* (Hawkins 1858). He also calculated the orbital elements of C/1861 J1 (Tebbutt), and these were published in the same Sydney newspaper (see Hawkins 1861a, b). This was all part of Hawkins' commitment to popularising astronomy (see Orchiston 1997a).

One of Hawkins' official Lands Department tasks was to supply field surveyors with lists of stars suitable for latitude and longitude observations at first-order trig stations (First Annual Report … 1882, p. 50). In order to assist these colleagues, and others, with their observations, Hawkins published *Practical Astronomy as Applied to Land Surveying* in 1876. This was reprinted in 1878, and a revised, extended, edition appeared posthumously, in 1890.

During its short-lived existence, Hawkins was active in Section A of the Royal Society of New South, and at the November 1879 meeting he presented a paper on 'Compiled catalogue of latitude stars, epoch 1880' (Orchiston and Bhathal 1991). This catalogue arose because of the deteriorating relations between the Department and Sydney Observatory during the late 1870s and throughout the 1880s, partly because of Henry Russell's reluctance to supply precise positions for reference stars observed at first order trig stations. In frustration, Hawkins eventually used published lists from Melbourne Observatory and the Royal Observatory at the Cape of Good Hope to produce his own catalogue containing positions for 920 stars. This subsequently was published by the Government (Hawkins 1881).

In 1883, Hawkins was awarded an M.A. degree in Mathematics by the University of Sydney, and he died less than 5 years later, on 31 March 1888, while still in the employ of the Lands Department (Venn and Venn 2011, p. 294). Some of the foregoing biographical information is contained in successive New South Wales *Blue Books*, and his avocational astronomical activities while with the Lands Department are discussed in Orchiston (1987).

Hawkins concluded his newspaper article by suggesting that the Earth *may* pass through the tail on 31 June, but by the time this news reached his readers it was all too late. So what of the grand event? In his diary, Tebbutt noted that on the evening of 30 June

> … I observed a peculiar whitish light throughout the sky, but more particularly along the eastern horizon; this could not have proceeded from the moon, but was probably caused by the diffused light of the comet's tail, which we are very near to just now. (Tebbutt 1859–63).

Tebbutt (1910) was later to describe this glow as aurora-like, and Hind, observing in England, noted a similar phenomenon:

> … on the evening of 30 June, while the comet was so conspicuous in the northern heavens, there was a peculiar phosphorescence or illumination of the sky, which he attributed at the time to an auroral glare; it was remarked by other persons as something unusual …
>
> Mr. Lowe, of Highfield House, confirmed Mr. Hind's statement of the peculiar appearance of the heavens on June 30. The sky, he says, had a yellow auroral glare-like look, and the Sun, though shining, gave but feeble light. (Chambers 1877, p. 321).

Clerke (1893) described the peculiar glare as inconspicuous and noted that although no 'unusual electrical excitement' was noticed by the instruments at Greenwich they '… were, indeed, disturbed on the following night; but it would be rash to infer that the comet had art or part in their agitation'.

Even when the comet was lost to Australian observers and had moved into northern skies, this did not stop the animated interchanges in both *The Empire* and the *Sydney Morning Herald*—even though the former newspaper published a letter by 'The Shade of Pollak' (1861) on 1 July expressing the hope that the protagonists '… will now close their controversies—which must by this time have ceased to interest the general public …'.

Nonetheless, given the level of public debate, Scott decided to offer his positional observations of the comet to anyone wishing to calculate the orbital elements for themselves, and on 2 July he published data for 27 and 30 May, and 10, 14, 16, 17, 22, 25 and 26 June in *The Empire* (Scott 1861d). Hawkins did not take up this option although he did submit a revised set of orbital elements, corrected for aberration and parallax, to the *Sydney Morning Herald* (see Table 6.1), and these were published on 5 July (Hawklins 1861b).

Meanwhile, a letter by Diogenes (1861c) appeared in the 3 July issue of *The Empire*, responding to Tebbutt's letter published on 28 June where Diogenes was accused of arriving at 'absurd conclusions.' Tebbutt (1861e) felt impelled to reply to Diogenes yet again and re-expose his inaccuracies, and this letter appeared in *The Empire* on 6 July. By now, the public—and presumably both editors—had endured more than enough of this ongoing 'cometic comedy', and C/1861 J1 (Tebbutt) disappeared from the pages of the Sydney newspapers, except for one transient reappearance.

Scott finally decided to bring together all of his own observations plus details of Tebbutt's discovery, and he wrote all this all up in two research papers. One was presented at a meeting of the Philosophical Society of New South Wales in 1861 (Scott 1861o), and the other was published in *Monthly Notices of the Royal Astronomical Society* (Scott 1861m). Scott's Sydney paper was reproduced in its entirety in the *Sydney Morning Herald* (see Philosophical Society 1861) and in *The Empire* on 18 July, much to Tebbutt's later delight:

> The vindications thus referred to of the correctness of my predictions afforded considerable amends for the ridicule to which I have been subjected by anonymous incompetent critics, and here I may add that I am deeply indebted to the old *Empire* for many notes of encouragement which I received from it during the early years of my career. (Tebbutt 1908, p. 21).

Australian astronomy was still in its infancy in 1861 (see Haynes et al. 1996; Orchiston 1988a), and apart from Sydney Observatory the only other professional institution in the country was Williamstown Observatory, near Melbourne (Fig. 3.4). Staff there possibly first heard of the new comet via a letter Scott (1861e) sent Ellery (the Director) on 25 May 1861, the very day that the news broke in the Sydney press.

The Assistant Astronomer at Williamstown, E.J. White (Bio Box 6.3), carried out positional observations of the comet over 6 nights in June and used these to compute the orbital elements. It was only in August that he forwarded his suite of observations and the result of his orbital computations (Table 6.1) to Tebbutt (White 1861b), and this valuable little data set subsequently was published in *Astronomische Nachrichten* (Ellery 1861). Surprisingly, these observations were omitted from the Observatory's own monograph, *Astronomical Observations*, which spanned the critical interval, 1861–1863 (see Ellery 1869).

**Bio Box 6.3: Edward John White**

Edward John White (Fig. 6.8) was born in Bristol, England, on 8 December 1831, and after completing his schooling trained as a mechanical engineer (About people 1913).

In January 1853, White emigrated to Australia and for the next 7 years worked in the Victorian goldfields. During this period, '… he superintended the erection of some of the earliest mining machinery in Bendigo, Victoria' (Tebbutt 1913).

**Fig. 6.8** Edward John White (*Courtesy* State Library of Victoria, Picture Collection)

(continued)

White joined Ellery on the staff of Williamstown Observatory in 1860 as Assistant and transferred to Melbourne Observatory when this new facility was established. Thus, like Ellery, he was a nice Australian example of the 'Amateur-turned-professional Syndrome' as discussed in Orchiston (2015). When White '… commenced duty at the [Williamstown] Observatory … He was [found to be] … a born computer, and much of the mathematical work he did showed unusual gifts' (Death … 1913).

White remained at Melbourne Observatory for the remainder of his working life, retiring as Chief Assistant in 1892 (Tebbutt 1913). While Ellery was overseas in 1875–1876 White served as Acting Government Astronomer of Victoria (About people 1913). He also '… had charge of the party sent to Hobart in 1882 to observe the transit of Venus … compiled the General Star Catalogue of the Melbourne Observatory, and wrote [many] sceentific [*sic*] papers …' (ibid.). Tebbutt (1913) has even higher praise for White: 'It is well-known that this [i.e. Melbourne] observatory held for many years the premier place among Australian Observatories, and it was mainly to the devotion of our departed friend that such was the case'. I wonder if Ellery would have felt a little short-changed by this assessment had he still been alive in 1913!

Apart from astronomical research, White also was committed to popularising astronomy, and he mainly did this through reports in the local newspapers, starting from when he was in Bendigo. Then, soon after moving to Melbourne, he '… became a regular contributor of "Astronomical Notes" to "The Australasian." His monthly contribution was eagerly looked for, and kindled interest in many of the outback homesteads and huts of Australia …' and it was 'Only advancing years [that] caused him to sever his connection with this journal a few years ago' (Death … 1913).

White was a Fellow of the Royal Astronomical Society from 1869 and was active in the Royal Society of Victoria, serving as President in 1902–1903 (About people 1913).

Edward John White died on 2 August 1913 and was described in one obituary notice as '… an old colonist and well known citizen …' (ibid.).

Apart from Tebbutt, the only other internationally-visible non-professional astronomer in Australia during the 1860s was Francis Abbott of Hobart (see Bio Box 6.4). Media coverage first directed Abbott's attention to Tebbutt's comet in early June, and he carried out observations on eight different mornings between 6 and 20 June. On the 20th, he forwarded a note on this work, including crude positions, to *Monthly Notices of the Royal Astronomical Society* and this included a description of the comet:

> The tail, which is quite straight and about 10 degrees in length, points in the direction of star Achernar, the nucleus forming nearly a right angle with that star and Canopus. The nucleus of the comet, measured with a Cavallo's micrometer in an eyepiece of 50, was 33″ in diameter;

> the coma or nebulosity surrounding the nucleus, as seen with a comet eye-piece of 27 in a 5-feet telescope, was 30 minutes in diameter. The breadth of nebulosity at the apex of the tail 50 minutes, where it becomes very diffuse. (Abbott 1861).

Abbott's right ascension and declination positions were too imprecise to be used for orbit computations.

**Bio Box 6.4: Francis Abbott**

Francis Abbott (Fig. 6.9) was born in Derby, England, in August 1799, and in 1825 he married Mary Woolley, with whom he had four children. He ran a successful business manufacturing clocks, watches and astronomical equipment until 1844 when he was found guilty of obtaining two watches under false pretences and sentenced to penal servitude in Hobart, Tasmania. After 4 years, he obtained his ticket-of-leave, and set up as a watch- and clock-maker in Hobart. Despite this less than auspicious start, by the time he died, in early 1883, Abbott was regarded as a respected member of Tasmanian society. One son, Francis Jnr., became Director of the Hobart Botanical Gardens, and his other two sons were accomplished early Tasmanian photographers.

While he was still a convict Abbott was given access to Rossbank (Geomagnetic) Observatory, and took part in its meteorological programme. When the Observatory closed at the end of 1854 he immediately established a private observatory at his home and continued the meteorological observations, over the next 25 years publishing monthly reports and 6 different monographs. His observatory also included 3 small refracting telescopes, a transit telescope and an astronomical clock, and for nearly 30 years he provided Hobart with a local time-service.

Although Abbott observed aurorae, meteors, sunspots, two transits of Mercury, a lunar occultation of Jupiter, meteors and the open cluster k Crucis, his main focus was on comets and the variable star η Carinae (or η Argus, as it was then known).

Abbott independently discovered the Great Southern Comet of 1865 although he generally is not credited with its discovery (see Orchiston 1997b). However, it was his observations of η Carinae that brought him greatest international exposure—and subsequent notoriety (see Frew and Orchiston 2003). His 1860s claim that the nebulosity surrounding this star had changed in shape and size since Sir John Herschel's observations of the 1830s ran counter to the prevailing wisdom and brought forth howls of protest from Herschel and other distinguished European astronomers, including the Astronomer Royal. Eventually Richard Proctor (1871) was called in to adjudicate, and from Abbott's point of view his report was damning. Over the years, Abbott published 37 different research and review papers on his astronomical work in the *Astronomical Register*, *Monthly Notices of the Royal Astronomical Society* and *Papers and Proceedings of the Royal Society of Tasmania*, nearly half of them about η Carinae.

(continued)

**Fig. 6.9** Francis Abbott seated in front of a doorway ca. 1860 (*Courtesy* Allport Library and Museum and Fine Arts, Tasmania Archives and Heritage Office, Hobart, 607375)

In addition to observational astronomy, Abbott was committed to astronomical education, and between 1878 and 1880 he published three booklets (Abbott 1878, 1879, 1880) that summarised recent astronomical developments (including the 'new astronomy', astrophysics). He also was a long-time member and Councillor of the Royal Society of Tasmania, a frequent presenter at its meetings and a regular contributor to its journal.

From 1855 to 1880 Francis Abbott served as Tasmania's *de facto* Government Astronomer and Meteorologist, and it is no coincidence that it was only after advanced years forced him to abandon his scientific work that the Tasmanian Government moved to establish a colonial observatory. Francis Abbott died in Hobart on 18 February 1883, and after Tebbutt he would rank as Australia's foremost nineteenth century independent astronomer. A detailed account of his work is presented in Orchiston (1992).

## 6.4 A Celestial Spectacle: The Comet in Northern Skies

In the early 1860s, before the advent of the telegraph, there was no way of rapidly communicating important news from Australia to England, the Continent or the United States, so no advance warning could be given of the new comet which had become a spectacular object in southern skies. Thus, the following formal announcement of Tebbutt's discovery, written by William Scott (1861n) on 21 June and submitted to *Astronomische Nachrichten*, only was published *after* the comet first became visible to northern hemisphere astronomers on 30 June (the very night that the Earth passed through the outer reaches of the tail):

> On 13th May a Comet was discovered by Mr. Tebbutt, an amateur Astronomer of this Colony, a few minutes of arc West of 1250 B.A. Catalogue. I was not aware of the discovery until May 21st. At that time my small Equatoreal [*sic*] was dismounted and preparations were being made for mounting the 7inch refractor by Merz which had lately arrived; this circumstance and some cloudy weather prevented me from obtaining any good observations until the 27th. The Comet was at that time but just visible to the naked eye …
>
> [By] June 8th before sunrise … [the comet] was very conspicuous to the naked eye fully equal to a Star of the fourth magnitude. It has continued to increase in brilliancy …
>
> The nucleus is bright, round well defined but I cannot detect any peculiarities or changes in appearance; the tail subtended an angle of about 18° pointing about 15° West of South; one narrow stream of light extending more than twice as far as the rest.

If Comet C/1861 J1 (Tebbutt) (as it is now known) had appeared impressive from Australia, it put on even more of a spectacle for northern hemisphere viewers. In his synthesis of the various published observations, Vsekhsvyatskii (1964) gives a magnitude equal to Saturn (i.e. 0.0) for the nucleus at the end of June. On 2 July, its magnitude was estimated as 1, but by 13 July it had faded to 3.

Perhaps the most notable feature of the nucleus was the splendid assemblage of envelopes (Rahe et al. n.d.), first seen surrounding it on 30 June and described by the British Oxford-trained clergyman-astronomer Reverend Thomas William Webb (1807–1885) '… as though a number of light, hazy clouds were floating around a miniature full moon. Portions of six could be more or less distinctly traced …' (Webb 1862). In the course of the next week, the configuration of the envelopes changed continuously (see Fig. 6.10), and by the eighth they had disappeared.

Between 2 and 8 July other observers described a fan-shaped 'luminous sector' apparently radiating from the nucleus (ibid.), its angular extent varying with time. This luminous sector is shown in Fig. 6.11.

But more than any other feature, it was the tail which set this comet apart. At the end of June, Keil Observatory's Theodor Brorsen (1819–1895) and Director of the Vatican Observatory Father Angelo Secchi (1818–1878) recorded a 90–120° long tail, with a maximum width of 6° (Vsekhsvyatskii 1964). At about the same date, the British astronomer George Williams drew a number of additional jets up to 40° or 50° in length (Fig. 6.12), making this one of the most interesting cometary tails seen in many years, and on 2 July the British researcher and soon-to-be highly successful populariser of astronomy Richard Proctor (1837–1888) also noted a fan-like tail (see Fig. 1.9; Weiss 1888), very reminiscent of Williams' drawing (see Proctor 1926).

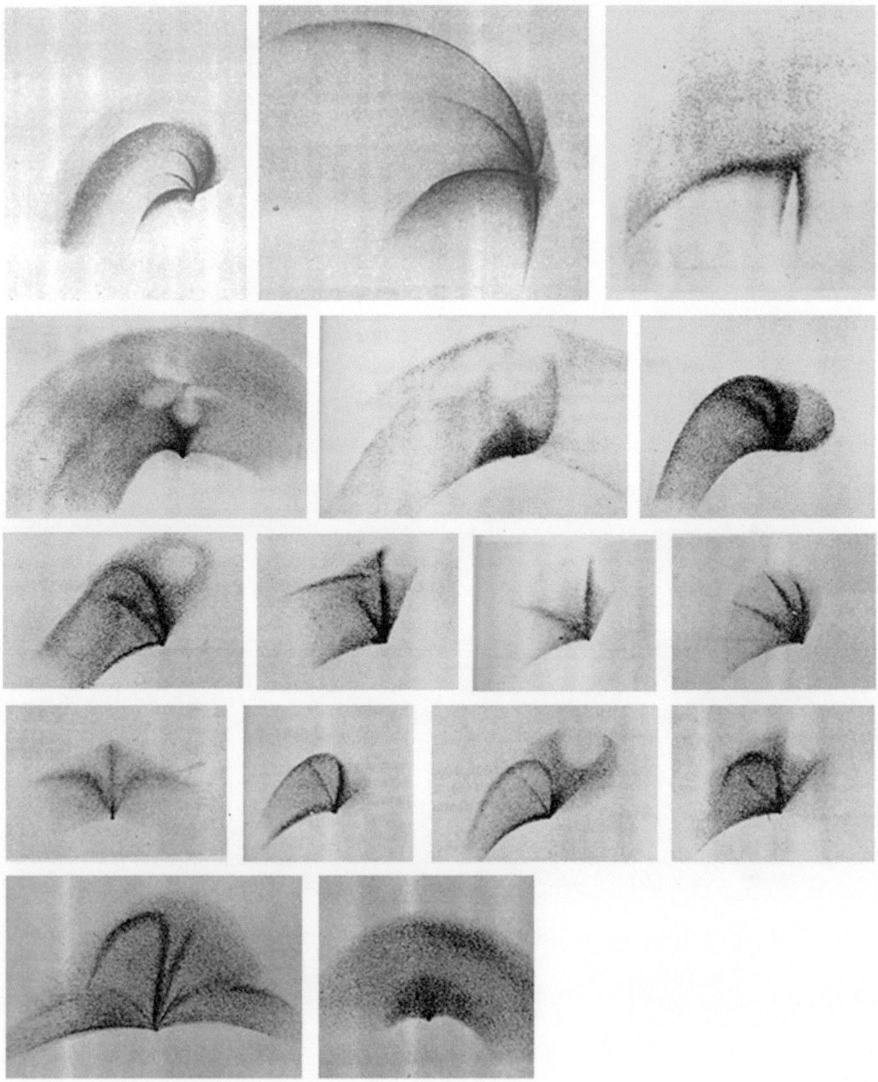

**Fig 6.10** Envelopes drawn by Julius Schmidt (1863) of Athens Observatory (from *top left*) on July 01.884, 02.119, 02.922 July 03.822, 03.893, 04.068 July 04.806, 04.822, 04.906, 04.920 July 05.092, 06.818, 06.840, 06.888 July 07.021, 07.795 (no visible envelopes) (after Rahe et al. n.d.)

Much later, in February 1910—and as a prelude to the upcoming appearance of Comet 1P/Halley—some astronomers questioned whether a number of rays really were visible in the tail of C/1861 J1 between 30 June and 2 July, as all they recalled was a long single tail, but Irish-born British cometary authority, Andrew Claude de la Cherois Crommelin (1865–1939) pointed out this probably was a short-term perspective effect:

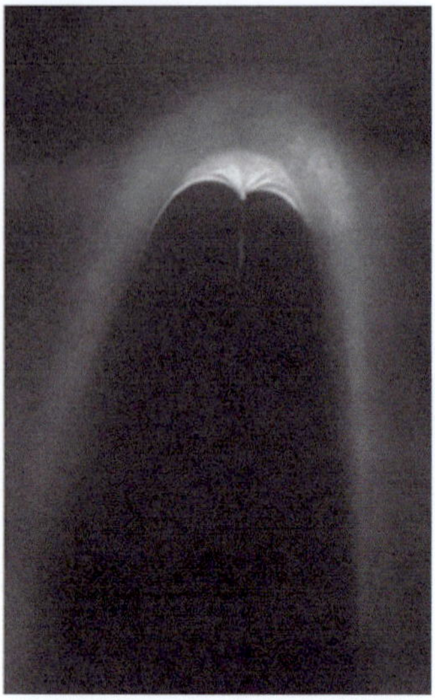

**Fig. 6.11** A drawing of the head of the comet showing the 'luminous sector' (after Guillemin 1866, Plate XXIII)

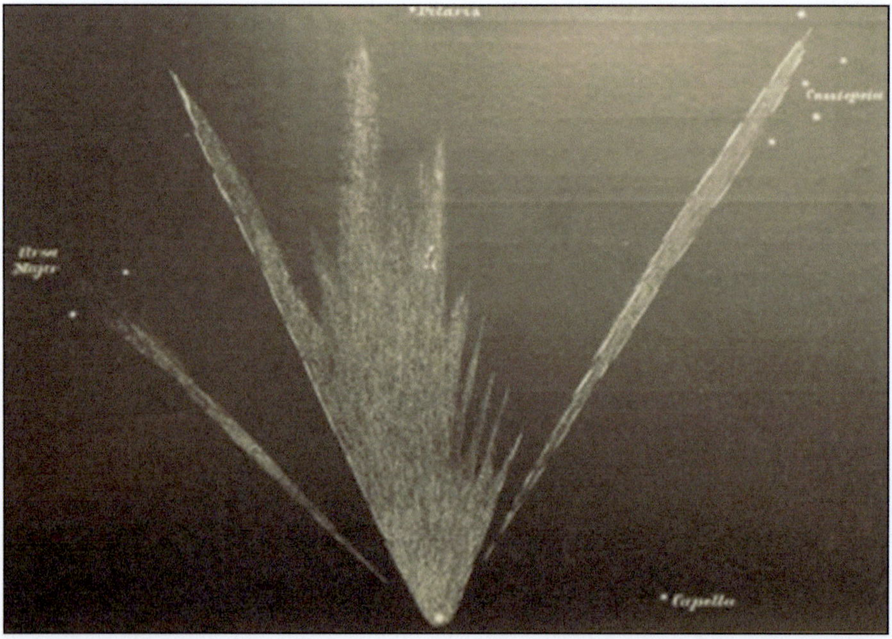


**Fig. 6.12** A drawing of the Great Comet made by G. Williams on 30 June, about the time when the Earth passed through the tail (after Chambers 1877, Plate XIX)

> … it must be remembered that, at the time when the comet's tail was exceedingly near the Earth, a short interval of time might produce a very different perspective effect. He [Crommelin] remembered that Mr. Proctor … [also] gave a picture … with the rays, at about the time, he believed, when the Earth was about to pass through the tail. It might well be, then, that some such presentment might last for a few hours and then pass away. He knew that in 1861, at the time when the Earth passed through the comet's tail, some people talked of having seen actually a visible motion across the sky, and it was quite likely that something of the kind might be seen. (Report of the Meeting … 1910, p. 231).

However, this explanation would only apply to Proctor's drawing if we accept that it was made on 30 June, not on 2 July.

On 1–2 July, Secchi mentioned only a single tail, and gave its length as 118° (Vsekhsvyatskii 1964). On 2 July, the Cambridge-educated British clergyman-astronomer Reverend Robert Main (1808–1878) recorded two tails, the larger more distinct one being at least 70° in length (Main 1861), while fellow-clergyman Webb (1862) confirmed the existence of two tails, noting the 'main tail' as about 80° long. This same day, the Director of Vienna Observatory Karl Ludwig Edler von Littrow (1811–1877) gave a value of 97° (Vsekhsvyatskii 1964). On the 3rd, Main (ibid.) mentioned only the main tail, about 60° in length, and on the 4th this could be traced for about 44°. By 8 July, according to Main (ibid.) the main tail had decreased to about 27°. Von Littrow provided a figure of 31° on this evening, as against 57° given by the German-born Athens Observatory Director Johann Friedrich Julius Schmidt (1825–1884; Sheehan and Dobbins 2014). Von Littrow also recorded the presence of a fainter wide fan-like tail on 8 July (Vsekhsvyatskii 1964). On 12 July, the main tail extended over 30°, and the wide (fan-tail) tail over 21° (ibid.). According to Main (1861), the main tail had decreased to 15° or 16° on 14 July, and on 16 July it was 12–13°. After this, both tails continued to rapidly decrease in size and to fade. At its best, Comet Tebbutt 1861 was one of the most spectacular-looking comets of the century, and Webb (1862) described its tail as a '… magnificent appendage'.

Apart from descriptions of the head and tail, many astronomers published positional observations of the nucleus. To indicate the geographical range of these, papers published in *Monthly Notices of the Royal Astronomical Society* were from Armagh Observatory, Cambridge University Observatory, Dudley Observatory, Harvard College Observatory, Liverpool Observatory, Pulkowa Observatory, Radcliffe Observatory (Oxford), the Royal Observatory (Greenwich) and the U.S. Naval Observatory, while most European observatories sent their papers to *Astronomische Nachrichten*.

In addition to astronomers, the general public also was enthralled by the comet. The following poem, 'The Way of the Comet' (1941), actually was written about a quite different object, but it also typifies the public reaction to the Great Comet of 1861:

> So very rarely do we see a Comet
> That People get a strange Excitement from it.
> Why, Some will scramble out of Bed at Night
> To see that far, mysterious, drifting Light.
> Astronomers debate
> About the Place and Date,

And measure on their own peculiar Scale
Diameter of Head and Length of Tail.

Brighter than Stars, the Sky's distinguished Guest,
The Comet sails serenely self-possess'd.
It needs no Propaganda Stunt to try
To bring itself before the Public Eye;
For we, who dumbly stare
Into the midnight Air,
Are only following the Way of Man
Since searching the heavenly Spheres began.

## 6.5 Recovery of the Comet: Its Reappearance in Australian Skies

After its brilliant northern showing, the comet returned to grace southern skies in early August and was again observed by Tebbutt (see Tebbutt 1908), but at first unknowingly. On the evening of 2 August, he wrote in his diary:

> I have this evening discovered what I strongly suspect is another comet: it first attracted my attention at about 7h 25m per clock. It was barely visible to the naked eye and was seen through a smoke proceeding from a fire under the horizon. I thought I could detect a slight appearance of a tail with the telescope. I with difficulty got the following [three] distances with the sextant … I have sent notice of the discovery to the astronomer [Mr. Scott] by post to-night. (Tebbutt 1859–63).

The following morning Tebbutt 'did his homework': using Scott's orbital elements published back on 27 June, he computed the position of 'his' comet on 2 August and discovered—to his utter surprise, and presumably delight—that rather than scoring a second comet discovery he had independently recovered his first one! Late that same morning he telegraphed Scott accordingly, and that evening obtained a further point for the comet (ibid.). He also advised Melbourne Observatory that his comet had returned, but after trying to observe it close to the horizon, White (1861a) responded to Tebbutt's telegram: 'I have become very indifferent however about the Comet, now that I know that my observations can have no real value. I scarcely think it worth my while to turn my Telescope to it'. This was a strange reaction given his access to the largest refracting telescope in Australia, furnished, moreover, with a micrometre, and after he had already computed realistic orbital elements (Scott 1861d) and submitted papers with his micrometric observations to *Astronomische Nachrichten* (Scott 1861o) and *Monthly Notices of the Royal Astronomical Society* (Scott 1861n). I suspect that by this time Scott simply may have had enough of comet C/1861 J1 (Tebbutt).

However, White's negative attitude did not influence Tebbutt, but cloudy weather did and he was only able to resume observations on 9 August when a small portion of the tail was indistinctly visible in his telescope (Tebbutt 1859–63). Further observations followed on 11 and 12 August. By this latter date, the comet was no longer visible to the naked eye, and 'Even in the telescope it was faint, and there was not

**Fig. 6.13** Tebbutt's sketch in his diary showing the motion of the comet relative to reference stars between 23 August (the bottom of the 'Track') and 2 September (position 'c'). Positions 'a' and 'b' mark the comet's location on 30 and 31 August, respectively (after Tebbutt 1859–63; *Courtesy* Mitchell Library)

the slightest appearance of a tail; it was however brighter than when I discovered it on the 13th of May last'. (ibid.).

In contrast to his run of clear evenings in June, cloudy skies continued to plague Tebbutt in August and his next observation of the comet was on the 23rd when it was very faint in the telescope and comparable to a star of about magnitude 7. Tebbutt noted (ibid.) that since 2 August it had moved about 4.5°. Further observations followed on 24, 30 and 31 August and on 2 and 5 September (see Fig. 6.13). By 5 September it was very faint, and Tebbutt surmised, correctly as it turned out, that 'I think I shall not be able to find it any more (ibid.). Thus ended Tebbutt's observations of the Great Comet of 1861 although it was observed by others through to May 1862 (Vsekhsvyatskii 1964).

## 6.6 Rightful Recognition: Compensating for the Passage of Time and Tyranny of Distance

By the 1870s, knowledge of Tebbutt's initial discovery of the Great Comet of 1861 was widespread amongst northern hemisphere astronomers, thanks largely to Scott's actions (e.g. see Tebbutt 1875), but not so well known were the facts that he was the first to compute the comet's orbital elements and to forecast the Earth's passage

**Fig. 6.14** Heinrich Kreutz (https://en.wikipedia.org)

through the tail. Because documentation of these aspects was preserved in Sydney newspapers rather than in more conventional outlets (astronomical or scientific journals), Tebbutt thought it prudent to pen a note for *Monthly Notices of the Royal Astronomical Society*, and this was published in 1878. He was particularly forthright in his priority claims, as the following quote demonstrates:

> … I claim to be not only the first observer of the comet, but also the first calculator of the elements of its orbit … I announced the discovery to the Sydney Observatory on May 21, and in the *Sydney Morning Herald* of May 25. My approximate determination of the orbit … appeared … in the *Herald* of June 15 …
>
> In the same communication I pointed out the near approach of the Earth to the comet's tail for June 29, and also that the comet itself would probably become visible in full daylight. This announcement was followed by a more complete prediction by me in the *Sydney Empire* of June 22. (Tebbutt 1878b).

Another version of this note was published in *Astronomische Nachrichten* (Tebbutt 1878a).

A further reason for documenting his original discovery was to counter a claim which appeared in Rambosson's *Astronomy* that that author, and others at Réunion (in the Indian Ocean), had made naked eye observations of the comet at the very time when Tebbutt made his discovery and that it was sporting an 18° tail. Tebbutt (1878c) was quick to point out that the comet was a telescopic object at the time and that it lacked a tail!

Not long thereafter, the distinguished German astronomer, Professor Heinrich Kreutz (1854–1907; Fig. 6.14), carried out a definitive investigation of all known micrometric observations of Comet C/1861 J1 (Tebbutt) and published the resulting orbital elements (Kreutz 1880). As Table 6.1 indicates, the values derived by Tebbutt, Scott and Hawkins on the basis of much more limited data sets are in close agreement with Kreutz's results. Kreutz (ibid.) derived a period of 409.4 years for the comet.

In his *Astronomical Memoirs*, Tebbutt (1908, p. 25) was later to state: 'My sole regret in connection with this comet is the circumstance that my instrumental means did not admit of my accurately observing it [the comet] and thus contributing to the data employed in Dr. Kreutz's investigation'. While this is regrettable, Tebbutt had no real grounds for concern. Given his lack of formal training in mathematics and the fact that he had recourse only to a simple marine telescope and a sextant, it is remarkable that he was able to make the observations that he did and compute realistic orbital elements.

## 6.7 Concluding Remarks

Comet C/1861 J1 (Tebbutt), better known to many as the 'Great Comet of 1861', was one of the most impressive comets seen during the nineteenth century and was discovered by Tebbutt more than 6 weeks before its unheralded and spectacular appearance in northern skies. At this time, Tebbutt was already widely known in New South Wales, but his name was totally unknown to the astronomers of Europe and North America. He was young and dedicated, had considerable mathematical acumen and extensive observational experience, but was furnished with only the simplest of instruments: a sextant and a small marine telescope.

Tebbutt used these instruments to diligently track the comet for close on four months (though with a hiatus in July) and employed his self-taught mathematical skills to compute the orbital elements and to correctly forecast the passage of the Earth through the tail of the comet. Although he lacked precise micrometric positions for the comet, his orbital elements proved to be remarkably accurate.

Even before he had confirmed the discovery of the comet, Tebbutt informed his friend and colleague, William Scott (the Government Astronomer of New South Wales), of his suspicions, and he then published a succession of reports in *The Empire* and the *Sydney Morning Herald* newspapers in order to share his discovery and ongoing observations and calculations with interested members of the general public. To his chagrin, some of these communications were to involve him in considerable public controversy.

When he discovered the comet, Tebbutt was not a member of any astronomical society, and he had not published any research papers. Consequently, his various newspaper accounts were the only documentation of his work and as such served as a substitute for formal publication. At a later date, after he had acquired an international reputation, he sought to remedy this by belatedly publishing short papers about his discovery and observations in both *Astronomische Nachrichten* and *Monthly Notices of the Royal Astronomical Society*.

In the absence of telegraphic links at this time between Australia and Europe or North America, it was not possible to forewarn northern hemisphere astronomers of the comet's existence, and its sudden appearance made a considerable impact. Clerke (1893, p. 395) has commented on this:

> Many of our readers must remember the unexpected disclosure, on the withdrawal of the sun below the horizon on that evening, of an object so remarkable as to challenge universal attention. A golden-yellow planetary disc, wrapt in dense nebulosity, shone out while the

> June twilight of these latitudes was still in its first strength. The number and complexity of the envelopes surrounding the head produced … a magnificent effect … As the sky darkened the tail emerged to view … [and] its dimensions proved to be extraordinary. It reached upwards beyond the zenith when the head had already set … Most remarkable, however, was the appearance of two widely divergent rays, each pointing towards the head …

Studies of the head and tail of this comet led to an improved understanding of cometary structure and behaviour. Pasachoff et al. (1996) have documented that the first comet to be successfully photographed was C/1858 L1 (Donati) by the English painter and photographer William Usherwood (ca. 1820–1916) and Harvard College Observatory astronomer George Phillips Bond (1825–1865), so it is no surprise that Britain's Warren de la Rue (1815–1889) attempted to photograph C/1861 J1 (Tebbutt), but apparently without success (Moore 1973, p. 61; de Vaucouleurs 1961, p. 41).

The discovery of the Great Comet of 1861 first brought Tebbutt's name before the professional astronomers of the Northern Hemisphere, and it inspired him to purchase an astronomical telescope, establish an observatory and embark on an independent astronomical career that ultimately would bring him international renown. Comet C/1861 J1 certainly was a magnificent career catalyst!

**Acknowledgements** I wish to thank Gerard Hayes (State Library of Victoria), Professor Nick Lomb (University of Southern Queensland, Toowoomba), the late Dr. Harley Wood (formerly Sydney Observatory), and staff at the Mitchell Library (Sydney) and the New South Wales Lands Department (Sydney) for their assistance in various ways, and the Mitchell Library, State Library of Victoria (Melbourne) and the Tasmanian Archives and Heritage Office (Hobart) for kindly supplying Figs. 6.2, 6.3, 6.6, 6.8, 6.9 and 6.13.

## References[1]

Abbott, F. (1861). [The Great Comet of 1861]. *Monthly Notices of the Royal Astronomical Society, 21,* 260.

Abbott, F. (1878). *Modern Astronomy* … Hobart: printed for the author.

Abbott, F. (1879). *Bibliographical Contribution to the Resume on Modern Astronomy*. Hobart: printed for the author.

Abbott, F. (1880). *Sidereal Systems of Modern Astronomy* … Hobart: printed for the author.

About people. *Examiner,* 6 August 1913, page 5.

Bhathal, R. (1991). Henry Chamberlain Russell – astronomer, meteorologist and scientific entrepreneur. *Journal and Proceedings of the Royal Society of New South Wales, 124,* 1–21.

Cable, K.J. (1976). Scott, William (1825–1917). In G. Serle & R. Ward (Eds.), *Australian Dictionary of Biography. Volume 6: 1851–1890. R-Z* (pp. 97–98). Melbourne: Melbourne University Press.

Chambers, G.F. (1877). *A Handbook of Descriptive Astronomy*. Oxford: Clarendon Press.

Clerke, A. (1893). *A Popular History of Astronomy during the Nineteenth Century*. London: Adam and Charles Black.

[1] The following abbreviation is used: SL = Sydney Observatory Outward Letterbooks. TL = Letters to J. Tebbutt, bound manuscript letters in the Mitchell Library, Sydney.

Comet of 1861 (1861). The comet's grievance. *Sydney Morning Herald,* 26 June.
Dawson, T. (1861). The comet. *The Empire,* 26 June.
de Vaucouleurs, G. (1961). *Astronomical Photography: From the Daguerreotype to the Electron Camera*. London: Faber & Faber.
Death of Mr. E.J. White. *The Australian,* 9 August 1913, page 5.
Diogenes (1861a). The comet. *The Empire,* 25 June.
Diogenes (1861b). The comet. *The Empire,* 28 June.
Diogenes (1861c). The comet. *The Empire,* 3 July.
Doyle, G.E. (2008). *Reverend William Scott M.A. First Government Astronomer and First Director of Sydney Observatory*. Sydney: published by the author.
Ellery, R.J. (1861). Beobachtungen des Cometen II. 1861. *Astronomische Nachrichten, 56,* 53–54.
Ellery, R.J. (1869). *Astronomical Observations Made at the Williamstown Observatory in the Years 1861, 1862, and 1863*. Melbourne: Government Printer.
FCB (1861a). The comet. *Sydney Morning Herald,* 24 June.
FCB (1861b). The comet. *Sydney Morning Herald,* 27 June.
First Annual Report of the Department of Lands, Being for the Year 1880. In *New South Wales Votes and Proceedings of the Legislative Assembly during the Session of 1881 ..., Volume 3* (pp. 1–126). Sydney: Government Printer (1882).
Frew, D.J. & Orchiston, W. (2003). Nineteenth century Australian observations of η Carinae: international controversy and astrophysical implications. Paper presented at the IAU General Assembly, Sydney.
G (1861). The comet. *The Empire,* 28 June.
Glass, I. (2013). *Le Caille Nicolas-Louis de la Caille Astronomer and Geodesist*. Oxford: Oxford University Press.
Gledson, J.J. (1861). The comet. *Sydney Morning Herald,* 29 May.
Guellimin, A. (1877). *The World of Comets*. London: Low, Marston, Searle & Rivington.
Hawkins, H.S. (1858). The comet. *Sydney Morning Herald,* 16 November.
Hawkins, H.S. (1861a). The comet. *Sydney Morning Herald,* 1 July.
Hawklins, H.S. (1861b). Letter to the Editor. *Sydney Morning Herald,* 5 July.
Hawkins, H.S. (1876). *Practical Astronomy as Applied to Land Surveying*. Sydney: George Loxton.
Hawkins, H.S. (1881). *Compiled Catalogue of 920 Latitude Stars for the Epoch 1880*. Sydney: Government Printer.
Haynes, R., Haynes, R., Malin, D. & McGee, R. (1996). *Explorers of the Southern Sky. A History of Australian Astronomy*. Cambridge: Cambridge University Press.
Houghton, T.H. (1917). Presidential address. *Journal and Proceedings of the Royal Society of New South Wales, 51,* 1–70.
Kreutz, H. (1880). *Untersuchungen uber die Bahn des grossen Kometen von 1861 (1861, II.)*. Bonn.
la Caille, M. l'Abee de (1755). Nebulae of the southern sky. Observed by M. l'Abbe de la Caille, from the Cape of Good Hope. *Memoirs of the Royal Academy,* 194–199.
Lynn, W.T., 1904. The first director of Sydney Observatory. *The Observatory, 27,* 371–373.
Macdonnell, W.J. (1904). Letter to John Tebbutt, dated 30 September. In TL.
Maiden, J.H. (1918). A contribution to the history of the Royal Society of New South Wales, with information in regard to other New South Wales societies. *Journal and Proceedings of the Royal Society of New South Wales, 52,* 215–361.
Main, R. (1861). Observations of Comet II, 1861, made with the heliometer at the Observatory, Oxford. *Monthly Notices of the Royal Astronomical Society, 22,* 50–57.
Moore, P. (1973). *The Comets, Visitors from Space*. Shaldon: Reid.
Orchiston, W. (1987). Contribution of the Lands Department to the development of astronomy in New South Wales during the nineteenth century. *Australian Journal of Astronomy, 2,* 65–74.
Orchiston, W. (1988a). The emergence of professional astronomy in nineteenth century Australia. *Search, 19,* 76–81.
Orchiston, W. (1988b). From research to recreation: the rise and fall of Sydney Observatory. *Vistas in Astronomy, 32,* 49–63.

Orchiston, W. (1992). The contribution of Francis Abbott to Tasmanian and Australian astronomy. *Vistas in Astronomy, 35,* 315–344.

Orchiston, W. (1997a). The role of the amateur in popularising astronomy: an Australian case study. *Australian Journal of Astronomy, 7,* 33–66.

Orchiston, W. (1997b). The "tyranny of distance" and Antipodean cometary astronomy. *Australian Journal of Astronomy, 7,* 115–126.

Orchiston, W. (1998a). Illuminating incidents in Antipodean astronomy: John Tebbutt and the Great Comet of 1861. *Irish Astronomical Journal, 25,* 167–178.

Orchiston, W. (1998b). Mission impossible: William Scott and the first Sydney Observatory directorship. *Journal of Astronomical History and Heritage, 1,* 21–43.

Orchiston, W. (2015). The amateur-turned-professional syndrome: two Australian case studies. In W. Orchiston, D.A. Green & R. Strom (Eds.), *New Insights from Recent Studies in Historical Astronomy: Following in the Footsteps of F. Richard Stephenson* (pp. 259–350). Cham: Springer.

Orchiston, W. & Bhathal, R. (1991). Illuminating incidents in Antipodean astronomy: Section A of the Royal Society of New South Wales. *Australian Journal of Astronomy, 4,* 49–67.

Orion (1861). The new comet. *Sydney Morning Herald,* 19 June.

Pasachoff, J., Olson, R.J.M. & Hazen, M.L. (1996). The earliest comet photographs: Usherwood, Bond and Donati 1858. *Journal for the History of Astronomy, 27,* 129–145.

Philosophical Society. *Sydney Morning Herald,* 18 July 1861.

Proctor, M. (1926). *The Romance of Comets*. New York: Harper.

Proctor, R. (1871). Note on Mr. Abbott's imagined discovery of great changes in the Argo nebula. *Monthly Notices of the Royal Astronomical Society, 32,* 62.

Rahe, J., Donn, B. & Wurm, K. (n.d.). *Atlas of Cometary Forms. Structure near the Nucleus.* Washington, U.S. Government Printing Office (NASA SP-198).

Report of the Meeting of the Association, held on February 23, 1910, at Sion College, Victoria Embankment, E.C. *Journal of the British Astronomical Association, 20,* 229–242 (1910).

Robertson, W.H. (1985a). Obituaries. Harley Weston Wood. *Quarterly Journal of the Royal Astronomical Society, 26,* 225–228.

Robertson, W.H. (1985b). Obituary: Harley Wood. *Proceedings of the Astronomical Society of Australia, 6,* 111–112.

Russell, H.C. (1892). *Observations of the Transit of Venus, 9 December 1874, Made at Stations in New South Wales*. Sydney: Government Printer.

Schmidt, J. (1863). Astronomische Beobachtungen über Kometen. *Publications de l'Observatoire d'Athenes*, 1re Series, Tome 1.

Scott, W. (1861a). The comet. *The Empire,* 17 June.

Scott, W. (1861b). The comet. *Sydney Morning Herald,* 24 June.

Scott, W. (1861c). The comet. *Sydney Morning Herald,* 27 June.

Scott, W. (1861d). The comet. *The Empire,* 2 July.

Scott, W. (1861e). Letter to R.J.L. Ellery, dated 25 May. In SL.

Scott, W. (1861f). Letter to the Editor. *Sydney Morning Herald,* 27 May.

Scott, W. (1861g). Letter to the Editor. *Sydney Morning Herald,* 31 May.

Scott, W. (1861h). Letter to the Editor. *Sydney Morning Herald,* 1 June.

Scott, W. (1861i). Letter to Professor Peters, dated 21 June. In SL.

Scott, W. (1861j). Letter to John Tebbutt, dated 25 June. In TL.

Scott, W. (1861k). Letter to J. Maclear, dated 18 July. In SL.

Scott, W. (1861l). Letter to G.B. Airy, dated 8 September. In SL.

Scott, W. (1861m). Observations of Comet II, 1861, at the Sydney Observatory. *Monthly Notices of the Royal Astronomical Society, 21,* 254–257.

Scott, W. (1861n). Observations of Comet II, 1861, at the Sydney Observatory. *Astronomische Nachrichten, 56,* 51–54.

Scott, W. (1861o). On the Sydney Observatory and Tebbutt's Comet. Paper presented at a meeting of the Philosophical Society of New South Wales.

Scott, W. (1874). The transit of Venus as observed at Eden. *Journal and Proceedings of the Royal Society of New South Wales, 8,* 113–116.
Sheehan, W. (1999). *Christian Heinrich Friedrich Peters.* Biographical Memoir. Washington: National Acadeemy of Sciences.
Sheehan, W. & Dobbins, T.A. (2014). Schmidt, Johann Friedrich Julius. In Hockey et al., (Eds.), *Biographical Encyclopedia of Astronomers. 2nd Edition* (1944–1947). New York: Springer.
Tebbutt, J. (1859–63). Diary. MS, Mitchell Library, Sydney (A3785).
Tebbutt, J. (1861a). The comet. *The Empire,* 22 June.
Tebbutt, J. (1861b). A comet is visible [Letter to the Editor]. *Sydney Morning Herald,* 25 May.
Tebbutt, J. (1861c). Letter to the Editor. *Sydney Morning Herald,* 15 June.
Tebbutt, J. (1861d). Letter to the Editor. *The Empire,* 28 June.
Tebbutt, J. (1861e). Letter to the Editor. *The Empire,* 6 July.
Tebbutt, J. (1875). The Windsor Observatory. *Sydney Morning Herald,* 27 March.
Tebbutt, J. (1878a). Note on the Great Comet of 1861. *Astronomische Nachrichten, 93,* 47.
Tebbutt, J. (1878b). Note on the Great Comet of 1861. *Monthly Notices of the Royal Astronomical Society, 38,* 412.
Tebbutt, J. (1878c). Rambosson and the Great Comet. *Sydney Morning Herald,* 23 April.
Tebbutt, J. (1908). *Astronomical Memoirs.* Sydney: printed for the author.
Tebbutt, J. (1910). Halley's Comet. *Windsor and Richmond Gazette,* 30 April.
Tebbutt, J. (1913). Death of an Australian astronomer. *Windsor and Richmond Gazette,* 16 August, page 6.
The Shade of Pollok (1861). The Comet of 1861. *Sydney Morning Herald,* 1 July.
The Way of the Comet. *Southern Stars, 7,* 31–31 (1941).
Trigonometrical Survey (Cost and Calculations, &c.). In *New South Wales. Votes and Proceedings of the Legislative Assembly during the Session of 1875 ..., Volume 4* (pp. 929–933). Sydney: Government Printer (1876).
Venn, J. & Venn, J.A. (eds.) (2011). *Alumni Cantabrigienses. A Biographical List of all Known Students, Graduates and Holders of Office at the University of Cambridge, from the Earliest Times to 1900. Volume 2: From 1752 to 1900. Part 3: Gabb–Justamond.* Cambridge: Cambridge University Press.
Vsekhsvyatskii, S.K. (1964). *Physical Characteristics of Comets.* Jerusalem: Israel Program for Scientific Translations.
Webb, T.W. (1862). On the Great Comet of 1861. *Monthly Notices of the Royal Astronomical Society, 22,* 305–314.
Weiss, E. (1888). *Bilderatlas der Sternewelt: Eine Astronomie für Jedermann.* Stuttgart: J.F. Schreiber.
White, E.J. (1861a). Letter to John Tebbutt, dated 10 August. In TL.
White, E.J. (1861b). Letter to John Tebbutt, dated 16 August. In TL.
Wilkinson, J.H., 1861. The comet. *Sydney Morning Herald,* 27 June.
WM (1861). The comet. *The Empire,* 22 June.
Wood, H. (1958). *Sydney Observatory 1858 to 1958.* Sydney: Government Printer (Sydney Observatory Papers No. 31).
XX (1861). The comet. *Sydney Morning Herald,* 2 July.

# Chapter 7
# An Urge for Independence: The Sydney Observatory Directorship

## 7.1 Introduction

In the mid-nineteenth century, professional observatories worldwide were concerned with positional astronomy, which focussed on recording the appearance of celestial objects and plotting their positions (and as a corollary determining their motions in space). Positional astronomy was primarily a descriptive discipline, and independent astronomers blessed with the requisite instruments and mathematical skills could participate on equal terms with many of their professional colleagues. Because of this, the boundary between professional and non-professional astronomy was blurred, and many independent astronomers made the transition to professional ranks (e.g. see Orchiston 2015 for some Australian examples).

The southern sky was crying out for detailed examination, and this challenge was answered by the Cape Observatory in South Africa (Glass 2015); small National Observatories in Argentina, Brazil and Chile (see Evans 1988 for a summary) and the Sydney, Melbourne and Adelaide Observatories in Australia (Haynes et al 1996).

Sydney Observatory opened in 1858 and grew from the ashes of Parramatta Observatory which had closed in 1847 (see Saunders 1990). The founding Director and Government Astronomer of New South Wales was the Reverend William Scott (see Bio Box 6.1), but in early 1862—less than 6 years after accepting the post—he offered his resignation. Well before this occurred, Scott tried to tempt Tebbutt to join the staff of the Observatory, and he also recommended to the Government that Tebbutt was the right man to replace him. Tebbutt refused on both counts.

In this chapter we expand on an earlier, preliminary, account published in 1988 (Orchiston 1988c), and draw on a more recent biography of Scott (Orchiston 1998). We summarise the early history of Sydney Observatory, factors leading to Scott's resignation, and relations between Scott and Tebbutt. We then examine Tebbutt's 'track record' in astronomy from 1853 to 1862 and his suitability for the vacancy,



W. Orchiston, *John Tebbutt*, Historical & Cultural Astronomy,
DOI 10.1007/978-3-319-44521-2_7

and consider his reasons for declining it. We conclude by discussing developments at Sydney Observatory under Smalley—Scott's successor—and Tebbutt's continuing association with that institution.

## 7.2 Trials and Tribulations: The First Sydney Observatory Directorship

Sydney Observatory was the second of the Australian colonial astronomical observatories to be founded, after Williamstown in Victoria (Haynes et al. 1996), and aspects of its history have been documented by Baracchi (1914), Ellery (1901), Lomb (2011), Nangle (1930), Orchiston (1988a), Russell (1882) and Wood (1958, 1982, 1983).

After Parramatta Observatory was closed in 1847, Captain Phillip Parker King (Bio Box 7.1) campaigned relentlessly for the establishment of a government observatory in Sydney, but only met with success following the appointment of Sir William Denison (1804–1871; Fig. 7.2) as Governor of New South Wales in January 1855 (see Orchiston 1988d; Wood 1958).

Denison (Currey 1972; Obituaries 1872) was a career politician with a serious interest in science, including astronomy (e.g. see Denison 1858). Earlier he had established an observatory while an instructor in the Royal Engineers and later was an observer at the Royal Observatory, Greenwich (Moyal 1976). He saw the advantages to the young colony of a professional observatory and called on the expertise of the Astronomer Royal, George Airy (Fig. 3.1; Airy 1896), who selected as founding Director an Englishman named William Scott (see Bio Box 6.1). In April 1856, Scott accepted the Sydney post and spent much of the next 2 months at the Royal Observatory, Greenwich

> … where he received every assistance from the Astronomer Royal in completing his knowledge of practical Astronomy, and in making himself familiar with the most recent improvements in the construction and use of astronomical instruments. (Wood 1958, p. 7).

When Scott arrived in Sydney in October 1856 he immediately set to work. He confirmed the site of the Observatory and advised on its plans; proceeded to set up a network of country meteorological stations (Scott 1858d, 1859a); commenced corresponding with astronomers from throughout the world, in order to establish a library, and to gain ideas about suitable astronomical research projects; and succeeded in forming the Observatory Board (Wood 1958).

He also gathered together the remaining instruments from Parramatta Observatory (then in storage), and on noting their limitations set in train the purchase of a new 7.25-in. (18.4-cm) Merz refractor (Fig. 6.5). In addition, he arranged for a 'computer' to be added to the Observatory staff. The appointee was Henry Chamberlain Russell (Fig. 7.3; Bhathal 1991), a recent B.A. graduate from the University of Sydney, and although he was their top student in mathematics and physics Scott (1858c) felt '… he will have much to learn before he can be considered fully official …' Writing to a clerical colleague in England the following year, Scott (1859a) described Russell as '… rather a slow coach but steady and hard working'. Further biographical material on Russell will be found in Chap. 12.

**Bio Box 7.1: Phillip Parker King**

Phillip Parker King (Fig. 7.1) was the son of Philip Gidley King (1758–1808), the Governor of New South Wales. He was born on Norfolk Island on 12 December 1793, and entered the British Navy in 1807 where his progress was rapid. By 1815 he was a Lieutenant and was very familiar with the principles of nautical astronomy, and particularly interested in the problems associated with the calculation of longitude. Between 1817 and 1830, King spent much of his time conducting cartographic surveys of the coasts of Australia and South America, and during this period was promoted to Captain and elected a Fellow of the Royal Society.

At the end of the South American survey, King retired from the Royal Navy and returned to New South Wales, in 1832 settling at Dunheved (west of Parramatta). In 1839 he moved to Tahlee at Port Stephens, but returned to Duneheved in 1849, where he died on 26 February 1856.

King erected simple observatories at Dunheved and Tahlee, which housed a transit telescope, repeating circles, a small equatorially-mounted portable refracting telescope, an astronomical clock and a number of meteorological instruments. The astronomical instruments were used to observe comets, lunar occultations, Jovian satellite phenomena, an occultation of Jupiter, a transit of Mercury, and solar and lunar eclipses, and King reported his observations in eight different papers in *Monthly Notices of the Royal Astronomical Society* and in two privately-printed pamphlets. He also summarised his 1843–1847 meteorological observations in the 1849 issue of the *Tasmanian Journal of Natural Science*.

During the 1840s, the Admiralty Board in England and Astronomer Royal, Airy, were concerned about the dilapidated state of Parramatta Observatory and its astronomer, James Dunlop (Bio Box 11.3), and the New South Wales Government was moved to establish a three-man Commission of Enquiry, one of whom was P.P. King. They visited the Observatory on 14 June 1847 and in their report of 28 June recommended its immediate closure. The Government acted on this recommendation, and the instruments and library were placed in storage pending establishment of a new astronomical observatory. This is something that King relentlessly lobbied for over the ensuing decade, despite equally-relentless opposition from Airy, and what began as a simple time-ball station, expanded into a combined time-ball and meteorological station and eventually evolved into a fully-fledged astronomical observatory—with ever-increasing architectural sophistication, and associated expenditure! Only with Denison's appointment as Governor and a certain element of luck did King's

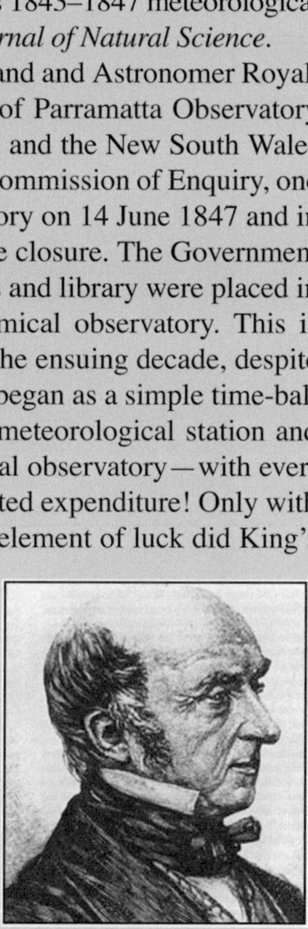

**Fig. 7.1** Phillip Parker King (https://en.wikipedia.org)

(continued)

dream become a reality, but tragically, he was to die (on 26 February 1856) not long before this impressive new facility was completed.

During the 1840s and early 1850s, the Dunheved and Tahlee facilities were the only functioning astronomical observatories in Australia, and King served an important purpose by maintaining Australia's international astronomical visibility and bridging that vital gap between the demise of Parramatta Observatory and the eventual emergence of Australia's next generation of distinguished independent astronomers, led by Abbott and Tebbutt. For a detailed account of King's astronomical work, and his involvement in the closing of Parramatta Observatory and founding of Sydney Observatory, see Orchiston (1988d).

**Fig. 7.2** A photograph of Sir William Denison taken in 1863, just 2 years after leaving Sydney to become the Governor-General of Madras in India (https://en.wikipedia.org)

**Fig. 7.3** A hand-tinted studio portrait of H.C. Russell taken by an unknown photographer around the time he was appointed Computer at Sydney Observatory (*Courtesy* Museum of Applied Arts and Sciences, Sydney, 95/239/1)

The original Sydney Observatory building is shown in Fig. 7.4 and comprised an equatorial dome, transit room, time-ball tower and astronomer's residence. It was completed in mid-1858, and Scott was able to initiate the time service. In December 1858 the old Parramatta Observatory meridian circle arrived back in Sydney after repairs in England, and this immediately was assigned to a survey of stars near the zenith and the south celestial pole. It was also used for positional observations of comets in 1860 and 1861, for lunar occultation studies and for observations of the 1861 solar eclipse and the transit of Mercury that occurred in that year.

Denison saw the role of Sydney Observatory as '… to provide service, including education, to the community and to engage in a program of work as a contribution to astronomical science' (Wood 1958, p. 30), and Scott addressed the education element by offering classes in astronomy for the general public. Planning these merely added to his already heavy workload.

As if astronomy, timekeeping and meteorology were not enough, Scott arranged with the Harbour Master for tidal readings and seawater temperature measurements to be taken from 1860 onwards.

In April 1860, Scott (1860d) wrote that he was much bothered by colds '… brought on by observing at night in a draught which is sometimes in winter enough to cut one in two …', and during the year his health continued to deteriorate. By 1861 he was totally disillusioned. Apart from his health and the impossible work-load, other matters troubled him deeply. For instance, he was bitterly disappointed when no interest was shown in his astronomy classes (Scott 1861b), and also resented public criticism that was directed towards him and the Observatory:

**Fig. 7.4** A drawing of Sydney Observatory made soon after it was completed in 1857 (Orchiston collection)

> [I] … am continually attacked in the public papers for not making more a plaything of the Observatory. The late Comet's appearance called forth attack innumerable especially for not being the first to observe it. (Scott 1861d).

The comet he was referring to was the one discovered by Tebbutt in May 1861 (C/1861 J1), and some of these attacks were discussed in the previous chapter. Scott also experienced continuous problems with the meteorological service: there were many instrument breakages, and his unskilled observers seemed expert in sending him inconsistent results. All this involved him in endless worry and incessant correspondence (Wood 1958).

Scott also became immersed in the petty bureaucracy which can become so much a part of Government. In early 1862, he discovered that payment for the new Merz telescope had been disallowed because the invoice was presented in the year after the vote was allocated (Scott 1862a). This led him to worry about other financial allocations to the Observatory, and over the next 3 months several letters were dispatched to the Colonial Secretary and the Auditor General, some of them marked 'Great Urgency'. These must have been trying times for Scott.

When it is remembered that Scott was a university academic, not an observer, and indeed had never even made a serious astronomical observation before coming to Sydney (see Scott 1861c, f), it is not surprising that he experienced considerable difficulty making the transition. He seemed ill-at-ease with the Observatory's instruments (Scott 1861e), although the arrival of the Merz refractor did, for a short while, inspire him with new spirit (Scott 1861d).

Following on from this, Scott was not altogether happy with the staffing situation at the Observatory. Although he found Russell to be an adequate computer, he would have preferred someone who also was an experienced observer (Scott 1862b). Russell was, after all, in his own mould, an academic, and had come to the Observatory straight out of university. What was needed, Scott felt, was a 'practical astronomer' who also possessed mathematical skills.

Finally, Scott (1861c) sorely felt his geographical isolation from the main centres of astronomical activity, in Europe, Britain and the United States. The only oasis in this astronomical desert called Sydney was John Tebbutt from nearby Windsor, and as we saw in the previous chapter the two men struck up an instant friendship. Scott (1860b) provided Tebbutt with advice on possible observational programmes, while Tebbutt furnished Sydney Observatory with detailed eclipse and occultation predictions (see Scott 1861a). Scott (1860a) also welcomed '… at any time suggestions or information that you may be inclined to send; as my time is so much occupied that I find little time for reading, and many important novelties may escape my notice'. The relationship was obviously a mutually-beneficial one, and for Scott was to some degree a compensation for the apparent lack of interest in astronomy shown by the citizens of Sydney. In later years, Tebbutt (1875) was to acknowledge the extent to which he valued Scott's assistance at this time:

> … I must say that, during his superintendence of the Sydney Observatory, I received from him many valuable suggestions, which encouraged me in no small degree in my studies. Any one isolated, as an amateur astronomer must necessarily be in a young colony like this, knows the value of such encouragement; and I am sure I feel grateful for it.

Notwithstanding his friendship with Tebbutt, Scott found the job untenable, and on 9 May 1862 he tendered his resignation to the Colonial Secretary:

> I have the honour to inform you that four years of close application to my duties have so affected my sight, as to convince me that I could not much longer continue to perform those duties in a manner satisfactory to myself. I have therefore thought it my duty to resign my appointment as soon as I should be able to do so without altogether ruining myself by so doing. Such an opportunity has just now occurred and … I shall be unable to attend fully to my official duties later than the end of July next. (Scott 1862c).

The search was now on to try to find a new director within the next 2 months.

## 7.3 John Tebbutt of Neighbouring Windsor: The Ideal New Director?

As early as April 1860 Scott was thinking of premature retirement, and he saw Tebbutt as part of his plans. He was clearly impressed with Tebbutt's knowledge, enthusiasm and mathematical skills (Scott 1860c, 1861a), and prophesied a brilliant future for the young Windsor astronomer:

> With your enthusiastic love of Astronomy mathematical ability and industry you might become one of the distinguished Astronomers of the age in fact Australia's first Astronomer … Why not come to Sydney give some proof of your powers at the University; work as an amateur at the Observatory (The Equatorial would be at your disposal) and take possession of my place when I give it up, which I hope to do in a few years? (Scott 1860b).

Just 1 year and 3 months later circumstances prompted Scott to take this scenario one step further, and on 12 December 1861 he wrote Tebbutt:

> Will you now allow me to ask you in confidence these two questions which you may be assured I would not ask without reason.
>
> 1. Would you be prepared, four or five years hence, to accept, if offered, the appointment which I now hold.
> 2. Would you object to take the office and Salary (£300) of Computer, on a more independent footing than that which the present computer occupies, with the understanding that you should succeed to my post when vacant.
>
>    I should be greatly obliged if you will explain to me as candidly as you think proper what are your feelings with regard to these two questions. (Scott 1861g).

In order to forestall an obvious question from Tebbutt, Scott explained (1862b) that in order to accomplish this staffing change, Russell would be offered employment elsewhere in the Public Service, in a post equivalent to the one he then held.

It is much to be regretted that the 'Tebbutt Collection' in the Mitchell Library does not contain copies of Tebbutt's outward letters, and that some of his letters have been culled from the Sydney Observatory inward correspondence files. However, in the light of later events, it is apparent that Tebbutt turned down Scott's offers as he preferred '… an unofficial life …' when it came to astronomy (N.S.W. 1893). He was not to know that the office of Government Astronomer would be vacant in a mere 5 months time, not the 5 years forecast by Scott!

Despite this response, when the position did become available in May 1862 Scott formally recommended Tebbutt's appointment, but Tebbutt once again declined the offer. In an account of the proceedings of the New South Wales Legislative Council for 29 October 1862 published in *The Empire*, funding was approved for Sydney Observatory salaries. The Colonial Secretary, Charles Cowper (1807–1875), was then quizzed about Scott's resignation and the search for a new Government Astronomer of New South Wales. After comments by Daniel Cameron Dalgleish (1827–1870) about Tebbutt's observations and discoveries, Cowper responded: 'As to Mr. Tebbutt, he had only to say the Government would be happy to give that gentleman the position; but he had refused it over and over again' (Parliament of New South Wales 1862). When pressed further by Dalgleish on whether it was the office of Astronomer or Assistant Astronomer that Tebbutt had been offered, Cowper elaborated: '… he had repeatedly refused the offer of assistant, and had as much as said that he would not accept the office of astronomer' (ibid.).

Others, besides Scott (1862d), tried to talk Tebbutt into changing his mind. One of these was the Italian-born scientific instrument-maker, Angelo Tornaghi (Maguire 1985), who had only arrived in Sydney in 1855 and later would provide Tebbutt's Windsor Observatory with a transit telescope and other equipment:

> I wish you would accept the place of Colonial Astronomer, as you are the only only [*sic*] compitent for it is well known. trusting you will axept the ofar. I know the Revd Scott is very anseous you should and so am I many others besides. (Tornaghi 1862).

While Tornaghi may win no prizes for spelling and punctuation, his sentiments are clear. They did not, however, bring about the desired result.

At a later date doubt was cast on whether Tebbutt really was offered the vacant directorship (Tebbutt 1908), but Scott's own public testimony put the matter beyond dispute:

> Mr. Tebbutt always had my friendly sympathy in his work, but the only occasion on which I had an opportunity of encouraging him was when I resigned my office of Astronomer in 1862, and recommended the Colonial Secretary to appoint him in my place. This recommendation, no doubt, would have been acted on had not Mr. Tebbutt communicated to Mr. Cowper, through me, his unwillingness to accept the appointment. (Scott 1875).

So what, precisely, were Tebbutt's credentials for the position? Was he more than your 'average' independent astronomer, who also dabbled in mathematics, and had the good fortune to have discovered a remarkable comet?

In fact, Tebbutt was a first class candidate. In the 10 years that he had been actively involved in astronomy he had made naked eye observations of meteors, the zodiacal light, aurorae, solar and lunar eclipses, and solar and lunar halos. Meanwhile, he gained extensive observational experience with his modest astronomical instruments (a small marine telescope, a sextant, and from November 1861 a portable 3.25-in. (8.3-cm) refractor), applying them to sunspots, a transit of Mercury, lunar occultations, Jupiter, Saturn, Jovian satellite phenomena, comets, variable stars and searches for the postulated intra-Mercurial planet, Vulcan (Tebbutt 1853–59, 1859–63). In 1860 and 1861, some of these observations were made in the company of the Reverend C.F. Garnsey, using his telescope. English-born Charles Frederick Garnsey (1828–1894) had moved to Windsor in 1860 and set up a successful 'collegiate school' (Cable 1972). Garnsey married Emma Stiles, who was the daughter of the Reverend H.T. Stiles, one of the teachers from whom Tebbutt acquired an interest in science (Tebbutt 1908). Garnsey also succeeded Stiles as the rector at St. Matthew's, the church which Tebbutt attended (Steele 1916). We do not know the aperture of Garney's telescope, although it was certainly greater than that of Tebbutt's small marine telescope (see the 4, 16 and 21 April 1860 entries in Tebbutt 1859–63).

Table 7.1 lists Tebbutt's annual astronomical targets and illustrates his observational diversity over a 10-year interval. His first recorded observations of Jupiter, a lunar and planetary conjunction (the Moon, Venus and Jupiter), a comet, a double star, meteors and fireballs were made in 1853; a meteor shower, sunspots and the zodiacal light in 1854; a lunar occultation in 1856; a solar eclipse and a variable star in 1857; a lunar eclipse in 1858; an aurora and a lunar halo in 1859; $\eta$ Argus, a solar halo and Jovian satellite phenomena in 1860; and a transit of Mercury in 1861. Tebbutt also began systematic comet-searches in 1861, as we noted in the previous chapter, and in 1860 he carried out what was to be the first of several searches for the ever-elusive Vulcan.

**Table 7.1** Astronomical observations by Tebbutt 1853–1862[a]

| Year/object | 1853 | 1854 | 1855 | 1856 | 1857 | 1858 | 1859 | 1860 | 1861 | 1862 |
|---|---|---|---|---|---|---|---|---|---|---|
| Aurorae | | | | | | | × | × | | |
| Comets | × | | | | | × | × | × | × | × |
| Double stars | × | | | | | | | | × | |
| Lunar eclipse | | | | | | × | | | | |
| Lunar occultations of stars | | | | × | | | | | × | |
| Meteors | × | × | × | | | × | × | × | | |
| Planets and satellites | × | | | | | × | | × | × | × |
| Sunspots | | × | | | × | × | × | × | × | × |
| Solar eclipses | | | | | × | | | | × | |
| Transit of Mercury | | | | | | | | | × | |
| Variable stars | | | | | × | × | | × | | × |
| Vulcan searches | | | | | | | | × | | × |
| Zodiacal light | | × | | | | | | | | |

[a]Up to and including April only

The 1856 lunar occultation (Table 7.1) was Tebbutt's first by accident rather than design, for as we saw in Chap. 2 he had attempted to observe an occultation of the star 8 Scorpii on 17 September 1855 but fogged up his eyepiece at the critical time.

What Table 7.1 does not do is indicate the frequency of Tebbutt's observations, and these varied considerably from year to year as the following accounts demonstrate.

It was Comet C/1853 G1 (Schweizer) which was the catalyst that awakened Tebbutt's deep but slumbering interest in astronomy, and he observed this impressive naked eye object on five different evenings in May (Fig. 7.5), commenced an astronomical diary (Tebbutt 1853–59) and recorded his observations in it, along with field sketches. In May and June 1853, he also made 20 drawings of the positions of the Jovian satellites. A second naked eye comet (C/1853 L1) attracted his attention, on 11 September 1853, and nine observations of it were made in the course of the next month. During its presence, Tebbutt journeyed to Sydney with John Stiles (ibid.), a former Royal Navy midshipman and brother of his old teacher (Tebbutt 1908), and purchased a sextant and a copy of Norie's *Epitome of Navigation*. He then used the sextant to take three readings of the position of the comet (Tebbutt 1853–59). September, October and November all greeted Tebbutt with meteors and fireballs, rounding out what had proved to be an exciting year in which to initiate a regular astronomical observing programme.

The following year, 1854, was less demanding. In July, Tebbutt observed the zodiacal light on five separate evenings, while fireballs and meteors attracted further attention during the first 8 months of the year. One of these, witnessed on 27 February, was a particularly interesting object and is described in Tebbutt's diary:

> A very beautiful meteor appeared at 10h 30m, p.m. the most beautiful for brilliancy, I think, I had ever seen. It passed from the zenith towards Procyon and the constellation Gemini, having a long double train of white light behind it. The meteor was of oblong figure apparently about half a degree in length. (Tebbutt 1853–59).

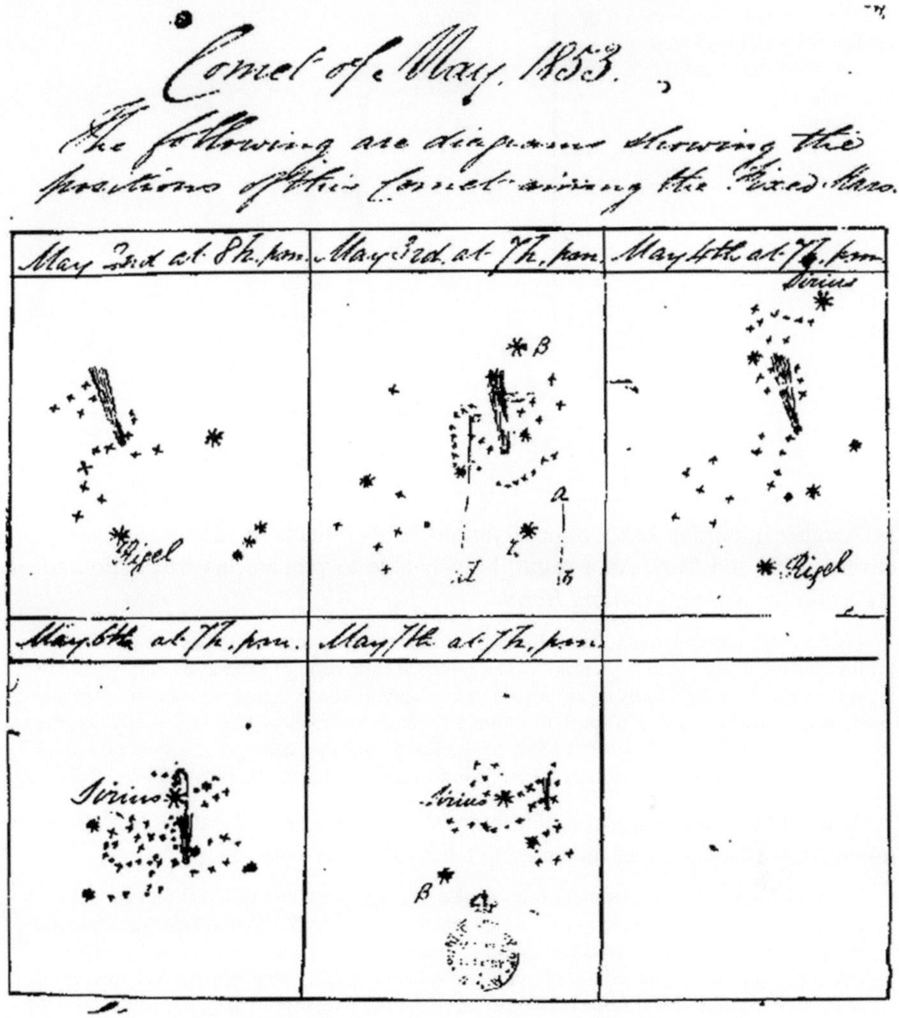


**Fig. 7.5** Tebbutt's observations of Comet C/1853 G1 (Schweizer) in his observing diary (after Tebbutt 1853–59; *Courtesy* Mitchell Library)

Although this was certainly impressive, undoubtedly Tebbutt's most notable achievement in 1854 was to record a minor meteor shower. In the course of about one and a half hours observing on the evening of 26 July 1854, five meteors were recorded, and when the paths of these were projected back four of them met at a point a few degrees to the west of Jupiter at about R.A. = 19 h 20 m, and Dec. = −22°. It is interesting to note that these co-ordinates do not coincide with the radiant positions of any of the major annual showers, but they are similar to radiant No. 221 in McIntosh's (1935) extensive catalogue of southern meteor showers, whose right ascension and declination are, respectively, 19 h 28 m and −23°. McIntosh lists 1–7 August as the period when this shower is active.

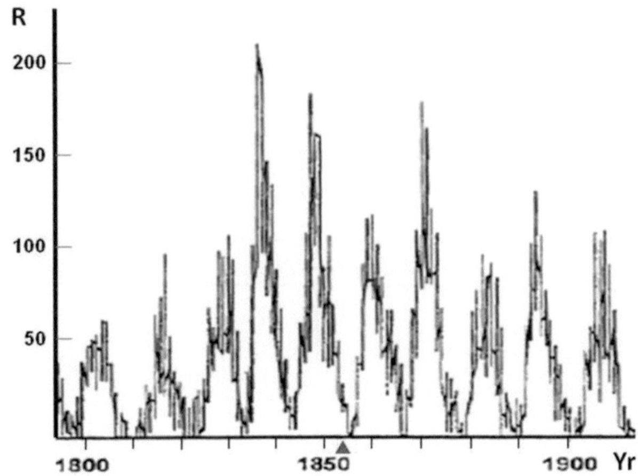


**Fig. 7.6** Sunspot Numbers (R) for 1795–1915; the *red triangle* indicates April–May 1854 (*Plot* Wayne Orchiston)

Another important series of observations Tebbutt made in 1854 was of sunspots, during April and May. As a result, he was able to prepare his first astronomical report for the *Sydney Morning Herald*:

> I beg to call attention to an unusually large group of solar spots which has been traversing the sun's disk for some days past … The group is composed of several nuclei, which are surrounded more or less by an umbra or fainter shade. By observations of these spots it may be ascertained that the sun revolves on his axis in about 25 days 12½ hours; and that the inclination of his equator to the plane of the Earth's orbit is nearly 8″ … Large tracts of faculae were also to be seen near the sun's limb a few days ago. (Tebbutt 1854).

This report arose as a result of Tebbutt's curiosity to empirically establish astronomical parameters rather than merely taking them as stated fact:

> I had read in popular works on astronomy that the sun rotated on its axis in about twenty-five days, and I proceeded to verify the statement. By means of the small marine telescope properly mounted, I projected very good pictures of the sun on a sheet of white paper … Having in this way obtained a series of sketches of the sun's disc and its spots I succeeded in deducing a fairly good value for the period of the sun's rotation. Thus I had actual experimental evidence of the truth which I had learned from popular works, which evidence, of course, gave great satisfaction. (Tebbutt 1908, p. 11).

What is particularly remarkable about these observations is that they were possible at all given that in 1854 the Sun was very close to a minimum in its sunspot cycle. This is clearly shown in Fig. 7.6.

Establishing the latitude of The Peninsula Homestead became Tebbutt's total pre-occupation in 1855, and in order to achieve this he carried out sextant observations of selected stars from February through to October, and positional observations of the Sun during May, June and July. Apart from the observations associated with this demanding programme, he recorded just one meteor and one fireball (Tebbutt 1853–59).

By contrast, 1856 was a very quiet year, the only astronomical observation of note being a lunar occultation, on 3 November (ibid.).

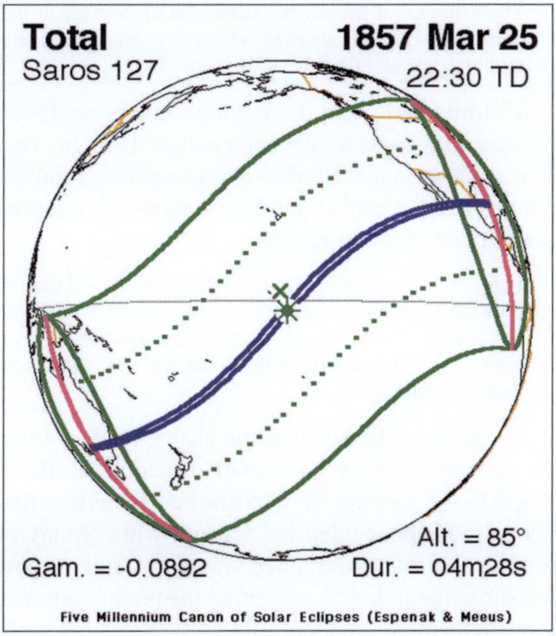


**Fig. 7.7** Espenak and Meeus' map showing the path of totality of the 25 March 1857 total solar eclipse (https://en.wikipedia.org)

While the last quarter of 1857 was to prove most demanding, it was the total solar eclipse of 26 March (Fig. 7.7) which initially attracted Tebbutt's attention. Although an annular eclipse was visible in Sydney in February 1851 (see Clarke 1851), this was the first total solar eclipse visible from the Sydney region since the European settlement of Australia, and as such it generated considerable public interest. From a scientific viewpoint, these eclipses were important events in that they offered albeit all-too-brief opportunities for astronomers to study the Sun's lower atmosphere, the chromosphere (with its accompanying prominences), and the nebulous outer atmosphere, the corona (although in 1857 there still was open debate about whether the corona was a solar, lunar or terrestrial phenomenon—see Ranyard 1879).

While Tebbutt was eager to carry out a scientific study of the March eclipse, he was also well aware of the spectacular nature of such events for the general populace, and so made a point of publicising the event through the *Sydney Morning Herald*. In a letter published on 12 March he reported that at sunrise the eclipse would already be mid-way through, with the Sun's disk totally hidden by the Moon, and that the end of the event would occur at 7:54 a.m. (Tebbutt 1857d). Five days later, he prepared a follow-up letter:

> The intense interest which this extraordinary phenomenon must naturally produce in the minds both of scientific persons and of the community in general, will, I trust, afford a sufficient apology for my once more troubling you with a few remarks concerning it. The eclipse, as seen from Windsor, Richmond, Pitt Town, and Wilberforce will be more interesting than as seen from Sydney… I find by a calculation, conducted with great exactness, that the eclipse will be almost central at Windsor … The line of central and total eclipse lies about two miles south of Windsor, and extends in the direction of Broken Bay.

> The diameter of the moon's shadow at the surface of the earth will be about eighty miles: consequently all places lying within forty miles north or south of Windsor will have the sun totally eclipsed. (Tebbutt 1857c).

Unfortunately, fate did not look kindly on Tebbutt's sole opportunity during his lifetime to witness a total solar eclipse from his very own place of domicile. On the day in question the Windsor sky was almost totally clouded over, the only clear area being a narrow strip about three-quarters of a degree deep extending above the eastern horizon. Tebbutt reports:

> In order to catch a glimpse of the sun as he passed through this clear opening, I took up my position a few minutes before sunrise. At twelve minutes past six o'clock the sun appeared, and a thrill of satisfaction and delight passed through my mind as I beheld the sun's disc with a large indentation in the upper limb. He did not remain more than three minutes in view. (Tebbutt 1857a).

By the time the Sun became visible totality was over, so Tebbutt devoted most of the newspaper report to a detailed account of the changing appearance of the landscape in the vicinity of Windsor and the effect that the eclipse had upon his farm animals. He concluded his account with: 'Many of the townspeople rose early to witness the interesting phenomenon. A few persons thought the affair a hoax, but the subsequent darkness verified the predictions of the dispised star-gazer, and convinced the most skeptical' (ibid.).

Nor were those in Sydney treated any better. Reverend William Branwhite Clarke (1798–1878), the well-known geologist, records that 26 February dawned cloudy, and therefore

> … was most unfavourable for observations of the eclipse … [nevertheless, there was a] crowd of persons on foot, or horseback, and in carriages, congregated or passing along the light-house ridge … [but] I speedily saw, by the shortness of their stay, and the quickness of their movements, that the disappointment respecting the weather was quite as great to them as to persons more inland. (Clarke 1857, p. 40).

Clarke (1857, p. 44) ended up with the following sobering reflection: despite the cloudy skies, the eclipse made people feel that something unusual was taking place, '… something that makes a man feel that there is a power in Nature above man's philosophy or man's indifference'.

Fortunately, those further to the west, in Bathurst, near the end of the path of totality, were graced with clear skies and saw a total eclipse.

Collectively, these accounts demonstrate the 'pulling power' that impressive astronomical events or phenomena had (and continue to have), even in country towns, but it also reveals Tebbutt's far from popular image in the eyes of some locals. For much of his life, Tebbutt was to be preoccupied with his status, as perceived by the Australian public and by professional and independent astronomers at home and abroad.

The Sun remained the focus of Tebbutt's observational attention during the remainder of 1857. On 11 September he commenced a series of solar observations, and in the course of the next 5 weeks made 13 different full-disk drawings showing the distribution of sunspots (ibid.). These distinctive objects were particularly abundant, despite the fact the Sun was just past sunspot minimum (see Fig. 7.6).

**Fig. 7.8** Comet C/1858 L1 (Donati) over Paris (after Weiss 1888)

September also offered Tebbutt a special treat—a second solar eclipse in the 1 year (but this time a partial one), and he lost no time in forewarning the public, again through the good offices of the *Sydney Morning Herald*: '[Although] … it will not be so remarkable as that which occurred in March last, it will be interesting, as one-half of the Sun's diameter will be obscured to Sydney and its vicinity' (The coming eclipse 1857). The article contained details of the starting and mid-eclipse times of this 18 September event; the end occurred after sunset. Once again clouds intervened on the day in question, although Tebbutt was able to successfully observe the first half of the event before this occurred (Tebbutt 1857b).

In November 1857, Tebbutt commenced a series of naked eye observations of the bright variable star Algol, and these continued into early 1858. Sunspots also proved an attraction in March 1858 (see Tebbutt 1858b), and a lunar eclipse in August, but from 12 October it was Donati's Comet (C/1858 L1) which commanded Tebbutt's attention. An impressive object while in northern skies (e.g. see Fig. 7.8), this naked-eye comet captivated Tebbutt (1858a), Scott (1858a, b) and other local astronomers (Clarke 1858; Hawkins 1858; Severn 1858a, b), and the general public (e.g. Bull 1858; Ionaramus 1858; Lector 1858; The Comet 1858). When first viewed from Sydney, the British-born independent astronomer Henry Augustus Severn (1833–1883; Orchiston 2016, Chap. 21) reported that it

> … presented a very fine nucleus, and a well-defined tail, which measured about ten degrees … the tail along its upper surface is very well defined and even, but the lower, or under side, it seemed crooked and uneven. (Severn 1858a).

Tebbutt observed this comet on 31 different occasions between 12 October 1858 and 4 January 1859 and made a number of drawings in his observing diary (Tebbutt 1853–59). He also recorded its position with his sextant on numerous occasions but, as Scott (1858a) cautioned, such measurements were not precise enough to furnish accurate orbital elements. On 18 October, Tebbutt (1853–59) recorded a 5–6° tail, despite bright moonlight; by 17 November the comet had all but lost its majesty and was barely visible to the naked eye (ibid.).

Ever since his observations of the two comets in 1853, Tebbutt had been teaching himself the mathematics required for the computation of cometary orbits, and in spite of Scott's warning about the limitations of sextant observations, he decided to test his computational competence on this comet (Tebbutt 1853–59). His first attempt at this challenging task was a success, which prompted him (in September 1859) to compute a 'rough orbit' for the comet of May 1853 (ibid.). Tebbutt proudly reports in his *Astronomical Memoirs*:

> These were the first attempts by me to calculate a cometary orbit, and I well remember the pleasure which the results afforded me. I took courage and persevered with my studies, much to the disapprobation of my good father who often expressed his fears that I should ruin my health by my intense application. (Tebbutt 1908, p. 14).

After the excitement of Comet C/1858 L1 (Donati), 1859 proved a relatively quiet year. From June on, Tebbutt carried out further sextant observations for latitude, and during September recorded sunspots. On 29 August 1859 he and Scott independently observed an *aurora australis* for the very first time, and both were so impressed that they forwarded accounts to the *Sydney Morning Herald* (Scott 1859b; Tebbutt 1859).

In contrast, 1860 was to prove a demanding year, observationally. During March and April Tebbutt participated in an international search for Vulcan, and in the process recorded sunspots. In his diary entry of 17 March he writes:

> I was not aware till yesterday of the existence of such a planet. I am indebted to the Rev. C.F. Garnsey for the information, who also kindly furnished me with certain particulars which appeared in the Sydney Empire relative to the planet. (Tebbutt 1859–63).

On 28 March, Tebbutt recorded an aurora, the first since August 1859, and the following evening he and Garnsey successfully observed Jovian satellite phenomena (in order to determine the longitude of Windsor). Acting on Scott's suggestion, on 4 May Tebbutt made his first magnitude estimate of the famous variable star η Argus (or η Carinae as it is now known). The comparison stars he used in order to make the magnitude estimates are shown in Fig. 7.9.

Two months later, on 9 July 1860, Tebbutt wrote in his observing diary: 'I discovered a comet at 6 h 31 m, p.m., per clock, in the constellation Leo. The nucleus was tolerably distinct to the naked eye, but its train was very short' (ibid.). Tebbutt immediately notified Scott and Quaife (his old tutor) of the discovery only to hear back from the latter that he too had independently 'discovered' this comet. In fact, Comet C/1860 M1 (also known as the Great Comet of 1860) was first detected on 18 June 1860 (Marsden 1982), and put on a spectacular performance while visible to northern hemisphere observers (see Fig. 7.10). Tebbutt continued to follow the comet until 19 August, making observations on 25 different evenings. In late July, he used his sextant observations to compute initial orbital elements for the comet

**Fig. 7.9** The chart that Tebbutt prepared so that he could make magnitude estimates of η Argus (which is located towards the *bottom* of the chart in the centre, as star number *1*) (*Courtesy* Mitchell Library)

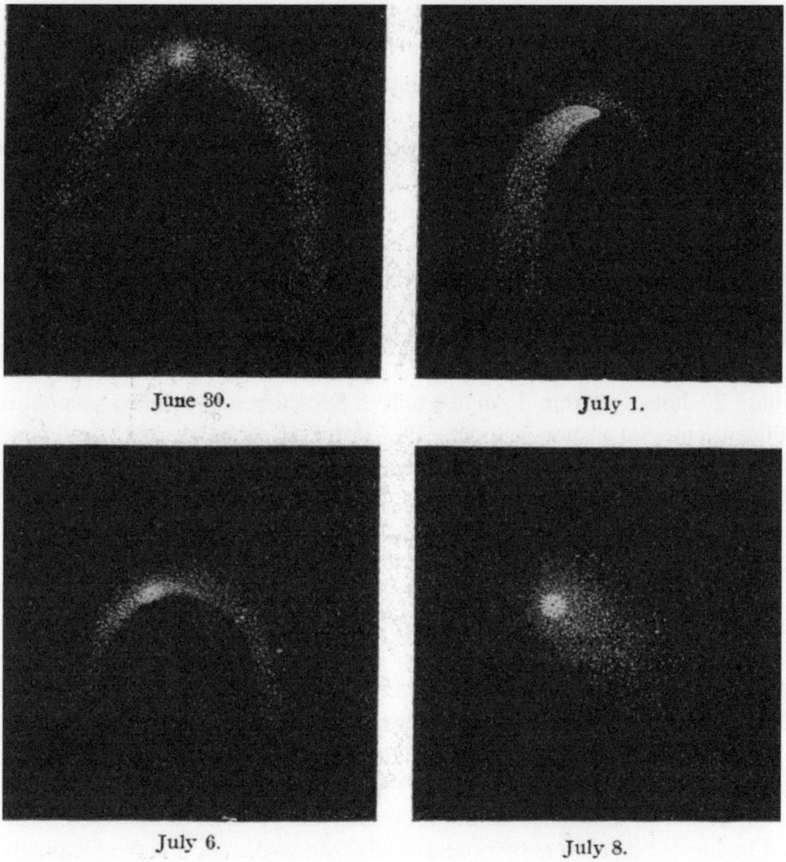


**Fig. 7.10** During June and July 1860 the head of Comet C/1860 M1 put on a spectacular display (after Phillips 1923, p. 418)

**Fig. 7.11** A modern photograph of M5 taken with a DSLR camera. It is easy to see how Tebbutt could have mistaken this for a faint comet when he saw it through his little marine telescope (https://en.wikipedia.org)

(Tebbutt 1860a), and after further observations came to hand was able to send the *Sydney Morning Herald* '… as accurate elements as I can hope to obtain …' (Tebbutt 1860b). Tebbutt (1859–63) was gratified to discover that his elements were similar to those calculated by Scott. Further sunspots, in October 1860, brought Tebbutt's very busy observational year to a close.

A partial solar eclipse on 11 January 1861 heralded in the new year, and given that 60 % coverage was expected, Tebbutt (1861a) made a point of promoting the event through the *Sydney Morning Herald*. On this occasion the weather was co-operative, but Tebbutt was moved to report: '… I made a few observations on the occasion, but they are of no value except to give your readers some idea of the accuracy with which eclipses can be predicted' (Tebbutt 1861b). Between April and November (inclusive), he made five observations or attempted observations of lunar occultations, and on 12 November recorded his first transit of Mercury, noting the presence of sunspots at the same time (Tebbutt 1859–63).

Undoubtedly Tebbutt's greatest astronomical coup during 1861 was his discovery of the Great Comet of that year (C/1861 J1), which he followed religiously from 13 May until 27 June and from 2 August until 5 September. A detailed account of this magnificent celestial visitor was presented in the previous chapter so we shall not dwell upon it here. This discovery was the result of a systematic search programme, which persisted throughout the year. On 23 September, Tebbutt thought he had found yet another comet, but when he and Garnsey investigated this using the latter's telescope, it proved to be the globular cluster M5 or NGC 5904 (ibid.). This object (Fig. 7.11) is described in *Nortons Star Atlas* as 'A fine globular cluster, 15′ in diameter, composed of 11th–15th mag. stars, with much-compressed centre (Norton 1959).

Another important milestone in Tebbutt's emerging astronomical 'career' occurred on 14 November 1861 when he purchased a 3.25-in. (8.3-cm) refractor, and with this substantially-increased light grasp a wider range of new 'prey' came within his reach. On 18 November he observed Jupiter, Saturn and the first magnitude double stars α Crucis and α Centauri (the latter was only discovered to be a triple system during the twentieth century), and noted in his observing diary: 'Before purchasing the telescope I had tested its powers by observing the Moon, Venus, and the great nebula in Orion' (Tebbutt 1859–63). In December, the Reverend Garnsey was delighted to be

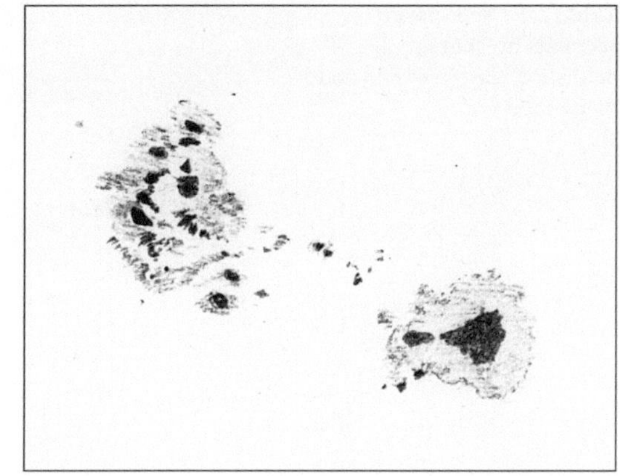

**Fig. 7.12** As well as making full-disk solar drawings, John Tebbutt also liked to sketch individual sunspot groups, as in this example, which dates to 29 July 1862 (*Courtesy* Mitchell Library)

able to join Tebbutt in exploring the wonders of the southern night sky with the new telescope. What had proved to be a very busy year, observationally, ended with unsuccessful telescopic searches for Encke's Comet on 25, 26 and 27 December.

During the 4 months in 1862, before he was officially offered the Sydney Observatory post, Tebbutt used his new telescope to carry out sunspot observations (see Fig. 7.12) on almost every clear day, and combined this with another unsuccessful search for Vulcan. He finally located Comet 2P/Encke on 16 February 1862, observing it on three further evenings, and recorded Jovian satellite phenomena on 1 and 14 February. On 23 January, he was intrigued to see Saturn with its rings edge-on. Saturn and Jupiter were observed on 12 February, and Tebbutt (ibid.) described the latter planet in his observing diary: 'The belts on Jupiter's disk were beautifully distinct. Four dark belts were visible, the broad one (see annexed figure) was very distinct, as also two yellow belts'. The magnitude of η Argus was estimated on 23 and 26 January, and the small marine telescope was used for comet-searches on 11 different evenings (ibid.).

In summary, over the 10 years or so leading up to his Sydney Observatory offer, Tebbutt had become an experienced observational astronomer and was adept in the use of scientific instruments. Furthermore, the types of observations associated with the objects and events listed in Table 7.1 were those required of a professional astronomer committed to positional astronomy. Although self-taught, Tebbutt was also an accomplished mathematician, and by 1862 had computed the orbital elements of four different comets, in each case relying solely on his own observations for positional data. This was a considerable achievement: in those halcyon days before computers, pocket calculators or even slide rules, the amount of effort involved in such calculations was prodigious.

Moreover, Tebbutt was thorough, systematic and totally professional in his approach, all qualities which Scott (1860c) admired. One aspect of this was his comet-searching, which in 1861 had netted him what turned out to be one of the Great Comets of the nineteenth century. Through this discovery, Tebbutt's name became known to the astronomers of the northern hemisphere (Tebbutt 1908, p. 14), an important point since at the time he was only a household name in Sydney and perhaps throughout New South Wales. For it to prosper, Sydney Observatory

**Table 7.2** Newspaper accounts by Tebbutt published between 1853 and April 1862

| Year | Number of accounts | |
|---|---|---|
| | *Sydney Morning Herald* | *The Empire* |
| 1853 | 0 | 0 |
| 1854 | 1 | 0 |
| 1855 | 0 | 0 |
| 1856 | 1 | 0 |
| 1857 | 5 | 0 |
| 1858 | 6 | 0 |
| 1859 | 2 | 0 |
| 1860 | 2 | 0 |
| 1861 | 7 | 9 |
| 1862 | 3 | 0 |

Based on the listing in Tebbutt (1887, pp. 38–39)

needed an internationally-recognised director. The only other Australian independent astronomer with any international visibility at this time was Francis Abbott of Hobart (see Bio Box 6.4), but he was not a contender for the Sydney position.

Another of Tebbutt's obvious talents was in the popularising of astronomy (Orchiston 1997), something to which Scott was equally devoted (Orchiston 1998). Tebbutt's technique was to bring astronomy to the general public through newspaper articles, reports and 'Letters to the Editor'. His first such contribution appeared in the *Sydney Morning Herald* on 13 May 1854 and more followed, all in a style and with a vocabulary that was palatable to the general public. By the time the Sydney Observatory appointment arose, Tebbutt had 36 publications of this kind (see Table 7.2). They dealt with an aurora, comets, lunar occultations, solar eclipses, sundials, sunspots and a transit of Mercury.

Finally, Tebbutt was a mature young man (about to turn 28) who potentially had a long and illustrious career ahead of him. Through his farming activities, business commitments, and family's relative affluence, he had gained the requisite skills in administrative and financial matters, and all he had to learn were the vagaries of the New South Wales Public Service and the Government—but this was best done on the job.

In summary, Tebbutt was an obvious contender for the Directorship, in that he possessed the right combination of experience, enthusiasm and intellect. Scott's recommendation that the Government should offer him the post was therefore justified. It is interesting, in this regard, that Scott chose to overlook his own assistant, Russell, even after Tebbutt chose not to accept the position.

If he was to all intents and purposes the ideal candidate, why then did Tebbutt refuse the position? The directorship, after all, offered him certain obvious advantages: the chance to work full-time in the service of science (and in fields other than just astronomy, if he so chose); to draw a regular salary, regardless of the state of the economy or flood-prone Windsor; to exchange his newly-arrived 3.25-in. Jones altazimuth refractor for access to an equatorially-mounted 7.25-in. instrument by the celebrated German firm, Merz; and to popularise astronomy in an official capacity.

But there were obvious disadvantages. Tebbutt saw Scott work himself into ill health, immersed in and exploited by the Public Service system, and he noted the attacks in the press by irate taxpayers or those with a political or anti-science 'axe to grind'. He must have questioned whether this would provide him with the right environment in which to raise his young family (a wife, and three daughters aged not quite 4, 2 and 4 months), especially if it meant loosening ties with family and friends in Windsor and moving to Sydney. Nor was the option of a regular salary an instant attraction, for when times were good farming would prove far more lucrative than astronomy. However, Tebbutt was very aware that although Windsor boasted some of the most fertile land in the colony, droughts and floods were frequent '... and the hopes of the farmer are often disappointed' (N.S.W. 1893).

But despite this, as an effective 'only child' (for his younger sister had died in infancy), Tebbutt had a family obligation to maintain the Peninsula Estate which his parents had worked so hard and long to establish. Neither parent was getting any younger. His father was already 68 years of age, and was not destined to live another 6 years, while his mother, just 2 years younger, was to die in 1866.

Under these circumstances, Tebbutt was not overly concerned about the limitations of the instruments at his disposal. He planned to remedy this situation as the opportunity arose, but believed the attitude and ability of the observer rather than telescope size *per se* were the critical factors:

> The best instrument in the world was of little use if an intelligent observer was not at the eye-piece, and an inferior instrument might be the means of doing good work if handled by a skilful observer, who was really the soul of his instrument. If real work was to be done, a good deal of self-denial must be exercised. (Reports of the Branches 1895).

There was ample work to be done with his 3.25-in. refractor, on comets, Jovian satellite phenomena, lunar occultations and variable stars, not to mention 'occasional phenomena' such as eclipses and transits, without recourse to a larger telescope (see Tebbutt 1908).

Tebbutt also contemplated the freedom of the independent astronomer, as clearly portrayed to him at a much later date by Sydney Observatory's H.A. Lenehan (Bio Box 7.2):

> ... you have a great advantage over the Sydney Observatory or other public institutions in so far that being the only one to be consulted and [can] arrange your work without the usual and annoying routine of a large establishment, where you are at the beck and call of your staff—and the persistent interruption of the public on matters irrelevant to your desires. You can to a certain extent lock yourself away from these troubles and devote your time and close attention to any particular object in view. (Lenehan 1906).

In view of Scott's troubles and the various personal factors mentioned earlier, Tebbutt ultimately decided that the directorship was too much of a challenge and was not worth the effort. And perhaps, like Sir John Herschel (1792–1871), he '... saw independence of means as intimately connected with independence of mind' (Chapman 1993, p. 75).

**Bio Box 7.2: Henry Alfred Lenehan**

Henry Alfred Lenehan (Fig. 7.13) was born in Sydney on 29 August 1843, 9 years Tebbutt's junior, and after completing his schooling worked in a number of positions in New South Wales and Queensland before becoming a draftsman with the Queensland Railways (Obituaries 1909a). He returned to Sydney in 1870 and was appointed an Assistant at Sydney Observatory (Macdonnell 1908). Despite his lack of an astronomical background Lenehan excelled in his new post, and 'For 32 years … worked under the forthright and uncompromising Russell and was responsible for cataloguing accurate star positions, using the transit instrument' (Wood 1986). He participated in the Observatory's ambitious 1874 transit of Venus programme (Lomb 2011; Orchiston 2004); became the Observatory's Deputy Director and served as Acting Government Astronomer in 1875, 1877 and 1901 while Russell was away.

Lenehan was active in the Royal Society of New South Wales for many years, serving as President during the 1905–1906 session. Earlier, in 1894, he had been elected a Fellow of the Royal Astronomical Society (Obituaries 1909a). He also joined the New South Wales Branch of the British Astronomical Association after this group was formed in 1895 and served on the Committee in 1897 (Orchiston 1988b). He always supported serious research-active independent astronomers (e.g. see Lenehan 1904) and strove to foster close professional–non-professional relations in Australian astronomy. He also was committed to popularising astronomy, '… and at all times was ready to impart his knowledge to others. Consequently, he was extremely popular with all sections of the community …' (Macdonnell 1908).

When Russell went on sick leave in 1903 Lenehan again assumed the role of Acting Government Astronomer, and on 1 January 1907—not long before Russell's death—he was appointed Government Astronomer of New South Wales and Sydney Observatory's fourth Director. His first year in office was an eventful one: the Federal Government extracted the meteorological services from the state observatories, but Sydney Observatory was to benefit astronomically as Lenehan convinced the New South Wales Government to approve the

**Fig. 7.13** Henry Alfred Lenehan (adapted from Russell 1892, Frontispiece)

(continued)

purchase of a new 15-in. (38.1-cm) Grubb refractor (Macdonnell 1908). Sadly, Henry Lenehan never lived to witness the arrival of this instrument as he died suddenly on 2 May 1908, a little over 1 year after taking office as Director (Obituaries 1909a). Thanks to C.J. Merfield's intervention, the order for the Grubb refractor then was cancelled.

## 7.4 Concluding Remarks

With Tebbutt's refusal, the search for a new Government Astronomer was on in earnest and the New South Wales Government once more turned to England for help. In due course Airy and Sir John Herschel arranged for a mathematician and astronomer named Smalley to replace Scott. George Robarts Smalley was born at Banbury, England on 18 April 1822, and after graduating with a B.A. degree from Cambridge University in 1845 accepted a position as Third Assistant at the Cape Observatory in South Africa. In 1851 he moved to a Chair in Mathematics at the South African College (which later became the University of Cape Town), but returned to England in 1854 to take up a Lectureship in Mathematics at King's College, London (Wood 1976).

Smalley arrived in Sydney on 7 January 1864 and faced a daunting assortment of tasks assigned him by Airy. Among the 19 on his list (see Wood 1958) were:

1. Prepare a catalogue of southern stars.
2. Undertake a trigonometrical survey of New South Wales.
3. Carry out magnetic observations.
4. Carry out meteorological observations.
5. Maintain a time service.
6. Establish and superintend tide gauges.
7. Carry out observations of Mars and Venus for parallax, and general observations of eclipses, lunar occultations, minor planets, southern comets, double stars, nebulae, etc.
8. Promote astronomy education.

This listing demonstrates the wide-ranging focus of government 'astronomical observatories' during the nineteenth century, and as a corollary, a rather catholic interpretation of the role of a professional astronomer was called for.

Smalley was very well aware of this, but soon after his arrival in Sydney felt obliged to point out that '… the various suggestions of Professor Airy, valuable as they are in every point of view, can only be fully carried out gradually as the occasion may arise and with the aid of additional assistance to the funds and the staff of this establishment' (cited in Wood 1958, p. 11). Given the reality of the situation, he decided to undertake the magnetic, tidal and trigonometrical work and to continue the time service and the meteorological programme that Scott had initiated. He also determined to follow southern comets and observed Comet 1864 R1 (Donati) in 1864 and Comet 2P/Encke in 1865 (e.g. see Smalley 1862–65).

**Fig. 7.14** Fort Denison on the *left*, is sited on a small island in Sydney Harbour, and tidal reading were taken there for Sydney Observatory (*Photograph* Darunee Lingling Orchiston)

In January and February 1865, Smalley began the magnetic survey, taking readings at 12 different stations, and further fieldtrips followed (Wood 1958). Russell was left in charge of the Observatory on these occasions. Meanwhile, systematic tidal readings were commenced in 1866 using automatic tide gauges located at Fort Denison (see Fig. 7.14), and in order to facilitate this programme Smalley served on the Harbour Commission (ibid.).

Initially, Smalley decided to continue the network of country meteorological stations established by Scott, but towards the end of 1869 he resolved to expand the programme. As Wood (1958, p. 12) has documented, 'This led immediately to the establishment of additional stations of which 43 are listed in a paper laid before the Legislative Assembly in March, 1870'. Although he was not a part of this network, Tebbutt submitted his own meteorological observations to Sydney Observatory, where they were combined with data provided by the country stations. This led to trouble when Smalley's 1865 Annual Report to the Sydney Observatory Board of Visitors included Windsor as one of its country stations. Tebbutt's response was to immediately 'set the record straight' by means of a letter which was published in the 8 September 1866 issue of the *Sydney Morning Herald*:

> There has, however, been no meteorological station at Windsor under the superintendence and control of the Government Astronomer since the year 1863. A meteorological observatory was established here by me in December, 1862, and regular observations three times a day commenced, of the same character as those made at the Sydney Observatory. I was then asked by Mr. Russell … if I would favour him with monthly copies of my observations, and thereby enable him to remove the instruments from the Windsor telegraph office to a new station. I at once acceded to the request, and my observations, after complete reduction, were forwarded monthly to the Sydney Observatory during the years 1863, 1864, and 1865. Considering these circumstances, I was rather surprised to find that my observatory was

> regarded in Mr. Smalley's first report as one merely attached to the Government department. I was still more surprised to find that an abstract of my work for 1863 was published in the Government volume for that year without any acknowledgement of the source from which it was obtained. I am sorry to find that the present report does not correct the erroneous impression; it is my intention, however, to publish my observations. (Tebbutt 1866).

This reaction clearly demonstrates Tebbutt's disdain for anything that suggested belittlement, and to his way of thinking Smalley's Annual Report threatened his independence and downgraded the status of his observatory and his own observations.

This incident contrasted markedly with the particularly amiable relations that had initially ensued between Smalley and Tebbutt following the former's Sydney appointment. In a letter dated 30 April 1864, Smalley (1864a) wrote Tebbutt offering 'Many thanks for your interesting and helpful communication received this morning. Nothing could be better than the arrangement you describe'. Later in the same year he wrote that Tebbutt could telegraph matters of scientific importance to Sydney Observatory free of charge using 'OHMS' (Smalley 1864b). But by 1866, Tebbutt was beginning to be publicly critical of Sydney Observatory staff, although it was a matter of misrepresentation that he resented rather than the actual programmes undertaken by the Observatory. This was to come later.

The other major project that Smalley initiated was the trigonometrical survey of the colony, which proved to be a challenging undertaking. In January 1867, Smalley was instructed by the Government to go ahead with the measurement of a baseline for triangulation, and after conferring with the Surveyor General and obtaining reports from surveyors in various localities he selected a site for it near Lake George (Wood 1958, p. 12). The length of the base line was 5 miles 64 chains and 72 links. Smalley devoted much of the next 3 years to the survey, which proved to be extremely demanding and at times frustrating, and kept him away from the Observatory for long periods of time. It is likely that the privations of fieldwork contributed to the illness that prematurely took Smalley's life on 11 July 1870 at the comparatively youthful age of 48 (Wood 1976).

After Smalley's death, Henry Chamberlain Russell was appointed Government Astronomer of New South Wales, and he went on to play a leading role in the development of Australian astronomy and meteorology. He also became a major force in the evolution of scientific societies in New South Wales (see Bhathal 1991). Soon after his appointment, Russell wrote to Tebbutt and complained that because Smalley had devoted himself so fully to the trigonometrical survey, little of an astronomical nature had been achieved at the Observatory during his tenure: no new instruments were added apart from a chronograph, and no astronomical results were published (Russell 1870a). Obviously Russell had forgotten Smalley's paper on Comet 2P/Encke! Tebbutt's reply must have been even more critical of Smalley, for Russell's response to it contained a vigorous defence of the trigonometrical survey work and outlined the innumerable problems that Smalley had encountered (Russell 1870b).

Having refused the Sydney Observatory post, Tebbutt determined to pursue astronomy as an independent astronomer (see N.S.W. 1893), and despite this apparent impediment was able to achieve an international reputation for himself and thus fulfil Scott's prophesy. In this regard, it is important to note that the range of objects

**Fig. 7.15** Melbourne Observatory (*Courtesy* Peter Hingley)

and events that he observed between 1853 and 1862 was to prove pivotal in his subsequent 'career' as an astronomer.

From the start, Tebbutt welcomed the open co-operation that was offered by his professional colleagues at both the Sydney and Melbourne Observatories, the only government astronomical observatories in existence in Australia at the time. In October 1862, Russell wrote that 'It gives me great pleasure to be able to comply to your request' (Russell 1862), and provided Tebbutt with cometary information. In January 1863, he told Tebbutt that he was very happy to compare barometric readings from the Sydney and Windsor Observatories (Russell 1863).

Those at Melbourne Observatory (Fig. 7.15) were even more forthright. On 7 July 1865, Robert Ellery (Bio Box 7.3), the Government Astronomer of Victoria, supplied Tebbutt with requested transit places of stars, and ended his letter with the following pledge: 'In conclusion I would assure you it will always give us great pleasure to give you any information or assistance in our powers that you may require' (Ellery 1865a). On 31 August 1865 he sent Tebbutt more star positions, plus observations of the Great Southern Comet of 1865 (Ellery 1865b), and further observations, made by White (Bio Box 6.3) with the 4.5-in. (11.4-cm) refractor, followed on 31 October (Ellery 1865c). In 1868 Ellery wrote:

> I am very glad indeed to find our Volume of Observations has been invaluable to you … Hoping to hear of the successful determination of your latitude & with the assurance to you that no apology is needed for asking anything astronomical of us. (Ellery 1868).

For the most part, this healthy attitude of professional–non-professional co-operation was to persist in Australia through into the twentieth century.

**Bio Box 7.3: Robert Lewis John Ellery**

Robert Lewis John Ellery (Fig. 7.16) was born at Cranleigh, Surrey, on 1827 July 14, the son of a surgeon (W.T.L. 1908). After completing his schooling he studied medicine, but through friends at the Royal Observatory, Greenwich, developed an interest in astronomy (Gascoigne 1992). In 1851 fabulous stories of gold discoveries in Victoria lured him to Australia, and 2 years later he was appointed Director of the Colonial Government's new Williamstown Observatory in Melbourne (Andropoulos 2014). In 1863 both the Williamstown and Flagstaff Observatories were closed, and the astronomical, timekeeping, meteorological, geomagnetic and trigonometrical survey functions of the two institutions were centralised in a larger facility, Melbourne Observatory. As the Government Astronomer of Victoria, Ellery assumed the directorship of this new institution and he continued in that capacity until his retirement in 1895. During this time, Melbourne Observatory snared the 'Great Melbourne Telescope' (see Gillespie 2011; Warner 1982), and Ellery built his institution into the most prestigious government observatory in Australia (see Gascoigne 1992).

In addition to astronomy, Ellery was particularly active in meteorology and trigonometrical survey work (he was Director of the Geodesic Survey from 1858 to 1874). He also was a leading force in the Royal Society of Victoria, serving on its Council from 1863 to 1905 and as President between 1867 and 1884 (Annual Report … 1908). He had earlier (in 1859) been elected a Fellow of the Royal Astronomical Society, and in 1873 was honoured when appointed a Fellow of the Royal Society (Obituaries 1909b). In addition to his formal duties at the Observatory, Ellery organised and for a time commanded the Victorian Torpedo Corps, and retired in 1889 with the rank of Lieutenant-Colonel (ibid.).

Robert Lewis John Ellery's '… kindly nature and fund of humour made him a universal favourite …' (Annual Report … 1908), and he passed away on 14 January 1908. For a detailed account of Ellery's work see Gascoigne (1992).

**Fig. 7.16** Robert Lewis John Ellery (after Gascoigne 1992)

**Acknowledgements** I am grateful to Iwona Hetherington (Museum of Applied Arts and Sciences, Sydney), the late Peter Hingley (Royal Astronomical Society, London), Professor Nick Lomb (University of Southern Queensland, Toowoomba) and staff at the Mitchell Library (Sydney) for their assistance in various ways, and to the late Peter Hingley, Darunee Lingling Orchiston, the Mitchell Library and the Museum of Applied Arts and Sciences for kindly supplying Figs. 7.3, 7.5, 7.9, 7.12, 7.14 and 7.15.

## References[1]

Airy, W. (Ed.) (1896). *Autobiography of Sir George Biddell Airy, K.C.B.* Cambridge: Cambridge University Press

Andropoulos, J. (2014). *The Astronomical Publications of Melbourne Observatory*. Ph.D. Thesis, James Cook University, Townsville, Australia.

Annual Reports: Mr. Tebbutt's Observatory … *Monthly Notices of the Royal Astronomical Society, 68,* 287 (1908).

Baracchi, P. (1914). Astronomy and geodesy in Australia. In G.H. Knibbs (Ed.), *Federal Handbook Prepared in Connection with the Eighty-fourth Meeting of the British Association for the Advancement of Science, Held in Australia, August, 1914* (pp. 326–390). Melbourne: Government Printer.

Bhathal, R. (1991). Henry Chamberlain Russell – astronomer, meteorologist and scientific entrepreneur. *Journal and Proceedings of the Royal Society of New South Wales, 124,* 1–21.

Bull, J. (1858). The comet. *Sydney Morning Herald,* 21 October.

Cable, K.J. (1972). Garnsey, Charles Frederick (1828–1894). In N.B. Nairn, A.G. Serle & R.B. Ward (Eds.), *Australian Dictionary of Biography. Volume 4: 1851–1890. D-J* (pp. 232–233). Melbourne: Melbourne University Press.

Chapman, A. (1993). An occupation for an independent gentleman: astronomy in the life of John Herschel. *Vistas in Astronomy, 36,* 71–116.

Clarke, W.B. (1851). Observations made at Paramatta during the eclipse of the Sun on Saturday, February 1, 1851. *Monthly Notices of the Royal Astronomical Society, 11,* 223–225.

Clarke, W.B. (1857). Observations made at St. Leonard's, North Shore, Sydney, during the eclipse of the Sun, March 26th, 1857. *Monthly Notices of the Royal Astronomical Society, 18,* 39–44.

Clarke, W.B. (1858). The six comets of 1858. *Sydney Morning Herald,* 29 October.

Currey, C.H., 1972. Denison, Sir William Thomas (1804–1871). In N.B. Nairn, A.G. Serle & R.B. Ward (Eds.), *Australian Dictionary of Biography. Volume 4: 1851–1890. D-J* (pp. 46–53). Melbourne: Melbourne University Press.

Denison, Sir William, 1858. The Moon's rotation. *The Sydney Magazine of Science and Art, 1,* 43–44.

Ellery, R.J. (1865a). Letter to John Tebbutt, dated 7 July. In TL.

Ellery, R.J. (1865b). Letter to John Tebbutt, dated 31 August. In TL.

Ellery, R.J. (1865c). Letter to John Tebbutt, dated 31 October. In TL,

Ellery, R.J. (1868). Letter to John Tebbutt, dated 1 April. In TL.

Ellery, R.J. (1901). A brief history of the beginnings and growth of astronomy in Australasia. In *Report of the Eighth Meeting of the Australasian Association for the Advancement of Science* (pp. 1–17). Melbourne: AAAS.

Evans, D.S. (1988). *Under Capricorn. A History of Southern Hemisphere Astronomy*. Bristol: Adam Hilger.

Gascoigne, S.C.B. (1992). Robert L.J. Ellery, his life and times. *Proceedings of the Astronomical Society of Australia, 10,* 170–176.

[1] The following abbreviation is used: SL = Sydney Observatory Outward Letterbooks. TL = Letters to J. Tebbutt, bound manuscript letters in the Mitchell Library, Sydney.

Gillespie, R. (2011). *The Great Melbourne Telescope*. Melbourne: Museum Victoria.
Glass, I.S. (2015). *The Royal Observatory at the Cape of Good Hope. History and Heritage*. Cape Town: Mons Mensa.
Hawkins, H.S. (1858). The comet. *Sydney Morning Herald,* 16 November.
Haynes, R., Haynes, R., Malin, D. & McGee, R. (1996). *Explorers of the Southern Sky. A History of Australian Astronomy*. Cambridge: Cambridge University Press.
Ionaramus (1858). The comet. *Sydney Morning Herald,* 29 October.
Lector (1858). The comet. *Sydney Morning Herald,* 21 October.
Lenehan, H.A. (1904). Letter to John Tebbutt, dated 19 January. In TL.
Lenehan, H.A. (1906). Letter to John Tebbutt, dated 17 February. In TL.
Lomb, N. 2011. *Transit of Venus: 1631 to the Present.* Sydney: NewSouth & Powerhouse Museum.
Macdonnell, W.J. (1908). Obituary. H.A. Lenehan, F.R.A.S. *Journal of the British Astronomical Association, 18,* 385–386.
Maguire, R. (1985). Angelo Tornaghi. An inventive Italian of the nineteenth century. *The Australian Antique Collector, 29,* 42–46.
McIntosh, R.A. (1935). An index of southern meteor showers. *Monthly Notices of the Royal Astronomical Society, 95,* 709–718.
Marsden, B.G. (1982). *Catalogue of Cometary Orbits*. Cambridge (Mass.): Minor Planet Center.
Moyal, A. (1976). *Scientists in Nineteenth Century Australia: A Documentary History*. Sydney: Cassell Australia.
Nangle, J. (1930). *The Sydney Observatory: Its History and Work*. Sydney: Sydney Technical College.
Norton, A.P., (1959). *Norton's Star Atlas and Reference Handbook. Fourteenth Edition*. Edinburgh: Gall & Inglis.
N.S.W. (1893). A self-taught astronomer. *The Australasian,* 26 August.
Obituaries. Lieut.-General Sir William Denison, K.C.B., R.E. *Monthly Notices of the Royal Astronomical Society, 32,* 119–120 (1872).
Obituaries. Henry Alfred Lenehan. *Monthly Notices of the Royal Astronomical Society, 69,* 248 (1909a).
Obituaries. Robert Lewis John Ellery. *Monthly Notices of the Royal Astronomical Society,* 69, 245–246 (1909b).
Orchiston, W. (1988a). From research to recreation: the rise and fall of Sydney Observatory. *Vistas in Astronomy, 32,* 49–63.
Orchiston, W. (1988b). Illuminating incidents in Antipodean astronomy: formation of the New South Wales Branch of the British Astronomical Association. *Journal of the British Astronomical Association, 98,* 75–84.
Orchiston, W. (1988c). Illuminating incidents in Antipodean astronomy: John Tebbutt and the Sydney Observatory directorship of 1862. *Australian Journal of Astronomy, 2,* 149–158.
Orchiston, W. (1988d). Illuminating incidents in Antipodean astronomy: Phillip Parker King and the founding of Sydney Observatory. *Vistas in Astronomy, 32,* 285–301.
Orchiston, W. (1997). The role of the amateur in popularising astronomy: an Australian case study. *Australian Journal of Astronomy, 7,* 33–66.
Orchiston, W. (1998). Mission impossible: William Scott and the first Sydney Observatory directorship. *Journal of Astronomical History and Heritage, 1,* 21–43.
Orchiston, W. (2004). The nineteenth century transits of Venus: an Australian and New Zealand overview. *Journal of Astronomical Data, 10,* 219–308.
Orchiston, W. (2015). The amateur-turned-professional syndrome: two Australian case studies. In W. Orchiston, D.A. Green & R. Strom (Eds.), *New Insights from Recent Studies in Historical Astronomy: Following in the Footsteps of F. Richard Stephenson* (pp. 259–350). Cham: Springer.
Orchiston, W. (2016). *Exploring the History of New Zealand Astronomy: Trials, Tribulations, Telescopes and Transits*. Cham: Springer.

Parliament of New South Wales. Legislative Council, Wednesday, October 29. *The Empire,* 30 October 1862.
Phillips, T.E.R. (Ed.) (1923). *Hutchinson's Spendour of the Heavens. Volume 1.* London: Hutchinson & Co.
Ranyard, A.C. (1879). Observations made during total solar eclipses. *Memoirs of the Royal Astronomical Society, 41,* 1–792 (plus plates).
Reports of the Branches. New South Wales Branch. *Journal of the British Astronomical Association, 5,* 343–345 (1895).
Russell, H.C. (1862). Letter to John Tebbutt, dated 28 October. In TL.
Russell, H.C. (1863). Letter to John Tebbutt, dated 8 January. In TL.
Russell, H.C. (1870a). Letter to John Tebbutt, dated 23 November. In TL
Russell, H.C. (1870b). Letter to John Tebbutt, dated 29 November. In TL.
Russell, H.C. (1882). *The Sydney Observatory. History and Progress.* Sydney: Government Printer.
Russell, H.C. (1892*). Observations of the Transit of Venus, 9 December 1874, Made at Stations in New South Wales.* Sydney: Government Printer.
Saunders, S.D. (1990). *Astronomy in Colonial New South Wales: 1788 to 1858.* Ph.D. Thesis, Department of General Philosophy, University of Sydney, Sydney, Australia.
Scott, W. (1858a). The comet. *Sydney Morning Herald,* 23 October.
Scott, W. (1858b). The comet. *Sydney Morning Herald,* 26 October.
Scott, W. (1858c). Letter to the Colonial Secretary, dated 20 December. In SL.
Scott, W. (1858d). On the meteorology of New South Wales, No. 2. *The Sydney Magazine of Science and Art, 2,* 118–121.
Scott, W. (1859a). Letter to the Reverend J.M. O'Neill, dated 11 February. MS, Mitchell Library, Sydney (ML DOC 50).
Scott, W. (1859b). Southern aurora. *Sydney Morning Herald,* 30 August.
Scott, W. (1860a). Letter to John Tebbutt, dated 5 April. In TL.
Scott, W. (1860b). Letter to John Tebbutt, dated 18 April. In TL.
Scott, W. (1860c). Letter to John Tebbutt, dated 3 August. In TL.
Scott, W. (1860d). Letter to the Reverend J.M. O'Neill, dated 15 April. MS, Mitchell Library, Sydney (ML DOC 50).
Scott, W. (1861a). Letter to John Tebbutt, dated 22 January. In TL.
Scott, W. (1861b). Letter to the Colonial Secretary, dated 2 February. In SL.
Scott, W. (1861c). Letter to the Secretary, Royal Astronomical Society, dated 19 February. In SL.
Scott, W. (1861d). Letter to J. Maclear, dated 18 July. In SL.
Scott, W. (1861e). Letter to John Tebbutt, dated 10 August. In TL.
Scott, W. (1861f). Letter to G.B. Airy, dated 8 September. In SL.
Scott, W. (1861g). Letter to John Tebbutt, dated 12 December. In TL.
Scott, W. (1862a). 1862a. Letter to the Colonial Secretary, dated 23 January. In SL.
Scott, W. (1862b). Letter to the Colonial Secretary, dated 11 April. In SL.
Scott, W. (1862c). Letter to the Colonial Secretary, dated 9 May. In SL.
Scott, W. (1862d). Letter to John Tebbutt, dated 19 June. In TL.
Scott, W. (1875). The Windsor Observatory. *Sydney Morning Herald,* 24 March.
Severn, H.A. (1858a). The comet. – No. 1. *Sydney Morning Herald,* 26 October.
Severn, H.A. (1858b). The comet. No. 2. *Sydney Morning Herald,* 16 November.
Smalley, G.R. (1862–65). The theory of Encke's Comet. *Transactions of the Philosophical Society of New South Wales,* 330–338.
Smalley, G.R. (1864a). Letter to John Tebbutt, dated 30 April. In TL.
Smalley, G.R. (1864b). Letter to John Tebbutt, dated 12 August. In TL.
Steele, J. (1916). *The Early Days of Windsor.* Sydney: Tyrrells Steele.
Tebbutt, J. (1853–59). *Observations made … at the Peninsula, near Windsor, in the Colony of New South Wales.* MS, Mitchell Library, Sydney (A3745$^{-1}$).
Tebbutt, J. (1854). Solar phenomena. *Sydney Morning Herald,* 13 May.
Tebbutt, J. (1857a). The eclipse. *Sydney Morning Herald,* 1 April.
Tebbutt, J. (1857b). The eclipse. *Sydney Morning Herald,* 2 October.

Tebbutt, J. (1857c). The forthcoming eclipse. *Sydney Morning Herald,* 17 March.
Tebbutt, J. (1857d). The solar eclipse. *Sydney Morning Herald,* 12 March.
Tebbutt, J. (1858a). The comet. *Sydney Morning Herald,* 29 October.
Tebbutt, J. (1858b). Letter to the Editor. *Sydney Morning Herald,* 17 March.
Tebbutt, J. (1859). *The aurora australis. Sydney Morning Herald*, 1 September.
Tebbutt, J. (1859–63). Diary. MS, Mitchell Library, Sydney (A3785).
Tebbutt, J. (1860a). The comet. *Sydney Morning Herald,* 23 July.
Tebbutt, J. (1860b). The comet's orbit. *Sydney Morning Herald,* 3 August.
Tebbutt, J. (1861a). Eclipse of the Sun. *Sydney Morning Herald,* 4 January.
Tebbutt, J. (1861b). The solar eclipse. *Sydney Morning Herald,* 15 January.
Tebbutt, J. (1866). The Government Astronomer's report. *Sydney Morning Herald,* 8 September.
Tebbutt, J. (1875). The Windsor Observatory. *Sydney Morning Herald,* 27 March.
Tebbutt, J. (1887). *History and Description of Mr. Tebbutt's Observatory, Windsor, New South Wales*. Sydney: printed for the author.
Tebbutt, J. (1908). *Astronomical Memoirs*. Sydney: printed for the author.
The Comet (1858). The comet. *Sydney Morning Herald,* 23 October.
"The coming eclipse." *Sydney Morning Herald,* 5 September 1857.
Tornaghi, A. (1862). Letter to John Tebbutt, dated 31 May. In TL.
Warner, B. (1982). The large Southern Telescope: Cape or Melbourne. *Quarterly Journal of the Royal Astronomical Society, 23,* 505–514.
Weiss, E., 1888. *Bideratlas der Sternenwelt. Eine Astronomie für Jedermann*. Stuttgart: Schreiber.
Wood, H. (1958). *Sydney Observatory 1858 to 1958*. Sydney: Government Printer (Sydney Observatory Papers No. 31).
Wood, H.W., (1976). Smalley, George Robarts (1822–1870). In G. Serle & R. Ward (Eds.), *Australian Dictionary of Biography. Volume 6: 1851–1890. R-Z* (pp. 136–137). Melbourne: Melbourne University Press.
Wood, H.W. (1982). *Sydney Observatory 1958 to 1981*. Sydney: Government Printer.
Wood, H.W. (1983). Sydney Observatory 1858 to 1983. *Proceedings of the Astronomical Society of Australia, 5,* 273–281.
Wood, H.W. (1986). Lenehan, Henry Alfred (1843–1908). In B. Nairn & G. Serle (Eds.), *Australian Dictionary of Biography. Volume 10: 1891–1939. Lat-Ner* (pp. 73–74). Melbourne: Melbourne University Press
W.T.L. (1908). Mr. Ellery, C.M.G., F.R.S. *Journal of the British Astronomical Association, 18,* 177.

# Chapter 8
# From Humble Beginnings: The Development of Windsor Observatory

## 8.1 Introduction

John Tebbutt's astronomical 'career' began in May 1853 with the appearance of a conspicuous naked eye comet. At that time, he was furnished with nothing more than the family's excellent Scottish 8-day grandfather clock (Tebbutt 1908, p. 11), and a small telescope described as '… a very good glass by Lynch & Co. of 36, Westmoreland Street, Dublin, constructed after the fashion of the ordinary navy telescope. Its clear aperture is 15/8 in. [4.1 cm]' (Tebbutt 1873). This telescope is shown in Fig. 8.1. Abrahams (2002) documents the existence of three successive James Lynches who were actively involved in scientific instrument-making in Dublin during the eighteenth and nineteenth centuries. James Lynch II died in 1833 and subsequently 'Lynch & Co.' (presumably under James Lynch III), operated in 1841–1842, and in 1845–1846 was replaced by 'James Lynch & Co.' On the basis of this rather limited evidence, Mollan (1995) dates the manufacture of Tebbutt's marine telescope to around 1841–1842, but it could have been a year or 2 later. In an interesting paper titled 'Telescopes for land and sea', Deborah Jean Warner (1998) explores the small portable telescopes that were made during the eighteenth and early nineteenth centuries for terrestrial and maritime use, and she identifies the following types: perspective glasses, pocket telescopes and day and/or night telescopes. Tebbutt's marine telescope is an example of a 'day and night telescope' designed for hand-held use, but Tebbutt (1879) later states that it frequently was used on a rough mounting, although he does not elaborate on this.

In September 1853, Tebbutt went to Sydney (Tebbutt 1853–59) and purchased a sextant of unknown provenance and manufacture. Much later, he was to describe this as a 7½-in. sextant (Tebbutt 1902).

Over the next 8 years Tebbutt would train his marine telescope, and from time to time the sextant, on comets, the planet Jupiter and its satellites, lunar occultations and eclipses, double stars, variable stars and the Sun. He also carried out naked eye observations of meteors, the zodiacal light and aurorae (Tebbutt 1853–59, 1859–63).



W. Orchiston, *John Tebbutt*, Historical & Cultural Astronomy,
DOI 10.1007/978-3-319-44521-2_8

**Fig. 8.1** A 31 October 1981 photograph of John Tebbutt's Lynch & Co. marine telescope, which is still owned by the Tebbutt Family (*Photograph* Wayne Orchiston)

Fate allowed his growing local reputation to attain international proportions in 1861, when Tebbutt discovered his first comet. Later known as the Great Comet of 1861, Comet C/1861 J1 (Tebbutt) turned out to be one of the most magnificent celestial visitors of the century and it was this discovery more than any other event which emphasised Tebbutt's urgent need for proper instrumentation if he wished to make any further contribution to astronomy. Thus, it was that he decided to purchase an astronomical telescope and to erect an observatory.

This chapter draws on material in Orchiston (1988c, 2001b) and documents the succession of astronomical instruments that Tebbutt acquired in the course of his long life as an astronomer. It also traces the development of Windsor Observatory from the first simple building constructed in 1863 to the full complex of buildings that graced the grounds of the Peninsula Estate in 1894 following the erection of the new 'equatorial house' for his 8-in. (20.3-cm) Grubb refractor. But telescopes and associated accessories merely provided Tebbutt with the means to achieve his observational goals, so we also examine the changing pattern of research that accompanied the growth of Windsor Observatory. To provide a more realistic picture of the entire research enterprise, there are also brief accounts of Tebbutt's meteorological and tidal instruments.

In its heyday, Windsor Observatory had an international reputation and was one of Australia's most productive astronomical institutions (Haynes et al. 1996). It has been likened by Ashbrook (1984, p. 66) to a '... one man Greenwich Observatory in the southern hemisphere'.

## 8.2 A Serious Commitment to Observational Astronomy: Purchase of the Jones Refractor

Immediately after discovering the 1861 comet, Tebbutt approached William Scott (Bio Box 6.1) at Sydney Observatory to see if he could obtain the old 3.25-in. (8.3-cm) refractor that had originally been at Parramatta Observatory. He knew that Scott was busy bringing the new 7.25-in. (18.4-cm) Merz refractor into operation and must have presumed that the old refractor was no longer required. But this was not so, for Scott wrote: 'I am sorry to tell you that the old Equatorial Instrument is the only one I am unable to part with as it is necessary to have some Instrument to observe phenomena hidden from the Dome by the Time Ball Tower' (Scott 1861). So Tebbutt was forced to persevere with his sextant and small marine telescope but on 14 November 1861, just 3 months after farewelling 'his' comet, he purchased his very own 3.25-in. (8.3-cm) refractor from the Sydney firm of Brush and Macdonnell. He has provided a useful description of this telescope:

> This instrument is the work of Chas. Jones of Strand Street, Liverpool, England; the clear aperture of the object glass is 3¼ inches [8.3 cm] and its focal length about 48 inches [1.22 m]… The telescope is provided with a brass tripod stand with altitude and azimuth movement … The telescope has a finder of about 11 inches [27.9 cm] focal length and $^{7}/_{8}$ inch [2.2 cm] clear aperture. It is also furnished with a terrestrial eyepiece with extra tube, [and] six negative eyepieces with powers varying from 30 to 120 … (Tebbutt 1873).

The telescope tube was made of brass (Tebbutt 1879).

At this time the leading manufacturers of refracting telescopes were Merz of Germany and the two English firms of Cooke, and Troughton and Simms (see van Helden 1984). In his 'Cyclopaedia of Telescope Makers', Andrews (1993) lists two quite different firms of astronomical telescope-makers with the name Jones who were active at about this time, but both were based in London. There is no 'Chas. Jones' of Liverpool listed, so he certainly was not one of the 'name' manufacturers. Consequently, his telescopes would have been better priced than more favoured models, and although this may have influenced Tebbutt's decision it also is likely that he was forced to select from the stock available at Brush and Macdonnell's shop in Sydney and the Jones refractor was the best option. At the time of the acquisition, Tebbutt (1859–63) described it as a splendid telescope, and in later years he never voiced any regrets regarding the purchase. To the contrary, he spoke of the objective as '… one of undoubted excellence' (Tebbutt 1873).

It is important to place the Jones refractor in a national and an international context. At the end of 1861, the equal-largest refracting telescopes in the world were the 15-in. (38-cm) refractors at Pulkovo and Harvard College Observatories (Welther 1984), while Australia's largest refractor was the 7.25-in. Merz (Fig. 6.5) just acquired by Sydney Observatory (see Wood 1958). Prior to this, Scott had been forced to rely on the old 3.25-in. Parramatta Observatory refractor. The only other Australian independent astronomer of note at this time was Francis Abbott of Hobart (Bio Box 6.4), and his main instrument was a 4-in. (10.2-cm) refractor (Orchiston 1992).

The Jones telescope may have been small, but it was capable of valuable work and to add to its usefulness Tebbutt had the Sydney instrument-maker Angelo Tornaghi (Bio Box 8.1) construct a ring micrometer so that he could carry out accurate positional observations (for a history of micrometer development see Brooks 1991). On 21 February 1862 Tornaghi (1862) advised Tebbutt that the micrometer was ready and that Scott had examined it and found it to be correct. Tornaghi also included an invoice which revealed that Tebbutt paid £3/5/0 for the micrometer and associated eyepiece. The new appliance arrived in Windsor on the 23rd, and Tebbutt (1859–63) noted in his diary

**Bio Box 8.1: Angelo Tornaghi**

Angelo Tornaghi (Fig. 8.2; Maguire 1985) was born in Milan, Italy, in 1831, and emigrated to Sydney in 1855. There he served as the local agents for the London firm of Negretti & Zambra (Barker 2013), but he also designed and manufactured a range of scientific instruments, taking out a number of patents. During the 1860s, Tornaghi sold various scientific instruments to Sydney Observatory, the New South Wales Lands Department and Windsor Observatory. By May 1864 he '... was running a highly respected business *importing and making* scientific instruments ... [and] Just 2 months later his instruments were praised by the Royal Society of New South Wales as being equal to those of European manufacture' (Barker 2010; my italics).

**Fig. 8.2** Angelo Tornaghi After Digby 1889–1890, p. 226; (*Courtesy* Research Library, Museum of Applied Arts and Sciences, Sydney)

At this time, Tornaghi '... gave his occupation as Scientific Instrument Maker, Optician and Clockmaker depending on the current emphasis of his work' (Maguire 1985, p. 43), and although his scientific instruments generally were of high quality, they tended to be '... plain and practical and precise rather than decorated or ornamental ...' (Maguire 1985, p. 44).

However, Tornaghi took business risks and by mid-1866 had crippling debts, even owing Negretti & Zambra £2,000 (Maguire 1985). Consequently, he was forced to sell all of the stock in his Sydney shop, which included clocks, watches and astronomical, mathematical, optical and surveying instruments. After this, he focussed his instrument-making activities on the construction and/or maintenance of public clocks throughout New South Wales (Barker 2013). In the 1887 Sands Directory, Tornaghi is only listed as a clockmaker (Maguire 1985).

Apart from supplying instruments to Sydney and Windsor Observatories, and encouraging Tebbutt to accept the Sydney Observatory Directorship when Scott vacated the position, Tornaghi's astronomical interests extended to the 1874 transit of Venus, and he was a member of the Sydney Observatory party that went to Goulburn and successfully observed the event (Russell 1892).

Four years later Tornaghi entered local politics, and the following year became the Mayor of Hunters Hill in suburban Sydney (Barker 2013). Prior to his death in Sydney in 1906 he was awarded the Cross of Italy by his country of birth for his services to the Italian community in Sydney (Barker 2010). Roslyn

(continued)

Maguire (1985, p. 46) concludes her biographical essay on Tornaghi with the following evaluation: 'Angelo Tornaghi deserves recognition for his various and inventive contributions to the colony not only as a successful importer, merchant, local politician and philanthropist but as a nineteenth century scientist, inventor and skilled instrument and clock maker'. For us, we should remember him for his contribution to nineteenth-century New South Wales astronomy.

that '... it seems to be well executed'. In 1866 Tebbutt ordered a second ring micrometer from Tornaghi (1866a), and although there were few delays on this occasion, when it arrived in November 1866 Tebbutt found it faulty and returned it (Tornaghi 1866b). One week later it was back in Windsor and working perfectly (Tornaghi 1866c).

Armed with his new telescope and the first micrometer, Tebbutt was soon able to carry out his first micrometric observations, as revealed in his *Astronomical Memoirs*:

> I soon afterwards found an opportunity for turning my modest equipment to account. In the beginning of September [1862] a comet discovered by Swift in the United States on July 15, came south and was a conspicuous object to the naked eye. My telescope was as yet only mounted on a tripod stand without circles of any kind, its location being under the verandah of my father's residence ... The comet was compared with stars by means of one of the ring micrometers. The comparison stars were identified by observing their transits across the diaphragm of the telescope in connection with other stars plainly visible to the naked eye, these latter being themselves identified by sextant observations. The difficulties thus surmounted can well be appreciated by practical astronomers. (Tebbutt 1908, p. 26).

The positions he obtained were used to calculate the orbital elements of what later became known as periodic Comet 109P/Swift-Tuttle, and in a classic case of publication overkill Tebbutt submitted his results in five separate research papers to *Astronomische Nachrichten*, *Monthly Notices of the Royal Astronomical Society* and *Transactions of the Philosophical Society of New South Wales*. Those in the first two journals were his very first scientific papers to be published in professional international journals and, to use his own words, '... constituted my first attempt to bring my work before the professional astronomers of the northern hemisphere' (Tebbutt 1908, p. 27). As such, they must have brought him a great deal of satisfaction. However, Tebbutt had one reservation: the Editor of the German journal described him as the Director of Windsor Observatory, which he found very annoying considering that his 'astronomical observatory' was merely a portable one (Tebbutt 1863). He should have been flattered by the compliment, even if it was unintended.

Tebbutt now sought to extend his work beyond astronomy into meteorology—as was the custom with the Australian colonial observatories at this time (see Orchiston 1988a)—and towards the end of 1862 he ordered a set of meteorological instruments. Other instruments arrived in 1863, and by the end of that year he owned a barometer, a hygrometer, ten different thermometers, a rain gauge and two evaporometers. Most of these (see Fig. 1.3) were installed near the new observatory referred to below in Sect. 8.3. A second rain gauge was purchased in 1871 or 1872. Detailed descriptions of all of these instruments are provided in Tebbutt (1873). Tebbutt quickly brought these instruments into operation and at the end of January 1863 was pleased to be able to release his first monthly meteorological report (see Russell 1863a).

## 8.3 'Wholly the Work of My Own Hands': The First Windsor Observatory Building

From mid-1863 Tebbutt had been giving serious thought to the adequate housing of his astronomical equipment, and what was needed was an observatory. Up to that time, he had persevered with the Jones refractor, using it on the verandah of the Tebbutt Homestead or outdoors—often under temporary shelter (Tebbutt 1879)—but in these situations it was subject to the vagaries of wind and light pollution. There was also the problem of stability that is associated with all telescopes featuring portable mountings, and the basic nuisance of having to take out the telescope and set it up every time observations were to be made. Clearly, what was needed was a permanent home, and even more so since he wished to also install a transit telescope and establish a local time service (see Stott and Hughes 1987). Thus, it was that Windsor Observatory came into existence.

Donnelly (1973, p. 56) points out that by the end of the eighteenth century '… the development of astronomical instruments of great sensitivity compelled architects to recognize the necessity of stable mounting and controlled interior climate was well as the differentiation of rooms for the various instruments'. Although no architect, Tebbutt was aware of these factors and took them into account when designing his first Windsor Observatory building. This was constructed towards the end of 1863, a little west of the Tebbutt Homestead (see Fig. 1.9). It became operational in January 1864, and '… was wholly the work of my own hands, and in accomplishing it I combined in my own person the handicrafts of the carpenter, the bricklayer and slater' (Tebbutt 1908, p. 29). Always a practical man when it came to new challenges, Tebbutt went and purchased two useful books before he began this project: *The Carpenter's Manual* and *Masons, Bricklayers, Plasterers, & Slater's Assistant* (see Tebbutt n.d.).

The resulting observatory comprised a 'transit room', and above it an 'equatorial room' which was accessed via stairs which led up from the transit room. Detailed descriptions of this observatory are provided in Tebbutt (1873) and (1879), but as Cremin (1987) has pointed out they are somewhat misleading in that they imply a single phase of construction. In fact, the observatory was built in two stages, in 1863 and 1867, with a 'prime vertical room' being added in the latter year. Let us first examine the original Windsor Observatory building.

The larger of the two functional spaces was the transit room, which was rectangular in shape with the long axis running east–west. Internally it measured 20 ft (6.1 m) by 10 ft (3.05 m), and the ceiling height was 7 ft 2 in. (2.2 m). This room was primarily designed for the transit telescope (when it arrived), and originally contained a single (eastern) pier which was constructed in May–June 1864 (Tebbutt 1864–69). This extended from a clay foundation well below floor level and was made of brick and capped with stone, ready to accommodate the metal frame of a transit telescope.

The transit room was also designed to accommodate Tebbutt's small but growing astronomical library and to serve as a study (Tebbutt 1879). The western wall, therefore, featured two long narrow vertical windows, similar in placement and positioning to the transit slits on the adjacent walls. Apart from these, the building had no other windows.

The equatorial room rose in the form of an octagonal tower from the centre of the transit room and was topped by a conical dome. The walls of the equatorial room were 4 ft 4 in. (1.32 m) high and around the top was a 10-ft (3-m) diameter circular wooden plate capped with iron, upon which the dome rotated by means of eight small iron wheels. The octagonal dome was made of tongue-and-groove Oregon pine covered with canvas, and one of its segments comprised a pair of shutters which opened outwards to allow access to the sky. Tebbutt (1873) thought this arrangement very good, as it effectively excluded the rain. The equatorial room was constructed specifically for the 3.25-in. Jones refractor, which was supported by a heavy well-braced wooden platform that extended up from the clay beneath the building (Tebbutt 1879). This provided the necessary rigidity to minimise vibration. Attached to this platform was a rather crude English equatorial mounting which Tebbutt (1873) constructed for the telescope.

The newly-constructed Windsor Observatory acquired its first instrument when the Jones refractor was installed in the tower in early 1864, but on 24 August 1866 Tebbutt first reported an unforeseen design problem:

> In the usual position of the equatorial I could not bring the telescope to point quite to Jupiter owing to one of the rafters supporting the pin on which the revolving roof turns. By shifting the south pivot of the polar axis out of its V I, however, continued to follow the planet … (Tebbutt 1864–69).

In this way he was able to observe the Jovian satellite event. Further events were successfully observed from the tower observatory until the evening of 13 September 1866, when he was obliged to remove the Jones telescope from its equatorial mounting and place it back on its tripod in order to make the observation (ibid.). Again the offending rafter was the problem. From this date on, an area of the ecliptic seems to have been inaccessible from the tower observatory, and on numerous occasions Tebbutt had to remove the telescope from the equatorial mounting in order to secure the observations (e.g. see Tebbutt 1864–69). But what is equally obvious when one peruses Tebbutt's 1864–69 and 1870–73 observing journals is that only some observations were affected and that on many occasions the tower observatory functioned perfectly. Thus, there is little support for Cremin's (1987) claim that the telescope basically served as a portable instrument and that for reasons of his own Tebbutt '… chose not to emphasise this point, which may be perceived as affecting the credibility of his observations'. To Tebbutt, the accuracy of the observations was paramount, regardless of where the telescope was located.

In addition to the offending rafter, Tebbutt also began encountering problems with the platform that supported the equatorial mounting, and in April 1870 he finally cut away part of the ceiling of the transit room, so that it would not interfere with the platform (Tebbutt 1870–73).

On 11 April 1864, the scientific instruments at Windsor Observatory were added to when Tebbutt (1879) purchased an 8-day box chronometer from Dawsons of Sydney. Manufactured by Parkinson and Frodsham of London (Fig. 8.3), this turned out to be a very reliable timepiece. Presumably this is the very same chronometer that Dawsons had for sale 1 year earlier at £35, delivery free of charge (Dawson 1863). Fifteen years after purchasing this chronometer, Tebbutt (1879) proudly declared that it '… has never once become deranged since I acquired it'. One can only wonder about the symptoms of a deranged chronometer!

**Fig. 8.3** A Parkinson and Frodsham box chronometer of slightly earlier vintage than the one purchased by Tebbutt in 1864 (www.rubylane.com/Item/1475848-CONSC15027/Antique-1820s-Parkinson-Frodsham-chronometer-Serial)

Although Windsor Observatory could now function successfully, Tebbutt still lacked a transit telescope, thanks to a chain of unforeseen events. Back in 1863, before construction of the observatory began, he had attempted to obtain a suitable instrument through Sydney Observatory. Initially this approach seemed promising when Henry Russell wrote:

> In reference to the Transit Instrument we have one here which I think would suit you. The focal length is 30 inches [76.2 cm] the aperture 2¼ inches [5.7 cm], but I am not in a position to say it is for sale, as Mr. Scott's successor may want it, but if I obtain authority to sell it, I will let you know. (Russell 1863b).

The instrument that Russell referred to was originally at Parramatta Observatory and was no longer used after the ex-Parramatta Observatory meridian circle was refurbished in England and returned to Sydney Observatory (Wood 1958). As we saw in the previous chapter, ill health brought on by chronic overwork had prompted Scott to resign on 9 May 1862, and when Russell wrote Tebbutt about the transit telescope he was still waiting to hear who the new Government Astronomer of New South Wales would be.

A little over 3 months after making his offer, Russell (1863c) advised that although he had no authority to sell the transit telescope Tebbutt should make an offer and thus give him something to act on. There is no evidence that Tebbutt actually did so, but on 19 November 1863 Russell (1863d) announced that Smalley had been appointed Government Astronomer and that the 'surplus' transit instrument was no longer available. Whether there was a connection between the two is not apparent.

**Fig. 8.4** The transit telescope at Uckfield Observatory with a stand similar to the one Tornaghi made for Tebbutt (after Chambers 1867, p. 653)

After this unfortunate delay, Tebbutt turned to Brush and Macdonnell but their reply of 12 March 1864 indicated that they did not have any suitable transit telescopes in stock and that Tebbutt would not find any in Sydney. However, they could order one in from overseas and prices ranged from £23 to £47/5/0 (Brush and Macdonnell 1864).

Instead of choosing this option, Tebbutt turned once more to Tornaghi who accepted the challenge, and in mid-1864 he advised that the new transit instrument was about half-finished (Tornaghi 1864a). By 8 July 1864 it was so near completion that Tornaghi (1864b) exhibited it at the monthly meeting of the Philosophical Society of New South Wales, but various unexpected delays ensued (see Tornaghi 1864c, d) and Tebbutt only took delivery of it on 8 September 1864 (Tebbutt 1864–69). Tebbutt (1873) has provided a description of this instrument:

> The telescope, which is of brass, is supported between the pivots and has an object-glass of 2.1 inches [5.3 cm] clear aperture and 20 inches [50.8 cm] focal length. The setting circle is of brass but the graduated limb is of silver about 5½ inches [14 cm] diameter. The divisions are each equal to 20′ and are subdivided by two opposite verniers to 20″. The pivots are of gun metal about $1^{1}/_{16}$ inch [1.7 cm] in diameter and the distance between them 12 inches [30.5 cm].

The telescope was supported by means of a heavy metal stand which Tebbutt (1887) likened to the one at Uckfield Observatory represented on page 653 in the 1867 edition of Chambers' *Descriptive Astronomy* (see Fig. 8.4, here), and this was bolted to the capping stone on the pier in the transit room (Tebbutt 1879).

Tebbutt soon came to realise that the existing arrangement did not allow him to successfully investigate the latitude of Windsor Observatory, and so he decided to add a prime vertical room to the Observatory and to remount the transit telescope on a new double pier in the western end of the transit room. This would leave the original transit frame free for use in the prime vertical room, and 'The same telescope will then be used both for observations in the meridian and the prime vertical, the former for time and the latter for latitude' (Tebbutt 1864–69, 16 November 1867 entry).

Construction of an 8 ft 6 in. (2.6 m) square prime vertical room abutting the southern wall of the transit room was commenced on 27 September 1867 and on 5 December Tebbutt (1864–69) recorded its completion in his journal. This room contained a single pier, '… constructed of well grouted brickwork coated with cement, painted, and capped with stone on which is secured the metal frame of the transit instrument' (Tebbutt 1887). There was a door at the southern end of the prime vertical room, but the main entrance to the observatory was via the eastern end of the transit room, where there was a small porch covered by a curved corrugated iron roof.

While the prime vertical room was under construction, a double transit pier was installed at the western end of the transit room. This consisted of:

> … a solid mass of brick-work well grouted rising from a clay foundation nearly to the floor of the building. On this mass stand two piers also of brick-work well grouted and surmounted with stone caps to which the Ys of the transit instrument are secured. The telescope moves between the piers in this arrangement… (Tebbutt 1873).

Beneath a trap door in the floor the main body of the pier supported a trough of mercury that was used in connection with a Bohnenberger eyepiece and a level for determinations of collimation error (Tebbutt 1887). The western piers were coated with cement and were painted.

Meanwhile, in September 1867 Tornaghi (1867a) was employed to fabricate the necessary lateral extensions to the telescope tube, and the housings for these (referred to as 'Ys') which would be attached directly to the tops of the two piers. The telescope would then pivot between the piers, with the eastern Y adjustable for azimuth and the western one for level (Tebbutt 1873). After the almost inevitable unforeseen delays (Tornaghi 1867b) the modifications were dispatched on 16 November 1867 (Tornaghi 1867c), and the transit telescope was operational 2 days later (Tebbutt 1864–69). On the last day of the month Tornaghi (1867d) sent his invoice for his work; the amount was £11/7/0.

Collimation of the transit telescope was achieved by means of a meridian mark which on 20 November 1867 was installed on the bank of South Creek, due south of the western pier. Tebbutt (1887) describes this:

> A stout iron-bark post is sunk perpendicularly into the ground to the depth of about five feet [1.53 m], its top being nearly level with the northern sloping surface of the bank. About six inches [15.2 cm] below the top a piece of white porcelain is let into the post and over this is secured a plate of copper painted black, with a circular hole in it about a quarter of an inch [0.6 cm] in diameter. The porcelain is thus seen through the perforation from the Observatory as a small white spot on a black ground.

The first Windsor Observatory building was rather attractive, and the final (1867) version is shown from the south-east in Fig. 8.5, where the roof shutters in the transit room and the prime vertical room are obvious (as are stands supporting some of the meteorological instruments). The frame of the building was timber, and tongue-and-groove Oregon pine was used for all of the outer walls, to line the various rooms and for the floors and ceilings. The roof was covered with slate tiles. Apart from the roof, the whole building was painted white, both inside and outside.

**Fig. 8.5** John Tebbutt posing in front of the meteorological instruments, with his first observatory building in the background (Orchiston collection)

Tebbutt's Windsor Observatory design was both simple and functionally effective. The transit and prime vertical areas were separate and both were divorced from the elevated equatorial room. In this regard, the basic precepts that he followed mirrored those first introduced into observatory design during the 1780s and 1790s with the erection of Dunsink Observatory (1785) and Armagh Observatory (1791) in Ireland (see Donnelly 1973).

Meanwhile, Tebbutt's use of simple materials and the resulting rather bland appearance of the building, devoid of any embellishment or decoration, was in keeping with Cassini's views on observatory architecture. He believed that

> … an observatory could not be a work of architecture because all ornament would be foreign or superfluous to it and should not be allowed unless it would in no way interfere with the use of the building or unless a public monument was intended for which no expense would be spared. (Donnelly 1973, p. 59).

Paris Observatory Director Jean Dominique Cassini (1748–1845) also was adamant that all possible funds should be devoted to the instruments within rather than to the building itself (Donnelly 1973), and judging from Tebbutt's 1863 design it would appear that he shared these views.

The basic plan that Tebbutt chose of a central elevated equatorial tower with dome, flanked on three sides by lower 'wings' with roof openings for meridian-type work, was the norm (with various permutations) during the nineteenth century (e.g. see Donnelly 1973, pp. 63–68), and although the materials of construction were different, the Hopkins Observatory (Donnelly 1973, p. 75; Pasachoff 1998), New Cincinnati Observatory (Lockyer 1878, p. 338) and the Western Reserve Academy Observatory (Donnelly 1973, p. 77), amongst others, all bear a basic resemblance to the 1863 Windsor building.

**Table 8.1** Publications in Tebbutt's library acquired before 1863 and containing information about observatories

| | Publication | | |
|---|---|---|---|
| Author(s) | Year | Title | Place |
| Barlow, P. | 1849 | *Plane Astronomy* | London |
| Chamber, G.F. | 1861 | *Handbook of Descriptive and Practical Astronomy* | London |
| Delambre, M. | 1814 | *Astronomie Theorique et Practique* (Three volumes) | Paris |
| Ferguson, J. and Brewster, D. | 1821 | *Astronomy, Explained Upon Sir Isaac Newton's Principles* | Edinburgh |
| Herschel, J.F.W. | 1851 | *Outlines of Astronomy* | London |
| Lardner, D. and Dunkin, E. | 1860 | *Handbook of Astronomy* | London |
| Loomis, E. | 1861 | *An Introduction to Practical Astronomy* | New York |
| Olmstead, D. | 1853 | *Mechanism of the Heavens* | London |
| von Humboldt, A. | 1849–1852 | *Cosmos* (Four volumes) | London |
| von Littrow, J.J. | 1821 | *Theoretische und Practische Astronomie* (Three volumes) | Wien |

*Note*: This is an updated version of Table 1 which appeared in Orchiston 1988c

It is difficult to know if Tebbutt was influenced in any way by overseas observatories when he designed the first Windsor Observatory building. We know from his writings that at least some elements of later Windsor Observatory buildings were inspired by what he saw as successful overseas prototypes (e.g. see Tebbutt 1879), but by the close of 1862 he had access to few books or journals which would have provided documentation on the design features of observatories. Those that he could consult are listed in Table 8.1 (after Tebbutt n.d.). It should be noted that the acquisition date of the Loomis volume is actually not recorded in Tebbutt's library catalogue, but this was one of his most valued reference books, along with the volumes by Chambers and Lardner and Dunkin. If the time delay between the publication of these latter two books and their purchase is anything to go by, then Tebbutt would also have taken delivery of Loomis's book in 1862, 1 year after its publication. Since Tebbutt read widely, some of the books listed in this table may have provided him with inspiration and guidance. Most likely candidates would be those by Chambers, Herschel (both tomes), and Lardner and Dunkin.

Certainly, Tebbutt did not look locally for observatories to copy for—as reference to Figs. 3.4 and 3.7 demonstrate—his first Windsor Observatory building was totally unlike either Sydney Observatory (see Wood 1958) or Williamstown Observatory (see Ellery 1869). Moreover, at the time he was designing his Windsor Observatory, Tebbutt had only visited Sydney Observatory on two occasions (see Tebbutt 1859–1863, entries for 1 October 1860 and 13 August 1862), and this was the only Australian observatory that he had visited. Nor did he draw on Scott's knowledge of overseas observatory design, as an 1875 newspaper article indicates:

> Now I have no wish to receive any honour which does not justly belong to me, I therefore beg to state that Mr. Tebbutt is not indebted to me in any way for his reputation as an astronomer, *nor for the building of the Windsor Observatory*. Mr. Tebbutt was an astronomer of some repute before I made his acquaintance. Whether or not the establishment of the Sydney Observatory influenced Mr. Tebbutt indirectly, and so led to his building one for himself, it is impossible for me to say. … (Scott 1875; my italics).

Another interpretation is possible: that Tebbutt came up with an independent, original, observatory design—without any outside help. If he simply responded to the diverse needs of the various instruments he planned to install, it is reasonable to assume that he would in any case have arrived at an observatory design that was in basic conformity with overseas thinking at the time, given the obvious evolution in observatory architecture during the eighteenth and nineteenth centuries. A newspaper article published in 1893 suggests just this: that the observatory was built '… after his own ideas …' (N.S.W. 1893).

Whatever the facts of the matter, with the completion of the prime vertical room and installation of the transit telescope on its new mounting in 1867, Tebbutt had created a research-grade private observatory and was now in a position to make a useful contribution to observational astronomy. Given the modest equipment at his disposal and his own particular interests, he resolved to focus on

> … systematic observations of occultations of stars by the moon, and of eclipses of Jupiter's satellites, and of positions of comets, with attention to phenomena of an occasional character. (Tebbutt 1908, p. 30).

Dewhirst (1985, pp. 147–148) has raised the thorny question of the utility of many of the types of observations carried out in the name of positional astronomy during the nineteenth century:

> Initially the need was clear … But somewhere along the way the initial purpose became lost, and it is a sad feature of the history of nineteenth century meridian astronomy that much of the investment of money and of observing time resulted in a large body of factual data but not in significant astronomical discoveries commensurate with the investment.

As we shall see in Chap. 11, when Tebbutt chose to focus on specific aspects of observational astronomy he was very aware of their particular value and of how they could contribute to the science of astronomy.

## 8.4 A New Home for a New Telescope: The Second Wooden Observatory

For a number of years Tebbutt had been looking for a larger telescope than the Jones refractor, and he was eventually successful on 13 June 1872 (Tebbutt 1870–73) when he purchased a 4.5-in. (11.4-cm) Cooke refractor from Anthony Hordern of Sydney (see Teale 1972) for the sum of £100 (Tebbutt 1879, 1887). Thomas Cooke (Fig. 8.6) was the founder of the British firm, Thomas Cooke (from 1868, Thomas Cooke & Sons), which manufactured refracting telescopes, transit telescopes and

**Fig. 8.6** Thomas Cooke (https://en.wikipedia.org)

other astronomical equipment for professional and non-professional astronomers (Andrews 1992; Gegen 1994; McConnell 1992).

The Horderns were members of a highly-influential and very successful Sydney business family, with a succession of Anthony Horderns managing various components of their business empire. Undoubtedly, the astronomer was either Anthony III (1842–1886) or his father, Anthony II (1819–1876), also known as Anthony junior to differentiate him from his father, Anthony senior (1788–1869). Be that as it may, it is apparent that whichever Hordern owned the telescope found little time to indulge any simmering astronomical interests, because it was sold just 2 years after Cooke manufactured it to order in 1870 (Hordern 1870).

This was a fine instrument, except that it lacked a drive, but it was comparatively modest by world standards. At that time, the 18.5-in. (47-cm) Alvan Clark refractor at Dearborn Observatory in the USA was the largest in the world, although there soon would be a quantum jump in aperture (see Fig. 8.7) when Cooke completed a 25-in. (63.5-cm) instrument for the affluent Scottish-born British-based independent astronomer, Robert Stirling Newall (1812–1889). This was to be Cooke's largest telescope, and for 2 short years it was the world's largest refractor until surpassed in 1873 by the 26-in. (66-cm) Clark refractor at the U.S. Naval Observatory (see Welther 1984).

Meanwhile, in 1872 the largest refractor in Australia was still the Sydney Observatory 7.25-in. Merz, although it would be eclipsed in 1874—thanks to the upcoming transit of Venus—with the arrival of the 11.5-in. (29.2-cm) Schroeder at Sydney Observatory (Fig. 8.8) and the 8-in. (20.3-cm) Cooke (Fig. 3.10) and Troughton and Simms (Fig. 3.11) instruments at Adelaide and Melbourne Observatories, respectively (Orchiston 1988a).

Tebbutt (1887) provides a useful description of the 4.5-in. Cooke refractor:

> Its object glass has a clear aperture of 4½-inches [11.4 cm] and a focal length of 70 inches [1.78 m]. It is mounted according to the Fraunhofer [German Equatorial] method and is adjustable to any latitude. When it came into the possession of the writer it was mounted on a firm wooden tripod … At the lower or northern end of the polar axis is the hour circle, five inches [12.7 cm] in diameter. It has a double set of divisions on its edge, and when clamped

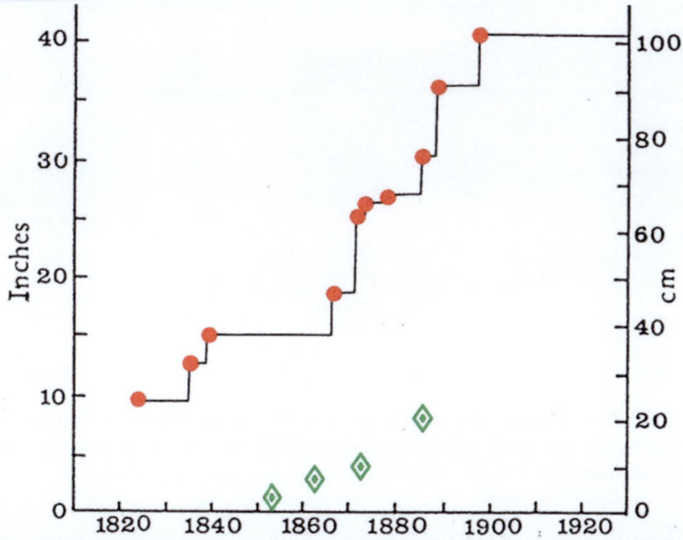


**Fig. 8.7** A plot showing the increase in aperture of the world's largest refracting telescopes between 1820 and 1900. By comparison, the *green diamonds* show (left to right) John Tebbutt's marine telescope and his Jones, Cooke and Grubb refractors (*Plot* Wayne Orchiston)

**Fig. 8.8** The 11.5-in. Schroeder refractor at Sydney Observatory (*Courtesy* Dr. Harley Wood)

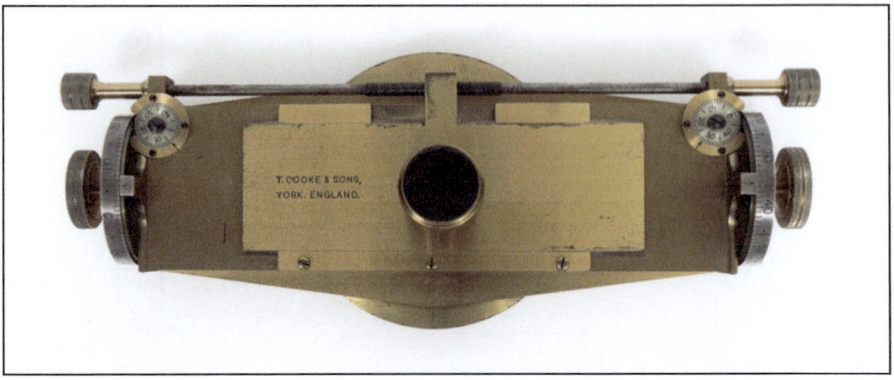


**Fig. 8.9** A late 1800s Thomas Cooke & Sons filar micrometer similar to the one that Tebbutt acquired in 1879 for his 4.5-in. Cooke telescope (http://utsic.escaltor.utoronto.ca/home/blog/instrument/filarmicrometer/)

> revolves with the polar axis, and when unclamped it revolves on the axis by means of a milled head … The hour circle is divided on brass to two minutes of time, and these divisions are again read off by a microscope and vernier to five seconds. The declination circle is five and a half inches [14 cm] in diameter and is divided on its brass edge to 30′ and subdivided by two verniers to single minutes. The telescope is provided with the usual arrangements for clamping and slow motion in right ascension and declination.

Tebbutt (1873) also indicates that there was a finder of $1^3/_8$ in. (3.5 cm) aperture and 15 in. (38.1 cm) focal length, and that the telescope was furnished with a dew cap, two diagonals and six standard eyepieces (ranging from 55× to 400×), as well as transit and comet eyepieces, a terrestrial eyepiece and solar and lunar filters (Tebbutt 1879). As we have already noted, the only major drawback was the lack of a drive (Tebbutt 1896–98).

The two ring micrometers that had been used with the 3.25-in. refractor were transferred to the new telescope, and on 25 April 1879 were joined by a new position filar micrometer that was constructed expressly for the telescope by Thomas Cooke & Sons (cf. Fig. 8.9). This was supplied by the Sydney firm of Flavelle Bros & Roberts, which adapted the micrometer to fit the telescope, and the all-up cost was £33/15/0 (ibid.).

The tripod was immediately dispensed with and the new telescope was installed on a solid wooden stand in the 'equatorial tower', but this facility proved far too small for it (Tebbutt 1873). Clearly, a new observatory was called for.

An observatory specifically designed to accommodate the 4.5-in. refractor was completed 2 years later, in July 1874, and took the form of a free-standing circular building close to and due south-east of the original observatory. It was 12 ft (3.66 m) in internal diameter and was made of pine timber (Tebbutt 1887). Around the top of the wall was a circular wooden plate, and on the upper side of this was a groove faced in iron which accommodated the eight 4-in. (10-cm) diameter iron wheels on which the dome ran.

As in the original observatory, the new dome was also made of wood and covered with canvas, but on this occasion the 14-in. (35.6-cm) wide shutters extended from horizon to horizon (ibid.). The conical dome was constructed in two sections

**Fig. 8.10** An 1880 woodcut showing the three Windsor Observatory buildings that existed at that time (Orchiston collection)

of different pitch, each with eight segments, as shown in Fig. 8.10, where this observatory is the middle one of the three. Judging from Tebbutt's comments (1879), the inspiration for this design came from the dome of the Cambridge University Observatory that housed the 'Northumberland Refractor' (see Fig. 8.11). On a more modest scale, a dome almost identical to Tebbutt's had been constructed at a slightly earlier date by the English independent astronomer John Drew (1809–1857) for a 3.75-in. (9.5-cm) refractor (see Fig. 293 in Andrews 1996b).

The Cooke telescope was installed in its new home on 1 August 1874 (Tebbutt 1873–79) and supported by an elaborately constructed pier consisting of

> … a stout post of iron bark sunk eight feet [2.44 m] in the ground. This post is stayed by four wooden buttresses under ground in the planes of the meridian and the prime vertical, and above ground by two buttresses in the plane of the meridian. The space between the post and the upper buttresses is built in with bricks and cement in order to destroy vibration. (Tebbutt 1879).

The mounting of the telescope was bolted to the top of the post (ibid.).

When the Cooke refractor was first installed in the tower observatory in 1872, the Jones refractor was remounted on its original tripod and was frequently used outdoors (Tebbutt 1890–92). After the Cooke telescope transferred to its own observatory in 1874, the tower observatory remained vacant until 1878 when it was '… occupied by an anemograph of my own design and chiefly of my own construction for recording both the direction and velocity of the wind' (Tebbutt 1879). The 'Tebbutt Letters' in the Mitchell Library reveal that Tornaghi (1878a) initially constructed this instrument for Tebbutt, but the lengths of the arms were incorrect (Tornaghi 1878b). Given the earlier quote and the lack of any further letters from Tornaghi on the matter, we must assume that Tebbutt carried out the modifications himself. Whatever the situation, this was the first addition to the stable of meteorological instruments since the founding of Windsor Observatory.

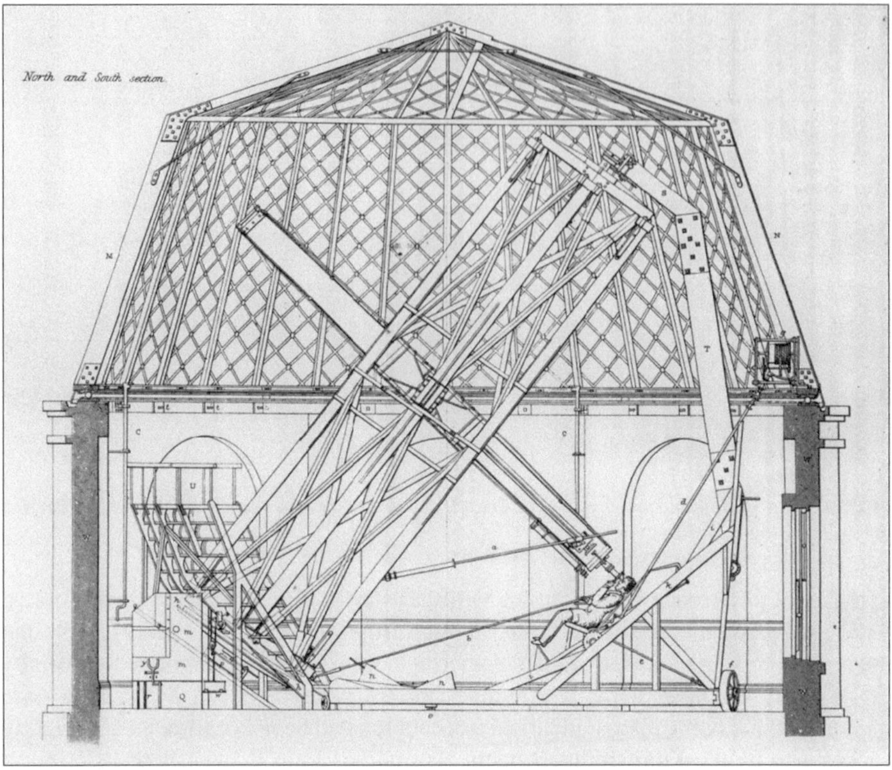


**Fig. 8.11** Schematic cut-away cross-section showing the Northumberland Refractor and its dome (after Challis 1879, Plate V)

Part of this anemograph is plainly visible in Fig. 8.10 protruding above the tower observatory.

It is interesting that, even when it was constructed, Tebbutt (1887) saw the 1874 observatory as no more than a temporary home for the Cooke refractor, which would indicate that he already must have been contemplating the construction of a larger more substantial observatory building. This came to pass just 5 years later.

## 8.5 A Proper Astronomical Centre: The Substantial Brick Building of 1879

One of Tebbutt's major achievements listed in his *Astronomical Memoirs* for the year 1879 was the erection of a substantial new brick observatory a few yards south-west of the old building. This also is shown on the right in Fig. 8.10 and has considerable visual appeal and architectural merit; to quote Tebbutt (1879), it is '... rather

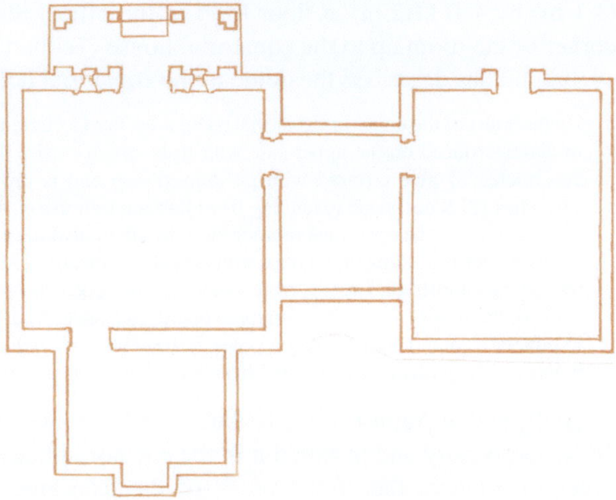

**Fig. 8.12** The floor plan of the 1879 brick building, showing (left to right) the office (*top*) and prime vertical room (*bottom*); transit room and the library; north is at the top (*Plan* Wayne Orchiston)

ornamental in external appearance'. Austin et al. (n.d.) have carried out a detailed architectural appraisal of the building and conclude that it is

> … an example of classical Georgian with its small entrance portico and well proportioned plan. John Tebbutt being associated with geometry has undoubtedly left this inherent with his buildings … Where any contradictions occur such as the small fenestrations and roof form over the equatorial room, these are purely functional elements derived from the needs of the building. The whole complex is the outcome of function and this is typical Georgian architecture.

A floor plan of this building is given in Fig. 8.12, which reveals the following components:

- An office/computing room at the western end of the building
- An 'equatorial house' directly above the office
- A small prime vertical room to the south of the office
- A transit room to the east of the office
- A large library room to the east of the transit room

Although it may be no more than a coincidence, it is interesting that this basic design, with the main dome at one end, meridian and transit rooms leading off this, and an office/library facility beyond that, is identical to the floor plan of Leipzig University Observatory constructed in 1861 (although Tebbutt's new observatory was, of course, on a much more diminutive scale). Donnelly (1973, p. 100) has suggested that the Leipzig design may have been inspired by an institution much closer to Windsor, namely Sydney Observatory, but in fact the resemblances are superficial.

Let us now examine each room of the 1879 Windsor Observatory building in turn. The office-cum-computing room was 15 ft (4.6 m) square and was entered from a north-facing ornamental porch. The centre of the room was interrupted by a pyramid-shaped pier of brick and cement which extended from the clay foundation beneath the building up to the equatorial house. This pier measured 3 ft 6 in.

(1.1 m) by 4 ft (1.2 m) at floor level in the office. Stairs led from the south-east corner of the room up to the equatorial house (Tebbutt 1879).

Tebbutt has described the dome in the equatorial house:

> On the walls of the upper or equatorial room is secured a circular plate of wood 15 feet [4.6 m] in diameter faced on the upper side with iron, which carries the revolving roof. The roof is constructed of pine covered with galvanised iron and is provided with an opening about 11 inches [27.9 cm] wide extending from horizon to horizon. The shutters six in number are hung on hinges; the upper one on each side opening outwards and the other four inwards. The revolving roof … turns on ten cast iron wheels 10 inches [25.4 cm] in diameter. The roof is moved by handles on two opposite sides … The spaces between the circular plate and the walls of the room are covered with pine boards and sheet lead, the rain running east and west to gutters … and then to … down pipes at the south east and south west corners of the main building. The gutters are concealed from view by means of ornamental facia boards. (ibid.).

At the end of August 1879, Tebbutt (1887) removed the 4.5-in. refractor from the 1874 observatory and installed it in the equatorial house in the new observatory. In order to improve the effectiveness of the telescope, in 1882 he commissioned Thomas Cooke & Sons to make him a square bar micrometer. From a letter dated 12 April 1882, he learnt that fabrication of the micrometer was proving very troublesome, hence the delay in dispatching it from England. It finally reached Windsor at the end of April 1882 (Flavelle Bros and Roberts 1882), and from that time on was used with the 4.5-in. telescope, replacing the old ring micrometers made by Tornaghi during the 1860s (Tebbutt 1887).

Leading off the office was the prime vertical room, which measured 10 ft 6 in. (3.2 m) by 7 ft 3 in. (2.2 m). It contained a 10-in. (25.4-cm) wide shuttered slit in the galvanised iron roof and in the walls. The roof shutters '… slide by means of pullies, and the openings in the walls are closed by tongued and grooved boards sliding vertically in grooves, by means of which the opening can be diminished in case of wind' (Tebbutt 1879). In the centre of this room was a pier to accommodate the transit telescope when latitude observations were required. In March 1881, Tebbutt (1879–81) removed the frame for the original transit telescope from the 1863 observatory and installed it in the new meridian room.

To the east of the office was the transit room, which measured 10 ft 6 in. (3.2 m) by 8 feet (2.4 m) and contained a

> … meridian opening from 8½ [21.6 cm] to 10 inches [25.4 cm] in width [which] is closed by shutters in the roof opening outwards, the southern one capping the other to exclude the rain. The openings down the north and south walls are closed by sliding boards similar to those in the prime vertical room. (Tebbutt 1879).

A solid single brick pier faced with cement extended from the clay foundation below the observatory building, and at its top measured 4 ft (1.2 m) by 1 ft 4 in. (0.4 m) (ibid.). In 1878 a new Thomas Cooke & Sons transit telescope of 3-in. (7.6-cm) aperture was ordered through Flavelle Bros & Roberts, and this was received on 6 September 1879 and 3 days later was installed in the transit room (Tebbutt 1879). Tebbutt (ibid.) describes this instrument:

> The instrument is of that form which has the telescope between the pivots. The telescope has a clear aperture of three inches [7.6 cm] and a focal length of about 36 inches [91.4 cm]. The pivots which are about 1¼ inch [3.2 cm] in diameter rest on segmental supports and are

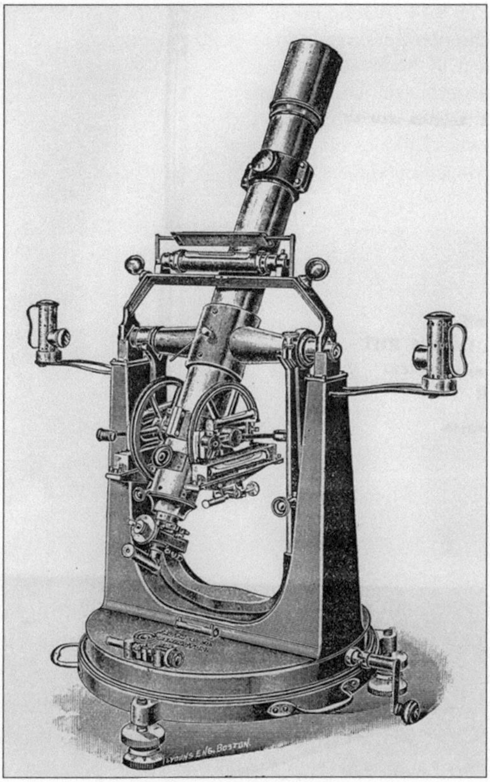


**Fig. 8.13** A 3-in. Cooke transit telescope similar to the one that Tebbutt acquired in 1879 (after Andrews 1996a)

> covered by moveable caps to exclude dust & … The instrument is clamped to an arm or bar which revolves on one end of the axis, a pin on this bar falling into a notch in a piece of moveable brass attached to the heavy iron frame which forms the support of the instrument.

It would appear from a letter written by Flavelle Bros & Roberts on 30 March 1880 that Tebbutt had some reservations about the new transit telescope, but he does not elaborate on these anywhere. Although he never provided a photograph of this instrument, Fig. 8.13 shows a Cooke transit telescope that was probably of similar vintage and design.

Initially, the old Parkinson and Frodsham chronometer was used to keep local time in the transit room, but in November 1882 it was joined by one manufactured by John Poole of London (see Fig. 8.14). In the mid-nineteenth century Poole was one of the leading chronometer-makers, but he died in 1867 (Mercer 2004) and his business was continued by his brother, James using the name 'James Poole & Co.' The fact that Tebbutt refers specifically to a 'John Poole' chronometer suggests that he acquired a second-hand instrument made in the 1860s. Later, at the end of 1885, while the John Poole chronometer continued to show local time the old Parkinson and Frodsham chronometer began keeping sidereal time (Tebbutt 1887).

Once the Cooke transit telescope was installed, Tebbutt began using the old Tornaghi transit telescope as a collimator for the new instrument. For this purpose the smaller transit telescope '… is mounted on a substantial pier of brick and cement

**Fig. 8.14** A John Poole chronometer thought to date to the 1860s and therefore similar to the one purchased by Tebbutt (www.antiques-marine.com/chronometers/1183/)

with stone cap a few feet north of the transit-room, and is protected from the weather by a small wooden building covered with galvanised iron' (Tebbutt 1890–92). In a later journal, Tebbutt (1896–98) states that the small wooden building is about 36 ft (11 m) north of the Cooke transit telescope. There being no further use for the original observatory, in late 1879 or in 1880 the old transit room was turned into a carpenter's shop (ibid.) and later, in 1890, the old prime vertical room was converted into a workshop and the instrument pier was removed (Tebbutt 1888–90).

The only room in the 1879 observatory not yet discussed is the library, which lay to the east of the transit room and measured 16 ft 6 in. (5 m) by 12 ft 6 in. (3.8 m). This room was accessed from the transit room or from outside through an entrance door located in the middle of the northern wall. Tebbutt was well aware that he possessed an outstanding astronomical research and reference library (see Tebbutt 1908, pp. 112–117), so he went to some pains to protect this by making the library wing fireproof. Consequently, the doors and roof were constructed of iron and the floor of cement (Tebbutt 1879). Tebbutt (ibid.) also used the library '… as an office for the transactions of my ordinary business'. From time to time it also served as a second computing room.

In just 16 years, Tebbutt saw Windsor Observatory grow from a single building to a cluster of 3 (Fig. 8.15), and the aperture of his refracting and transit telescopes increases significantly (see Fig. 8.7). Although they were still modest by professional standards, Tebbutt more than made up for this with his dedication and application. Between 1864 and 1885 (inclusive), he used the 3.25-in. telescope and from 1872 the 4.5-in. telescope for an impressive range of observational programmes (see Table 1.1). While he frequently recorded sunspots prior to the establishment of the Windsor Observatory (see Table 7.1), he came to regard such activity as 'amateurish' (Turner 1918) once he was furnished with instruments capable of more 'serious' work. Likewise, naked eye targets such as aurorae, meteors and the zodiacal

**Fig. 8.15** A view of Windsor Observatory, looking north, showing (left to right) the 1879 main building, the original 1863/1867 building with the tower observatory and the 1894 equatorial house (after Tebbutt 1986, p. 121)

light no longer featured in his observing repertoire. I am a little surprised though, that he did not bother to observe the 1866 Leonid meteor storm, which even from the southerly latitude of Windsor must have been an impressive sight.

Tebbutt's observing programme from 1864 to 1885 resulted in the appearance of 185 different research publications: 75 in *Astronomische Nachrichten*, 61 in *Monthly Notices of the Royal Astronomical Society*, 22 in *The Observatory*, 14 in *Journal of the Royal Society of New South Wales*, 9 in the *Astronomical Register*, 3 in *Transactions of the Philosophical Society of New South Wales* and 1 in *Memoirs of the Royal Astronomical Society* (after Tebbutt 1887, 1908). In addition, there were two chapters of books dealing with astronomy in New South Wales, one research paper on Australian storms and four volumes of meteorological observations (Tebbutt 1868b, 1874a, 1877c, 1882). However, the total of 185 is a little misleading, for Tebbutt was in the habit of sometimes publishing the very same paper in more than one outlet—a practice which in those days was acceptable but today is frowned on. Thus, of the 185 papers listed, 35 different papers were published in both *Astronomische Nachrichten* and *Monthly Notices*. Despite this complication, we are dealing with a prodigious publications output for a single individual, indulging in astronomy (and meteorology) at his own expense and in his spare time.

In addition to this research focus, Tebbutt used Windsor Observatory to actively popularise astronomy, and between 1864 and 1885 (inclusive) he produced 206 articles for the Sydney and local newspapers, plus 20 on aspects of meteorology and 1 on seismology (see Tebbutt 1887).

**Fig. 8.16** One of the bright comets Tebbutt observed (and published eight papers on) was C/1874 H1 (Coggia), but unlike other astronomers—who focussed on its tail and ever-changing head (as shown here)—all Tebbutt did was make positional measurements of the nucleus (after Phillips 1923, p. 421)

When we examine all of Tebbutt's publications for this period his preoccupation with comets is immediately apparent (see Fig. 8.16), but there is evidence of a more catholic astronomical interest. Also represented are observations of solar and lunar eclipses, lunar occultations, planets, minor planets, transits of Mercury, Jovian satellite phenomena, double stars and, finally, variable and suspected variable stars. Of greatest interest, though, are his observations of the 1874 transit of Venus; his involvement in the international searches for the intra-Mercurial planet, Vulcan; and his discovery of a nova (although he did not realise, at the time, that this is what it was).

As we saw in Chap. 3, transits of Venus are rare events, and the pair of transits in 1874 and 1882 were the first to occur since the 1760s and Cook's first voyage to the South Seas (see Orchiston 2005, 2017; van Helden 1995). By positioning observers at widely-differing latitudes and accurately timing the ingress and egress contacts as Venus traversed the Sun's disk, it was hoped that a more precise value of the Astronomical Unit (the distance from the Earth to the Sun) could be computed (see Forbes 1874; Grant 1874). Consequently, a number of nations dispatched observing teams to the far corners of the globe for the 1874 event (Janiczek and Houchins 1974), and two American teams were stationed in Tasmania (see Dick 2003, pp. 257–260; Orchiston and Buchanan 1993, 2004; Orchiston et al. 2015, for accounts of the Hobart and Campbell Town parties). The locations of these stations are shown in Fig. 8.17.

In mainland Australia, as Fig. 8.17 also indicates, a number of Government observatories established observing stations (e.g. see Lomb 2011; Orchiston 2004; Russell 1892), but Tebbutt was the most prominent non-professional astronomer to carry out independent observations. This was so important an event that Tebbutt accepted the assistance of a local man, and an amusing incident occurred the day before the great event, while they were practising their observing procedure:

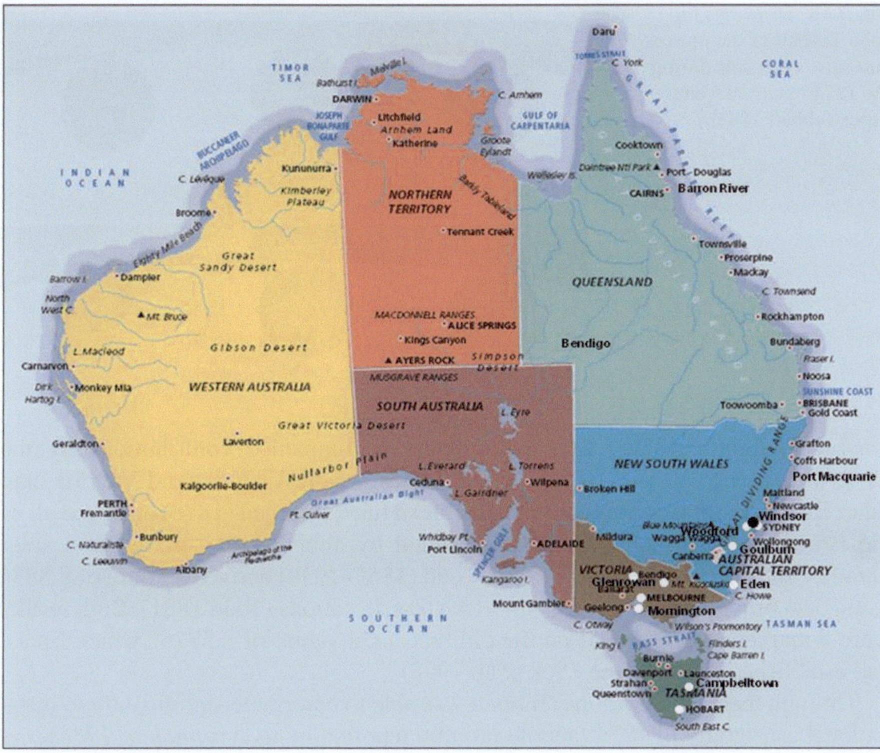


**Fig. 8.17** The *white dots* show the locations of the two American transit stations in Tasmania; Melbourne Observatory in Victoria and its transit stations at Mornington, Bendigo and Glenrowan; and Sydney Observatory in New South Wales and its transit stations at Eden, Goulburn and Woodford. Windsor Observatory is marked by the *black dot* (*Map* Wayne Orchiston)

> While we were so engaged … a visitor from the town entered the open door of the observatory and walked up to the place where I was stationed with my telescope. On completing the counting I looked up and found my visitor to be a resident of Windsor whose attentions were usually devoted to Bacchus rather than to Urania. He accosted me in the following terms:- "I have come, Mr. Tebbutt, to ask you if I can be of any service to you in your observation of the transit of Venus to morrow." I thanked him for his offer of assistance, and suggested as politely as possible that if his visit were made at the moment of the great event the result would probably be that I should loose the observation for which I had been waiting so many years. (Tebbutt 1908, pp. 39–40).

Tebbutt should have been flattered that an astronomical event could have so sobering an effect upon one of his neighbours, but the experience was a timely warning and he had the gates to the Estate locked to prevent any further visitors. The day of the transit, 9 December, was fine and sunny, and Tebbutt succeeded in obtaining the contact observations using the 4.5-in. telescope, but his assistant was less successful with the 3.25-in. refractor.

Soon after the event, Tebbutt wrote up his observations for *Astronomische Nachrichten* (Tebbutt 1875), and in 1883 he published a fuller account, complete with contact diagrams (see Fig. 8.18), in the *Memoirs of the Royal Astronomical*

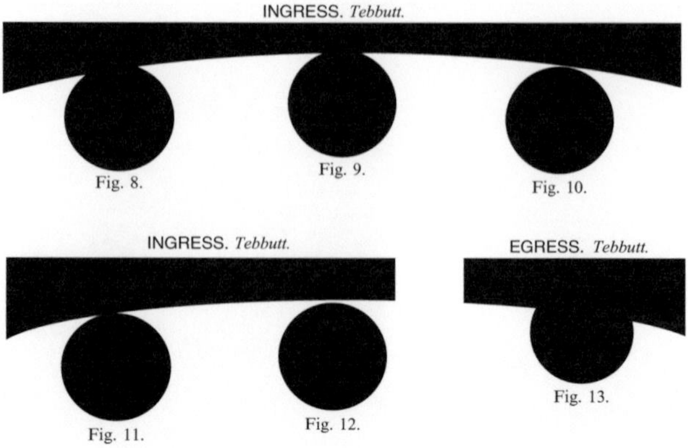


**Fig. 8.18** Drawings by John Tebbutt of the ingress and egress contacts during the 1874 transit of Venus (after Tebbutt 1883b)

*Society* (Tebbutt 1883b). This latter paper accompanied contributions by the Government Astronomers of New South Wales, South Australia and Victoria, and other British Empire astronomers. In the mean time, Tupman (1878) analysed all of the British results (including those provided by Tebbutt) and arrived at a solar parallax of 8.8455″, while David Peck Todd (1855–1939) derived a figure of 8.883″ using data provided by the eight American transit stations (Todd 1881). Both results were a marked improvement on the old 'accepted value' of 8.5776″ which Encke had established in 1824 (see Dick et al. 1998).

Through these observations, Tebbutt was able to contribute significantly to international astronomy in 1874 (and he proudly reported in his *Astronomical Memoirs* that Tupman assigned his contact timings double weight) and his observations and publications are discussed in more detail in Orchiston (2004), but he was not so fortunate with the 1882 transit of Venus for Windsor and other New South Wales stations were clouded out on the vital day.

Another fascinating international programme with which Tebbutt was associated was the search for a planet which was thought to exist between Mercury and the Sun. During the 1850s the unpopular Paris Observatory Director Urbain Jean Joseph Leverrier (1811–1877; Fig. 8.19; Lequeux 2016) developed the theory that irregularities in Mercury's orbit resulted from perturbations of this missing planet, which he named 'Vulcan' (Chambers 1877, Chap. 3). Between 1860 and 1877 a succession of international solar monitoring programmes was organised to detect Vulcan during calculated transits (see Baum and Sheehan 1997). Tebbutt participated in four of these, in 1860, 1862, 1873 and 1877. Although the first search was carried out with the small marine telescope and the first two searches occurred prior to erection of Windsor Observatory, all four are dealt with here for reasons of thematic continuity.

On 16 March 1860 Tebbutt for the first time learnt of the supposed existence of Vulcan from the Reverend C.F. Garnsey (Tebbutt 1859–63). From time to time, Tebbutt had used Garnsey's small telescope for observing, prior to purchasing the 3.25-in. Jones refractor. In his journal entry for 17 March 1860, Tebbutt (ibid.) wrote:

> I propose to examine the sun's disc frequently during the next three or four weeks in order if possible to detect M. Lescarbault's intra-Mercurial planet, which is expected to transit the disc sometime between the 25th instant and the 10th proximo.

**Fig. 8.19** U.J.J. Leverrier (https://en.wikipedia.org)

He successfully recorded sunspots on twenty different days between 17 March and 13 April (inclusive), but there was no sign of Vulcan (ibid.). In 1862, Tebbutt conducted another unsuccessful Vulcan search during April (ibid.).

Fontenrose (1973, pp. 147–148) reports that by the mid-1860s '… the question of Vulcan had become extremely confused. While several textbooks went ahead and listed the new planet among the others, most astronomers preferred to remain non-committal'. After unsuccessful searches in 1869, 1870, 1871 and 1873 enthusiasm waned, and Tebbutt too must have been disappointed as he was one of those involved in the 1873 programme. On 20 and 21 March he searched the Sun's disc for any sign of Vulcan, but all he saw were sunspots (Tebbutt 1870–73).

International interest in Vulcan was revived after a reported sighting in April 1876, and a further search was arranged for October of that year (which proved unsuccessful) and another for March 1877. Tebbutt became involved in the 1877 search when Airy sent a telegram to Russell, soliciting Australian support for this programme. The critical dates in Australia were 22 and 23 March, and on both days Tebbutt used projection to examine the solar disk at intervals throughout the day. He used a magnification of 55× with the 4.5-in. refractor. On 22 March he made 32 separate observations between 10:03 and 18:08 local time, and on the following day 21 observations between 08:45 and 17:34. For good measure, he also made seven observations on 24 March (Tebbutt 1877d). Neither he nor any other observer saw Vulcan.

With Leverrier's death in 1877 interest in Vulcan diminished, although further searches—but not on the scale of previous years—were mounted almost annually from 1878 to 1885. By 1886, to quote Fontenrose (1973, p. 153), '… Vulcan's stock had reached what was to be a permanent low'. Thirty years later, Albert Einstein introduced his General Theory of Relativity, and we now know that by substituting relativistic physics for Newtonian physics Mercury's orbital anomaly is resolved. There was no need for Vulcan after all!

The third case study we shall examine is Tebbutt's discovery of a nova, which is a brilliant example of astronomical detective work. A nova is a star that increases enor-

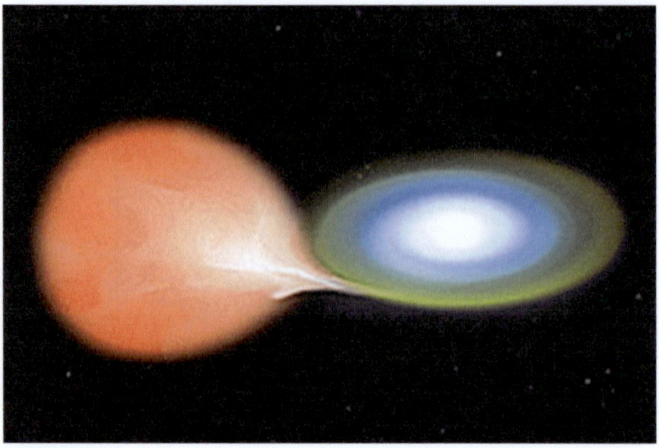

**Fig. 8.20** An artist's impression of the mass transfer that leads to a nova (https://en.wikipedia.org)

mously in brightness over just a few hours and then slowly fades usually over several months. It is believed that novae are associated with close binary stars, where one component is a white dwarf and the other a red giant. The two stars are so close together that matter can flow from the giant to the dwarf star and after this has occurred for a time the accumulated material is ignited explosively and some of it is blasted into space. When this happens we observe a nova (see Fig. 8.20; Zeilik 1985).

Tebbutt discovered a nova quite by chance, years after the event. He recounts that in 1877

> With a view to the determination of more accurate places of Swift's comet of 1862, I made a thorough examination of the original observations for the determination of the places of the stars of comparison. I found that I had fixed … the positions of the two stars of comparison for October 4, namely Brisbane 5754 and 5799 relative to a conspicuous star of the 5th magnitude, whose place was determined on October 9th [1862] by means of sextant distances from θ Scorpii, Altair, Antares and ε Sagittarii. This fifth magnitude star was remarked as being about a degree north-east of δ Arae and somewhat more conspicuous to the naked eye. (Tebbutt 1908, p. 43).

Various catalogues had found their way into Tebbutt's library between 1862 and 1877, but he could not find a naked eye star near δ Arae listed in any of them, so

> My curiosity being aroused, I turned the 4½-inch Equatorial on the spot in the evening of the 13th [November 1877] instant, and was surprised to find that the field contained stars of the 10th and 11th magnitude only. One star in particular attracted my attention: it is extremely faint, and could only be observed with great difficulty. (Tebbutt 1878a).

This was in the precise position of the bright star seen in 1862, and Tebbutt could only conclude that it was a variable star.

With all due modesty, Tebbutt reported his discovery in the 19 November 1877 issue of the *Sydney Morning Herald*:

> I have the honour to announce my discovery of a new and highly interesting variable star … An account of the steps by which I became aware of the variable character of the star would be too voluminous for your columns, and indeed would hardly be understood by ordinary readers. I therefore propose to lay the circumstances of the discovery before the Astronomical Section of the [local] Royal Society. (Tebbutt 1877a).

This paper was duly presented and was subsequently published (Tebbutt 1877b). In it, Tebbutt raised the possibility that the star would exhibit another outburst in future years. It was only much later, in 1901, that R.T.A. Innes (Bio Box 8.2) suggested to Tebbutt that his 'variable star' was in fact a nova. He went further: 'But should you not announce that this star should be included in the list of "New Stars"? I think if you do not care to do so, I will feel justified in doing so after I examine the region' (Innes 1901). In fact, neither Innes nor Tebbutt published an announcement amending the original identification, but Tebbutt must have accepted Innes's interpretation for he subsequently described the missing star as '… a temporary one' (i.e. a nova) in his *Astronomical Memoirs* (Tebbutt 1908).

**Bio Box 8.2: Robert Thorburn Ayton Innes**

Robert Thorburn Ayton Innes (Fig. 8.21) was born in Edinburgh on 10 November 1861, but went to school in Dublin (Obituaries 1933). In 1890 he emigrated to Sydney and set up in business as a successful wine and spirit merchant. Innes displayed an early interest in astronomy, showing such aptitude for mathematical astronomy that he was elected a Fellow of the

**Fig. 8.21** Robert Thorburn Ayton Innes at his desk at Union Observatory (*Courtesy* South African Astronomical Observatory)

(continued)

**Bio Box 8.2** (continued)

Royal Astronomical Society at the tender age of 17 years 2 months (Obituaries 1933b). He began publishing papers on mathematical astronomy soon after settling in Sydney.

In 1894 the well-known Sydney independent astronomer, Walter Gale, loaned Innes an old 6.25-in. (15.9-cm) refractor, and in 1896 Innes supplemented this with a 16.5-in. (42-cm) reflector of his own. After the 18-in. (46-cm) reflectors owned by Gale and Hans Frandsen Madsen (1843–1937; Madsen 1886; Orchiston 2003a; Orchiston and Bembrick 1995), Innes' 16.5-in. was the largest reflecting telescope in Sydney at that time. Innes then became enamoured with observational astronomy and used these instruments mainly to search for new double stars and to monitor known variable stars. At the same time he continued his mathematical investigations, and he published a succession of research papers on mathematical astronomy and on his observations in *Astronomische Nachrichten*, *Journal of the British Astronomical Association* and *Monthly Notices of the Royal Astronomical Society*.

In addition to his research activities and involvement in societies, Innes contributed to the popularisation of astronomy in Sydney. For example, between 1890 to early 1896 he submitted articles to the *Australian Star*, *Daily Telegraph*, *Echo* and the *Sydney Morning Herald* on a wide range of topics (Orchiston 1997a).

Early in 1896 Innes moved to a clerical position at the Royal Observatory, Cape of Good Hope, and continued double star work in his spare time, eventually publishing his authoritative *Reference Catalogue of Southern Double Stars* (Innes 1899). With strong support from Sir David Gill, he was appointed founding Director of the Transvaal (later Union) Observatory in 1903, where he continued to focus on double stars, minor planets and lunar occultations. One of his most notable discoveries was Proxima Centauri (see Glass 2008). In 1923, Leiden University recognised his lifelong astronomical work by awarding him an honorary D.Sc., and by the time he died from a heart attack on 13 March 1933 he was recognised as an international figure in double star astronomy.

Throughout his life, Innes was well known for his unconventional views, and '… his unaffected manner of expressing them made Innes a charming companion in daily life …' (Obituaries 1933a). His sociability—and wonderful way with words—is revealed in a letter written to Tebbutt in 1895 about an incorrect report that appeared in one of the Sydney newspapers: 'Whilst the Reporter was in my cellar we found a severe local drought and irrigated accordingly and I fear between the wine, astronomy & meteorology the poor reporter got mixed up' (Innes 1895).

For further information on Innes's astronomical activities in Australia and South Africa, see Orchiston (2001a, 2003b).

More recently, both Ashbrook (1984b) and White (1979) have discussed this star, which is now designated Nova V728 Scorpii. In 1979, White (ibid.) postulated that the star in question should now '… be red, about 16th magnitude and probably not showing any ejecta shell'. He searched the ESO-B and SRC-J sky surveys but could not identify a suitable candidate, which may indicate that Tebbutt misidentified the progenitor star back in 1877. Nonetheless, we must applaud Tebbutt's scholarly approach, which allowed him to prove the existence of a nova fully 15 years after the event.

After these observational diversions, let us now move forward to 1886 and resume our narrative of the development of Windsor Observatory.

## 8.6 Accommodating the Eight Inch: The Demand for a New Brick Observatory

In mid-1886, astronomy at Windsor entered a new era when Tebbutt purchased a larger telescope, the 8-in. (20.3-cm) Grubb refractor (see Fig. 1.12) which was formerly the property of Dr. William Bone (Bio Box 8.3) of Castlemaine, Victoria (see Orchiston 1986, 1987b, 1997b). The 1880s and 90s saw the emergence of the 'giant refractors' when, as Fig. 8.7 shows, apertures jumped from 26 in. (66 cm) to 40 in. (1 m) (see Welther 1984), so by world standards Tebbutt's 8-in. Grubb was extremely modest. Yet it was the largest refractor owned by an Australian independent astronomer, and it was equal in aperture to the 8-in. instruments at Adelaide and Melbourne Observatories and exceeded only by the 11.5-in. Schroeder at Sydney Observatory. As such, it certainly was capable of performing useful work in positional astronomy.

Tebbutt has provided the following detailed description of this telescope:

> This consists of a fine equatorial refractor of eight inches [20.3 cm] clear aperture, and 9 feet 7 inches [2.92 m] focal length … The pillar supporting the equatorial consists of two massive hollow cylinders of cast iron, one standing on the other. The lower casting is securely bolted to a pier … The upper casting is bolted to the lower one, and admits at the same time of a small azimuthal adjustment by means of a tangent screw. To the upper cylinder is secured another massive casting containing the socket for the polar axis: it also contains the clock, whose driving weight is within the cylindrical pillar. This casting consists of two parts fitted in such a way as to admit of the usual adjustment for latitude within a very wide range. The mounting is that known as the Fraunhofer or German. The telescope tube is in three parts and of rolled steel. The hour circle is at the upper extremity of the polar axis, and the declination circle next to the telescope cradle, being respectively 8.6 [21.8 cm] and 13.0 [33 cm] inches in diameter. They are divided, the former to 2 minutes of time and read off to 10 seconds by the vernier, and the latter to 10′, read off by opposite verniers to 30″. The latter is very conveniently read off from the eye end of the telescope by means of a tube with lenses and a prism. The illuminating lamp is suspended at the counterpoise extremity of the declination axis. The light passing up this hollow axis and falling on a set of reflectors in the telescope cradle, serves for illuminating the micrometer threads, the graduated limb of the declination circle and a transparent position circle at the eye end of the telescope. The telescope has a filar micrometer of the best description … a Dawes solar eyepiece and four negative eyepieces whose magnifying powers are 131, 175, 224 and 340. (Tebbutt 1887, pp. 14–15).

All of the micrometers used with the 4.5-in. Cooke were also modified so that they could be used with the new telescope.

With this acquisition, Tebbutt (1908, p. 65) felt that '… the observatory was now in a position to do work equal to that of many of the observatories in the northern hemisphere'. As might be anticipated given Tebbutt's commitment, a great deal of 'work' was indeed accomplished with this telescope, and a detailed discussion of this is presented in Chap. 11.

**Bio Box 8.3: Dr. William Bone**

William Bone (Fig. 8.22) was born in Penzance, Cornwall, in 1836, and studied at King's College, University of London, before transferring to St. Andrews University in Scotland where he graduated as a doctor (Death of Dr W. Bone 1885).

**Fig. 8.22** Dr. William Bone and his 8-in. Grubb telescope, which passed to John Tebbutt in 1886 (Orchiston collection)

In 1863 he migrated to Tasmania (Bone 1882b), and about 1869 settled in Castlemaine, Victoria (Death of Dr W. Bone 1885). It is apparent that Bone was well liked in Castlemaine: '… it is noticeable in his career as a physician that he gradually rose in favour by the keenness of his observations, and skill in diagnosis and treating diseases … Whenever care was required he was so unrelaxing in his attention that he would often sit up all night by the bedside of patients in straitened circumstances and bring them presents from his own home of such food, delicacies, or wines that were deemed requisite by him for their nourishment' (ibid.). He was also '… strongly embued with a love of science, especially astronomy, & endeavoured to impregnate others with his own ardour' (ibid.).

(continued)

Bone acquired this interest in astronomy from his father who '… used to explain the planetary motions to me at breakfast time by utilising the plates and cups and saucers etc. for Sun, planets etc. to the horror and consternation of my mother and sister' (Bone 1882c). Once settled in Castlemaine, Dr. Bone erected an observatory for his $4^5/_8$-in. (11.9-cm) Wray refractor. In 1883 he acquired a new 8-in. refractor from Grubb (Death of Dr W. Bone 1885).

In 1880, he was one of several Australians to independently discover the Great Southern Comet of that year, C/1880 C1 (see Orchiston 1997c; Russell 1880), and he followed this magnificent object (Fig. 4.1) religiously from February 2 to 17. He obtained micrometric positions on four evenings and on one of these also tried to observe the spectrum of the comet but found that the light grasp of his telescope was not sufficient for this purpose. Bone then proceeded to prepare a four-page report, which was published in *Monthly Notices of the Royal Astronomical Society* (Bone 1880). Bone went on to publish papers on the 7–8 November 1881 transit of Mercury (Bone 1882a) and Comet C/1881 K1 (Tebbutt) (Bone 1882d).

Bone died suddenly in 1885 before he had much of a chance to make full use of his 8-in. Grubb telescope (Death of Dr W. Bone 1885). Further details of Dr. Bone and Castlemaine Observatory are provided in Orchiston (1986, 1987b).

The Grubb telescope arrived in Windsor on 20 May 1886, and the following day Tebbutt and four others temporarily mounted the instrument outdoors, near the main entrance to the 1863 observatory building. The next evening, Tebbutt was able to test the performance of his expensive new acquisition:

> … I pointed the telescope on Jupiter and Mars. The definition was excellent. Jupiter was a magnificent sight … I could observe the occultation and reappearance from eclipse of Jupiter's first satellite. I accordingly observed these two phenomena with much greater accuracy than I could possibly obtain with the 4½ inch equatorial. (Tebbutt 1884–86).

The little 1874 observatory was then prepared for the new telescope. Tebbutt later reported in his journal (ibid.) that a substantial pier of brick and cement whose foundation was 8–9 ft (2.4–2.7 m) below the ground surface was prepared for the telescope, and that the dome was increased slightly in height. On 18 June 1886 the Grubb telescope was installed in its new home (ibid.), but it did not take Tebbutt long to realise that this was far too small for so large an instrument.

Despite the obvious shortcomings, Tebbutt persevered for a while, and it was only in 1891 that he decided a new observatory was essential (Tebbutt 1892, p. 3). By 1893 the situation had become desperate, and he was moved to act:

> The proposed enlargement of the circular chamber, accommodating the 8-inch equatorial refractor, is now in progress. The old wooden structure is not only exceedingly inconvenient as regards size, but it will soon become useless in consequence of the inroads of the white ant. A substantial building of brick is now in course of erection, with an internal diameter of eighteen feet [5.5 m]. The telescope will be retained in its present position, and

**Fig. 8.23** This 1906 photograph shows the 1894 equatorial house on the left and behind it the 1879 observatory building (Orchiston collection)

> the old structure of wood will, of course, be removed as soon as the new building is completed. (Tebbutt 1894, p. 3).

With the continuing action of the white ants, the survival of the building was at stake and with it the proper protection of the Grubb refractor. Size, therefore, was merely one of the constraints that demanded a new observatory.

In the 22 February 1895 entry in his Journal, Tebbutt announces completion of the new building:

> Mr. Thomas Collison and his son, who were employed in the construction of the revolving roof of the new equatorial chamber … completed their work on the 19th instant, and likewise the removal of the old structure of wood … This new structure has a circular brick wall nine inches [22.9 cm] in thickness, with an external diameter of 19 feet 5 inches [5.9 m] … (Tebbutt 1894–96).

The ten-sided dome turned on ten 13.5-in. (34.3-cm) diameter wheels that ran on a circular iron track provided with a rail to keep the wheels on course (Tebbutt 1895). The dome framework was constructed of pine timber and covered by galvanised iron. There was an 18-in. (45.7-cm) slit which extended from horizon to horizon and was closed by two shutters which were very conveniently opened and closed from the floor (Tebbutt 1894–96).

Tebbutt (ibid.) thought the new observatory was very commodious, and believed that it '… will add greatly to my comfort in observing'. With his increasing years, this was becoming an important factor. A view of the completed observatory, as it appeared in 1906, is shown in the foreground in Fig. 8.23.

## 8.7 Concluding Remarks

With the erection of the 'equatorial house' in 1894, Windsor Observatory reached its maximal size (see Fig. 8.24). In constructing four separate yet adjacent observatory buildings between 1863 and 1894, Tebbutt was actually following the dictates of observatory design that became more commonplace in overseas observatories after the late 1870s—of moving away from centralising all activities and instruments in the one building to dispersing them. This was especially so of those institutions boasting large refractors, and an 'astronomical precinct' rather than a single building became the norm (see Donnelly 1973).

Yet this was an architectural rather than a functional precinct at Windsor, for completion of the 1879 building allowed Tebbutt to centralise most of those astro-

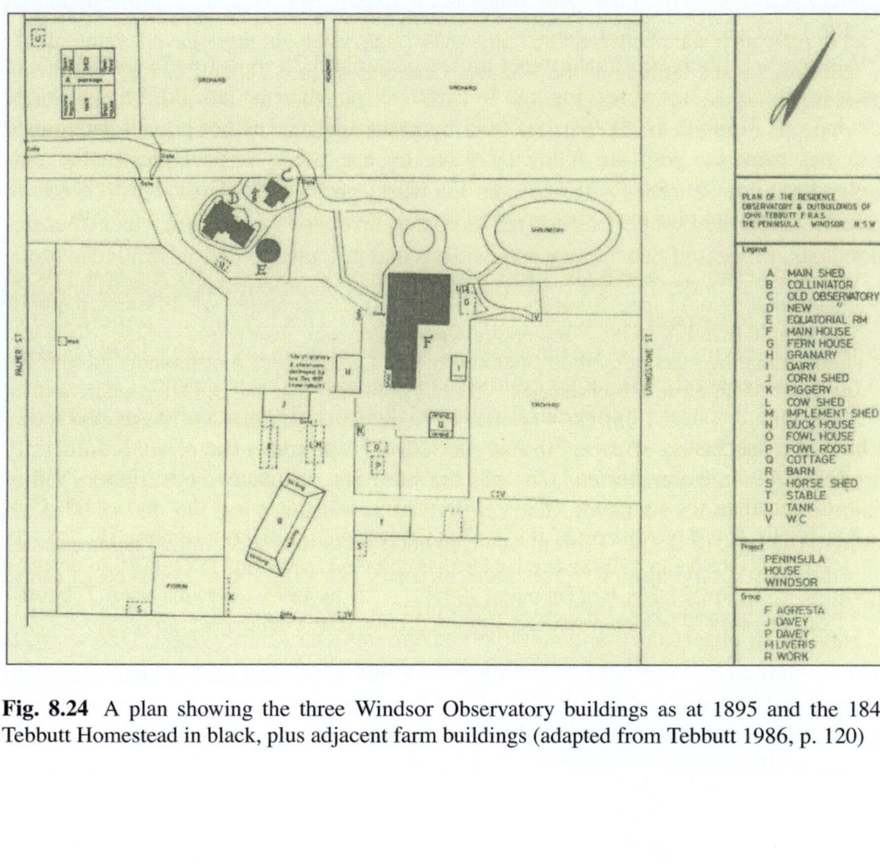


**Fig. 8.24** A plan showing the three Windsor Observatory buildings as at 1895 and the 1845 Tebbutt Homestead in black, plus adjacent farm buildings (adapted from Tebbutt 1986, p. 120)

**Fig. 8.25** Old Cincinnati Observatory with its distinctive drum dome (https://en.wikipedia.org)

nomical activities which had previously been dispersed between his homestead and the 1863 and 1874 observatories.

Although Tebbutt continued to set up the portable 3.25-in. refractor outdoors and use it for variable star observing and to examine the Moon's dark limb for probable occultations (Tebbutt 1890–92), he could now carry out the rest of his observing programmes from the comfort of his 1879 observatory. This was where he also conducted literature searches and wrote up his work, prepared lectures, wrote letters to and articles for newspaper editors, responded to an ever-increasing number of correspondents, processed new library acquisitions and attended to administrative matters. To have all this under the one roof was rather convenient, but this situation changed in 1886 with the arrival of the 8-in. Grubb telescope. This became the preferred telescope for observing and, as we have seen, it was installed in the nearby 1874 observatory and subsequently in a larger building constructed on the same site.

While the disparate architectural styles of the four different buildings erected at Windsor Observatory over a 30-year period are apparent, there was conformity throughout on one element and that was the selection of a domed observatory rather than a roll-off roof or a turret (also referred to as a drum—see Fig. 8.25). When it came to the type of dome, Tebbutt opted in each instance for a conical design rather than the more conventional hemispherical form. For craftsmen unskilled in dome construction this was certainly an easier option.

Having an observatory and suitable instruments was simply part of the story. If Tebbutt wanted to contribute observations of value to science then he also needed to determine the precise location of Windsor Observatory. What should, at first glance, have been a simple exercise turned into a complex ongoing saga—as the following account will indicate.

Determining latitude was a comparatively-straightforward matter, and involved the observation of particular stars, with precisely-known positions, as they approached the zenith. Tebbutt (1887) notes that originally he obtained the latitude from numerous sextant observations of the meridian altitudes of the Sun and selected stars, but once Windsor Observatory was in existence he installed the original transit telescope in the prime vertical room and between 1867 and 1874 (inclusive) used this to observe nine 'clock stars'. These gave a value of 33° 36′ 28.9″ S for the latitude of Windsor Observatory (see Tebbutt 1884). After the construction of the brick observatory in 1879 and purchase of the new Cooke transit telescope, he repeated the exercise in 1881 by observing transits of six well-determined stars and came up with the following value for the prime vertical pier in the new observatory: 33° 36′ 30.8″ S (ibid.). He was later to note in his *Astronomical Memoirs* that this was the latitude that finally was adopted for the observatory (Tebbutt 1908, p. 59).

In stark contrast to latitude, there was considerable debate about the best way of establishing longitude (see Baracchi 1914, pp. 366–373) as a number of different methods were available (Challis 1879). One option was to carry out observations of the Moon and Moon-culminating stars, a method favoured by the Government Astronomers at both Sydney and Melbourne Observatories. Tebbutt also used this technique in 1866 and 1867, deriving values of 10 h 03 m 19.1 s E and 10 h 03 m 23.1 s E, respectively, but he also arrived at a figure of 10 h 03 m 15.7 s E, based upon knowledge of the longitude of Sydney Observatory, and initially decided to adopt this figure (Tebbutt 1868a). However, Tebbutt eventually had greater faith in the use of lunar occultations for longitude determinations, and from 1864 he made systematic observations of these events and published a succession of papers in *Astronomische Nachrichten* and *Monthly Notices of the Royal Astronomical Society*. In 1880, Tebbutt used his occultation series from 1864 to 1876 to calculate the longitude of his Observatory, and published the result, 10 h 03 m 21.8 s E, in *Monthly Notices of the Royal Astronomical Society* (Tebbutt 1880b). In 1884, Arthur Auwers (1838–1915; Fig. 8.26; Tenn 1990) used Tebbutt's published data from 1873 to 1876 to derive a figure of 10 h 03 m 21.25 s E (Tebbutt 1908, p. 42), and in 1896 Dr. Hugo Clemens

**Fig. 8.26** Arthur Auwers (adapted from https://en.wikipedia.org)

from Gottingen University used Tebbutt's data from 1864 to 1870, and produced a value of 10 h 3 m 20.60 s E (Tebbutt 1908, p. 35).

Another approach to the longitude problem was to arrange a telegraphic exchange of time signals between Windsor Observatory and some other observatory with a well-determined longitude. The problem was identifying a reliable reference observatory. In terms of proximity, Sydney was the obvious choice, but Tebbutt did not place much faith in its longitude figure (see Tebbutt 1883a, 1897). Indeed, in 1878 he went so far as to publish a paper in the *Journal of the Royal Society of New South Wales* suggesting a correction to its 'accepted' longitude (Tebbutt 1878b) and followed this up with a further paper 2 years later (Tebbutt 1880a). Both of his derived figures differed from that reported earlier by Russell (1878) and, to use Tebbutt's (1891) own words, '... were treated with indifference'. In 1884, Auwers produced what was seen as a reliable figure for this Observatory, and that gave Tebbutt hope of obtaining a realistic figure for his own Observatory. In 1887, he went ahead and planned a telegraphic transfer of time signals between these two institutions and Melbourne Observatory, but this exchange was only achieved after a good deal of frustration. Once he had made the necessary arrangements with Ellery (Bio Box 7.3) in Melbourne and Lenehan (Bio Box 7.2) in Sydney (Russell was overseas at the time), Tebbutt wrote twice to the Superintendent of Telegraphs explaining the nature and importance of the exercise, but he received no reply. Eventually, his friend Lenehan intervened and

> ... I received a verbal reply through the window of the Windsor telegraph office, that the lines would be at my service at convenient opportunities. I thought that sufficient interest might have been taken in the matter to have allowed of my observatory being temporarily connected with the telegraph wires about a quarter of a mile distant, thus enabling me to give and receive signals of the actual star transits. Being, however, only an amateur and not an official, it was of course too much for me to expect. In order to transmit and receive signals I therefore myself carried a mean time chronometer to the telegraph station, distant only half a mile west of my observatory. (Tebbutt 1908, pp. 67–68).

The result was a difference of only 1 m 29.39 s between Sydney and Windsor Observatories.

On seven separate occasions, Tebbutt also journeyed in to Sydney with the chronometer—no mean feat in the days before the automobile—and compared his timekeeper with Sydney Observatory chronometers, which was another way of determining longitude. The mean difference, at 1 m 29.49 s, was remarkably similar to that obtained using the telegraphic exchange. The resulting longitudes from these two investigations were 10 h 03 m 20.21 s E and 10 h 03 m 20.11 s E.

Although these figures were consistent with earlier results, Tebbutt must have been contemplating further telegraphic transfers in 1899 because he received strong counsel against this from T.F. Furber (1899):

> [I] ... thought it my duty to urge the futility of continually tinkering at the adopted values of Australian longitudes derived in this way, until some great improvement is made in that method.

**Bio Box 8.4: Thomas Frederick Furber**

Thomas Frederick Furber (Fig. 8.27) was born in England on 13 May 1855, and moved to Sydney with his family while a young boy (Baracchi 1914; Obituary 1925; Orchiston 1987a; Wood 1981). He trained as a surveyor in the Lands Department, and rose from Draftsman and Computer in the Trigonometrical Survey Branch to Chief Computer and eventually Director of the Branch.

**Fig. 8.27** Thomas Frederick Furber (https://en.wikipedia.org)

Very early in his career, Furber '… displayed wonderful mathematical ability …' (Anderson 1925), and this was applied to his astronomical interests. In 1882 Furber observed the transit of Venus, and for many years he lectured in geodesy and astronomy at the University of Sydney. His published *piece de resistance* was a 62-page paper which was primarily about the trigonometrical survey of New South Wales (Furber 1898).

Furber was elected an FRAS in 1896, and at various times served as President, Vice-President, Secretary and Editor of the New South Wales Institution of Surveyors (Mr. T.F. Furber 1904). He also was a stalwart of the NSW Branch of the British Astronomical Association and served on the Committee for many years—including as President and Vice-President—but he rarely participated in observing programmes (Orchiston 1988b).

Thomas Frederick Furber died on 7 October 1924.

At the time Thomas Frederick Furber (Bio Box 8.4) was Chief Computer at the Lands Department and a prominent independent astronomer in Sydney.

Yet another means of determining the longitude of an observatory was to link it directly to the local trigonometric survey, which in the case of New South Wales had made considerable progress by 1887 (see Furber 1898). Tebbutt (1888) reports that in this year, he

> … made a trigonometrical connection between my transit instrument and the north terminal of the Government base-line, about three miles nearly due west of my observatory … A

**Bio Box 8.5: Joseph Brooks**

Joseph Brooks (Fig. 8.28) was born at Stockport, England, on 2 August 1847, and emigrated to Adelaide at the age of 9 (Dun 1919). In 1864 he joined the South Australian Department of Survey and Crown Lands as a draughtsman, and in 1875 was transferred to the field staff as a trigonometrical surveyor (Wood 1979).

Brooks joined the staff of the New South Wales Lands Department as a surveyor in 1877 (Dun 1919), and worked in the Trigonometrical Survey Branch. In 1886, he was promoted to the new position of 'Field Astronomer', the only person outside of Sydney Observatory to be officially employed by the colonial Government at this time as a professional astronomer (see Orchiston 1987a). Eventually he became Superintendent of the Trigonometrical Survey Branch.

Expedition (after McClean et al. 1910, Plate 31).

**Fig. 8.28** Joseph Brooks, is shown front row on the extreme right, during the Port Davey solar eclipse

Over the years, Brooks carried out numerous latitude and longitude determinations in order to pinpoint the positions of first-order trig stations (Wood 1979) and was described as '... a born observer with considerable mechanical ability, which enabled him to use his instruments to the best advantage' (Dun 1919, p. 4). In addition to surveying instruments, he also had access to an excellent 4.75-in. (12.1-cm) Schroeder refractor, which was acquired by the Lands Department in 1880.

(continued)

In 1881, Brooks and W.J. Conder (Bio Box 10.2), also from the Lands Department, observed a transit of Mercury from Bathurst, as part of a programme organised by Sydney Observatory (see Orchiston 1987a). In later years, Brooks was one of Tebbutt's strongest supporters, assisting with his occultation programme by computing the details of forthcoming events (see Tebbutt 1908).

Brooks is described as having a somewhat difficult temperament, and he could be impatient with the shortcomings of others (Wood 1979). Initially, he worked closely with H.C. Russell from Sydney Observatory, but as the relationship between the Observatory and the Lands Department deteriorated during the 1880s the two men became bitter enemies and by 1891 were not even on speaking terms (see Orchiston 1987a).

Throughout his life Brooks was active in astronomical groups and societies. He was a member of the short-lived Section A of the Royal Society of New South Wales (Orchiston and Bhathal 1991), and served as Acting Chairman and Acting Secretary on several occasions in 1881. He was elected a Fellow of the Royal Astronomical Society in 1890 and the Royal Geographical Society, but refused to join the NSW Branch of the British Astronomical Association when this group was formed because of his intense dislike of one of the leading members.

After his retirement, in 1906, Brooks was able to pursue his interest in solar eclipses, and between 1908 and 1911 (inclusive) he took part in expeditions to Flint Island (in the Pacific Ocean), Port Davey (Tasmania) and Vavau, near Tonga (see Lockyer 1912; McClean et al. 1908, 1910). Joseph Brooks died at his home in Sydney on 9 May 1918 (Dun 1919).

comparison of these results gives 1 m 29 s.77 as the longitude of my transit instrument west of the Sydney transit-circle.

Although this was a similar result to that obtained by the telegraphic transfer of time signals, the Lands Department's 'Field Astronomer', Joseph Brooks (see Bio Box 8.5), must have felt that it could be improved on, for in 1896 he offered to come to Windsor Observatory and carry out a triangulation using two visible trig points (Brooks 1896). It is not clear whether in fact he did this, but we do know that by the turn of the century the latitude and longitude of Windsor Observatory were known with considerable precision, and that this private establishment joined with Sydney and Melbourne Observatories in providing the fundamental reference framework for all Australian astronomical observatories. A summary of the various aforementioned attempts to pin down the longitude of Windsor Observatory is presented in Table 8.2.

Although no further telescopes or astronomical equipment were acquired for the Windsor Observatory after 1886 (when the 8-in. Grubb refractor arrived), one major non-astronomical instrument was added and this was a tide gauge. Because of its proximity to the Hawkesbury River, South Creek was tidal in the vicinity of the Peninsula Estate, and in 1889 Tebbutt decided to install a tide gauge. This device

**Table 8.2** A chronological listing of the various longitude values derived for Windsor Observatory

| Technique | Year | Longitude value | Reference |
|---|---|---|---|
| Telegraphic exchange of time signals with Sydney Observatory | 1865 (Smalley and Tebbutt) | 10 h 03 m 15.7 s | Tebbutt (1868a) |
| Moon culminations | 1866 (Tebbutt) | 10 h 03 m 19.1 s | Tebbutt (1868a) |
| Moon culminations | 1867 (Tebbutt) | 10 h 03 m 23.1 s | Tebbutt (1868a) |
| Lunar occultations (1864–1876) | 1880 (Tebbutt) | 10 h 03 m 21.8 s | Tebbutt (1880b) |
| Lunar occultations (1873–1876) | 1884 (Auwers) | 10 h 03 m 21.25 s | Auwers (1884) |
| Telegraphic exchange of time signals with Sydney and Melbourne Observatories | 1887 (Tebbutt Lenehan and White) | 10 h 03 m 20.21 s | Tebbutt (1888) |
| Transporting a Windsor Observatory chronometer to Sydney Observatory | Several times prior to 1888 (Tebbutt) | 10 h 03 m 20.11 s | Tebbutt (1888) |
| Reference to the Trig Survey | 1887 (Tebbutt) | 10 h 03 m 20.59 s | Tebbutt (1888) |
| Lunar occultations (1864–1870) | 1896 (Clemens) | 10 h 03 m 20.60s | Tebbutt (1908) |

> … was constructed in the usual way. By means of a float and pencil the daily tide traces were obtained on a sheet of paper wrapped on a cylinder turned at a rate of about three-quarters of an inch per hour by a clock. The instrument was mounted in a tide house on the bank of the South Creek at a point due south of the observatory. (Tebbutt 1908, p. 74).

Tebbutt (ibid.) reports that between 26 March 1889 and 1 February 1890, 331 high tides and 337 low tides were recorded and their mean heights were calculated. The mean difference between high and low water was 1.62 ft (49.4 cm), significantly less than on the nearby Hawkesbury River. Tebbutt was also able to draw on his own tidal data and prepare a short paper on the high tides of 15–17 June 1889 (Tebbutt 1889), and 4 years later he published a further paper on high tides (Tebbutt 1893).

Back in the second half of the nineteenth century, South Creek originated on rural land to the west of suburban Sydney, and after flowing 43 miles (70 km) and dropping 308 ft (94 m) it flowed into the Hawkesbury River. Because South Creek and its 17 tributaries covered a substantial catchment area, South Creek near Windsor was prone to flooding during and immediately after heavy rain and substantial areas around Windsor were liable to be inundated (see Fig. 1.15). Fortunately, Tebbutt maintained a record of all floods and freshes (minor floods) and published these in his meteorological monographs (see Tebbutt 1877c, 1882, 1886b, 1891, 1898, 1916). Following the installation of the tide gauge, he was able to establish the relative height of the barometer cistern back at the Observatory (51.98 ft, or 15.8 m), and this was then used as the local datum for measuring floods and freshes.

Tebbutt found it essential to keep a rowing boat in his barn, and he and his family used this to commute to Windsor when South Creek rose more than 28.5 ft (8.7 m) and water completely surrounded 'Observatory Hill' (as in Fig. 8.29). On these occasions, when the Tebbutt Homestead and Windsor Observatory were totally isolated from the town, they presented a picturesque image which one journalist likened to '… an island home and suggesting the idea of a miniature Uraniburg' (Clericus n.d.). He was, of course, referring to Tycho Brahe's Uraniborg (i.e. 'Castle

**Fig. 8.29** An undated photograph of a flood on South Creek where the Tebbutt Homestead and Windsor Observatory are isolated and can only be accessed by boat (Orchiston collection)

of the Sky') on the island of Hven between Zealand (Denmark) and Norway. A list of all those floods that isolated the Tebbutt Homestead and observatory buildings in this way is given in Table 2.3.

Table 2.3 shows that there were 28 floods that isolated Windsor Observatory during this 48-yr period, and that these were not distributed randomly. In some years there was more than one major flood, whereas there were no such floods at all between September 1879 and May 1889. Furthermore, some months were more flood-prone than others: 50 % of all of the floods listed in Table 2.3 occurred in just the 3 mid-year months of May, June and July, and the consecutive 6 months, February–July account for 82 % of all floods.

Serious flooding had a major effect on Tebbutt's farming operations and drastically disrupted his daily routine, thereby impacting on his daytime *and* evening astronomical activities. However, the only flood that actually entered and damaged the Windsor Observatory was that of 23 June 1867. Tebbutt wrote in his *Astronomical Memoirs*:

> … the observatory was almost completely submerged. The instruments had been removed to a higher position for safety, and were replaced after the subsidence of the inundation. The flood at the maximum was ten feet five inches [3.18 m] deep on the ground floor of my father's residence, and exactly reached the lower ceiling. Up to the date [1908] of this

publication, this is the only flood which has entered the residence or the observatories. (Tebbutt 1908, p. 33).

Over the years, Tebbutt supplied information on local tides and floods to a number of researchers, particularly those from the Harbours and Rivers Branch of the Department of Public Works (e.g. see Moriarty 1869). One such individual was Joshua Percy Josephson (1852–1911; Maiden 1912) who carried out an accurate trigonometrical survey of the Hawkesbury River and published a valuable paper on its floods (Josephson 1885). Tebbutt also prepared a number of newspaper articles about local floods (e.g. Tebbutt 1867, 1874b, 1886a) based upon his own records.

**Acknowledgements** I am grateful to Dr. Aedeen Cremin (University of Sydney), Dr. Steve Dick (formerly U.S. Naval Observatory, Washington DC), Iwona Hetherington (Museum of Applied Arts and Sciences, Sydney), Julian Holland (formerly Macleay Museum, Sydney), Professor Nick Lomb (University of Southern Queensland, Toowoomba), the late Dr Harley Wood (Sydney Observatory) and staff at the Mitchell Library (Sydney) for their assistance, and to the Museum of Applied Arts and Sciences and the South African Astronomical Observatory (Cape Town) for kindly supplying Figs. 8.2 and 8.21.

## References[1]

Abrahams, P. (2002). The telescope in Ireland: Obscure makers & marks. Irish telescope makers and Irish signatures on telescopes. (http://home.europa.com/~telscope/tsireland.txt).

Anderson, C. (1925). Presidential address. *Journal and Proceedings of the Royal Society of New South Wales, 59,* 1–34.

Andrews, A.D. (1992). Cyclopaedia of telescope makers. Part 1 (A–F). *The Irish Astronomical Journal, 20,* 102–183.

Andrews, A.D. (1993). Cyclopaedia of telescope makers. Part 2 (G–J). *The Irish Astronomical Journal, 21,* 1–82.

Andrews, A.D. (1996a). Cyclopaedia of telescope makers. Part 5 (Sae–Sim). *Irish Astronomical Journal, 23,* 57–117.

Andrews, A.D. (1996b). Cyclopaedia of telescope makers. Part 6 (Sin–Syk). *Irish Astronomical Journal, 23,* 215–242.

Ashbrook, J. (1984). John Tebbutt, his observatory, and a probable nova. In J. Ashbrook, *The Astronomical Scrapbook. Skywatchers, Pioneers and Seekers in Astronomy* (pp. 66–71). Cambridge (Mass.): Sky Publishing Corporation.

Austin, Tollis, Weiss & Youden (n.d.). *Tebbutt's Observatory Peninsular Estate Windsor*. Sydney: School of Architecture and Building, University of New South Wales. Copy in the Mitchell Library (PXD 110-1).

Auwers, A. (1884). Bestimmung eines fundamentalen Meridians für Australien durch absolute Methoden. *Astronomische Nachrichten, 110,* 289.

Baracchi, P. (1914). Astronomy and geodesy in Australia. In G.H. Knibbs (Ed.), *Federal Handbook Prepared in Connection with the Eighty-fourth Meeting of the British Association for the*

[1]The following abbreviation is used: TL = Letters to J. Tebbutt, bound manuscript letters in the Mitchell Library, Sydney.

*Advancement of Science, Held in Australia, August, 1914* (pp. 326–390). Melbourne: Government Printer.

Barker, G. (2010). Science underground – Angelo Tornaghi: instrument maker and entrepreneur. Sydney, Powerhouse Museum (www.powerhousemusuem.com/insidethecollection/2010/08/science-underground-angelo-tornaghi-instrument-maker-and-entrepreneur/)

Barker, G. (2013). Angelo Tornaghi – scientific instrument maker Sydney – 1831–1906. Sydney, Powerhouse Museum (http://geoffsleftfoot.blogspot.com/2013/08/angelo-tornaghi-instrument-maker-and.html)

Barlow, P. (1849). *Plane Astronomy*. London: John Joseph Griffin and Co.

Baum, R. & Sheehan, W. (1997). *In Search of Planet Vulcan. The Ghost in Newton's Clockwork Universe*. New York: Plenum

Bone, W. (1880). Observations of the Great Southern Comet I (1880) made at Castlemaine, Victoria. *Monthly Notices of the Royal Astronomical Society, 41,* 85–88.

Bone, W. (1882a). Computed and observed times of contacts of the transit of Mercury at Castlemaine, Victoria … *Monthly Notices of the Royal Astronomical Society, 42,* 129.

Bone, W. (1882b). Letter to John Tebbutt, dated 5 May. In TL.

Bone, W. (1882c). Letter to John Tebbutt, dated 13 June. In TL.

Bone, W. (1882d). On an object seen near Comet b, 1881, on June 10, 1881. *Monthly Notices of the Royal Astronomical Society, 42,* 105–106.

Brooks, J. (1896). Letter to John Tebbutt, dated 25 May. In TL.

Brooks, R.C. (1991). The development of micrometers in the seventeenth, eighteenth and nineteenth centuries. *Journal for the History of Astronomy, 22,* 127–173.

Brush and Macdonnell (1864). Letter to John Tebbutt, dated 12 March. In TL.

Challis, J. (1879). *Lectures on Practical Astronomy and Astronomical Instruments*. Cambridge: Deighton, Bell and Co.

Chambers, G.F. (1861). *Handbook of Descriptive and Practical Astronomy*. London: Murray.

Chambers, G.F. (1867). *Descriptive Astronomy*. Oxford: Clarendon Press.

Chambers, G.F. (1877). *A Handbook of Descriptive Astronomy*. Oxford: Clarendon Press.

Clericus (n.d.). Tour to the Hawkesbury and Colo. Newspaper Clipping, Volume 78, page 234, Mitchell Library, Sydney (Q981 N).

Cremin, A. (1987). Report on the Excavation of Tebbutt's First Observatory at Windsor, NSW May and July 1987. Unpublished Report, Department of History, University of Sydney.

Dawson, M. (1863). Letter to John Tebbutt, dated 5 April. In TL.

Death of Dr W. Bone. *Mount Alexander Mail,* 2 September 1885.

Delambre, M. (1814). *Astronomie Theorique et Practique*. Three volumes. Paris: Courcier.

Dewhirst, D.W. (1985). Meridian astronomy in the private and university observatories of the United Kingdom: rise and fall. *Vistas in Astronomy, 28,* 147–158.

Dick, S. (2003). *Sky and Ocean Joined. The U.S. Naval Observatory 1830–2*000. Cambridge: Cambridge University Press.

Dick, S., Orchiston, W. & Love, T. (1998). Simon Newcomb, William Harkness and the nineteenth century American transit of Venus expeditions. *Journal for the History of Astronomy, 29,* 221–255.

Digby, E. (Ed.) (1889–1890). *Australian Men of Mark. Volume 2. Sydney*: C.F. Maxwell.

Donnelly, M.C. (1973). *A Short History of Observatories*. Eugene: University of Oregon Press.

Dun, W.S. (1919). Presidential address. *Journal and Proceedings of the Royal Society of New South Wales, 53,* 1–28.

Ellery, R.J. (1869). *Astronomical Observations Made at the Williamstown Observatory in the Years 1861, 1862, and 1863*. Melbourne: Government Printer.

Ferguson, J. & Brewster, D. (1821). *Astronomy, Explained Upon Sir Isaac Newton's Principles. Second Edition*. Edinburgh: Stirling and Slade.

Flavelle Bros & Roberts (1880). Letter to John Tebbutt, dated 30 March. In TL.

Flavelle Bros & Roberts (1882). Letter to John Tebbutt, dated 28 April. In TL.

Fontenrose, R. (1973). In search of Vulcan. *Journal for the History of Astronomy, 4,* 145–158.

Forbes, G. (1874). *The Transit of Venus*. London: Macmillan.

Furber, T.F. (1898). The trigonometrical survey of New South Wales, with mention of similar surveys in the other Australian colonies. In *Report of the Seventh Meeting of the Australasian Association for the Advancement of Science* (pp. 176–237). Sydney: AAAS.

Furber, T.F. (1899). Letter to John Tebbutt, dated 29 March. In TL.

Gegen, D. (1994). Thomas Cooke … the English Alvan Clark. *Journal of the Antique Telescope Society, 5,* 5–7.

Gingerich, O. (Ed.) (1984). *The General History of Astronomy. Volume 4. Astrophysics and Twentieth Century Astronomy to 1950: Part A*. Cambridge: Cambridge University Press.

Glass, I.S. (2008). *Proxima, the Nearest Star (other than the Sun)*. Cape Town: Mons Mensa.

Grant, R. (1874). *The Transit of Venus in 1874*. Glasgow: Maclehose.

Haynes, R., Haynes, R., Malin, D. & McGee, R. (1996). *Explorers of the Southern Sky. A History of Australian Astronomy*. Cambridge: Cambridge University Press.

Herschel, J.F.W. (1851). *Outlines of Astronomy.* London: Longman, Brown, Green & Longman.

Hordern, A. (1870). Letter to John Tebbutt, dated 6 July. In TL.

Innes, R.T.A. (1895). Letter to John Tebbutt, dated 13 September. In TL.

Innes, R.T.A. (1899). Reference catalogue of southern double stars. *Annals of the Cape Observatory, 2* (2).

Innes, R.T.A. (1901). Letter to John Tebbutt, dated 21 November. In TL.

Janiczek, P.M. & Houchins, L. (1974). Transits of Venus and the American expedition of 1874. *Sky and Telescope, 48,* 366–371.

Josephson, J.P. (1885). History of floods in the Hawkesbury River. *Journal of the Royal Society of New South Wales, 19,* 97–107.

Lardner, D. & Dunkin, E. (1860). *Handbook of Astronomy*. London: Walton & Maberly.

Lequeux, J. (2016). *Le Verrier—Magnificent and Detestable Astronomer*. New York: Springer.

Lockyer, J.N. (1878). *Stargazing: Past and Present*. London: Macmillan & Co.

Lockyer, W.J.S. (1912). *Report of the Solar Eclipse Expedition to Vavau, Tonga Islands, April 29, 1911 (Eastern Date)*. London: His Majesty's Stationery Office.

Lomb, N. (2011). *Transit of Venus: 1631 to the Present.* Sydney: NewSouth & Powerhouse Museum.

Loomis, E. (1861). *An Introduction to Practical Astronomy*. New York: Harper.

McClean, F.K., et al. (1908). *Report of the Solar Eclipse Expedition to Flint Island January 3, 1908*. London: printed for the authors.

McClean, F.K., et al. (1910). *Report of the Solar Eclipse Expedition to Port Davey, Tasmania May, 1910*. London: printed for the authors.

McConnell, A. (1992). *Instrument Makers to the World. A History of Cooke, Troughton and Simms*. York: William Sessions.

Madsen, H.F. (1886). Notes on the process of polishing and figuring 18 inch glass specula by hand. *Journal and Proceedings of the Royal Society of New South Wales, 20,* 79–92.

Maguire, R. (1985). Angelo Tornaghi. An inventive Italian of the nineteenth century. *The Australian Antique Collector, 29,* 42–46.

Maiden, J.H. (1912). Presidential address. *Journal and Proceedings of the Royal Society of New South Wales, 46,* 1–73.

Mercer, T. (2004). *Chronometer Makers of the World*. Revised Edition. London: NAG Press.

Mollan, C. (1995). *Irish National Inventory of Historic Scientific Instruments*. Dublin: Samton.

Moriarty, S. (1869. Letter to John Tebbutt, dated 19 November. In TL.

Mr. T.F. Furber. *The Surveyor, 17,* 221–223 (1904).

N.S.W. (1893). A self-taught astronomer. *The Australasian,* 26 August.

Obituaries. R.T.A. Innes. *Journal of the British Astronomical Association, 43,* 260–262 (1933a).

Obituaries. Robert Thorburn Ayton Innes. *Monthly Notices of the Royal Astronomical Society, 94,* 277–281 (1933b).

Obituary: Thomas Frederick Furber. *Monthly Notices of the Royal Astronomical Society 85,* 310 (1925).

Obituary. R.T.A. Innes: 1861–1933. *Journal of the Astronomical Society of South Africa, 3,* 125–127 (1933).

Olmstead, D. (1853). *Mechanism of the Heavens*. London: T. Nelson & Sons.

Orchiston, W. (1986). *Observatories of Victoria. 1: Dr William Bone and the Castlemaine Observatory*. Melbourne: Astronomical Society of Victoria (Historical Notes No.3).

Orchiston, W. (1987a). Contribution of the Lands Department to the development of astronomy in New South Wales during the nineteenth century. *Australian Journal of Astronomy, 2,* 65–74.

Orchiston, W. (1987b). Dr William Bone, and the role of the amateur observatory in Australian astronomy. *Southern Stars, 32,* 111–128.

Orchiston, W. (1988a). The emergence of professional astronomy in nineteenth century Australia. *Search, 19,* 76–81.

Orchiston, W. (1988b). Illuminating incidents in Antipodean astronomy: formation of the New South Wales Branch of the British Astronomical Association. *Journal of the British Astronomical Association, 98,* 75–84.

Orchiston, W. (1988c). Illuminating incidents in Antipodean astronomy: John Tebbutt and the founding of the Windsor Observatory. *Journal of the British Astronomical Association, 98,* 287–293.

Orchiston, W. (1992). The contribution of Francis Abbott to Tasmanian and Australian astronomy. *Vistas in Astronomy, 35,* 315–344.

Orchiston, W. (1997a). The role of the amateur in popularising astronomy: an Australian case study. *Australian Journal of Astronomy, 7,* 33–66.

Orchiston, W. (1997b). The role of the large refracting telescope in Australian amateur astronomy: an historical perspective. *Australian Journal of Astronomy*, 7, 89–114.

Orchiston, W. (1997c). The "tyranny of distance" and Antipodean cometary astronomy. *Australian Journal of Astronomy*, *7,* 115–126.

Orchiston, W. (2001a). From amateur astronomer to observatory director: the curious case of R.T.A. Innes. *Publications of the Astronomical Society of Australia, 18,* 317–327.

Orchiston, W. (2001b). 'Sentinel of the Southern Heavens': The Windsor Observatory of John Tebbutt. *Journal of the Antique Telescope Society, 21,* 11–23.

Orchiston, W. (2003a). Amateur telescope making in Australia. An historical perspective. In W.J. Cook (Ed.), *The Best of Amateur Telescope Making Journal. Volume 2* (pp. 208–239). Richmond, Willmann-Bell.

Orchiston, W. (2003b). Passion and patronage: the Afro-Australian astronomy of Robert Thorburn Ayton Innes. *Monthly Notices of the Astronomical Society of Southern Africa, 62,* 4–13.

Orchiston, W (2004). The nineteenth century transits of Venus: an Australian and New Zealand overview. *Journal of Astronomical Data, 10,* 219–308.

Orchiston, W. (2005). James Cook's 1769 transit of Venus expedition to Tahiti. In D.W. Kurtz (Ed.), *Transits of Venus: New Views of the Solar System and Galaxy* (pp. 52–66). Cambridge: Cambridge University Press.

Orchiston, W. (2017). Captain Cook and the 1769 transit of Venus: the Australian connection. *Records of the Queensland Museum*, in press.

Orchiston, W. & Bembrick, C. (1995). The role of the large reflecting telescope in amateur astronomy: an Australian case study. *Australian Journal of Astronomy, 6,* 53–74.

Orchiston, W. & Bhathal, R. (1991). Illuminating incidents in Antipodean astronomy: Section A of the Royal Society of New South Wales. *Australian Journal of Astronomy, 4,* 49–67.

Orchiston, W. & Buchanan, A. (1993). Illuminating incidents in Antipodean astronomy: Campbell Town, and the 1874 transit of Venus. *Australian Journal of Astronomy, 5,* 11–31.

Orchiston, W. & Buchanan, A., (2004). 'The Grange', Tasmania: survival of a unique suite of 1874 transit of Venus relics. *Journal of Astronomical History and Heritage, 7,* 34–43.

Orchiston, W., Buchanan, A. Price, G. & Sprent, T. (2015). Highlighting our international astronomical heritage: Tasmanian relics of the 1874 transit of Venus. *Journal of Astronomical History and Heritage, 18,* 149–160.

Pasachoff, J. (1998). Williams College's Hopkins Observatory: the oldest extant observatory in the United States. *Journal of Astronomical History and Heritage, 1,* 61–78.

Phillips, T.E.R. (Ed.) (1923). *Hutchinson's Splendour of the Heavens. Volume 1.* London: Hutchinson & Co.
Russell, H.C. (1863a). Letter to John Tebbutt, dated 13 February. In TL.
Russell, H.C. (1863b). Letter to John Tebbutt, dated 10 July. In TL.
Russell, H.C. (1863c) Letter to John Tebbutt, dated 21 October. In TL.
Russell, H.C. (1863d). Letter to John Tebbutt, dated 19 November. In TL.
Russell, H.C. (1878). On the longitude of Sydney Observatory. *Journal and Proceedings of the Royal Society of New South Wales, 12,* 222–225.
Russell, H.C. (1880) Observations of the Great Southern Comet, 1880, made at the Observatory, Sydney. *Monthly Notices of the Royal Astronomical Society, 40,* 379.
Russell, H.C. (1892). *Observations of the Transit of Venus, 9 December, 1874; Made at Stations in New South Wales*. Sydney: Government Printer.
Scott, W. (1861). Letter to John Tebbutt, dated 4 June. In TL.
Scott, W. (1875). The Windsor Observatory. *Sydney Morning Herald,* 24 March.
Stott, C. & Hughes, D.W. (1987). The amateur's small transit telescope in the nineteenth century. *Quarterly Journal of the Royal Astronomical Society, 28,* 30–42.
Teale, R. (1972). Hordern, Anthony (1819–1876). In R.B. Nairn, A.G. Serle & R.B. Ward (Eds.), *Australian Dictionary of Biography. Volume 4: 1851–1890. D-J* (pp. 423–424). Melbourne: Melbourne University Press.
Tebbutt, J. (n.d.). *Catalogue of the Observatory Library of John Tebbutt*. MSS, Mitchell Library (A3781).
Tebbutt, J. (1853–59). *Observations made … at the Peninsula, near Windsor, in the Colony of New South Wales*. MS, Mitchell Library, Sydney (A3745$^{-1}$).
Tebbutt, J. (1859–63). Diary. MS, Mitchell Library, Sydney (A3785).
Tebbutt, J. (1863). Diary. MS, Mitchell Library, Sydney (A3786).
Tebbutt, J. (1864–69). *Journal. Observatory, Peninsula, Windsor, New South Wales*. MS, Mitchell Library, Sydney (A3747).
Tebbutt, J. (1867). The weather at Windsor. *Sydney Morning Herald,* 28 June.
Tebbutt, J. (1868a). Eclipses of Jupiter's satellites, observed at Windsor, New South Wales. *Monthly Notices of the Royal Astronomical Society, 28,* 215–217.
Tebbutt, J. (1868b). *Meteorological Observation Made at the Private Observatory of John Tebbutt, Junr., The Peninsula, Windsor, New South Wales, in the Years 1863, 1864, 1865, and 1866*. Sydney: printed for the author.
Tebbutt, J. (1870–73). *Journal of the Observatory. The Peninsula, Windsor, New South Wales*. MS, Mitchell Library, Sydney (A3748).
Tebbutt, J. (1873). Untitled description of the Windsor Observatory. In Tebbutt 1873–79.
Tebbutt, J. (1873–79). Untitled journal. MS, Mitchell Library, Sydney (A3783).
Tebbutt, J. (1874a). *Meteorological Observations Made at the Private Observatory of John Tebbutt, The Peninsula, Windsor, New South Wales, in the Years 1867, 1868, 1869, and 1870*. Sydney: printed for the author.
Tebbutt, J. (1874b). The periodic occurrence of heavy rains and floods. *Sydney Morning Herald,* 2 March.
Tebbutt, J. (1875). Beobachtung des Venusdurchganges in Windsor (N.-S.-Wales). *Astronomische Nachrichten, 85,* 173–176.
Tebbutt, J. (1877a). Discovery of a remarkable variable star. *Sydney Morning Herald,* 19 November.
Tebbutt, J. (1877b). On a new and remarkable variable star in the constellation Ara. *Journal of the Royal Society of New South Wales, 11,* 197–202.
Tebbutt, J. (1877c). *Results of Meteorological Observations Made at the Private Observatory of John Tebbutt, The Peninsula, Windsor, New South Wales, in the Years 1871, 1872, 1873, 1874, 1875 & 1875*. Sydney: printed for the author.
Tebbutt, J. (1877d). *Search for Vulcan*. MS, Mitchell Library, Sydney (A3783).
Tebbutt, J. (1878a). On a new variable in the constellation Ara. *Monthly Notices of the Royal Astronomical Society, 38,* 330–331.

Tebbutt, J. (1878b). Proposed correction to the assumed longitude of the Sydney Observatory. *Journal and Proceedings of the Royal Society of New South Wales, 12,* 63–69.

Tebbutt, J. (1879). Untitled description of the Windsor Observatory, dated 18 September. In the "Introduction" to Tebbutt, 1879–1881.

Tebbutt, J. (1879–81). *Journal for Astronomical Observations. The Peninsula, Windsor, N.S.Wales.* MS, Mitchell Library, Sydney (A3749).

Tebbutt, J. (1880a). On the longitude of Sydney Observatory. *Journal and Proceedings of the Royal Society of New South Wales, 14,* 19–21.

Tebbutt, J. (1880b). On the longitude of the Observatory, Windsor, N. S. Wales. *Monthly Notices of the Royal Astronomical Society, 40,* 440.

Tebbutt, J. (1882). *Results of Meteorological Observations Made at the Private Observatory of John Tebbutt, The Peninsula, Windsor, New South Wales, in the Years 1877, 1878, 1879, 1880, & 1881.* Parramatta & Sydney: printed for the author.

Tebbutt, J. (1883a). Note on Professor Newcomb's remarks on the Windsor observations of the transit of Venus in 1874. *Monthly Notices of the Royal Astronomical Society, 43,* 279–280.

Tebbutt, J. (1883b). Observations of the transit of Venus, 1874, December 8–9, made at Windsor, N.S.W. *Memoirs of the Royal Astronomical Society, 47,* 89–92.

Tebbutt, J. (1884). On the latitude of the Observatory, Windsor, N.S. Wales. *Monthly Notices of the Royal Astronomical Society, 44,* 101–103.

Tebbutt, J. (1884–86). *Journal for Astronomical Observations. The Peninsula. Windsor, N.S.Wales.* MS, Mitchell Library, Sydney (A3751).

Tebbutt, J. (1886a). The great flood at Windsor. *Sydney Morning Herald,* 26 April.

Tebbutt, J. (1886b). *Results of Meteorological Observations Made at the Private Observatory of John Tebbutt, The Peninsula, Windsor, New South Wales, in the Years 1882, 1883, 1884, & 1885.* Parramatta & Sydney: printed for the author.

Tebbutt, J. (1887). *History and Description of Mr. Tebbutt's Observatory, Windsor, New South Wales.* Sydney: printed for the author.

Tebbutt, J. (1888). On the difference of longitude between Mr. Tebbutt's Observatory, Windsor, New South Wales, and the Government Observatories at Sydney and Melbourne. *Monthly Notices of the Royal Astronomical Society, 48,* 313–316.

Tebbutt, J. (1888–90). *Journal for Astronomical Observations. The Peninsula, Windsor, New South Wales.* MS, Mitchell Library, Sydney (A3753).

Tebbutt, J. (1889). On the high tides of June 15–17th. *Journal and Proceedings of the Royal Society of New South Wales, 23,* 51–53.

Tebbutt, J. (1890–92). *Journal for Astronomical Observations. The Peninsula, Windsor, New South Wales.* MS, Mitchell Library, Sydney (A3754).

Tebbutt, J. (1891). *Results of Meteorological Observations Made at the Private Observatory of John Tebbutt, The Peninsula, Windsor, N.S. Wales, in the Years 1886, 1887, 1888, 1889, & 1890.* Sydney: printed for the author.

Tebbutt, J. (1892). *Report of Mr Tebbutt's Observatory, The Peninsula, Windsor, New South Wales, For the Year 1891.* Sydney: printed for the author.

Tebbutt, J. (1893). On the probability of extraordinarily high spring-tides about the December solstice of 1893. *Journal and Proceedings of the Royal Society of New South Wales, 27,* 359–361.

Tebbutt, J. (1894). *Report of Mr. Tebbutt's Observatory, The Peninsula, Windsor, New South Wales, For the Year 1893.* Sydney: printed for the author.

Tebbutt, J. (1894–96). Untitled journal. MS, Mitchell Library, Sydney (A3756).

Tebbutt, J. (1895). *Report of Mr. Tebbutt's Observatory, The Peninsula, Windsor, New South Wales, For the Year 1894.* Sydney: printed for the author.

Tebbutt, J. (1896–98). Untitled journal. MS, Mitchell Library, Sydney (A3757).

Tebbutt, J. (1897). The longitude of the Sydney Observatory. *The Australian Star,* 21 January; and *The Observatory, 20,* 169–171.

Tebbutt, J. (1898). *Results of Meteorological Observations Made at the Private Observatory of John Tebbutt, The Peninsula, New South Wales, in the Years 1891, 1892, 1893, 1894, 1895, 1896, and 1897.* Sydney: printed for the author.

Tebbutt, J. (1902). *Report of Mr. Tebbutt's Observatory, The Peninsula, Windsor, New South Wales, For the Year 1901*. Sydney: printed for the author.
Tebbutt, J. (1908). *Astronomical Memoirs*. Sydney: printed for the author.
Tebbutt, J. (1916). *Results of Meteorological Observations at Mr. Tebbutt's Observatory The Peninsula, Windsor, New South Wales, During the Period 1898–1915*. Windsor: printed for the author.
Tebbutt, J. (1986). *Astronomical Memoirs*. Windsor: Hawkesbury Shire Council (Reprinted edition).
Tenn, J.S. (1990). Arthur Auwers: the second Bruce Medalist. *Mercury, 19,* 49–50.
Todd, D.P. (1881). The solar parallax as determined from the American photographs of the transit of Venus, 1874, December 8–9. *American Journal of Science, 21,* 491–493.
Tornaghi, A. (1862). Letter to John Tebbutt, dated 21 February. In TL.
Tornaghi, A. (1864a). Letter to John Tebbutt, dated 20 May. In TL.
Tornaghi, A. (1864b). Letter to John Tebbutt, dated 8 July. In TL.
Tornaghi, A. (1864c). Letter to John Tebbutt, dated 26 July. In TL.
Tornaghi, A. (1864d). Letter to John Tebbutt, dated 23 August. In TL.
Tornaghi, A. (1866a). Letter to John Tebbutt, dated 15 October. In TL.
Tornaghi, A. (1866b). Letter to John Tebbutt, dated 8 November. In TL.
Tornaghi, A. (1866c). Letter to John Tebbutt, dated 17 November. In TL.
Tornaghi, A. (1867a). Letter to John Tebbutt, dated 25 September. In TL.
Tornaghi, A. (1867b). Letter to John Tebbutt, dated 7 November. In TL.
Tornaghi, A. (1867c). Letter to John Tebbutt, dated 18 November. In TL.
Tornaghi, A. (1867d). Letter to John Tebbutt, dated 31 December. In TL.
Tornaghi, A. (1878a). Letter to John Tebbutt, dated 30 September. In TL.
Tornaghi, A. (1878b). Letter to John Tebbutt, dated 18 October. In TL.
Tupman, G.L. (1878). On the mean solar parallax as derived from the observations of the transit of Venus, 1874. *Monthly Notices of the Royal Astronomical Society, 38,* 429–457.
Turner, H.H. (1918). Obituary: John Tebbutt. *Monthly Notices of the Royal Astronomical Society, 78,* 252–255.
van Helden, A. (1984). Telescope building, 1850–1900. In Gingerich: 40–58.
van Helden, A. (1995). Measuring solar parallax: the Venus transits of 1761 and 1769 and their nineteenth-century sequels. In R. Taton & C. Wilson (Eds.), *The General History of Astronomy. Volume 2. Planetary Astronomy from the Renaissance to the Rise of Astrophysics. Part B: The Eighteenth and Nineteenth Centuries* (pp. 153–168). Cambridge: Cambridge University Press.
von Humboldt, A. (1849–1852). *Cosmos.* Four volumes. London: Longman, Brown, Green and Longman.
von Littrow, J.J. (1821). *Theoretische und Practische Astronomie* (Three volumes). Wien: J. B. Wallishausser.
Warner, D.J. (1998). Telescopes for land and sea. *Rittenhouse, 12* (2), 33–54.
Welther, B.L. (1984). The world's largest telescopes, 1850–1950. In Gingerich, Ai–Avi.
White, G. (1979). John Tebbutt and the astronomy at the Windsor Observatory. *Proceedings of the Astronomical Society of Australia, 3,* 408–411.
Wood, H. (1958). *Sydney Observatory 1858 to 1958*. Sydney: Government Printer (Sydney Observatory Papers No. 31).
Wood, H. (1979). Brooks, Joseph (1847–1918). In B. Nairn & G. Serle (Eds.), *Australian Dictionary of Biography. Volume 7: 1891–1939. A-Ch* (pp. 157–158). Melbourne: Melbourne University Press.
Wood, H. (1981). Furber, Thomas Frederick (1855–1924). In B. Nairn & G. Serle (Eds.), *Australian Dictionary of Biography. Volume 8: 1891–1939. Cl-Gib* (pp. 599–600). Melbourne: Melbourne University Press.
Zeilik, M. (1985). *The Evolving Universe*. New York: Harper & Row.

# Chapter 9
# A Timely Spectacle: The Great Comet of 1881

## 9.1 Introduction

Some comets make an impact because of their remarkable tails, while others present observers with a cavalcade of observational treats as they throw off a succession of filaments and streamers from their activated nuclei. There are also 'Great Comets' that impress with their overall majesty. These are naked-eye spectacles, with long, prominent tails.

Occasionally a comet will appear at a fortuitous time, when important new instrumentation or a novel analytical technique is being adopted by or applied to astronomy. Thus, it was that Donati's Comet arrived in 1858, just as astrophotography was being pioneered (see Pasachoff et al. 1996). Only rarely do we witness the appearance of a Great Comet which coincides with major new advances in astronomy, but such was the case with the Great Comet of 1881 which arrived when important developments were taking place in *both* astronomical spectroscopy (Hearnshaw 2014) *and* astronomical photography (Hughes 2013; Lankford 1984). This comet is also known as Comet C/1881 K1 (Tebbutt) and was John Tebbutt's second comet (and, for that matter, second 'Great Comet'). In the twenty years between his two discoveries, he had observed seven different comets (but one of them, periodic comet 2P/Encke, on four separate returns) and had published a number of papers on them.

In this chapter we examine Tebbutt's discovery of the 1881 comet and trace his subsequent observations and the problems associated with his computation of the orbital elements. We also discuss the controversy that emerged over discovery priorities and outline ways in which Comet C/1881 K1 (Tebbutt) made significant contributions to both cometary photography and cometary spectroscopy. In so doing, I draw freely on material that was published in 1981 to mark the centennial of this comet's discovery (Orchiston 1981) and on a more recent study (Orchiston 1999; see, also, Orchiston 2010).

 
W. Orchiston, *John Tebbutt*, Historical & Cultural Astronomy,
DOI 10.1007/978-3-319-44521-2_9

## 9.2 A Successful Scan of the Southern Sky: Discovery of a New Comet

Although Tebbutt summarises the discovery in his *Astronomical Memoirs* (1908), more information is available in his unpublished journal of observations:

> On going out this evening [22 May 1881] at 6h 15m as I had done several times lately to scan the sky for comets with the naked eye I detected what appeared to be a nebulus [*sic*] object in the south west a few degrees above the place where I discovered the great comet of 1861. I saw at a glance that it was a new object as I am very familiar with that part of the heavens. (Tebbutt 1879–81)

Between 18:37 and 19:11 h local time, Tebbutt used his 4.5-in. Cooke refractor and the filar micrometer to obtain eight positional measurements of the object. For comparisons he used $\gamma^1$ and $\gamma^2$ Caeli, with visual magnitudes of approximately 4.5 and 5.5, respectively, which conveniently lay in the comet's telescopic field of view (ibid.). Tebbutt (ibid.) also noted '… a broad tail [which] can be seen in the telescope …', although this feature was not yet visible to the naked eye.

These telescopic observations confirmed the existence of a hitherto unrecorded comet, and the following day (23 May), Tebbutt lost no time in dispatching a telegram to the *Sydney Morning Herald* and notifying the government observatories in Sydney and Melbourne. Later the same day, he prepared a fuller public announcement of the discovery, but this only appeared in the *Sydney Morning Herald* on 26 May 1881 (one day after his 47th birthday):

> I announced to you this morning by telegraph that I discovered on the previous evening a comet between the constellations of Columba and Eridanus … it appeared to the unaided eye as a hazy patch, but without a tail. In the telescope, however, the tail could be distinctly seen … It is a remarkable circumstance that the date of discovery here is only nine days later than the anniversary of the discovery of the great comet of 1861, and that the present comet was found only about fifteen degrees east of the position of the great comet just mentioned. (Tebbutt 1881c)

Tebbutt's first report to the international astronomical fraternity was in *Astronomische Nachrichten* (Tebbutt 1881f), and this new object was eventually designated Comet C/1881 K1 (Tebbutt) and is also known as 'The Great Comet of 1881' (e.g. see Gill 1882).

However, Tebbutt's claim on the comet, as was often the case with nineteenth-century comet discoveries, was for some time in contention as he waited to see whether documentation of a legitimate prior discovery would emerge. Such was the tyranny of distance! And so, as letters and verbal messages of congratulation flooded into the Windsor Observatory, we find sentiments expressed like the following:

> I suppose you have not heard yet whether Professor Gould caught the first glimpse of the stranger. You certainly have the credit of being the first Australian astronomer to detect it—if not its actual discoverer. (Powys 1881a)

A similar fear is contained in T.H.F. Griffin's letter of 24 May, the first one Tebbutt received applauding his discovery: that the comet hopefully had not been located earlier by astronomers at the Cape of Good Hope or at some other southern observatory with professional staff and large telescopes (Griffin 1881a). There was something wonderfully romantic about a naked-eye comet to the people of colonial Sydney town, even more so if New South Wales and its renowned 'astronomer' could again lay valid claim to a new one!

Some were so carried away by the emotion of the moment that they simply ignored the issue of prior discovery altogether. The following delightful letter takes this line:

> And now my dear Sir let me congratulate you … You from your watch tower the sentinel of our southern heavens the first to challenge the celestial stranger, having no guard to him out equal to the occasion will have to grapple with the monster & record him his place and path. (Butterfield 1881)

This letter was from George Butterfield (see Bio Box 9.1), an independent astronomer with an interest in popularising astronomy, who at times had a tendency to employ rather flowery prose.

**Bio Box 9.1: George Butterfield**

George Butterfield was born in Royston, England, in March 1841 and emigrated to Sydney at the age of 12. Eager to work in astronomy, the closest he came was as a draughtsman with the New South Wales Lands Department (1876–1880), where there were a number of staff members with strong interests in astronomy (see Orchiston 1987). After this, permanent employment proved hard to find, but by the mid-1890s, he was working for a Sydney printer and an avid independent astronomer, C. Matthews (d. 1919; Orchiston and Bembrick 1995). Both were founding members of the New South Wales Branch of the British Astronomical Association (Orchiston 1988).

Butterfield's primary astronomical interest was in comets, and although he owned a crudely-mounted 7-in. (17.8-cm) speculum mirror reflector and a 3-in. (7.6-cm) refractor which were quite unsuited to systematic micrometric observation, between 1880 and 1907, he observed many of the comets within the range of his telescopes. From time to time, he also had access to the 8.5-in. (21.6-cm) Browning-With reflectors owned by Matthews and Ebeneler Reginald Morris (a colleague from the Lands Department—see Orchiston 1987) and equatorially-mounted refractors owned by fellow-members of the New South Wales Branch of the British Astronomical Association. His most memorable non-cometary observations were of the 1874 transit of Venus and the 1878 transit of Mercury, but clouds thwarted his attempt to join Tebbutt in observing the 1882 transit of Venus.

While Butterfield's other interest was in popularising astronomy through magazine articles and letters in the Sydney newspapers, he was merely one of a number of Sydney-based non-professional astronomers with this commitment during the last quarter of the nineteenth century (see Orchiston 1997a). However, his main claim to fame is that he produced Australia's earliest planispheres (Orchiston 2003b).

(continued)

(continued)

On the society front, Butterfield was a potential participant in the Australian Comet Corps (see Chap. 10), and as a founding member of the local Branch of the British Astronomical Association, he frequently attended meetings, where he discussed 'current comets' and displayed 3-D models of their orbits.

An inveterate letter writer, the 'Tebbutt Collection', in the Mitchell Library includes 70 letters from him, plus 3 telegrams and 8 postcards, and collectively these provide much of our current knowledge of the inimitable George Butterfield. Despite his obvious commitment to astronomy, limited access to suitable instruments and occasional *faux pas* in his newspaper reports prevented Butterfield from making the major impact on New South Wales astronomy that his dedication and considerable efforts promised.

George Butterfield died in Sydney on 25 January 1910 and was survived by his wife and five of his six children. The death certificate listed his occupation as 'Draftsman'.

In the end, most people were not to be disappointed, for Tebbutt was awarded his comet and they an eventual celestial spectacle. None of the other southern astronomers referred to by Sydney's A.O. Powys and bank manager T.H.F. Griffin (a friend from neighbouring Richmond) detected the comet before Tebbutt, although Archdeacon Arthur Stock (the Astronomical Observer at the Colonial Observatory in Wellington, New Zealand) detected it just one day after Tebbutt (see Orchiston 2016, Chap. 9) and Dr. Benjamin Apthorp Gould (1824–1896, Hodge 1971), founding Director of the Argentine National Observatory in Cordoba, independently discovered it 2 days later, on 25 May (Vsekhsvyatskii 1963).

But not everyone was happy, for prior discovery claims by non-astronomers also emerged. In a letter dated 25 May 1881, Charles Crouch of Parramatta writes:

> I beg to inform you that a Mr Clayfield saw the Comet that you discovered fully 10 Days ago he said that he saw it early in the Morning he being a very early riser … You must excuse my liberty in writing to you but hearing that you are always willing to receive any information in reference to our Planets I have sent you these few lines as I always take great interest in any astronomical observation (Crouch 1881a)

It appears from this letter that Crouch had a sincere case to put to Tebbutt on behalf of Mr. Clayfield and that no malice was intended. Tebbutt certainly took the matter seriously and sought further details, which were forthcoming in another letter from Crouch, dated 7 June: '… Mr. Clayfield says that he first saw the Comet on the 12th May between 5 & 6 Oclock in the Morning about Rising Due East. I also inform you that he Made Mention of it so he must have seen something A Mr. Mitchell Jn$^{r}$ of Liverpool also saw it on 22' (Crouch 1881b). It is clear from this account that Clayfield's claim must be dismissed since Comet C/1881 K1 (Tebbutt) would not have been of the naked-eye magnitude on 12 May, but given the very limited evidence presented, it is not possible to assess Mitchell's claim.

Evidence of a third 'discovery' claimant comes from a completed 'question sheet' that Tebbutt forwarded to a relative of his living near Bungonia. This document, bound together with manuscript letters in *Letters to John Tebbutt 1881* in the Mitchell Library, records the discovery of the comet in the morning sky between 4 and 5 a.m. on 15 May, a little west of south. The correspondent says that she 'Could see the tail distinctly it appeared to be rather long'. On the basis of this specific item of information, and the reported position of the object, Tebbutt (1881d) also felt justified in rejecting this discovery claim.

The Tebbutt correspondence in the Mitchell Library contains an even earlier discovery claim, which must also be dismissed. Writing from Camberwell, Melbourne, on 7 June 1881, a Mrs. Gaggery states that:

> Mr. & Mrs. Dennis Grant have much pleasure to state that the Comet which was foretold would appear about April 1st/81 was seen by Them on the Evenings of 2 and 3 of April a little after sun set in a northerly direction. The brightness of the Star quite eclipsed the other Stars, The Tail was only visible for a few seconds. (Gaggery 1881)

To the untrained mind, bright meteors, or bolides, could easily become 'comets', but this example, and the others presented above, illustrate how there were often competing public claimants when the glory of a new comet discovery was at stake.

For Tebbutt to be credited with the discovery of this new comet was only just, as the distinguished Scottish astronomer, Charles Piazzi Smyth, (Brück and Brück 1988) pointed out later in 1881:

> … I do not know anyone who better deserves on the highest and most appropriate grounds to be discoverer of such a comet, & to have it known by his name;—when you must have devoted yourself from early life, to Comet seeking, & Comet computing, & Comet publications with such persistency & ability. (Smyth 1881)

And just as occurred with Tebbutt's discovery of his first Great Comet in 1861, the professional astronomers of Australia were called on to explain why it was that they were not the first to detect the new comet. At the June 1881 meeting of the Royal Society of Victoria, E.J. White (Bio Box 6.3) from Melbourne Observatory was quick to remind his audience that:

> … the public were wrong in expecting that professional astronomers should be the first to discover the advent of comets and other unusual occurrences in the heavens: the fact being that the ordinary duties of professional astronomers prevented them from scanning the heavens fully, their attention being almost always concentrated on the particular spot under observation. (Proceedings of Societies …, 1881)

## 9.3 An Evolving Object: Subsequent Observations of the Comet

Details of Tebbutt's subsequent observations of the comet are contained in his journal. On the evening of 23 May 1881, the night after the initial discovery, Tebbutt made seven micrometric observations with the 4.5-in. telescope, using two nearby stars of about magnitude 7.5 for comparison. He described the appearance of the comet:

> The comet's nucleus when bisected under the illuminated red field [of the filar micrometer] appeared oval, the major axis being parallel to a meridian of right ascension; it was also of woolly appearance, and not starlike. Without illumination it was more star like and brilliant, and the outline of the coma was well defined. I could just detect traces of the tail with the unassisted eye. (Tebbutt 1879–81)

The evening of 24 May was overcast and no observing was possible, but on the 25th (his birthday), Tebbutt was presented with a beautiful clear night and the comet was clearly visible (ibid.). Seven comparisons were made, with the aid of one of the nearby 7.5 magnitude stars, and the tail was obvious in the telescope field and seemed to be gradually increasing in brightness. The nucleus, meanwhile, continued to display the bright starlike appearance of 2 nights previously. The Government Astronomer of New South Wales, H.C. Russell, obtained his first view of the comet this same evening. He pinpointed its position and commented on the complexity of the head (Russell 1881b). His description, provided for the general public, appeared in the following day's issue of the *Sydney Morning Herald*:

> It is an object quite conspicuous to the naked eye, and it can be very well seen with a binocular glass. The nucleus is condensed bright, and almost star-like; thence the tail may be traced fully 12 degrees; but the smoky haze … prevented me from seeing how long it was … Examined with the large equatoria[l] the nucleus presents a bright, nearly round condensation, in front of which is a sort of half-halo, something like the drawing of the comet of 1858 [Donati's Comet], but the state of the atmosphere prevented me from seeing the details which the telescope should have revealed. (Russell 1881a)

Observations of the comet made in Melbourne this same evening also appeared in the *Sydney Morning Herald* on 26 May: 'The nucleus was quite bright and stellar, and the tail—which was narrow and well defined, but somewhat faint—could be traced for 15 or 20 degrees' (The Comet 1881).

Tebbutt rose early the next morning (26 May) and, at 05:30 h local time, observed the comet about 5° above the south-eastern horizon. No tail was apparent to the naked eye, and he reports that 'Were it not that I was acquainted with the position of the comet I should not have recognized it as a comet with the unassisted eye' (Tebbutt 1879–81). While observing the comet, Tebbutt

> … was surprised by one of the most interesting spectacles I have ever witnessed. The sky was beautifully clear … and just above the eastern horizon were grouped together within a circle, whose diameter did not exceed seven degrees, the crescent moon, and the planets Venus, Jupiter and Saturn. The planets were all conspicuous among their brilliant surroundings, more especially Venus, which is now approaching the most favourable part of her orbit for such observation. A creeping snow-white fog, extending from the base of the Observatory Hill over the low lands to the hills in the distant horizon, also lent charm to this lovely picture. (Royal Society of New South Wales, 1881)

In his journal, Tebbutt (1879–81) described this as '… one of the most charming sights I have ever beheld'.

The evenings of 26 and 27 May and the morning of the 27th were all cloudy, and so Tebbutt's next observation of the comet occurred on the morning of the 28th. However, he had time for only a single comparison before the arrival of clouds. The tail still could not be detected with the naked eye, but the head was clearly visible. When viewed through the telescope, the nucleus was found to contain '… an appendage which made it appear to be elongated in the direction of east and west, quite different from the elongation on the 23rd' (ibid.).

No observations were possible from 28 to 31 May inclusive owing to cloud cover, and by the time Tebbutt viewed the comet again, on 1 June, it had become much brighter and the tail was conspicuous. Sextant observations showed the latter to be 8° 38′ long. From 12 telescopic observations, Tebbutt was able to pinpoint the comet's celestial position and allocate a visual magnitude of about 6.5 to the nucleus (ibid.). Sharing in the excitement of these observations was Tebbutt's friend, T.H.F. Griffin, who finally was able to indulge his avid interest in astronomy and take up Tebbutt's offer of 24 May to come and view 'the little stranger' at Windsor Observatory (see Griffin 1881a; Tebbutt 1879–81).

The comet was also visible on 2 June, but there was time for only a single observation before clouds intervened. Tebbutt (1879–81) did note, however, that the tail had not grown appreciably, if at all. That same evening the Sydney independent astronomer W.J. Macdonnell (see Bio Box 9.2) viewed 'the interesting stranger' with his 2.25-in. (5.7-cm) Wray refractor and noted that the coma had a jet-like form. He estimated the tail at 4–5° (Macdonnell 1881). Another Sydney observer that night was A.O. Powys, who commented that the comet '… was very distinguishable to the naked eye, & under a telescope of small power I could trace the tail, I think, for about 7 degrees' (Powys 1881b). Russell also briefly observed the comet from Sydney Observatory on this evening and noted that in the course of the preceding eight days '… the coma had increased very much, and spread out like a fan before the nucleus, the greater part of it being turned to the following side' (Russell 1881d).

**Bio Box 9.2 William John Macdonnell**

William John Macdonnell (Fig. 9.1) was born in Dublin on 28 November 1842 and emigrated to Sydney when just 5 years of age. After completing his schooling in Germany, he returned to Sydney and found employment in a bank, eventually rising to the post of Manager. He remained in the banking industry throughout most of his working life, first in Sydney and then in Port Macquarie on the New South Wales north coast.

It is likely that Macdonnell acquired a youthful interest in astronomy from his father, who imported telescopes and other scientific instruments. Although he used a succession of smaller telescopes, by the early 1880s, he owned a $3\frac{5}{8}$-in. (9.3-cm) Parkes refractor, and when he moved to Port Macquarie in 1882, he erected his first observatory (Macdonnell 1871–1885). Eventually, this housed a 6-in. (15.2-cm) Grubb refractor and a small transit telescope (Beattie 1910). In 1896 he moved back to Sydney and in 1908 constructed an observatory there for a 4.75-in. (12.1-cm) Parkes refractor (having previously been forced to sell the Grubb refractor during the depression of the early 1890s).

(continued)

(continued)

Macdonnell was involved in observational astronomy over many years and published a number of short descriptive papers in local and overseas journals on sunspots, transits of Mercury, lunar occultations of planets, Jovian satellite phenomena, comets and double stars. In addition, he participated in Sydney Observatory's successful 1874 transit of Venus programme and contributed to the report that Russell (1883) published in *Memoirs of the Royal Astronomical Society*.

Apart from observational astronomy, Macdonnell also was committed to popular astronomy. While living in Port Macquarie, he maintained a time service for local citizens and may also have run public viewing nights at his observatory. He was always ready to respond to those with astronomical inquiries, and in 1909 Merfield described him as '… an unofficial astronomer much above the ordinary who has always taken an active part in the progress of the science in the State of New South Wales. His advice and assistance have done much to help the fostering of a love for astronomy in the State …' (Merfield 1909).

During its short-lived existence, Macdonnell was very active in the Astronomy Section of the Royal Society of New South Wales, serving for a time as Secretary, and he was also a leading member of the New South Wales Branch of the British Astronomical Association, serving in various capacities on the committee—including two sessions as President. A feature of Macdonnell's home, in Sydney, was his large and valuable astronomical library, which '… was freely at the disposal of his friends for reference and use' (Obituaries 1911).

Between 1870 and 1910, Macdonnell made an important, multi-faceted, long-term contribution to New South Wales astronomy, and he would have to rank as the State's second most important independent astronomer (after Tebbutt). For further details, see Orchiston (2001).

Macdonnell died in Sydney on 22 September 1910 and was survived by his wife. At the time, the Reverend Dr. Thomas Roseby (1910) wrote: '… next to Mr. Tebbutt, we all looked up to him as the G.O.M. [Grand Old Man] of Australian Astronomy'.

Cloudy skies occurred on 3 June and observations were not possible, but as some measure of compensation, Tebbutt's paper (ibid.) on the discovery and observation of the comet was read before the Astronomical Section of the Royal Society of New South Wales that evening and was reproduced the next day in both the *Sydney Morning Herald* and the *Evening News*. In this newspaper account, Tebbutt recounts how:

> It has been my custom for some years past, and especially since the appearance of the great comet at the beginning of last year, to scan the western sky in the evening, and occasionally the eastern sky in the morning, with the unaided eye, in search of bright comets. (Royal Society of New South Wales, 1881)

**Fig. 9.1** W.J. Macdonnell during the 1880s (*Courtesy* Port Macquarie Museum, William Macdonnell's Photo Album, A57)

It was while conducting such a search that he discovered the new comet. In the article, Tebbutt continues by outlining how independent astronomers can made a useful contribution: 'I hardly need say that some of our amateurs may do good service by sketching the position of the tail among the stars on different evenings, paying particular attention to the place of its extremity' (ibid.).

On 4 June Tebbutt managed to obtain three positional observations, but these were not deemed ideal, because of intermittent cloud and the low position of the comet in the sky.

Russell (1881e) obtained some useful observations of the comet on the morning of 5 June and noted that there had been a further change in the appearance of the head since his previous observation on 2 June:

> … the bulk of the coma was turned like a Prince of Wales feather to the preceding side, the extremity being turned towards the tail, and a second branch on the following side was turned almost directly to the tail. (See Drawing A.) It seemed as if in the interval from June 2 to 5 the coma had divided into two branches, one turning to the preceding, and the other to the following side; and it presented all the characteristics of something which had been shot out of the nucleus, and carried far in front by its velocity, and then turned gracefully back to form part of the tail.

Russell's 'Drawing A' is reproduced here as Fig. 9.2a.

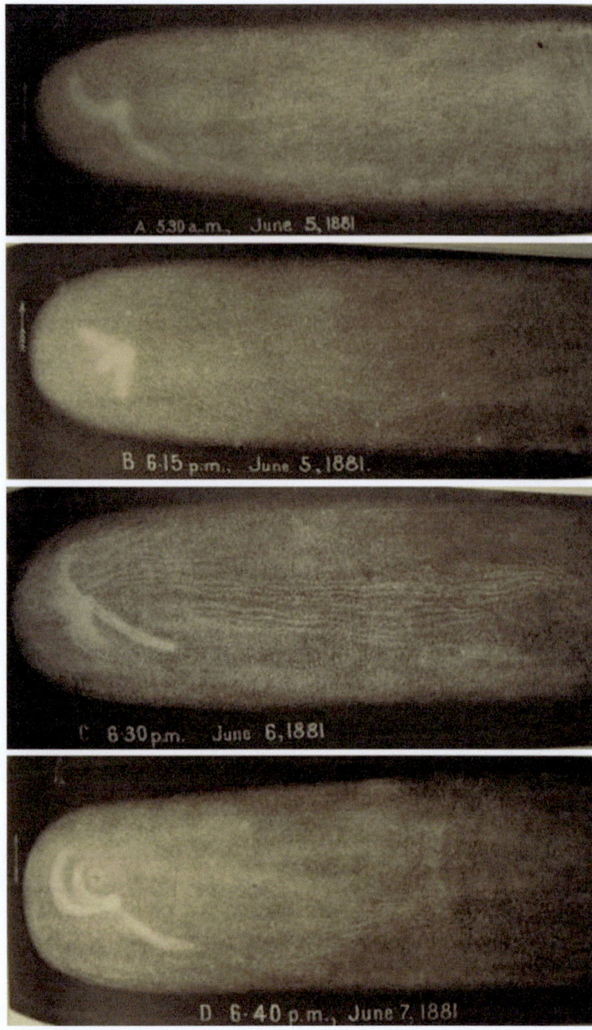


**Fig. 9.2** (**a–d**) Russell's drawings of the head and part of the tail of Comet C/1881 K1 on 5 June (**a**, **b**), 6 June (**c**) and 7 June (**d**) (after Russell 1881e)

The evening of 5 June was clear, and Tebbutt (1879–81) made six micrometric comparisons, while Russell (1881e) found the coma not so striking as that morning—although it still presented the two 'branches' (see Fig. 9.2b). While Russell was carrying out his observations, the comet occulted a ninth magnitude star, '… some of the brighter parts of the coma going over it without stopping any of its light, so far as I could see' (ibid.). This star is shown in Fig. 9.2b, near the nucleus.

The evening of 6 June was clear, and Tebbutt made eight comet comparisons with stars of magnitude 3 and between 6 and 6.5 in the same field. He noticed that the comet's head was plainly visible to the naked eye (Tebbutt 1879–81). Russell also carried out observations this evening and again noted changes to the coma:

> … this was the only occasion in which there seemed to be a shadow or dark cone in the rear of the nucleus. The coma was more conspicuous than ever, and presented the appearance of a bird with outstretched wings in front of the nucleus; and the branch first seen in the morning of the 5th was increasing, and turned directly to the tail. (Russell 1881e)

Figure 9.2c shows Russell's drawing of the comet on this evening.

Tebbutt (1879–81) made four further micrometric comparisons on the 7 June with the aid of stars of magnitude 3 and 6.5, and Russell (1881e) was only able to gain a glimpse of the comet between clouds but still managed to make a drawing (Fig. 9.2d) and note further changes:

> … the coma in front of the nucleus had divided into two shoots, which extended a great distance in front of the nucleus, and then arched gracefully until the tips turned right to the tail, thus forming two enormous hook-like appendages, and the branch on the following side was very much enlarged, but retained the same general direction. (See Drawing D.) Altogether, it presented a most striking appearance, which I shall not soon forget.

Tebbutt was also able to observe the comet on the evening of 8 June, when five comparisons were made. However,

> … owing to the prevalence of westerly winds and the low altitude at which the observations were made the comet and stars were blurred and very ill-defined. The stars appeared large and woolly and the comet's nucleus was not starlike as on the evening of the 6th. (Tebbutt 1879–81)

These westerly winds had also influenced the quality of observations made on 5 and 7 June.

On the basis of his June observations, Tebbutt prepared a popular account of the comet, and this appeared in the *Sydney Morning Herald* on 9 June: 'The comet now visible is obviously approaching the earth … It seems probable that … its tail will, as a consequence, become an imposing object in the course of a few days' (Tebbutt 1881d). Tebbutt concluded by noting that the one major outstanding chore now was to compute the comet's orbital elements, but he was forced to announce that '… this work, I regret to say, I shall be obliged to leave to others. Although urged by some to undertake the calculation, I cannot venture in my present weak state of health to add this labour to my other important avocations' (ibid.).

Clouds prevented any observations on the evenings of 9 and 10 June, and on the 11th, Tebbutt could do no more than catch a naked-eye glimpse of the comet for a few minutes through a break in the clouds. He only had time to note that the head was very brilliant (Tebbutt 1879–81).

Tebbutt rose early the next morning (12 June), and although the sky was clear, the diffused twilight and presence of a full Moon made observing difficult. Furthermore, no stars brighter than magnitude 7 were found near the comet. Consequently, he had to use Rigel for his five positional observations, and needless to say, the results were not entirely satisfactory. But by now the comet was spectacular: 'The comet's head was a very conspicuous object, and a portion of the tail was visible in the twilight' (ibid.).

A cloudy western horizon that same evening prevented further observations of the comet, but Tebbutt was successful the following night. However, the comet was too low in the sky for precise observation, and although one comparison was obtained, using two fourth-magnitude stars, all three objects appeared large and blurred (ibid.).

The evening of 13 June proved to be the last occasion upon which Tebbutt viewed the comet. On the 14th he went to Sydney for a well-earned rest (or 'change', as Tebbutt prefers to call it), prompted by the pleas of his friend, Griffin. On 2 June, after having observed the comet the previous evening from Windsor Observatory, Griffin wrote urgently to Tebbutt:

> I confess that I was much struck with your look of ill heath last night … I desire now to urge upon you with all the force of a sincere friend the strong necessity of a complete rest, and as long a change of air as you can possibly allow yourself. I do hope that in the interests of your health—not to speak of the strong claims that science has upon you—you may be prevailed upon at once to take the needed change, for I feel sure that if you do not, a serious illness will probably be the result of the continual strain your mental faculties are subjected to. (Griffin 1881b)

This letter suggests that Tebbutt was suffering from both mental fatigue and a physical affliction. Just 2 months later, Griffin (1881e) was to write to Tebbutt again, about his 'serious indisposition', while Powys (1881d) refers to 'pains in the back' and discusses rheumatic complaints. Perhaps rheumatism already plagued Tebbutt back in May and June when he was carrying out these frequent, precise, demanding observations of the comet and contributed to the 'look of ill health' that Griffin mentioned on 2 June.

Whatever the exact nature of the complaint(s), Tebbutt decided to take a break from observing (Griffin, 1881c), but he only spent 4 days in Sydney—at Manly (Griffin 1881d), presumably enjoying the sea air—before the combined pull of his observatory, wife and family (but not necessarily in that order!) drew him back to Windsor. He rose early on the morning of 19 June in the hope of viewing the comet, but the sky was overcast. Next morning, with a clear sky, he '… watched for the comet but did not see it. I waited till 6h 30m per chronometer, but it was obvious that the comet was so near the strong twilight when it rose as to be invisible to the naked eye' (Tebbutt 1879–81).

Thus ended Tebbutt's run of observations, or attempted observations, of Comet C/1881 K1. From the date of its discovery on 22 May until it disappeared from Windsor skies, Tebbutt made 68 positional measurements with the filar micrometer, on 13 different dates, and also recorded useful information about the tail and the head (but particularly the changing nature of the nucleus). He also determined the apparent visual magnitude of the comet on a number of occasions.

From 22 June the comet was visible to northern-hemisphere astronomers and was extensively observed (see Vsekhsvyakskii 1964). Its tail continued to draw attention (Fig. 9.3) and to grow in length, reaching 20° on 25 June, when the nucleus was about magnitude 1. At its best, the comet was an impressive sight (see Fig. 9.4) and was still a naked-eye object in August, but by month's end, the tail had all but disappeared (e.g. see the listing in Backhouse 1882). By the end of October, the integrated magnitude of the comet had dropped to 9 (note that Vsekhsvyakskii 1964 incorrectly lists some June dates as July on page 258). The last observations of this comet listed by Marsden (1982) were made on 15 February 1882.

While northern observers were following the comet, Tebbutt had decided to calculate the orbital elements after all (notwithstanding his earlier pronouncement to the contrary in the *Sydney Morning Herald*), and this became his major preoccupation

**Fig. 9.3** W.F. Denning's drawing of the comet on 27 June, when it showed '... an interesting series of more or less parallel bands' (after Phillips, 1923, p. 422)

during July. Towards this end, others from Australia and overseas forwarded him their own positional observations, in case these could be of value (e.g. see Biggs 1881; Ewer 1881; Joseph 1881; and Robinson 1881). Accordingly, he devoted himself to this mentally demanding exercise, spurred on no doubt by the possibility that Russell might beat him to the task (see Powys 1881b). As it transpired, it actually was White from Melbourne Observatory who was the first Australian to produce such results, late in June (Powys 1881c), and Tebbutt was not to complete his own orbital computations until the end of July. The calculations occupy almost 40 manuscript pages (Tebbutt 1881k), and although Tebbutt included a preliminary orbit in his paper read at the 5 August meeting of the Astronomical Section of the Royal Society of New South Wales (Tebbutt 1881g), it was in another paper, presented at the 7 September meeting, that the final orbital elements were announced (Tebbutt 1881i).

This latter paper showed a close orbital resemblance between Comet C/1881 K1 (Tebbutt) and the Great Comet of 1807 (C/1807 R1), as Table 9.1 indicates, but a definitive study published later by Riem (1894) revealed that they had to be two totally different comets since the 1881 comet had a period of somewhere between 2409 and 2446 years (see Tebbutt 1908).

In addition to the aforementioned research papers and newspaper accounts, Tebbutt also forwarded reports on his comet to *Astronomische Nachrichten* (Tebbutt 1881a, f, j), *Monthly Notices of the Royal Astronomical Society* (Tebbutt 1881e), and *The Observatory* (Tebbutt 1881b, h). This was yet another classic case of 'publication overkill'!

**Fig. 9.4** A chromolithograph of a painting by Étienne Léopold Trouvelot (1827–1895) of the Great Comet of 1881 on the night of 25–26 June (https://en.wikipedia.org)

**Table 9.1** A comparison of the orbital elements of the Great Comets of 1807 (C/1807 R1) and 1881 (C/1881 K1)

| Orbital element | 1807 comet | 1881 comet |
|---|---|---|
| Longitude of perihelion | 271° 57′ | 264° 56′ 15.5″ |
| Longitude of the ascending node | 267° 49′ | 270° 54′ 0.2″ |
| Inclination of the orbit | 63° 10′ | 63° 27′ 14.0″ |
| Perihelion distance (in AU) | 0.6461 | 0.7357 |
| Heliocentric motion | Direct | Direct |

## 9.4 An Enduring Legacy: The Great Comet's Contribution to Astronomy

In her monumental and masterful *A Popular History of Astronomy during the Nineteenth Century*, the English chronicler of astronomical history, Agnes Mary Clerke (1842–1907; Fig. 9.5; Brück 1994, 2002), summarises the 'service to science' of Comet C/1881 K1 (Tebbutt) as follows:

> It afforded, when compared with the comet of 1807, the first undeniable example of two such bodies travelling so nearly in the same orbit as to leave absolutely no doubt of the existence of a generic tie between them. Cometary photography came to its earliest fruition with it; and cometary spectroscopy made a notable advance by means of it. (Clerke 1893, p. 429)

Let us now critically evaluate each of the components of this statement.

In fact, Tebbutt was not the first to note the orbital resemblance between the Great Comets of 1807 and 1881, as Gould calculated the orbital elements of Comet C/1881 K1 before either White or Tebbutt (ibid.), noticed the similarity in elements and telegraphed London on 1 June prematurely announcing the return of the 1807 comet! However, this was not the first case of two quite different comets with very similar orbits. Just one year earlier, the Great Comet of 1880 (C/1880 C1—see Fig. 4.1) was found to possess an orbit virtually identical to those of the major comets of 1668 (C/1668 E1) and 1843 (C/1843 D1), and later the Great Comet of 1882 (C/1882 R1) was added to this group. Another member of the group was also discovered in 1887 in the form of Comet C/1887 B1 (Chambers 1909). This is the best documented of all comet groups, but by 1888 another half dozen groups had been identified (Young 1888). Marsden (1989) has provided an account of current thinking on the sungrazing comet groups. We see, then, that Clerke is in error, for the 1807–1881 comet group was in fact not the first to be identified. The idea of comet groups, meanwhile, had an even earlier origin. It saw its genesis with the Danish astronomer Thomas Clausen (1801–1885) in 1831 and was further developed by the Dutch astronomer Martin Hoek (1834–1863), from 1865 (Clerke 1893).

**Fig. 9.5** Agnes Clerke in about 1905 (https://en.wikipedia.org)

**Fig. 9.6** A photograph of Andrew Ainslie Common (https://en.wikipedia.org)

Clerke's second claim is that Comet C/1881 K1 (Tebbutt) was the first for which a satisfactory photograph was obtained, and this is indeed the case if we define 'satisfactory' as producing an image of the head *and* tail. Certainly, there were earlier cometary photographs. As we noted in Chap. 6, Pasachoff et al. (1996) have documented how the English professional photographer William Usherwood and the Harvard College Observatory astronomer, George Phillips Bond, successfully photographed the head of Donati's Comet in 1858 (cf. Hughes 2013; Proctor 1926).

However, only with the advent of the more sensitive dry gelatine plate and improvement in telescope drives did it become possible to successfully record the comparatively faint light from the tail of even the brightest of comets (see Clerke 1893). The Great Comet of 1881 arrived at just the right moment in this regard, and three different astronomers successfully took advantage of this. On 24 June, the British independent astronomer, Andrew Ainslie Common (1841–1903; Fig. 9.6; Obituaries 1904), took a 20 min exposure but found

> … a result that I ought to have anticipated, but did not; this was that the rapid motion in declination, not being provided for, caused the image to be a trail on the plate some quarter of an inch long. As far as it went the picture was good—that is, it shows the nucleus, head, and part of the tail, more particularly that part, narrow and bright, that proceeded from the nucleus. (Common 1881)

In New York, Dr. Henry Draper (1837–1882; Fig. 9.7; Whitney 1971) also captured an image of the comet this same evening:

> I succeeded in photographing the Comet in Auriga on Friday night, June 24th, 1881. Since then I have taken several photographs of it. One made last night [date not specified] with an exposure of 2 hours and 42 minutes shows the tail about 10° long. There are many stars on the plate, some shining through the tail. (Draper 1881c; cf. Draper 1881b)

At Meudon Observatory near Paris, Pierre Jules Cesar Janssen (1824–1907; Fig. 9.8; Launay 2011) used a 51-cm f/3 reflector to obtain a 30-min exposure of the comet on 30 June (Janssen 1881). The resulting photograph, which showed '… the structure of the tail with beautiful distinctness to a distance of 2½° from the head'

**Fig. 9.7** Henry Draper in about 1871 (https://en.wikipedia.org)

**Fig. 9.8** An oil painting of Jules Janssen by Jean-Jacques Henner, now in the Musée d'Orsay in Paris (https://en.wikipedia.org)

(Clerke 1893, p. 428), was reproduced by de Vaucouleurs (1961) and is shown here in Fig. 9.9. However, in sending Tebbutt a copy on 4 March 1882, Smyth queried the authenticity of the Janssen photograph. He explains: '… M. Janssens negatives seem to have been too faint to print on paper; so he adopted the rather dangerous expedient of making a drawing from them, & then photographing that, which explains his stars being so round & his comet's head so sharp …' (Smyth 1882). Even a casual examination of Fig. 9.9 suggests that there are good grounds for accepting Smyth's explanation, but I will leave it to each reader to make up their own mind.

Common was subsequently awarded the Gold Medal of the Royal Astronomical Society, in 1884, for obtaining '… a photograph of the Comet b, which is probably the earliest successful photograph of any comet …' (Stone 1884), and given that he, Draper and Janssen, all produced photographs showing the head and tail, it is hard to understand Mary Proctor's claim (1926) that the first 'entirely satisfactory' comet photographs were taken the following year when the Great Comet of 1882 made its extremely spectacular appearance.

Over the last 120 years, the photographic plate has been a major research tool in cometary astronomy (surpassed only in comparatively recent times by the CCD camera). It has been used extensively not just for astrometric observations and the recording

**Fig. 9.9** Janssen's reputed photograph of Comet C/1881 K1 (Tebbutt) (after de Vaucouleurs 1961, Plate 8, Figure 1)

of minute detail in the structure of comets but also in the recovery of known periodic comets and in the discovery of new comets (e.g. see Roemer 1963). It is to Lick Observatory's Edward Emerson Barnard (1857–1923; see Fig. 10.7) that we owe the first photographic discovery of a comet, in 1892 (see Sheehan 1995, p. 225).

The third aspect of Clerke's statement is cometary spectroscopy's 'notable advance' with the Great Comet of 1881. What, precisely, is the basis of this claim? It is well known that Giovanni Battista Donati (1826–1873), Director of the Observatory in Florence, carried out the first visual observation of a comet's spectrum in 1864, when Comet C/1864 N1 (Tempel) appeared (see Olivier 1930), and between that year and the end of 1880, Clerke maintains (1893) that 18 different comets were observed spectroscopically. The brightest of these was Comet C/1874 H1 (Coggia), the principal comet of that year, and a very significant object from the viewpoint of cometary structure (see Fig. 8.16). So although still a fledgling specialisation, cometary spectroscopists had already built up a considerable pool of observational data by 1881.

Nevertheless, because of its majesty, Comet C/1881 K1 (Tebbutt) drew widespread attention once it became a northern-hemisphere object, and Vsekhsvyatskii (1964) lists the different observational teams or individuals that studied its spectrum.

One person he missed was Russell who was the first to observe this comet spectroscopically, on 6 June, while it was still a southern-hemisphere object. In a letter to Tebbutt dated 11 June, Russell writes:

> I also managed to get its spectrum which is rather different from that of other comets, but the difference seems to identify it still more than the others with [hydro]Carbon. I could not see any sign of other lines. I hope I may get another chance of finding the positions of the lines it is a work of considerable difficulty. (Russell 1881c)

Russell subsequently published his spectroscopic observations in the *Journal and Proceedings of the Royal Society of New South Wales*—which explains why Vsekhsvyatskii was not aware of them—so, given their chronological priority, it is worthwhile to reproduce part of his account here:

> Turning the telescope so that the slit of the spectroscope received light from the tail of the comet, I found it was too faint to give a visible spectrum, and I could not see any until parts near the head were brought upon the slit, when three bright lines became visible. Getting nearer the nucleus a faint continuous spectrum was visible, crossed by three bright lines, but in the faint grey continuous spectrum I could not see any dark lines. The three lines were not sharply defined, but were sufficiently so to admit of good measures to the centre of each. The middle line was by far the brightest; the next in brilliance was in the yellow, and the third and faintest was in the violet. The whole of this spectrum increased in brightness as the slit approached the nucleus, but when the nucleus itself was on the slit all the additional light seemed concentrated in the middle of the B line until it shone almost like a star, and quite as bright as the nucleus itself, proving that its light is monochromatic. Even in the brightest part of the comet I was unable to see any dark lines in the continuous spectrum. (Russell 1881e)

Russell also was able to briefly observe the comet's spectrum on 7 and 8 June.

Later observations by others showed that the spectrum of Comet C/1881 K1 (Tebbutt) was not anomalous. Vsekhsvyatskii (1964, p. 259) succinctly highlights the results of these later observations:

> 22 June the nucleus showed an intense continuous spectrum without dark lines (molecular emission); in the envelope, there were traces of bright bands; 25 June, diffuse green and blue bands were observed in the bright continuous background; 29 June, the three common bands were in the intensity ratio 1:5:2; the yellow band was the faintest. 30 June, a well developed band spectrum of three bands; two additional bands were also reported. Through a high light-power instrument all parts of the comet and the tail over a considerable distance clearly showed a band spectrum. Polarized light emitted by the head. 5 July, the continuous spectrum of the nucleus grew fainter; the bands were clearly visible.

In addition, Chambers (1909) records that Fraunhofer lines appeared in the continuous spectrum on 29 June.

Clearly distinguishable in the above account are the two major components of most cometary spectra:

1. The continuous spectrum, usually with Fraunhofer lines, which is produced by reflected sunlight.
2. The bright 'hydrocarbon bands' that are produced by the gases from the comet itself. Only comparatively recently has it been possible to assign particular molecules to the various bands (see Jackson 1982).

**Fig. 9.10** A photograph of Hermann Carl Vogel (https://en.wikipedia.org)

**Fig. 9.11** A.W. Wright photographed in the 1870s (https://en.wikipedia.org)

Given the rather typical nature of the spectrum of Comet C/1881 K1 (Tebbutt) (e.g. see Young 1888), what then is the basis of Clerke's claim for a 'notable advance'? There are, in fact, two significant aspects to consider. Firstly, June 1881 was only the second occasion upon which a spectrum was obtained for an *entire* comet, when Sydney's Henry Russell, the Director of Potsdam Astrophysical Observatory Hermann Carl Vogel (1841–1907; Fig. 9.10), Yale's Professor Arthur Williams Wright (1836–1915; Fig. 9.11) and Princeton's Professor Charles Augustus Young (1834–1908; Fig. 9.12) detected the bright hydrocarbon bands in the tail; indeed, Vogel was able to trace them to the very end of the tail. The last three observers also recorded a faint continuous spectrum (Chambers 1909; Clerke 1893). The first entire comet to be analysed spectroscopically was Comet C/1874 H1 (Coggia), but William Huggins found that:

> When the slit was brought past the nucleus on to the commencement of the tail, the gaseous spectrum [emission bands] became rapidly fainter, until, at a short distance from the nucleus, the continuous spectrum predominated so strongly that the middle band only, which is the brightest, could be detected on it. (quoted in Guillemin 1877, p. 262)

**Fig. 9.12** Professor C.A. Young (https://en.wikipedia.org)

**Fig. 9.13** Sir William Huggins in 1910 (https://en.wikipedia.org)

Comet C/1881 K1 (Tebbutt), therefore, was the first in which the hydrocarbon bands could be traced throughout the length of the tail, and according to Chambers (1909), it was only with the appearance of Comet C/1907 L2 (Daniel) and Comet C/1908 R1 (Morehouse) that research into the spectra of cometary tails continued.

The second, and perhaps more important, point to note is that Comet C/1881 K1 (Tebbutt) was the first comet for which successful photographs of a spectrum were obtained (all previous cometary spectroscopic observations having been performed visually). The photographs in question were taken by the British astronomical spectroscopy pioneer, Sir William Huggins (Fig. 9.13; Maunder [1913]), on 24 June and the American, Dr. Henry Draper (1881a, b), on an unspecified date after 24 June.

Sir William Huggins (1881a) obtained a 1-h exposure that revealed:

> … a very distinct spectrum of the comet … [consisting of] … a pair of bright lines in the ultra-violet region, and a continuous spectrum which can be traced from about F to some distance beyond H. The bright lines, a little distance beyond H … appear to belong to the spectrum of carbon … In the continuous spectrum … the dark lines of Fraunhofer can be seen.

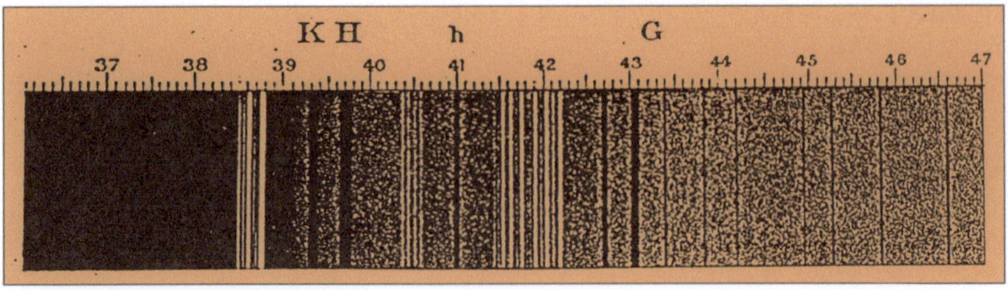


**Fig. 9.14** One of Huggin's spectrograms of Comet C/1881 K1 (Tebbutt) (after Huggins 1881b, facing page 2)

Huggins obtained a second successful spectrogram—using a 1½-hour exposure—the following evening (25 June).

In order to obtain his spectrogram of the comet, Draper employed the two-prism spectroscope that he had used in the past to photograph stellar spectra. He took three photographs:

> … with exposures of 180 minutes, 196 minutes and 228 minutes, and with a comparison spectrum on each. The continuous spectrum of the nucleus was plainly seen … [but] the most striking feature is a heavy band above H which is divisible into lines, and in addition two faint bands, one between G and h and another between h and H. (Draper 1881b)

One of Huggins' spectrograms is reproduced in Fig. 9.14, which shows the continuous spectrum and emission bands noted by both investigators. Some of these emission bands were previously unknown (Olivier 1930; Vsekhsvyatskii 1964). In the light of this achievement, and the new data presented on the spectra of cometary tails, Clerke is fully justified in attributing Comet C/1881 K1 (Tebbutt) a significant place in the history of cometary spectroscopy.

The foregoing analysis reveals that the Great Comet of 1881 has a special place in the history of astronomy. But this comet has a further 'claim to fame' which has not been adequately recognised: it contributed significantly to nineteenth-century knowledge and understanding of the structure and nature of comets by building on valuable data already provided by Comet C/1858 L1 (Donati), C/1861 J1 (Tebbutt) and Comet C/1874 H1 (Coggia). All of these comets were characterised by structurally-complex and rapidly-changing heads and intricate tails, and Comet C/1881 K1 (Tebbutt) continued in this tradition. The head of the 1881 comet, to quote Young (1888, p. 412), was '… always doing something outre, throwing off jets, breaking into fragments, and, in fact, continually exhibiting unexpected behaviour.' Sharpless and Philips (1882, p. 201) provide further relevant data:

> At one time … a separation in the nucleus led astronomers to think that it would follow the example of Biela's Comet, but the following night the smaller part had disappeared. Very frequent changes were observed by the telescope, in the shape of the jets and envelopes around the nucleus. Many drawings were made of it …

Meanwhile, eight drawings by the British independent astronomer Edward Ball Knobel (1831–1940) are reproduced here in Fig. 9.15, and they build on the earlier accounts provided by Russell when this comet was still a southern object (cf. Fig. 9.2).

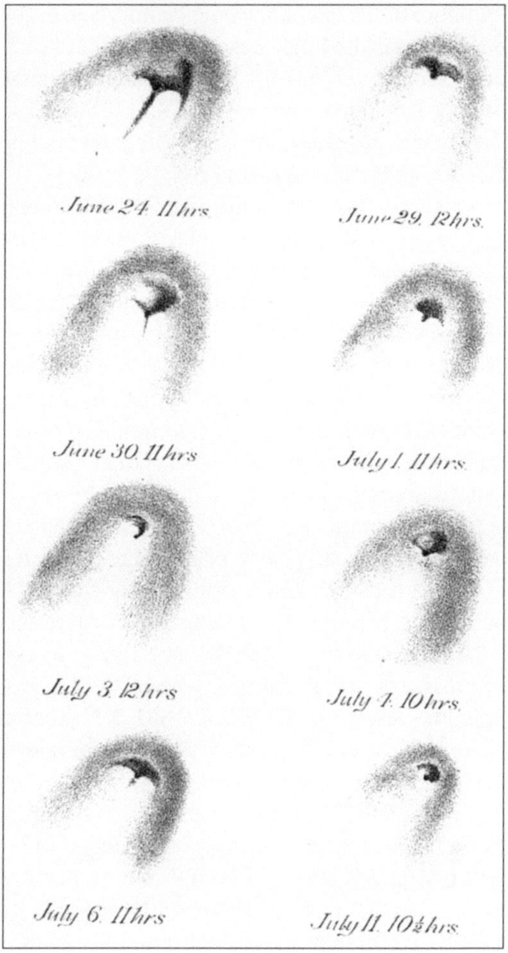


**Fig. 9.15** Drawings of the head of Comet C/1881 K1 (Tebbutt) between 24 June and 11 July (after Knobel, 1882, facing page 76)

The tail of Comet C/1881 K1 (Tebbutt) was equally impressive, when fully developed, as the following account demonstrates:

> … a curious dorsal spine of strong illumination formed the axis of the tail, which extended in clear skies over an arc of 20°. It belonged to the same "type" as Donati's great plume … the appendage was, for a few nights, and by two observers, perceived to be double. Tempel, on June 27, and Lewis Boss, at Albany (N.Y.), June 26 and 28, saw the long straight ray [and] … the curved train … It had vanished by July 1, but made a temporary reappearance July 22. (Clerke 1893, p. 426)

On every count, then, Comet C/1881 K1 (Tebbutt) was a remarkable astronomical phenomenon; indeed, it was, in the opinion of some, '… the finest object of the kind since 1861' (Clerke 1893, pp. 425–426). On the basis of the physical spectacle to the lay public, and its contribution to astronomical knowledge, it must rank with the major comets of 1843, 1858, 1861 and 1882 as one of the 'Great Comets' of the nineteenth century. Yet it has received scant recognition in much of the general astronomical literature of the late nineteenth century and early twentieth century. Of the 32 different books I consulted that were published between 1882 and 1930 and specifically mention 'great', 'interesting' or 'significant' comets of the 1850–1900 period, only 13 (40.6 %) include Comet C/1881 K1 (Tebbutt) (see Table 9.2). Moreover, three of these simply mention this comet in passing, without providing any account of it whatsoever. Many authors are happy to promote Comet C/1874 H1 (Coggia) and the major comet of 1880 (C/1880 C1), at the expense of Comet C/1881 K1 (Tebbutt).

It is interesting to speculate as to why the Great Comet of 1881 is not always given its rightful billing in the astronomical literature. One factor may be the number of major naked-eye comets seen during the nineteenth century. Although there are marked discrepancies in the numbers reported (cf. Jones 1946, p. 256 and Young 1888, p. 395), it is apparent that the second half of the century was blessed with an unexpected abundance of these objects. This situation may have made a subjective choice difficult, given the wide range of selection criteria available, and I suspect that most authors simply chose 'Great Comets' on the basis of their personal idiosyncratic impressions and their own observational involvement, rather than on the actual astronomical merits of each candidate. This is perhaps understandable in the case of those early writers in Table 9.2, given that they personally would have observed most, if not all, of the Great Comets about which they wrote, but it cannot be a contributing factor in the case of more recent authors.

Another factor which is very likely to have coloured some of these authors' impressions was the appearance of the Great Comet of 1882 (C/1882 R1) so soon after the advent of Comet C/1881 K1 (Tebbutt). This newcomer completely outclassed its notable predecessor in a number of ways, even becoming a daytime naked-eye object. The majesty of this comet may have unwittingly reduced the 'status' of the 1881 Comet in the eyes of some. These comments notwithstanding, it is to be hoped that future books and chapters of books dealing with 'Great Comets' will assign due space to the Tebbutt comet of 1881. In this context, it is heartening to see that Yeomans (1991) and Olson and Pasachoff (1998) pay due attention to C/1881 K1 in their authoritative works on cometary history.

Finally, there is one further reason why Comet C/1881 K1 (Tebbutt) was, in the eyes of some members of the lay public, perceived to be a 'Great Comet'. For centuries, comets were viewed by many as ill omens and portends of disaster, and various examples of major calamities and the almost simultaneous appearance of a comet are on record (e.g. see Chambers 1909; Proctor 1896). In this context, the following poem (cited in Smyth 1844, I, p. 214) is particularly apposite:

> The blazing star,
> Threat'ning the world with famine, plague, and war:
> To princes, death; to kingdoms, many crosses;
> To all estates, inevitable losses;
> To herdsmen, rot; to ploughmen, hapless seasons;
> To sailors, storms; to cities, civil treasons.

As chance would have it, the Great Comet of 1881 appeared just 8 weeks after the prophesied arrival of just such a visitor (see Gaggery 1881), and Russell could

**Table 9.2** 'Great Comets', 1850–1900

| Authority and reference | Comet | | | | | | | | | | |
|---|---|---|---|---|---|---|---|---|---|---|---|
| | 1858 | 1858 | 1860 | 1861 | 1862 | 1874 | 1880 | 1881 | 1882 | 1887 | 1892 |
| Baikie (1906) | × | | | × | × | × | × | | × | | |
| Ball (1893) | × | | | | | | | × | × | | – |
| Burgel (1911) | × | | | o | | o | | | × | | |
| Chambers (1909) | × | | | × | | × | × | o | × | × | |
| Clerke (1898) | × | | | × | | | × | × | × | | |
| Comstock (1901) | × | | | × | | × | | | × | | |
| Crommelin (n.d.) | × | | | × | | × | | | × | × | |
| Denning (1891) | × | | | × | | × | × | × | × | × | – |
| Dolmage (1909) | × | | | × | | | o | × | × | o | |
| Dunkin (1891) | × | | | × | | × | | | × | | – |
| Elson (1910) | × | | | × | | × | × | o | × | | |
| Flammarion (1903) | × | | | × | | × | | | × | | |
| Giberne (1910) | × | | | × | × | | | | | | |
| Gore (1909) | × | | | × | | | | | × | × | |
| Heath (1908) | × | | | × | | | | × | × | | |
| Hollis (n.d.) | × | | | × | | | | × | × | | |
| Howe (1897) | × | | | × | | | | | × | | o |
| Lewis (1921) | × | | | | | | × | × | × | | |
| Lynn (1909) | × | | | × | × | | × | | × | | |
| Macpherson (1906) | × | | | × | | o | × | | × | | |
| Main (1882) | × | | | × | × | × | | × | × | – | – |

(continued)

**Table 9.2** (continued)

| | Comet | | | | | | | | | | |
|---|---|---|---|---|---|---|---|---|---|---|---|
| Authority and reference | 1858 | 1858 | 1860 | 1861 | 1862 | 1874 | 1880 | 1881 | 1882 | 1887 | 1892 |
| Mee (1897) | × | | | × | | × | | | × | | |
| Moulton (1908) | × | | | × | | | × | | × | | |
| Neison (1886) | × | | | × | | | × | | × | – | – |
| Newcomb (1908) | × | × | × | | | | × | o | × | | |
| Olivier (1930) | × | | | × | | | | | | | × |
| Parkes (1888) | × | | | × | | | | | × | – | – |
| Peck (1890) | × | | | × | | | × | | × | | – |
| Proctor (1926) | × | | | × | | | × | | × | × | |
| Serviss (1909) | × | | | × | | | × | | × | | |
| Sharpless/Philips (1882) | × | | | × | | × | | × | × | – | – |
| Todd (n.d.) | × | | | × | | × | | × | × | | |

*Key*: × = 'Great' or major comet, o = mentioned in passing, but no details, – = not applicable not help but reflect on the effect this may have on some people. In a letter to Tebbutt dated 25 May 1881, he writes:

> For one reason I am rather sorry it [i.e. the comet] has come just now, and that is, because it imparts a seeming confirmation to one of those irresponsible and unsigned predictions that have been published in England of late, apparently with the object of frightening the multitude. (Russell 1881b)

Despite the advent of Comet C/1881 K1 (Tebbutt), the world survived!

## 9.5 Concluding Remarks

Tebbutt discovered the Great Comet of 1881 almost exactly 20 years after his 1861 find and at much the same position in the sky. Both comets were spectacular objects and contributed to our understanding of cometary structure and behaviour, but because of the emergent roles of spectroscopy and photography in 1881, comet C/1881 K1 was destined to play a key role in the evolution of cometary photography.

That Tebbutt succeeded in tracking his new discovery despite ill health and bad weather (for 39 % of all potential observing nights were cloudy) is a tribute to his dedication and perseverance. He then faced the onerous task of computing the comet's orbit, and given his precarious health, the expectations of his friends and a certain element of competition from some of his colleagues, it is very much to his credit that he was able to satisfactorily accomplish this feat.

In the midst of his own observing programme, Tebbutt was distracted by two unusual observations of the comet reported in *Astronomische Nachrichten* and *Monthly Notices of the Royal Astronomical Society*. In the first of these, B.A. Gould (Fig. 9.16) claimed to have observed a companion to Comet C/1881 K1 (Tebbutt) on 11 June 1881. Tebbutt refuted this interpretation, suggesting that:

> … he had observed by mistake, the two well known stars numbered 1592 and 1597 in the British Association Catalogue, which were in the immediate neighbourhood of the comet, and whose relative magnitudes and positions both agreed with his published observations of the supposed double comet. Both objects were excessively blurred in consequence of the thick haze and mists on the horizon. (Tebbutt 1908, p. 52)

Tebbutt (1882b) presented his interpretation in *The Observatory*, which was disputed by Gould (1882), and that brought a final rejoinder from Tebbutt (1882c). Later, he was to write that 'Any one reading the discussion in the Observatory and the Astronomische Nachrichten of the time must be satisfied of the correctness of my explanation …' (Tebbutt 1908, p. 53). Undoubtedly Gould was not 'satisfied', for to believe Tebbutt's interpretation was to raise doubts over the observational competence of the Director of the Argentine National Observatory. In the end it came down to a personal choice. Both were noted astronomers, even if one was a

**Fig. 9.16** Benjamin Apthorp Gould (https://en.wikipedia.org)

professional and the other a non-professional. The American astronomer, Peters, was forthright in his preference, assuring Tebbutt that his explanation of Gould's observation on 11 June:

> … is to me entirely satisfactory. But it is very difficult to convince people of the stamp of Dr. Gould, that they could have been mistaken. I had the same difficulty with Watson in regard to the alleged intra-mercurial planets. One can not use other arguments than those, that follow from internal evidence of the case; while there really enters also the character of the man as an element. And some people always like to publish something extraordinary, either for making a sensation, or for gaining thereby some cheap reputation with the ignorant general public. (Peters 1882).

It is not clear if Peters's last sentence was specifically meant to apply to Gould.

The second astronomer to publish contentious observations of Comet C/1881 K1 (Tebbutt) was Australia's Dr. William Bone (Bio Box 8.2), who reported that he too had seen a second object near the comet, but one day prior to Gould (Bone 1882b). On this occasion, Tebbutt was far less charitable than he had been when responding to Gould. He wrote:

> No sooner had Dr. Gould's mysterious observation of Comet b, 1881, been, as I conceive, satisfactorily explained, than another bone of contention is presented to astronomers … it must be confessed, I think, that the amateur astronomer at Castlemaine has strange experiences at times, as witness a note appended to his paper on the great Comet of 1880 … We must therefore not be surprised if a little unskilfulness on the part of the observer has caused him to mistake some bright but hazy fixed star for another comet. (Tebbutt 1882a)

Tebbutt (ibid.) went on to suggest that Bone's 'second' comet was actually 8 Leporis. When queried further on the matter, Bone (1882a) simply repeated what was written in his *Monthly Notices* paper and quoted White at Melbourne Observatory explaining the observation away as either the fortuitous appearance of a second comet in the field or else evidence of the splitting of the nucleus of Comet C/1881 K1 (Tebbutt). This observation clearly riled Tebbutt, and from this date on, he showed nothing but indifference towards Dr. Bone, even though the latter tried to maintain cordial relations (Orchiston 1987).

Unbeknown to Tebbutt, the Great Comet of 1881 ushered in a 'golden age' for Australian cometary discoveries. In December 1883, two Tasmanians, Clevers and Thirwall, discovered a naked-eye comet (see Orchiston 1983) which has not been included in the standard catalogues (e.g. Marsden and Williams 1996). The Melbourne astronomer, David Ross (Bio Box 9.3), discovered a comet in 1884 (see Orchiston and Brewer 1990), and a Queensland astronomer, J. Ewen Davidson (Bio Box 9.4), was successful in 1889 (Baracchi 1914; Darlington and Orchiston 2017). Between them, Australians accounted for >10 % of all new comets discovered worldwide during the period 1881–1889, inclusive, and it was Tebbutt who set the ideal example with the Great Comet of 1881. The only unfortunate aspect of these discoveries is that none of them was made during the year starting 1 February 1885 when Mr. H.H. Warner of Rochester, USA, was offering a prize of $US200 for any new comet discovery. Tebbutt inquired and found that Australian astronomers were eligible for this prize, and, in the absence of any existing astronomical society in Australia at that time, he made a point of communicating this fact (along with the discovery criteria) through the local newspapers (see Tebbutt 1885).

**Bio Box 9.3: David Ross**

David Ross (Fig. 9.17) was born in Ardgay, Scotland, on 26 February 1850 and migrated to Melbourne in 1876. Ten years later he joined the Brunswick Branch of The National Bank of Australasia Ltd. in suburban Melbourne. Ross remained with the branch the rest of his working life, and being a bachelor lived on the premises. Upon retiring in 1915 he moved to Yarra Glen, near Melbourne, living with his brother and sister-in-law (Orchiston and Brewer 1990).

**Fig. 9.17** David Ross (Orchiston collection)

(continued)

(continued)

Some time during the 1870s, Ross acquired a 3-in. (7.6-cm) refractor and from 1893 embarked on the manufacture of home-made reflectors, starting with a 9-in. (22.9-cm) instrument and culminating in the completion of a 12-in. (30.5-cm) equatorially-mounted Newtonian reflector in 1909 (see Fig. 4.12). Along with Robert W. Wigmore (ca. 1856–1945; Orchiston and Perdrix 2004), Ross was one of Victorian's pioneers in amateur telescope-making (see Orchiston (2003a). Initially his telescopes were mounted in the back yard of the bank in Brunswick, but after moving to Yarra Glen, the 3-in. refractor was housed in an observatory. Named 'The Peace Observatory' (Fig. 13.10), in anticipation of a rapid resolution of WWI, this wooden structure was unusual in that the whole building rotated (and not just the dome).

Ross had a special observational interest in comets. He discovered two comets (C/1884 A1 and C/1906 F1) and used his telescopes for systematic observations of these and known comets. As such, he was an obvious individual to target for membership of the Australian Comet Corps (see Chap. 10) as he had just become actively involved in observational astronomy when this group was formed. As if to highlight this fact, Ross discovered his first comet in early January 1884.

Ross occasionally observed other solar system objects or phenomena and published a number of short papers in the *Journal of the British Astronomical Association*. During the 1890s, Ross began experimenting successfully with astronomical photography, and among non-professional astronomers, he was one of Australia's pioneers in this field.

He also was one of the founders of the Victoria Branch of the BAA in 1897 and during its 10-year existence remained on the Committee and served at one time or another as Secretary and Vice President (Orchiston and Perdrix 1990, 2002). He was also their '… leading local role-model in the areas of cometary astronomy, astrophotography, and telescope-making' (Orchiston and Brewer 1990, p. 179).

By the end of the nineteenth century, David Ross was Victoria's leading independent astronomer, and he died at Yarra Glen on 20 May 1930 at the age of 80. For further details of his astronomical work, see Orchiston and Brewer (1990).

**Bio Box 9.4: John Ewen Davidson**

J. Ewen Davidson (Fig. 9.18) was born in London in 1841 and was educated at Harrow and then Oxford University. He graduated with a BA in 1863, specialising in science.

After examining sugar plantations in the West Indies and British Guiana, he came to Australia in 1865 and began working on a sugar plantation near Cardwell, in North Queensland, north of Townsville. In 1867 he moved to coastal Mackay (south of Townsville) where he teamed with Thomas Henry Fitzgerald (1824–1888) and established a sugar and cotton plantation on the fertile flood plain of the Pioneer River to the west of the city. Having proved that sugar-growing was a commercially-viable proposition, in 1868 they built the first sugar mill in the district. Davidson went on to become a 'sugar baron' and a prominent pioneer of the Australian sugar industry, and he also played a leading role in local politics and sporting activities (A sugar pioneer 1923; Mills 1981; Sugar pioneer … 2009).

From his youth Davidson had shown an interest in astronomy, but it was only when he was settled near Mackay that he was able to pursue this. In 1888 he purchased a 6-in. (15.2-cm) equatorially-mounted Cooke refractor from a Townsville astronomer and installed this in a roll-off roof observatory (see A peep … 1889; Orchiston 1997b; Russell 1889). This telescope was one of two Cooke refractors brought to Queensland by a British party for the 1882 transit of Venus (Darlington and Orchiston 2017; Haynes et al. 1993).

Davidson's primary observational interest was in comets, and on 21 July 1889, he discovered a new comet (Davidson 1890), which now has the designation C/1889 O1 (Davidson), and after moving into northern skies, it became yet another example of a comet that experienced nuclear splitting (Sekanina 1979). Details of Davidson's discovery and his subsequent observations of

**Fig. 9.18** John Ewen Davidson in about 1900 (https://en.wikipedia.org)

(continued)

(continued)

this comet, and observations made by other Australian astronomers and overseas astronomers, are provided by Darlington and Orchiston (2017).

Continuing his interest in cometary astronomy, Davidson discovered a second comet, 19P/Holmes, on 9 November 1892 (Russell 1892), but because he was only the fourth independent discoverer, his name was not assigned to this comet (see Darlington and Orchiston 2017). This comet was known colloquially as the 'Andromeda Comet' (e.g. see Ellery 1892).

As an active observational astronomer, one of the ways that Davidson tried to counter the geographical and intellectual isolation of living in country Queensland was to join international astronomical societies, such as the Astronomical Society of the Pacific and the British Astronomical Association (BAA) (where he was particularly active in the Comet Section). When the BAA formed a New South Wales Branch in Sydney in 1895 (see Orchiston 1988), Davidson joined, despite his distant domicile, and he sometimes sent reports that were read at the Branch's monthly meetings (e.g. see Davidson, 1897a, b).

Following the depression of the early 1890s, Davidson found himself in an untenable situation, with major changes to the Australian sugar industry planned by government (Mills 1981) and climatic extremes making farming an increasingly risky proposition. His response was to retire, and he moved back to England in 1900, settling in Oxford, where he died in 1923 at the age of 85 (a sugar pioneer, 1923). For further biographical details about Davidson and his astronomical activities in Queensland, see Darlington and Orchiston (2017).

**Acknowledgements** I am grateful to Professor Nick Lomb (University of Southern Queensland, Toowoomba), Clive Smith (Port Macquarie Historical Society), and the staff at the Mackay Historical Society, the Port Macquarie Museum and the Port Macquarie Library for their assistance in various ways and to the Port Macquarie Museum for kindly supplying Fig. 9.1.

## References[1]

A peep into Queensland houses. *The Sydney Morning Herald*, 12 September 1889, page 5.

A sugar pioneer. *Cairns Post,* 14 December 1923, page 9.

Backhouse, T.W. (1882). Observations of Comet b, 1881. *Monthly Notices of the Royal Astronomical Society, 42,* 413–421.

Baikie, J. (1906). *Through the Telescope*. London: Adam and Charles Black.

Ball, R. (1893). *The Story of the Heavens*. London: Cassell.

[1] The following abbreviation is used: RA = Royal Astronomical Society Letters Archives, London, TL = Letters to J. Tebbutt, bound manuscript letters in the Mitchell Library, Sydney.

Baracchi, P. (1914). Astronomy and geodesy in Australia. In G.H. Knibbs (Ed.), *Federal Handbook Prepared in Connection with the Eighty-fourth Meeting of the British Association for the Advancement of Science, Held in Australia, August, 1914* (pp. 326–390). Melbourne: Government Printer.

Beattie, E.H. (1910). Obituary. William J. MacDonnell. *Journal of the British Astronomical Association, 21,* 41–42.

Biggs, A.B. (1881). Letter to John Tebbutt, dated 5 August. In TL.

Bone, W. (1882a). Letter to John Tebbutt, dated 5 May. In TL.

Bone, W. (1882b). On an object seen near Comet b, 1881, on June 10, 1881. *Monthly Notices of the Royal Astronomical Society, 42,* 105–106.

Brück, H.A. & Brück, M.T. (1988). *The Peripatetic Astronomer. The Life of Charles Piazzi Smyth.* Bristol: Adam Hilger.

Brück, M.T. (1994). Agnes Mary Clerke, chronicler of astronomy. *Quarterly Journal of the Royal Astronomical Society, 35,* 59–79.

Brück, M.T. (2002). *Agnes Mary Clerke and the Rise of Astrophysics*. Cambridge: Cambridge University Press.

Burgel, B.H. (1911). *Astronomy For All*. London: Cassell.

Butterfield, G. (1881). Letter to John Tebbutt, dated 27 May. In TL.

Chambers, G.F. (1909). *The Story of Comets Simply Told for General Readers*. Oxford: Clarendon Press.

Clerke, A. (1893). *A Popular History of Astronomy during the Nineteenth Century*. London: Adam and Charles Black.

Clerke, A. (1898). The Solar System. In A.M., Clerke, A. Fowler & J.E. Gore (Eds.), *The Concise Knowledge of Astronomy* (Section III). London: Hutchinson.

Common, A.A. (1881). Photography as applied to Comet b 1881. *The Observatory, 4,* 232–233.

Comstock, G.C. (1901). *The Text-Book of Astronomy*. London: Hirschfeld.

Crommelin, A.C.D. (n.d.). *The Star World*. London: Collins.

Crouch, C. (1881a). Letter to John Tebbutt, dated 25 May. In TL.

Crouch, C. (1881b). Letter to John Tebbutt, dated 7 June. In TL.

Darlington, V. & Orchiston, W. (2017). North Queensland and the 1882 transit of Venus: our final role in the quest for the Astronomical Unit. *Memoirs of the Queensland Museum*, in press.

Davidson, J.E. (1890). Comet e, 1889. *The Observatory, 13,* 247.

Davidson, J.E. (1897a). Letter to John Tebbutt, dated 15 November. In TL.

Davidson, J.E. (1897b). Letter to John Tebbutt, dated 13 December. In TL.

de Vaucouleurs, G. (1961). *Astronomical Photography*. London: Faber and Faber.

Denning, W.F. (1891). *Telescopic Work for Starlight Evenings*. London: Taylor and Francis.

Dolmage, C.G. (1909). *Astronomy of Today*. London: Seeley.

Draper, H. (1881a). Note on photographs of the spectrum of Comet b 1881. *The Observatory, 4,* 252–253.

Draper, H. (1881b). On photographs of the spectrum of the comet of June, 1881. *American Journal of Science, 22,* 134–135.

Draper, H. (1881c). Photographs of Comet b 1881. *The Observatory, 4,* 239.

Dunkin, E. (1891). *Midnight Sky*. London: The Religious Tract Society.

Ellery, R.L.J. (1892). Letter to John Tebbutt, dated 22 November. In TL.

Elson, H.W. (1910). *Comets. Their Origin, Nature and History*. London: Sidgwick and Jackson.

Ewer, F.H. (1881). Memo to John Tebbutt, dated 27 July. In TL.

Flammarion, C. (1903). *Astronomy for Amateurs*. London: Nelson.

Gaggery, Mrs. (1881). Letter to John Tebbutt, dated 7 June. In TL.

Giberne, A. (1910). *Sun, Moon, and Stars*. London: Seeley.

Gill, D. (1882). On observations of Comets 1881 II. & III., of Wells's Comet, and of the Great Comet (b) 1882, made at the Royal Observatory, Cape of Good Hope. *Monthly Notices of the Royal Astronomical Society, 43,* 7–19.

Gore, J.E. (1909). *Astronomical Curiosities*. London: Chatto & Windus.

Gould, B.A. (1882). Dr. Gould's observation (June 11th) of Comet b, 1881. *The Observatory, 5,* 229–231.
Griffin, T.H.F. (1881a). Letter to John Tebbutt, dated 24 May. In TL.
Griffin, T.H.F. (1881b). Letter to John Tebbutt, dated 2 June. In TL.
Griffin, T.H.F. (1881c). Letter to John Tebbutt, dated 30 June. In TL.
Griffin, T.H.F. (1881d). Letter to John Tebbutt, dated 2 July. In TL.
Griffin, T.H.F. (1881e). Letter to John Tebbutt, dated 5 August. In TL.
Guillemin, A. (1877). *The World of Comets*. London: Sampson Low, Marston, Searle and Rivington.
Haynes, R.D., Haynes, R.A. & Kitson, W.S. (1993). The history of astronomy in Queensland. *Vistas in Astronomy, 36,* 231–252.
Hearnshaw, J.B. (2014). *The Analysis of Starlight. One Hundred and Fifty Years of Astronomical Spectroscopy. Second Edition.* Cambridge: Cambridge University Press.
Heath, T. (1908). *The Twentieth Century Atlas of Popular Astronomy*. Edinburgh: Johnston.
Hodge, P. (1971). Benjamin Apthorp Gould and the founding of the Argentine National Observatory. *The Americas, 28,* 152–175.
Hollis, H.P. (n.d.). *Chats on Astronomy*. London: Werner.
Howe, H.A. (1897). *Elements of Descriptive Astronomy. A Text-Book*. London: George Philip.
Huggins, W. (1881a). On the photographic spectrum of Comet b 1881. *The Observatory, 4,* 233–234.
Huggins, W. (1881b). Preliminary note on the photographic spectrum of Comet c 1881 *Proceedings of the Royal Society, 33,* 1–3.
Hughes, S. (2013). *Catchers of the Light. The Forgotten Lives of the Men and Women Who First Photographed the Heavens*. Two Volumes. Paphos: ArtDeCiel Publishing.
Jackson, W.M. (1982). Laboratory studies of photochemical and spectroscopic phenomena related to comets. In L.L. Wilkening (Ed.), *Comets* (pp. 481–495). Tuscson: University of Arizona Press.
Janssen, J. (1881). Photographie de la Comete 1881 III. *Vierteljahrsschrift der Astronomischen Gesellschaft, 16,* 308–311.
Jones, H.S. (1946). *General Astronomy*. London: Arnold.
Joseph, A.M. (1881). Letter to John Tebbutt, dated 20 June. In TL.
Knobel, E.B. (1882). Notes on sketches of Comet b, 1881. *Monthly Notices of the Royal Astronomical Society, 42,* 76–77.
Lankford, J. (1984). The impact of photography on astronomy. In O. Gingerich (Ed.), *The General History of Astronomy. Volume 4. Astrophysics and Twentieth Century Astronomy to 1950: Part A* (pp. 16–39). Cambridge: Cambridge University Press.
Launay, F. (2011). *The Astronomer Jules Janssen: A Globetrotter of Celestial Physics*. New York: Springer.
Lewis, I.M. (1921). *Splendours of the Sky*. London: Murray.
Lynn, W.T. (1909). *Remarkable Comets, A Brief Survey of the Most Interesting Facts in the History of Cometary Astronomy*. London: Bagster.
Macdonnell, W.J. (1871–1885). Untitled astronomical diary. MS, Mitchell Library, Sydney (B201).
Macdonnell, W.J. (1881). Letter to John Tebbutt, dated 3 June. In TL.
Macpherson, H. (1906). *A Century's Progress in Astronomy*. Edinburgh: Blackwood.
Main, R. (1882). *Rudimentary Astronomy*. London: Crosby, Lockwood.
Marsden, B.G. (1982). *Catalogue of Cometary Orbits*. Cambridge (Mass.): Minor Planet Center.
Marsden, B.G. (1989). The sun-grazing comet group. II. *Astronomical Journal, 98,* 2306–2321.
Marsden, B. & Williams, G.V. (1996). *Catalogue of Cometary Orbits 1996*. Cambridge (Mass.): Minor Planet Center.
Maunder, E.W., [1913]. *Sir William Huggins and Spectroscopic Astronomy*. London: Jack.
Mee, A. (1897). *Observational Astronomy: A Practical Handbook for Amateurs*. Cardiff: Western Mail.
Merfield, C.J. (1909). Letter to the Secretaries, dated 11 October. In RA.
Mills, J.A. (1981). Davidson, John Ewen (1841–1923). In B. Nairn & G. Searle (Eds.), *Australian Dictionary of Biography, Volume 8* (p. 228). Melbourne: Melbourne University Press.

Moulton, F.R., 1908. *An Introduction to Astronomy*. New York: Macmillan.
Neison, E. (1886). *Astronomy: A Simple Introduction to a Noble Science*. London: Ward, Lock.
Newcomb, S. (1908). *Astronomy for Everybody*. London: Pitman.
Obituaries: Andrew Ainslie Common. *Monthly Notices of the Royal Astronomical Society, 64,* 274–278 (1904).
Obituaries: William John Macdonnell. *Monthly Notices of the Royal Astronomical Society, 71,* 270–271 (1911).
Olivier, C.P. (1930). *Comets*. Baltimore: Williams and Wilkins.
Olson, R.J.M. & Pasachoff, J.M. (1998). *Fire in the Sky. Comets and Meteors, the Decisive Centuries, in British Art and Science*. Cambridge: Cambridge University Press.
Orchiston, W. (1981). Illuminating incidents in Antipodean astronomy: John Tebbutt and Great Comet of 1881. *Journal of the Astronomical Society of Victoria, 34,* 78–92.
Orchiston, W. (1983). Illuminating incidents in Antipodean astronomy: the mysterious Tasmanian comet of December 1883. *Proceedings of the Astronomical Society of Australia, 5,* 282–284.
Orchiston, W. (1987). Contribution of the Lands Department to the development of astronomy in New South Wales during the nineteenth century. *Australian Journal of Astronomy, 2,* 65–74.
Orchiston, W. (1988). Illuminating incidents in Antipodean astronomy: formation of the New South Wales Branch of the British Astronomical Association. *Journal of the British Astronomical Association, 98,* 75–84.
Orchiston, W. (1997a). The role of the amateur in popularising astronomy: an Australian case study. *Australian Journal of Astronomy, 7,* 33–66.
Orchiston, W. (1997b). The role of the large refracting telescope in Australian amateur astronomy: an historical perspective. *Australian Journal of Astronomy, 7,* 89–114.
Orchiston, W. (1999). C/1881 K1: A forgotten "Great Comet" of the nineteenth century. *Irish Astronomical Journal, 26,* 33–44.
Orchiston, W. (2001). William John Macdonnell and the development of astronomy in New South Wales. *Journal of the British Astronomical Association, 111,* 13–25.
Orchiston, W. (2003a). Amateur telescope making in Australia. An historical perspective. In W.J. Cook (Ed.), *The Best of Amateur Telescope Making Journal. Volume 2*. (pp. 208–239). Richmond: Willmann-Bell.
Orchiston, W. (2003b). Australia's earliest planispheres. *Journal of the British Astronomical Association, 113,* 329–332.
Orchiston, W. (2010). The Great Comet of 1881, John Tebbutt and poor old 'Jacko'. *Journal of the British Astronomical Association, 120,* 116–117.
Orchiston, W. (2016). *Exploring the History of New Zealand Astronomy: Trials, Tribulations, Telescopes and Transits*. Cham: Springer.
Orchiston, W. & Bembrick, C. (1995). The role of the large reflecting telescope in amateur astronomy: an Australian case study. *Australian Journal of Astronomy, 6,* 53–74.
Orchiston, W. & Brewer, A. (1990). David Ross and the development of amateur astronomy in Victoria. *Journal of the British Astronomical Association, 100,* 173–181.
Orchiston, W. & Perdrix, J. (1990). The British Astronomical Association in Australia: a centennial perspective. *Australian Journal of Astronomy, 3,* 163–179.
Orchiston, W. & Perdrix, J. (2002). A history of the British Astronomical Association in Australia: the fate of the Branches. *Journal of the British Astronomical Association, 112,* 68–77.
Orchiston, W. &, Perdrix, J. (2004). Who was Robert W. Wigmore? *Crux, 22* (5), 4–5.
Parkes, S.H. (1888). *Unfinished Worlds. A Study in Astronomy*. London: Hodder and Stoughton.
Pasachoff, J., Olson, R.J.M. & Hazen, M.L. (1996). The earliest comet photographs: Usherwood, Bond and Donati 1858. *Journal for the History of Astronomy, 27,* 129–145.
Peck, W. (1890. *A Popular Hand-Book and Atlas of Astronomy*. London: Gall and Inglis.
Peters, C.H.F. (1882). Letter to John Tebbutt, dated 8 December. In TL.
Powys, A.O. (1881a). Letter to John Tebbutt, dated 28 May. In TL
Powys, A.O. (1881b). Letter to John Tebbutt, dated 4 June. In TL.
Powys, A.O. (1881c). Letter to John Tebbutt, dated 25 June. In TL.

Powys, A.O. (1881d). Letter to John Tebbutt, dated 27 August. In TL.
Proceedings of Societies: Royal Society of Victoria. *Southern Science Record, 1,* 126–127 (1881).
Proctor, M. (1926). *The Romance of Comets*. New York: Harper.
Proctor, R. (1896). *Myths and Marvels of Astronomy*. London: Longmans, Green.
Robinson, G. (1881). Letter to John Tebbutt, dated 29 May. In TL.
Roemer, E. (1963). Comets, discovery, orbits, astronomical observations. In B.M. Middlehurst & G.P. Kuiper (Eds.), *The Solar System. IV The Moon, Meteorites and Comets* (pp. 527–549). Chicago: University of Chicago Press.
Roseby, T. (1910). Letter to W.H. Wesley, dated 25 November. In RA.
Royal Society of New South Wales [Report of a Meeting]. *Sydney Morning Herald*, 4 June 1881.
Russell, H.C. (1881a). The Comet. *Sydney Morning Herald*, 26 May.
Russell, H.C. (1881b). Letter to John Tebbutt, dated 8 February. In TL.
Russell, H.C. (1881c). Letter to John Tebbutt, dated 25 May. In TL.
Russell, H.C. (1881d). Letter to John Tebbutt, dated 11 June. In TL.
Russell, H.C. (1881e). Spectrum and appearance of the recent comet. *Journal and Proceedings of the Royal Society of New South Wales, 15,* 81–86.
Russell, H.C. (1883). Observations of the transit of Venus, 1874, December 8–9, made at stations in New South Wales. *Memoirs of the Royal Astronomical Society, 47,* 49–88.
Russell, H.C. (1889). Letter to W.H. Wesley, dated 17 September. In RA.
Russell, H.C. (1892). Letter to W.H. Wesley, dated 26 November. In RA.
Sekanina, Z. (1979). Relative motions of fragments of the split comets: III. A test of splitting and comets with suspected mutliple nuclei. *Icarus, 38,* 300–316.
Serviss, G.P. (1909). *Curiosities of the Sky. A Popular Presentation of the Great Riddle and Mysteries of Astronomy*. New York: Harper.
Sharpless, I. & Philips, G.M. (1882). *Astronomy for Schools and General Readers*. Philadelphia: Lippincott.
Sheehan, W. (1995). *The Immortal Fire Within. The Life and Work of Edward Emerson Barnard.* Cambridge: Cambridge University Press.
Smyth, P. (1881). Letter to J. Tebbutt, dated 28 August. In TL.
Smyth, P. (1882). Letter to J. Tebbutt, dated 4 March. In TL.
Smyth, W.H. (1844). *A Cycle of Celestial Objects*. Two volumes. London: Parker.
Stone, E.J. (1884). [Presidential] Address. *Monthly Notices of the Royal Astronomical Society, 44,* 221–223.
Sugar pioneer sighted a new comet in 1889. *The Daily Mercury*, 14 October 2009.
Tebbutt, J. (1879–81). Journal for Astronomical Observations. The Peninsula, Windsor, N.S.Wales. MS, Mitchell Library, Sydney (A3749).
Tebbutt, J. (1881a). Apparent places of Comet 1881 II [*sic*] *Astronomische Nachrichten, 100,* 333–334.
Tebbutt, J. (1881b). Appearance of another Southern Comet (Comet b 1881). *The Observatory, 4,* 239–240.
Tebbutt, J. (1881c). The Comet. *Sydney Morning Herald*, 26 May.
Tebbutt, J. (1881d). The Comet. *Sydney Morning Herald*, 9 June.
Tebbutt, J. (1881e). Observations and elements of Comet b 1881. *Monthly Notices of the Royal Astronomical Society, 41,* 442–443.
Tebbutt, J. (1881f). Observations of a southern comet 1881. *Astronomische Nachrichten, 100,* 171–174.
Tebbutt, J. (1881g). On Comet II [*sic.*], 1881. *Journal and Proceedings of the Royal Society of New South Wales, 15,* 87–91.
Tebbutt, J. (1881h). Orbit-elements of Comet b 1881. *The Observatory, 4,* 299.
Tebbutt, J. (1881i). The orbit-elements of Comet II [*sic.*], 1881. *Journal and Proceedings of the Royal Society of New South Wales, 15,* 393–395.
Tebbutt, J. (1881j). Orbit elements of Comet 1881. *Astronomische Nachrichten, 100,* 335–336.

Tebbutt, J. (1881k). Reduction of the Windsor Observations of Comet Tebbutt, 1881, with an approximate calculation of the orbit. MS, in Tebbutt, J. *Reduction of Observations. Windsor, Comets, 1862–1882*. Mitchell Library, Sydney (A3730).

Tebbutt, J. (1882a). Dr. Bone's supposed comet. *The Observatory, 5,* 199.

Tebbutt, J. (1882b). Dr. Gould's observation (June 11th) of comet b 1881. *The Observatory, 5,* 18–20.

Tebbutt, J. (1882c). Dr. Gould's observation of Comet 1881, III, on June 11. *The Observatory, 5,* 363–365.

Tebbutt, J. (1885). Astronomical prizes. *Sydney Morning Herald*, 24 March.

Tebbutt, J. (1908). *Astronomical Memoirs*. Sydney, printed for the author.

The comet. *Sydney Morning Herald*, 26 May 1881.

Todd, D.P. (n.d.). *A New Astronomy*. London: Sampson.

Vsekhsvyatskii, S.K. (1964). *Physical Characteristics of Comets.* Jerusalem: Israel Program for Scientific Translations.

Whitney, C.A. (1971). Draper, Henry. In C.C. Gillespie (Ed.), *Dictionary of Scientific Biography. Volume IV: DE-FIR* (pp. 178–181). New York: Scribner.

Yeomans, D.K. (1991). *Comets. A Chronological History of Observation, Science, Myth, and Folklore*. New York: Wiley.

Young, C.A. (1888). *A Text-book of General Astronomy for Colleges and Scientific Schools*. Boston: Ginn.

# Chapter 10
# An Abortive Association: The Australian Comet Corps

## 10.1 Introduction

Comets were an important aspect of nineteenth-century science, as astronomers sought to learn more about their orbits, trace temporal variations in the appearance of the heads and tails of individual comets, investigate the possibility of distinct 'comet families' and propose different 'types' of tails (see Chambers 1909; Clerke 1893; Guillemin 1877). Each new comet discovery was heralded with anticipation and delight by those addicted to cometary astronomy.

Consequently, comet-searching became a preoccupation for many nineteenth-century astronomers, and although some met with considerable success it was clear that new comets were being missed. What was needed was a co-ordinated search programme, and this led to the concept of a 'Corps of Comet-seekers' which was developed in 1881–1882 by the Boston Science Society.

In January and February of 1882, the Society sent out circulars to all known United States comet-seekers outlining the concept. Each member of the Corps would be assigned a specific region of sky and charged with systematically searching this for new comets. The circulars were succeeded in March by a 3-page pamphlet titled *A Systematic Search for Comets* (1882), and this was distributed widely within the United States as the scheme became public. The pamphlet outlined the obvious advantages of such a scheme: '… the possibility of discovering more comets, less liability to duplicate work, and the accumulation of data, showing what portions of the sky are neglected … the number of comets visiting our system, and their distribution'. It also listed those already involved in the 'American Comet Corps' (as I shall call it) or planning to participate, and these included Edward Emerson Barnard (1857–1923; Fig. 10.7), William Robert Brooks (1844–1921), John Martin Schaeberle (1853–1924) and Lewis A. Swift (1820–1913; Fig. 11.7), all now well-known names in cometary astronomy (see Chambers 1909; Marsden and Williams 1996).



W. Orchiston, *John Tebbutt*, Historical & Cultural Astronomy,
DOI 10.1007/978-3-319-44521-2_10

Later in 1882, Tebbutt received a letter and the 3-page pamphlet from the Society, suggesting that he might like to set up an 'Australian Comet Corps'. This unsolicited invitation offered a new direction for him, as he had never before been involved in the organisation of any astronomical group or society. Although he had been an active member of the Philosophical Society (later the Royal Society) of New South Wales since 1861 (Tebbutt 1908), and had presented papers at meetings and published in the Society's journal, he had never served on the Council. The closest he came in this regard was as a Committee member of the Society's short-lived Section A (effectively its Astronomical Section) in 1881, the year of its demise (see Orchiston and Bhathal 1991). By 1882, Tebbutt also was a Fellow of the Royal Astronomical Society but, as would be expected given his domicile, he played no organisational role in the affairs of that august body. To form an Australian Comet Corps therefore offered a considerable challenge, but one that Tebbutt responded positively to.

In this chapter, we trace Tebbutt's attempt to form the Australian Comet Corps, introducing along the way a number of others who were actively involved in Australian astronomy. We highlight the problems that he faced (or thought that he faced), document the failure of the Comet Corps after just 1 year of existence and present a post-mortem in an attempt to identify the reasons behind its demise. In so doing, I draw freely on a paper about the abortive 'Australian association of comet observers' (as I once called it) which was prepared to mark the centennial of Australia's first national astronomical group (Orchiston 1982), and on additional material that I included in a later paper, about Dr. William Bone of Castlemaine (Orchiston 1987b).

## 10.2 An Association of Observers: The Proposed 'Australian Comet Corps'

The seed of the idea for an Australian Comet Corps were planted in the letter that John Ritchie Jn. (1853–1930) of the Boston Scientific Society sent Tebbutt in 1882. Ritchie begins: 'Knowing of your interest in comets, and having received from you at different times, your publications, I am pleased to send to you the enclosed circular, which will not need any explanatory remarks' (Ritchie 1882). His reason for writing Tebbutt becomes obvious as one proceeds through his long letter. After explaining that the northern sky will be well covered by American observers, he directs his attention south:

> But good as may be our intentions, our position is not one that can command every point the care of the South polar regions must devolve upon some observer or observers in the Southern Hemisphere, or go undone, save in a desultory manner, as heretofore. It has seemed possible … that you might care to arrange your work as to provide for the sweeping of such zones as cannot be reached by us—once during each month … *It is quite possible that should the field seem a promising one to you that you may know enough astronomers in Australia who by joining work could sweep to the Equator*. (ibid.; my italics).

**Table 10.1** Tebbutt's scholarly publications on comets, 1862–1881

| Journal | Papers |
|---|---|
| *Astronomical Register* | 1 |
| *Astronomische Nachrichten* | 30 |
| *Journal of the Royal Society of New South Wales* | 3 |
| *Monthly Notices of the Royal Astronomical Society* | 19 |
| *The Observatory* | 6 |
| *Transactions of the Philosophical Society of New South Wales* | 2 |
| Total: | 61 |

His choice of Tebbutt was obvious, for between 1853 and the start of 1882 he had observed 13 different comets, and 1 of these, Comet 2P/Encke, on 4 different occasions (in 1862, 1865, 1875 and 1878). But more important was his record of publication: by 1882, he had presented his cometary observations in 61 research papers, published in 6 different journals, 4 of which appeared overseas (see Table 10.1).

Furthermore, Tebbutt had carried out some systematic comet-searching, having already discovered the Great Comets of 1861 (C/J11861; Orchiston 1998b) and 1881 (C/1881 K1; Orchiston 1999). If Ritchie was looking for an Australian to organise a local 'Comet Corps', then Tebbutt appeared to be the right man. However, Ritchie (ibid.) caused some confusion by mentioning in his letter that he also had written to Dr. William Bone of Castlemaine, Victoria. Presumably, Ritchie had seen Bone's (1880) *Monthly Notices* paper about the 1880 comet, and it was this that had prompted him to write.

Apparently Tebbutt saw considerable merit in Ritchie's suggestions, for he immediately wrote to the *Sydney Morning Herald*:

> … I have just received … a letter, requesting me to organize, if possible, a body of amateur observers for the purpose of cometary discovery … Can I prevail on any of our amateurs to join with me in the formation of a corps of comet-hunters? If some half-dozen would combine, a thorough periodic examination of the southern sky might be secured, and that with but little labour for each observer. (Tebbutt 1882a).

The same day he wrote this letter (8 May), Tebbutt began preparing his own circular about a mooted 'Corps of Amateur Comet-seekers in Australia', and the following day he forwarded this to astronomers he personally knew of who owned or had access to telescopes. He also submitted a copy to the *Sydney Morning Herald* (Tebbutt 1882b):

> Dear Sir,—I have received from the president of the Boston Scientific Society a letter requesting me to use my influence in organizing a corps of amateur comet-seekers in Australia … America has done well of late in the matter of cometary discovery … Australia can, I am sure, do something in this department. May I, therefore, respectfully inquire if your leisure and scientific means are such as to permit of joining in this scheme for systematic comet-seeking, and in the event of this being so, if you are willing to undertake a share in the work? Should half a dozen amateurs be found to combine, the work of sweeping would be very light for each, and need not occupy more than two mornings and two evenings in each month. Should a reply be received in the affirmative, I shall be obliged by

a description of the instrumental means which you have at your disposal for the work. State particularly the kind, aperture, magnifying powers, and mounting of your telescope, and what star-charts and lists of nebulae you have for reference. When the number is ascertained of observers that are likely to join in the work, the sky can conveniently be divided into zones, for sweeping, and a zone apportioned to each observer by mutual arrangement. Specimen forms will be supplied for the proper record of the sweeps. Should you know of any person in your immediate neighbourhood who would be likely to entertain the proposals contained in this circular, I shall be obliged by your communicating to me his address. As I am desirous of forwarding a progress report to the president of the Boston Scientific Society as soon as possible, an early reply is solicited to this appeal.

I am, dear sir, yours faithfully
JOHN TEBBUTT, F.R.A.S.

With the aid of supplementary names supplied through some of the astronomers he sent the circular to, Tebbutt ended up with 15 people whom he contacted directly or had others contact on his behalf, and these are listed in Table 10.2. In addition, he wrote to Archdeacon Arthur Stock, the Astronomical Observer at the Colonial Observatory in Wellington (see Orchiston 1985b, 1998c, pp. 99–101), and invited New Zealand astronomers to join the programme (see Stock 1882). Stock was the only professional astronomer in Australia or New Zealand whom Tebbutt approached about the local Comet Corps.

There was an interesting range of expertise and experience among those listed in this table but not all were potential members. Francis Abbott (see Bio Box 6.4) was in his twilight years, and although the concept probably appealed to him, he was not in any position to participate (and he died the following year at the age of 83). After Abbott, Launceston bank officer, Alfred Barrett Biggs (Bio Box 10.1), was Tasmania's foremost astronomer and was a Tebbutt protégé and a keen observational astronomer.

**Bio Box 10.1: Alfred Barrett Biggs**

Alfred Barrett Biggs (Fig. 10.1) was born in London on 10 April 1825, and after Francis Abott was Tasmania's other prominent nineteenth-century independent astronomer. He moved to Hobart in 1833 and chose school-teaching as a career. In about 1878 he moved to Launceston and a new position as accountant and head ledger-keeper with the Launceston Bank for Saving, and he retained that post until the end of 1900 when he retired. He died just 3 weeks later.

Biggs became addicted to astronomy after assisting the United States party that was based at Campbell Town for the 1874 transit of Venus (see Orchiston and Buchanan 1993; Smith and Jetson 1994). Soon after this event he obtained a small refractor and after moving to Launceston constructed an observatory. In 1885 he acquired an 8.5-in. (21.6-cm) Browning-With reflector and built a second observatory. These instruments were used mainly to observe comets and double stars. Under Tebbutt's guidance and encouragement he went on to publish a number of research papers in *Monthly Notices of the Royal Astronomical Society* and *Papers and Proceedings of the Royal Society of Tasmania.*

(continued)

(continued)

**Fig. 10.1** A.B. Biggs (Orchiston collection)

Like Tebbutt, Biggs also was committed to the popularisation of astronomy: he maintained a local time service and a meteorological station, and he and the Reverend Canon Brownrigg (see Condon, 1969) published monthly meteorological reports in the local press. He conducted frequent public viewing nights at one of his observatories, lectured on astronomy, wrote for the local newspapers (Orchiston 1997a), responded to public inquiries in astronomy, and was known locally to some as the 'Astronomer Royal' (Scott and Scott, 1935).

A remarkable non-professional scientist, Biggs was also one of Australia's earliest seismologists (see Giordano 1995). A detailed account of his astronomical activities is presented in Orchiston (1985a).

Dr. William Bone (Bio Box 8.3), a general practitioner, was anticipating the arrival of a new 8-in. (20.3-cm) Grubb refractor. Butterfield, Brooks and Conder all were employed by the New South Wales Lands Department (Orchiston, 1987a), Butterfield (Bio Box 9.1) as a draftsman, Brooks (Bio Box 8.5) as a surveyor and Conder (Bio Box 10.2) as Superintendent of the Trigonometrical Survey Branch.

**Bio Box 10.2: William Jacomb Conder**

William Jacomb Conder (Fig. 10.2) is somewhat of a mystery man, even though he attained some prominence in the New South Wales Lands Department during the nineteenth century. Nothing has been published about his early life, but he apparently trained as a surveyor, and in 1870 was employed by the Trigonometrical Survey Branch of the New South Wales

(continued)

(continued)

**Fig. 10.2** A photograph of W.J. Conder by David Scott (*Courtesy* State Library of New South Wales)

Lands Department. He must have displayed considerable talent for a decade later, in September 1880, he was appointed Superintendent of Trigonometrical Surveys (Orchiston 1987a).

Apart from conducting astronomical observations associated with determining latitude and longitude, Conder observed the 1881 transit of Mercury from Bathurst, along with the Lands Department surveyor Joseph Brooks (see Bio Box 8.5). They used the Department's excellent 4.75-in. (12.1-cm) Schroeder refractor and a 3.75-in. (9.5-cm) refractor loaned by Sydney Observatory. This was a collaborative project between the Department and the Observatory.

An active member of the short-lived Section A of the Royal Society of New South Wales, Conder joined Brooks on the Committee in its final year (1881), and at the August meeting he presented a paper on 'Some observations for longitude at Lambie'. Subsequently, this was published in the Society's journal (see Orchiston and Bhathal 1991). Soon after this, Conder was posted to the country, which effectively ended his direct involvement in astronomical activities taking place in Sydney.

It should be noted that W.C. Conder was the uncle of the well-known British-born Australian landscape painter, Charles Conder (1868–1909), who at the age of 16 was sent to the 'colonies'. For a while he worked as an apprentice surveyor, supervised by his uncle, but this did not prove a rewarding vocation and he soon gravitated towards art.

William Jacomb Conder died in December 1890, and his biography has yet to be written. This is an ideal project for someone with an interest in nineteenth-century New South Wales astronomy and the trigonometrical survey of the colony.

George Denton Hirst (Bio Box 10.3), a wine and spirit merchant, did not own a telescope at the time, but had been using Colyer's excellent equatorially-mounted 10.25-in. (26-cm) Browning reflector. Bank Manager, William Macdonnell (Bio Box 9.2) owned a 2.25-in. (5.7-cm) Wray refractor when he heard from Tebbutt (Macdonnell 1871–1885, 1881), but he soon replaced this with a 3.65-in. (9.3-cm) instrument (Macdonnell 1882) and moved to Port Macquarie.

**Bio Box 10.3: George Denton Hirst**

George Denton Hirst (Fig. 10.3) was born in Sydney on 7 April 1846, and although he became a successful wine and spirit merchant, his principal passion was astronomy.

By 1880, he was one of the most active independent astronomers in New South Wales after Tebbutt, even though he did not own a telescope at the time. Instead, he made use of the 11.5-in. (29.2-cm) Schroeder refractor at Sydney Observatory (Fig. 8.8) and the 10.25-in. (26-cm) reflector owned by his friend, J.U.C. Colyer. It was only in 1904, when Sydney Observatory disposed of a number of telescopes left over from its nineteenth-century transit of Venus programmes, that Hirst acquired a fine 4.25-in. (10.8-cm) Cooke refractor, complete with equatorial mounting and a filar micrometer (Lenehan 1904).

Hirst took part in Sydney Observatory's 1874 transit of Venus and 1881 transit of Mercury programmes (see Russell 1881, 1892), but his special interest was planetary astronomy, and 'As an astronomical draftsman Mr. Hirst had no equal in Australia. His drawings of Mars are marked by the same skill and delicacy as those of N.E. Green of England … A long series of fine drawings of Jupiter which he made in 1876 and 1878 were presented by him to the [Royal Astronomical] Society and are deposited in the library' (Obituaries 1916). In addition to Jupiter, Mars held a special fascination for him (ibid.), and once in possession of the Cooke refractor he also observed double stars

**Fig. 10.3** G.D. Hirst (adapted from Russell 1892, Frontispiece)

(continued)

(continued)

(Hirst 1909). Over the years, he published a number of papers on his work in Australian and overseas journals, but his Jovian observations of the 1870s were of particular interest to later researchers investigating the nineteenth-century reddening of the Great Red Spot (e.g. see Hockey 1992, 1999).

From its founding, Hirst was an active member of the short-lived Section A of the Royal Society of New South Wales and served as Secretary for two sessions. At a later date, he also was a stalwart of the New South Wales Branch of the British Astronomical Association (see Orchiston 1988), and his '... advice and practical help were always available to those in difficulties, and often on Sunday mornings a small gathering of amateurs met at his home and criticised each other's observations' (Greig-Smith 1916). He was elected a Fellow of the Royal Astronomical Society in 1895.

When Hirst died in 1915, '... amateur astronomy in New South Wales lost one of its ablest supporters' (ibid.).

Clergyman, George Martin (Bio Box 10.4) had a passion for astronomy and the telescope with which to pursue it. We know little about Charles Sydney Norris (1859–1935), except that he had an interest in astronomy and meteorology and produced monthly meteorological reports (Norris 1882a). When he received the Comet Corps circular he said that he was about to order a 4.5-in. (11.4-cm) or 5-in. (12.7-cm) Cooke refractor (Norris 1882b), but apparently nothing ever came of this (Darlington and Orchiston 2017).

**Bio Box 10.4: George Martin**

George Martin (Fig. 10.4) was born in Gainsborough, England, in about 1835 and trained as a Wesleyan minister. He came to Sydney in 1859 where he practised as a minister, dying in Sydney in 1905 (Martin 1907).

From his youth, Martin had a passion for knowledge, and a retentive memory allowed him to '... harvest the fruits of his reading'.

**Fig. 10.4** Reverend George Martin (after Martin 1907)

(continued)

(continued)

Martin had a strong interest in astronomy, and owned a magnificent 5-in. (12.7-cm) Cooke refractor (see Martin 1875). Although there is no evidence that he ever engaged in any systematic observing programmes, Martin '... knew nature in its more familiar aspects, but by skilled use of his telescope and microscopes he was able to look into its remote recesses and make himself acquainted with its otherwise unseen marvels and beauties. He saw nature with the eye of the scientist ...' (Martin 1907, p. xii).

From all accounts, Martin was a popular public speaker. He was known for his lectures (and not just on astronomy), which '... thrilled and delighted large audiences wherever they were delivered' (Martin, 1907, p. vi). Over the years, he gave many public lectures (ibid.; Orchiston 1997a). He was also an active member of Section A of the Royal Society of New South Wales throughout its short existence (see Orchiston and Bhathal 1991).

The last astronomer listed in Table 10.2, H.G.A. Wright (Bio Box 10.5) was another medical doctor with a strong astronomical interest and an excellent telescope.

**Bio Box 10.5: Dr. Horatio George Anthony Wright** Horatio George Anthony Wright (Fig. 10.5) was born in Maidstone, England, in 1827, and studied medicine in London. He emigrated to Sydney in 1853 and set up what was to become a flourishing practice in Hunter Street (Hoare 1976).

Wright had a strong interest in astronomy and was yet another in the colony who had the financial resources to purchase a Browning-With reflector. George With (1827–1904) was renowned for his mirrors (see Marriott 1996). Wright acquired his 8.5-in. (21.6-cm) telescope in 1870, and installed this in an observatory at his home in Wynyard Square in the centre of Sydney

**Fig. 10.5** Dr. Horatio Wright (adapted from Russell 1892, Frontispiece)

(continued)

(continued)

(Hordern 1870). In about 1873, under Russell's guidance Wright '... ground a reflector of 9 $7/8$ inches and the definition is very good indeed' (Russell 1873).

When time permitted, Wright carried out observational programmes, including monitoring of the η Carinae nebula in order to investigate the changes that had been suggested by Abbott (see Orchiston 1992). He also observed the 1874 transit of Venus and his report is included in Russell's 1892 monograph.

For many years Wright was a stalwart of the Royal Society of New South Wales, although his influence was '... "unobtrusive" at a time of marked improvements in its scientific standards and prestige ...' (Hoare 1976). He was particularly active in Section A of the Society, serving each year on the Committee (see Orchiston and Bhathal 1991).

Dr. Wright died in Sydney in 1901 (Hoare 1976), and his astronomical biography has yet to be written.

Of the others listed in Table 10.2, little if anything is known of their astronomical activities, which strongly suggests that they were not particularly active.

Nine of the fifteen astronomers in Table 10.2 replied, and their responses varied considerably. Some, like Biggs and Hirst, were particularly enthusiastic: 'I must say I feel with you that it is quite time that Australia took her share in this branch of Astronomical Work ...' (Hirst 1882; cf. Biggs 1882b). Butterfield (1882), on the other hand, somewhat arrogantly chided Tebbutt for not mentioning in his circular the anticipated rate of successful discoveries, if the group was formed.

Of the nine respondents, only five (Biggs, Bone, Butterfield, Collins and Suttor) were prepared to participate in the new Comet Corps. The remaining astronomers declined, for one reason or another: Parker (Biggs 1882c) and Wright (1882) complained of lack of time; Hirst (1882) stated that he had recently moved house and no longer had ready access to Colyer's reflector; and Cox (from the Royal Naval Depot) reported that had just been ordered back to England, and therefore was looking to sell his telescope (Cox 1882). Meanwhile, in New Zealand the mooted Comet Corps was raised at a meeting of the New Zealand Philosophical Society, but for Tebbutt (1887, p. 17) the outcome was disappointing: '... the discussion resulted merely in the suggestion that the Government should establish an astronomical observatory ...'

Even though approximately half of those approached agreed to join the Australian Comet Corps, most expressed some reservations about their instrumental means and/or their observational abilities. For example, after detailing the equipment in his little observatory, Biggs (1882b) concluded: 'If you think that with such, I can further the object you have in view it will give me great pleasure to co-operate with you and others'. Butterfield (1882), for his part, emphasised his severe time constraints.

**Table 10.2** Australian independent astronomers approached by Tebbutt in 1882 regarding the proposed Australian Comet Corps

| | | |
|---|---|---|
| F. Abbott (Hobart, TAS) | 4-in. Varley refractor, with Dallmeyer equatorial mounting with circles<br>3.5-in. Cooke refractor, altazimuth mounting<br>Micrometer<br>2-in. transit telescope | Abbott (1861, 1862, 1871a, b), White (1891), Orchiston (1992) |
| A.B. Biggs (Launceston, TAS) | 3-in. refractor, equatorial mounting with circles<br>Home-made micrometer<br>Small home-made transit telescope | Biggs (1881, 1882a, b), Orchiston (1985a, 2004) |
| Dr W. Bone (Castelmaine, VIC) | 8-in. Grubb refractor on order, equatorial mounting, with circles and drive<br>Micrometer<br>Transit telescope | Bone (1882a, c), Orchiston (1986, 1987b, 1997c) |
| J. Brooks (Sydney, NSW) | Access to 4.75-in. Schroeder refractor, equatorial mounting with circles<br>Micrometer<br>Access to transit telescope | Russell (1892), Orchiston (1987a) |
| G. Butterfield (Sydney, NSW) | 7-in. reflector (speculum mirror), altazimuth mounting<br>2-in. refractor | Butterfield (1882), Orchiston (1997a, 2003) |
| G. Collins (Launceston, TAS) | ~3.5-in. refractor, 'very poor mounting' | Biggs (1882b) |
| W.J. Conder (Sydney, NSW) | Access to 4.75-in. Schroeder refractor, equatorial mounting with circles<br>Micrometer<br>Access to transit telescope | Russell (1892), Orchiston (1987a) |
| F.F. Cox (Sydney, NSW) | 4-in. Wray refractor, equatorial mounting | Cox (1882) |

(continued)

**Table 10.2** (continued)

| | | |
|---|---|---|
| G.D. Hirst (Sydney, NSW) | Access to Colyer's 10.25-in. Browning reflector, equatorial mounting with circles Micrometer | Hirst (1882), Colyer (1888) |
| W. Macdonnell (Sydney, NSW) | 2.25-in. refractor | Macdonnell (1881), Orchiston (2001) |
| Rev. G. Martin (Sydney, NSW) | 5-in. Cooke refractor | Martin (1875), Orchiston (1997a) |
| C.S. Norris (Townsville, QUEENSLAND) | About to order a 4.5- or 5-in. Cooke refractor, equatorial mounting | Norris (1882b), Darlington and Orchiston (2017) |
| Mr. Parker (Launceston, TAS) | 4- or 4.5-in. Browning reflector | Biggs (1882b) |
| Hon. N.H. Suttor (Bathurst, NSW) | 3.5-in. refractor | Suttor (1882) |
| Dr. H.G.A. Wright (Sydney, NSW) | 8.5-in. Browning-With reflector, equatorial mounting<br>Micrometer | *Catalogue* … (1881), Orchiston (2004) |

*Note*: This Table is an update on the versions published in Orchiston (1982, 1987b), and draws on a list on a loose sheet of paper inserted in Tebbutt's Journal for 1882–1884

Of all those who responded to Tebbutt's circular letter, Dr. Bone was the only one who did not in some way doubt his own instrumental or observational prowess—and he had no reason to do so. Like Biggs, he was an experienced observer, and instrument-wise he was destined to be far better served than any other Australian independent astronomer of the day, Tebbutt included, given the impending arrival of the new 8-in. Grubb refractor. On 5 May 1882, Bone (1882a) wrote Tebbutt announcing that the telescope was due in Castlemaine in 2 months time, and later that month mentioned that that he was looking forward to immersing himself in the comet-search programme (Bone 1882b). However, he later felt obliged to remind Tebbutt that 'I have a much heavier work on hand … than comet seeking alone, (although that is work I am very fond of and take great interest in) …' (Bone 1882c).

## 10.3 In Search of Comets: Putting the Concept to the Test

On the basis of the replies received by early June 1882, Tebbutt decided to persevere with the Comet Corps concept, and he began planning search zones. However, he seems only to have allocated these to Bone, Biggs and himself. The fact that Bone had been contacted directly by the Boston Society initially caused some confusion, but Bone (1882a) eventually suggested to Tebbutt that 'You ought to dictate and I obey …' Tebbutt accepted this, much to Bone's subsequent regret, as his next letter indicates: 'For the present I will take the zone you assign 60–80° although it is not that which I had wished. But we cannot have all we wish for' (Bone 1882c). So much for Tebbutt's (1882b) plan to assign zones '… by mutual arrangement …', for Bone (1882b) had hoped to be assigned the zone from 70° to the Pole. But, as it happened, the non-appearance of his new telescope during 1882 blocked his attempts to participate in the comet-search programme, as he already had sold his previous telescope (Bone 1882a).

This left just Tebbutt and Biggs as the only viable participants in the Australian Comet Corps. Initially Biggs (1882b) was allocated the region from –40° to the Pole, but this enormous search zone was subsequently reduced to –40° to –60° (Biggs 1882c). However, fog and cloudy weather severely restricted his observing schedule during the ensuing winter and spring months (Biggs 1882d, e, f, g, i), though he did write Tebbutt on 22 June 1882 advising that 'My evenings have been completely absorbed lately, from Comet-time until sometimes past 12 o'clock with the Telescope …' (Biggs 1882f).

Just 6 days later Biggs wrote again, lamenting the little progress he had made: 'I have not done nearly as much as I expected in the time, although I have availed myself of every opportunity. I find it slower work than I expected, to do it thoroughly, as I do … I have done nothing in the morning yet' (Biggs 1882g). In fact, his actual 'track record' was not nearly as bad as his letter suggests, given the adverse sky conditions of Launceston, for on 13 September he forwarded Tebbutt a map showing

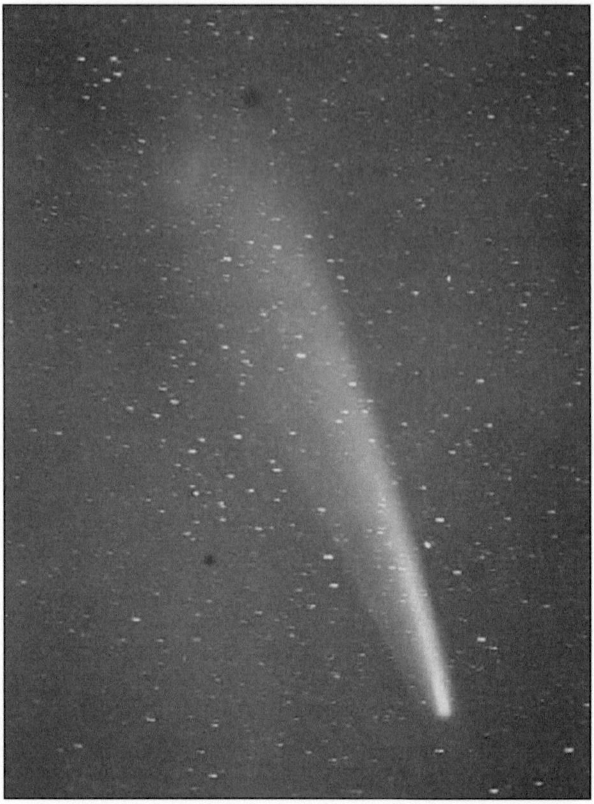

**Fig. 10.6** The Great Comet of 1882 (C/1882 R1) (https://en.wikipedia.org)

those areas he had searched and the dates involved. From this document we discover that Biggs found time for comet-searching on the following evenings that winter:

June: 8, 10, 14, 17, 21 and
July: 6, 17, 20 and
August: 10, 11, 13 and 14.

In the course of their correspondence, Tebbutt and Biggs discussed the appropriate action to be taken in the event of a comet discovery, and towards this end devised and refined a standardised telegraphic code to be used by all members of the Australian Comet Corps (see Biggs 1882g, h). Most of the refinements derived from Biggs, who proved to be a very practical and logical man.

Cloudy skies and 'extra duties' at the bank where he worked continued to plague Biggs (1882j) through into mid-October, to the point where he found hardly any time for comet-searching, let alone observing that brilliant new celestial visitor, the Great Comet of 1882, C/1882 R1 (see Fig. 10.6). From the middle of the month clearer skies prevailed, and he was rewarded with observations of the comet on three evenings (Biggs 1882k). These successes must have provided the incentive for a

**Fig. 10.7** E.E. Barnard, who discovered C/1882 R2 when it was located in Biggs' search zone (https://en.wikipedia.org)

return to comet-searching, and on 6 November he proudly announced: 'I have commenced a re-sweep of my comet zone; and I hope in future to be able to prosecute it more regularly' (ibid.).

But this was not to be. Instead, Biggs (1882i, m) spent much of his spare time preparing for the forthcoming 6 December 1882 transit of Venus, including the renovation of the small transit telescope owned by the Reverend Marcus Blake Brownrigg (1835–1890) which they planned to use in order to obtain accurate time (Biggs 1882l). Biggs (1883a) also continued to make regular observations of the Great Comet.

The final blow to Biggs's comet-search aspirations came when the U.S. astronomer E.E. Barnard (Fig. 10.7) discovered Comet C/1882 R2 while '… sweeping outside of his assigned search zone …' (Sheehan 1995, p. 54). The problem was that the comet was actually situated in Biggs' search zone! Understandably, Biggs was dismayed, lamenting that 'There seems to [be] a fatality against my prosecution of the work I have undertaken. I have made two or three sweeps lately, but without going over much ground' (Biggs 1882l). Although he was still very interested in comets, and continued to observe those that were visible from Launceston, Biggs does not seem to have resurrected his comet-search programme in 1883.

## 10.4 A Combination of Factors: The Demise of the Australian Comet Corps

Tebbutt took the disappointing outcome of these earnest endeavours philosophically, and in mid-1883 announced in the Sydney newspapers the demise of the Australian Comet Corps (see Morris 1883b; Tebbutt 1883). This was reinforced in another newspaper article, published in 1885:

> At the request of the Boston Scientific Society two or three years ago, I attempted to form a band of comet-seekers in the Australian colonies, but notwithstanding a good deal of time expended in the effort my labours met with disappointment. (Tebbutt 1885).

Two years later, Tebbutt provided further relevant comments:

> … in the year 1882, a request was made by the Scientific Society of Boston that he [i.e. Tebbutt] should … [form] … an Australian Corps of Comet Seekers … With kind words of encouragement from friends in England he proceeded to the task, but it was unfortunately one that ended in disappointment. Much time was spent in correspondence and in the preparation of regulations for the work, but although many gentlemen were written to, only two promised to join. These were residents of neighbouring colonies. One of them carried on the work for a short time, but the other does not appear to have done anything. Although there are several gentlemen in the Colonies possessed of telescopes adapted for ordinary amateur work, the idea of systematic observation, which alone can be of any benefit to science, does not present itself to their minds. (Tebbutt 1887, pp. 16–17; cf. Tebbutt 1908, pp. 58–59).

This presents a somewhat distorted interpretation of the facts. As we have seen in Sect. 10.2, there were actually more than just two astronomers prepared to join the Comet Corps, but Tebbutt ignored the rank beginners, Collins and Suttor, instead of supporting and encouraging them, and he did not take Butterfield seriously. It was his choice to proceed with just Biggs and Bone, knowing full well that Bone was without a telescope at the time.

While the concept of an Australian Comet Corps was a logical and laudable one, and on the surface Tebbutt appeared the obvious organiser, perhaps his very prominence was one of the reasons for its failure. The problem was that to all intents and purposes Tebbutt was viewed as a professional astronomer by most Australian non-professional astronomers, and the standards he set for members of the group were those that he aspired to. In reality, they were unattainable for the average independent astronomer. There was almost an element of intimidation here, albeit unintended, and this may explain why six of those he wrote to did not even bother to reply!

Tebbutt's excessive demands are apparent when we examine Biggs' activities during 1882. Although he actively supported the Comet Corps concept and was encouraged by Tebbutt (e.g. see Biggs 1882g), his letters contain a number of apologies for not providing enough of the specific sorts of details that Tebbutt had requested. This constant observational exertion, on top of a full-time day job, finally took its toll and by the end of 1882 his health was failing and he was obliged to take a well-earned holiday (Biggs 1883b).

But perhaps the comet-search programme may not have been too onerous a workload if Biggs and Tebbutt had not been faced with other pressing astronomical distractions: the 1882 December transit of Venus and three observable comets.

The simultaneous presence of so many observable comets in southern skies during 1882, and two of them major ones, diverted time from systematic comet-searching and thereby contributed to the downfall of the Comet Corps. Comet C/1882 F1 (Wells) was a conspicuous object during June–July and occupied a good deal of Biggs's and Tebbutt's time (see Table 10.3). For example, Tebbutt carried out observations on 32 different nights, but only found time on one of these to also

**Table 10.3** Observations of Comet C/1882 F1 (Wells) and The Great Comet of 1882 (C/1882 R1) by Biggs and Tebbutt, 1883–84 (after Tebbutt 1882–84)

| Comets and observers | Observations made | |
|---|---|---|
| | Year | Dates |
| Comet Wells | | |
| Biggs (1882g, h) | 1882 | June 16, 17, 18, 19, 20, 21, 24, 25, 26, 27, 30<br>July 1, 2, 5, 9, 15, 16, 19 |
| Tebbutt (1882–84) | 1882 | June 15, 16, 19, 20, 21, 22, 23, 24, 25, 26, 27, 28, 29, 30 |
| | | July 1, 2, 3, 4, 5, 6, 7, 9, 12, 13, 14, 17, 18, 19, 22, 23, 24, 25 |
| Great Comet of 1882 | | |
| Biggs (1883a, c) | 1882 | October 7, 9, 22, 28<br>November 9, 23, 30<br>December 15, 19, 29 |
| | 1883 | January 3, 5, 6, 8, 9, 12, 15<br>February 5, 13<br>March 2, 3 … through to April 7 |
| Tebbutt (1882–84) | 1882 | September 9, 10, 11, 17, 18, 19, 20, 21, 22, 24, 25, 26, 27, 28, 29, 30<br>October 1, 3, 4, 6, 7, 8, 11, 13, 14, 15, 17, 18, 19, 20, 23, 24, 27, 31<br>November 4, 6, 7, 8, 10, 11, 15, 19, 21, 23, 27<br>December 1, 2, 4, 9, 12, 13, 14, 15, 18, 19, 28, 29 |
| | 1883 | January 1, 2, 7, 9, 12, 14, 18, 26<br>February 6, 7, 12, 13, 15, 26<br>March 1, 2, 5 |

undertake a comet-search (Tebbutt 1882–84). Immediately after Comet Wells disappeared he began comet-searching again (ibid.), but from 9 September was preoccupied with the Great Comet of 1882 (C/1882 R1). He and Biggs then followed this remarkable comet (see Clerke 1893) through to March and April 1883, respectively (Table 10.3). Because of these commitments, between 10 May 1882 and 4 April 1883 (inclusive) Tebbutt (1882–84) was only able to devote 11 h 17 min to comet-searching, on just 14 different evenings and one morning.

In addition to these two comets, Tebbutt observed Comet C/1882 R2 (Barnard) on three evenings in December 1882 (Tebbutt 1882–84). Biggs (1882l) received an ephemeris from Tebbutt, but apparently did not succeed in detecting this comet, and the fact that he missed out on the initial discovery must have been demoralising. This factor, coupled with his precarious health, led Biggs to abandon systematic comet-searching.

So, in a sense, it could be said (with hind-sight), that Tebbutt's attempt to form the Australian Comet Corps in 1882 was extremely ill-timed, given the enormous observational commitment demanded by the three 1882 comets. Add to these the December transit of Venus, and it is clear that 1882 was a rather exceptional year and one of the worst ones imaginable in which to try and establish a comet-search group—especially in light of the very small number of active members.

**Table 10.4** Other Australian independent astronomers with telescopes in 1882

| Name | Instrument(s) | Reference(s) |
|---|---|---|
| W. Banerlen (Sydney, NSW) | 3-in. refractor | Banerlen (1882) |
| T. Brindley (Sydney, NSW) | 3-in. refractor | Orchiston and Bhathal (1991) |
| S.S. Browne (Sydney, NSW) | 2.5-in. refractor | Browne (1882) |
| G.R. Harrison (Sydney, NSW) | Small refractor | Butterfield (1882) |
| E.R. Morris (Sydney, NSW) | 8.5-in. With reflector, home-made equatorial mounting and drive Two micrometers | Morris (1884a), Orchiston (1987a, 2004) |
| D. Ross (Melbourne, VIC) | 3-in. refractor, equatorial mounting with circles | Orchiston and Brewer (1990) |
| W. Russell (Port Adelaide, SA) | 4.7-in. refractor, equatorial mounting with circles and drive | Orchiston (1987b), Waters (1980) |
| Mr. Scott (Sydney, NSW) | 3.25-in. refractor | Butterfield (1882), Willcox and Gibbs (1887) |
| Rev. C. Stead (Orange, NSW) | 'Telescope' | Streatfeild (1882) |
| W. Swindlehurst (Corowa, NSW) | 3-in. refractor, altazimuth mounting | Swindlehurst (1877a, b) |
| J. Taylor Campbell Town, TAS) | 8.5-in. Browning-With reflector, equatorial mounting | Biggs (1882b) |
| J.W. Ward (Sydney, NSW) | 6-in. Cooke refractor, altazimuth mounting | Ward (1883a, b), Orchiston (1997b) |
| D.T. Wiley (Penrith, NSW) | 3-in. refractor | Wiley (1879) |

However, Tebbutt could have succeeded nonetheless, had he started with a larger Comet Corps membership. What is surprising—even inexplicable—is that he chose not to approach some independent astronomers he knew of who owned telescopes and were potential members of the Comet Corps. These individuals are listed in Table 10.4.

Of these 13 individuals, Morris (Bio Box 10.6), Ross (Bio Box 9.3), Russell and Ward (Bio Box 10.7) all had useful instruments and observational experience.

**Bio Box 10.6: Ebeneler Reginald Morris**

Ebeneler Reginald Morris was a lithographic draftsman who was appointed to the New South Wales Lands Department in 1879 (Orchiston 1987a). Prior to that, he had lived in Melbourne, and in 1874 he observed the transit of Venus as a member of one of the Melbourne Observatory parties (Morris 1883b, c), but using his own telescope, an 8.5-in. (21.6-cm) With reflector with a home-made equatorial mounting and drive (Orchiston 2004).

(continued)

(continued)

In his first letter to Tebbutt, dated 3 October 1883, Morris states that this telescope '... is a very suitable one for comet seeking' (Morris 1883a). His apparent lack of interest in the mooted Comet Corps back in 1882 is explained in his next letter: 'I might state that I should have been willing but for not having my instrument in good working order at the time, *but if you thought of forming another I should be happy to take part. at present I sweep about in most places where I can get a clear view*' (Morris 1883b; my italics).

After his telescope became fully operational, in Sydney, Morris observed the Great Comet of 1882. On 12 November 1883 he wrote Tebbutt that he was keen to observe periodic Comet 12P/Pons-Brooks (Morris 1883d), which was expected to be accessible to southern observers in late 1883. Early in 1884, Morris (1884a) was busy observing both this comet and the newly-discovered Comet C/1884 A1 (Ross), and he also recorded an eclipse and a lunar occultation. He also planned to add double stars to his observing schedule (Morris 1884b).

Morris was a keen observer and obviously was sympathetic to the concept of an Australian Comet Corps. Not with standing the foregoing account, E.R. Morris still remains very much a 'mystery man', and it is to be hoped that archival evidence will emerge that documents his birth and death dates and throws further light on his astronomical activities, not to mention the fate of his telescope.

**Bio Box 10.7: J.W. Ward**

Little is known about J.W. Ward, who was a leather merchant in Sydney and seems to have been converted to astronomy by the 1874 transit of Venus.

In 1876 he acquired his first telescope, a 4.25-in. (10.8-cm) refractor, but wrote Tebbutt that he did not know how to use it properly (Ward 1876). This situation soon changed, for by 1882 he had purchased a comparatively new 6-in. (15.2-cm) altazimuth-mounted Cooke refractor (Butterfield 1882; Orchiston 1997b).

By 1883 Ward had developed a special interest in comets, and in September and October he was searching for periodic Comet 12P/Pons-Brooks, which Tebbutt had publicised in the *Sydney Morning Herald*. The notion of systematic comet-seeking also appealed to him, judging from a letter he wrote Tebbutt on 29 September 1883. In it, he makes the following perceptive comments: 'I remember your invitation to amatures [*sic*] but being a perfect stranger to you did not like to correspond, I often find myself stuck up for information; but do not like to apply to any one not being know[n] …' (Ward 1883a).

It is clear that Ward still owned the Cooke refractor in 1892 and continued to observe comets with it (Ward 1892a), while in 1891 he had observed the transit of Mercury and timed the contacts. However, he was somewhat peeved that Russell did not publish his observations, so he complained to Tebbutt: 'I was within 20″ of the true time of Contact and the other details were as correct.

(continued)

(continued)

Yet withall I was never publically credited though others who sent but a poor report were so acknowledged. *But Sir I am but a Bootmaker and cant place half the alphabit after my name*' (Ward 1892b; my italics). The masthead of this letter indicates that Ward wished officially to be known as a 'Leather and Grindery Merchant'.

Three years later, in 1895, the NSW Branch of the British Astronomical Association was formed in Sydney, but there is no evidence that Ward was ever a member of this vibrant group, even though it boasted a small pool of dedicated cometary observers (see Orchiston 1988).

Nor have I been able to discover where and when Ward died (let alone where and when he was born). It is to be hoped that as yet untapped archival or family records will one day reveal more about this independent astronomer, including the fate of his 6-in. telescope. It will be interesting to see if its actual fate tallies with any of the possibilities that I mention in Orchiston (1997b).

Meanwhile, of others listed in Table 10.4, Brindley (Bio Box 10.8) was destined to become a dedicated observer. In addition, Swindlehurst (Bio Box 10.9) was keen to make a useful contribution, even though he was still a relative newcomer to observational astronomy.

**Bio Box 10.8: Thomas Brindley**

Thomas Brindley was born in Sydney on 18 October 1855 (Obituary 1940), and in August 1878 wrote Tebbutt: 'Having taken up the study of the glorious and bewitching science of astronomy I have lately purchased a small telescope, as large as my means would allow …' (Brindley 1878). Soon he was actively observing (ibid.), and he also joined the short-lived Section A of the Royal Society of New South Wales (see Orchiston and Bhathal 1991).

Although he had no cometary observational experience, there was clear evidence of unbounded enthusiasm, and with appropriate encouragement Tebbutt may have been able to tempt him to join the new Comet Corps. Much later in life, Brindley was a stalwart of the New South Wales Branch of the British Astronomical Association, serving on Council for many years and as Director of the Saturn and Jupiter Observing Section for 13 years (Obituary 1940).

Thomas Brindley died in 1939 (ibid.).

**Bio Box 10.9: W. Swindlehurst**

When he first came to light, in an astronomical sense, Swindlehurst was a schoolteacher at Corowa in south-western New South Wales (Swindlehurst 1877b). On 18 January 1877 he wrote Tebbutt that 12 months earlier he had purchased a 3-in. (7.6-cm) refractor from W. Macdonnell in Sydney. He then proceeded to seek advice on astronomy books, sextants, observational programmes and the necessary supplementary instrumentation (Swindlehurst 1877a). In his next letter (following Tebbutt's reply) he resolved to save up for a better-quality telescope with an equatorial mounting, and to '… study the theoretical, & practical Mathematical parts of the science' (Swindlehurst 1877b). He also announced his observational intentions: 'I intend to prosecute the line of study you recommend viz:– to confine my observations to Variable Stars and sweep the sky for Comets' (*ibid.*). Three weeks later, Swindlehurst (1877c) was planning to sell his refractor and purchase an equatorially-mounted reflector, but more than a year later he still had not acted on this intention, when he sent Tebbutt his observations of the 1878 transit of Mercury (Swindlehurst 1878).

When Tebbutt next heard from Swindlehurst, in 1890, he was a minister of religion living in Newcastle. Although he still owned the original 3-in. refractor, he reported that he had just made an altazimuth-mounted 6-in. (15.2-cm) Newtonian reflector (Swindlehurst 1890).

Soon after this, he joined the newly-formed British Astronomical Association, contributing to the Star Colours Section, and began agitating for the formation of a Branch of the Association in Australia (Swindlehurst 1891). This came to pass a little over 3 years later.

Swindlehurst is yet another nineteenth-century New South Wales astronomer we know too little about.

It is apparent that had Tebbutt tried to establish the Australian Comet Corps in 1883 instead of 1882, or had he persisted with the embryonic group through into 1884 instead of giving up midway through 1883, then he would very likely have met with success. Not only were there fewer comets visible in 1883 to divert members of the Comet Corps from systematic searches, but he could definitely have counted on support from Bone, Morris and Ward. After unexpected delays, Bone's 8-in. Grubb refractor reached Castlemaine in late 1883, and as soon as it was installed in the observatory and ready for use Bone (1883) wrote to Tebbutt:

> As soon as possible I shall take up my zone of sweeps for the Boston Science Observer comet seeking society … Having splendid clockwork I presume the best plan will be to sweep in Dec within the limits of my zone … I am longing to get to work again.

In late 1883, a newcomer appeared on the Australian astronomical scene in the form of Edwin Norris of Townsville, the father of C.S. Norris who was mentioned in Table 10.2. On 17 September 1883, he wrote Tebbutt:

> In December last I purchased from the English Transit of Venus party to Queensland (Captn Morris & Lieut Darwin) their 6 in Cooke refractor Equitorial [*sic*] with Clock motion double parallel wire Micrometer Cookes prismatic illuminating apparatus a very fine instrument with all other necessary appliances. (Norris 1883).

Norris (ibid.) was about to erect an observatory, and planned to be engaged in active observing by the end of the year, or early in 1884 (see Darlington and Orchiston 2017). The Cooke refractor was ideal for comet-sweeping, and Tebbutt should have been able to entice Norris to join the Comet Corps had this group still been in existence at the end of 1883.

## 10.5 Concluding Remarks

The Australian Comet Corps is a story of lost opportunities and bad timing, but despite its short-lived existence and lack of cometary discoveries, it has a certain claim to fame as Australia's first formal national scientific group of any kind in that it predated by 2 years the formation of the Geographical Society of Australasia (see Branagan 1972; Hoare 1974). It therefore has a special place in Australian astronomical history. However, it is also clear that Tebbutt abandoned the Australian Comet Corps too early and too easily. A little patience and perseverance would have won the day.

In Australia, localised generalist astronomical groups struggled to survive in the 1880s, and only became firmly established during the 1890s (see Orchiston 1998a). In the 1880s a specialist national group like the Comet Corps was premature, and it is ironic that it took *exactly* one hundred years from the time of Tebbutt's abortive attempts before Australia would gain a national comet group for dedicated independent astronomers!

**Acknowledgements** I wish to thank Professor Nick Lomb (University of Southern Queensland, Toowoomba) and staff at the Mitchell Library (Sydney), the New South Wales Lands Department (Sydney) and the Royal Society Archives for their assistance in various ways, and the State Library of New South Wales for kindly supplying Fig. 10.2.

## References[1]

*A Systematic Search for Comets*. Boston: Boston Science Society (1882). Copy in Mitchell Library, Sydney, bound in TL.

Abbott, F. (1861). [The Great Comet of 1861]. *Monthly Notices of the Royal Astronomical Society, 21*, 260.

[1] The following abbreviation is used: TL = Letters to J. Tebbutt, bound manuscript letters in the Mitchell Library, Sydney.

Abbott, F. (1862). Observations of Comet II 1862. *Monthly Notices of the Royal Astronomical Society, 23,* 31–32.
Abbott, F. (1871a). Some further observations on the variable star η Argus, and its surrounding nebula. *Monthly Notices of the Royal Astronomical Society, 31,* 226–228.
Abbott, F. (1871b). [On η Argus]. *Monthly Notices of the Royal Astronomical Society, 31,* 230–231.
Banerlen, W. (1882). Letter to John Tebbutt, dated 18 June. In TL.
Biggs, A.B. (1881). Letter to John Tebbutt, dated 17 November. In TL.
Biggs, A.B. (1882a). Letter to John Tebbutt, dated 14 March. In TL.
Biggs, A.B. (1882b). Letter to John Tebbutt, dated 13, 15 May. In TL.
Biggs, A.B. (1882c). Letter to John Tebbutt, dated 2 June. In TL.
Biggs, A.B. (1882d). Letter to John Tebbutt, dated 6 June. In TL.
Biggs, A.B. (1882e). Letter to John Tebbutt, dated 12 June. In TL.
Biggs, A.B. (1882f). Letter to John Tebbutt, dated 22 June. In TL.
Biggs, A.B. (1882g). Letter to John Tebbutt, dated 28 June. In TL.
Biggs, A.B. (1882h). Letter to John Tebbutt, dated 22 July. In TL.
Biggs, A.B. (1882i). Letter to John Tebbutt, dated 13 September. In TL.
Biggs, A.B. (1882j). Letter to John Tebbutt, dated 14 October. In TL.
Biggs, A.B. (1882k). Letter to John Tebbutt, dated 6 November. In TL.
Biggs, A.B. (1882l). Letter to John Tebbutt, dated 2 December. In TL.
Biggs, A.B. (1882m). Letter to John Tebbutt, dated 12 December. In TL.
Biggs, A.B. (1883a). Letter to John Tebbutt, dated 16 January. In TL.
Biggs, A.B. (1883b). Letter to John Tebbutt, dated 16 March. In TL.
Biggs, A.B. (1883c). Letter to John Tebbutt, dated 27 September. In TL.
Bone, W. (1880). Observations of the Great Southern Comet I (1880) made at Castlemaine, Victoria. *Monthly Notices of the Royal Astronomical Society, 41,* 85–88.
Bone, W. (1882a). Letter to John Tebbutt, dated 5 May. In TL.
Bone, W. (1882b). Letter to John Tebbutt, dated 23 May. In TL.
Bone, W. (1882c). Letter to John Tebbutt, dated 13 June. In TL.
Bone, W. (1883). Letter to John Tebbutt, dated 14 November. In TL.
Branagan, D.F. (1972). Words, actions, people: 150 years of scientific societies in Australia. *Journal and Proceedings of the Royal Society of New South Wales, 104,* 123–141.
Brindley, T. (1878). Letter to John Tebbutt, dated 29 August. In TL.
Browne, S.S. (1882). Letter to John Tebbutt, dated 30 October. In TL.
Butterfield, G. (1882). Letter to John Tebbutt, dated 31 May. In TL.
*Catalogue of Astronomical Telescopes*. Sydney: W. MacDonnell & Co. (1881).
Chambers, G.F. (1909). *The Story of Comets Simply Told for General Readers*. Oxford: Clarendon Press.
Clerke, A. (1893). *A Popular History of Astronomy during the Nineteenth Century*. London: Adam and Charles Black.
Colyer, J.U.C. (1888). Letter to John Tebbutt, dated 3 October. In TL.
Condon, H.H. (1969). Brownrigg, Marcus Blake (1835–1890). In *Australian Dictionary of Biography. Volume 3*. Melbourne: Melbourne University Press (http://adb.anu.edu.au/biography/brownrigg-marcus-blake-3090).
Cox, F.F. (1882). Letter to John Tebbut, dated 13 May. In TL.
Darlington, V. & Orchiston, W. (2017). North Queensland and the 1882 transit of Venus: our final role in the quest for the Astronomical Unit. *Memoirs of the Queensland Museum,* in press.
Giordano, M. (1995). *Watcher of the Skies. The Story of Alfred Barrett Biggs. Pioneer in Astronomy, Seismology and Telecommunications*. Launceston: Regal Publications.
Greig-Smith, R. (1916). Presidential address. *Journal and Proceedings of the Royal Society of New South Wales, 50,* 1–81.
Guillemin, A. (1877). *The World of Comets*. London: Sampson Low, Marston, Searle and Rivington.

Hirst, G.D. (1882). Letter to John Tebbutt, dated 16 May. In TL.
Hirst, G.D. (1909). Letter to the Secretary, dated 18 October. In RAS Letter Archives.
Hoare, M.E. (1974). *Science and Scientific Societies in Eastern Australia 1820–1890*. Ph.D. Thesis, Australian National University, Canberra, Australia.
Hoare, M.E. (1976). Wright, Horatio George Anthony (1827–1901). In G. Serle & R. Ward (Eds.), *Australian Dictionary of Biography. Volume 6: 1851–1890. R-Z* (pp. 442–443). Melbourne: Melbourne University Press.
Hockey T.A. (1992). Seeing red: observations of colours in Jupiter's Equatorial Zone on the eve of the modern discovery of the Great Red Spot. *Journal for the History of Astronomy, 23,* 93–105.
Hockey T.A. (1999). *Galileo's Planet. Observing Jupiter before Photography*. Bristol: Institute of Physics.
Hordern, A. (1870). Letter to John Tebbutt, dated 6 July. In TL.
Lenehan, H.A. (1904). Letter to John Tebbutt, dated 19 January. In TL.
Macdonnell, W.J. (1871–1885). Untitled astronomical diary. MS, Mitchell Library, Sydney (B201).
Macdonnell, W.J. (1881). Letter to John Tebbutt, dated 3 June. In TL.
Macdonnell, W.J. (1882). Letter to H.C. Russell, dated 8 September. MS, Archives Office of New South Wales (Sydney Observatory Papers: 1882 Transit of Venus).
Marriott, R.A. (1996). The life and legacy of G.H. With, 1827–1904. *Journal of the British Astronomical Association, 106,* 257–264.
Marsden, B. & Williams, G.V. (1996). *Catalogue of Cometary Orbits 1996*. Cambridge (Mass.): Minor Planet Center.
Martin, G. (1875). Letter to John Tebbutt, dated 17 August. In TL.
Martin, G. (1907). *The Radiant Days of Life and Other Sermons and Addresses*. Sydney: Angus and Robertson
Morris, E.R. (1883a). Letter to John Tebbutt, dated 3 October. In TL.
Morris, E.R. (1883b). Letter to John Tebbutt, dated 8 October. In TL.
Morris, E.R. (1883c). Letter to John Tebbutt, dated 18 October. In TL.
Morris, E.R. (1883d). Letter to John Tebbutt, dated 12 November. In TL.
Morris, E.R. (1884a). Letter to John Tebbutt, dated 8 April. In TL.
Morris, E.R. (1884b). Letter to John Tebbutt, dated 24 April. In TL.
Norris, C.S. (1882a). Letter to John Tebbutt, dated 27 February. In TL.
Norris, C.S. (1882b). Letter to John Tebbutt, dated 7 May. In TL.
Norris, E. (1883). Letter to John Tebbutt, dated 17 September. In TL.
Obituaries. George Denton Hirst. *Monthly Notices of the Royal Astronomical Society, 76,* 261 (1916).
Obituary. Thomas Brindley. *Journal of the British Astronomical Association, 50,* 218 (1940).
Orchiston, W. (1982). Illuminating incidents in Antipodean astronomy: John Tebbutt and the abortive Australian association of comet observers. *Journal of the Astronomical Society of Victoria, 35,* 70–83.
Orchiston, W. (1985a). The contribution of Alfred Barrett Biggs to Australian and Tasmanian astronomy. *Records of the Queen Victoria Museum, 89,* 1–28.
Orchiston, W. (1985b). A survey of manuscript material in the Mitchell Library (Sydney) relating to the history of astronomy in New Zealand. *Southern Stars, 31,* 225–238.
Orchiston, W. (1986). *Observatories of Victoria. 1: Dr William Bone and the Castlemaine Observatory*. Melbourne: Astronomical Society of Victoria (Historical Notes No.3).
Orchiston, W. (1987a). Contribution of the Lands Department to the development of astronomy in New South Wales during the nineteenth century. *Australian Journal of Astronomy, 2,* 65–74.
Orchiston, W. (1987b). Dr William Bone, and the role of the amateur observatory in Australian astronomy. *Southern Stars, 32,* 111–128.
Orchiston, W. (1988). Illuminating incidents in Antipodean astronomy: formation of the New South Wales Branch of the British Astronomical Association. *Journal of the British Astronomical Association, 98,* 75–84.

Orchiston, W. (1992). The contribution of Francis Abbott to Tasmanian and Australian astronomy. *Vistas in Astronomy, 35,* 315–344.
Orchiston, W. (1997a). The role of the amateur in popularising astronomy: an Australian case study. *Australian Journal of Astronomy, 7,* 33–66.
Orchiston, W. (1997b). The role of the large refracting telescope in Australian amateur astronomy: an historical perspective. *Australian Journal of Astronomy, 7,* 89–114.
Orchiston, W. (1997c). The "tyranny of distance" and Antipodean cometary astronomy. *Australian Journal of Astronomy, 7,* 115–126.
Orchiston, W. (1998a). Amateur-professional collaboration in Australian science: the earliest astronomical groups and societies. *Historical Records of Australian Science, 12,* 163–182.
Orchiston, W. (1998b). Illuminating incidents in Antipodean astronomy: John Tebbutt and the Great Comet of 1861. *Irish Astronomical Journal, 25,* 167–178.
Orchiston, W. (1998c). *Nautical Astronomy in New Zealand. The Voyages of James Cook.* Wellington: Carter Observatory.
Orchiston, W. (1999). C/1881 K1: A forgotten "Great Comet" of the nineteenth century. *Irish Astronomical Journal, 26,* 33–44.
Orchiston, W. (2001). William John Macdonnell and the development of astronomy in New South Wales. *Journal of the British Astronomical Association, 111,* 13–25.
Orchiston, W. (2003). Australia's earliest planispheres. *Journal of the British Astronomical Association, 113,* 329–332.
Orchiston, W. (2004). The nineteenth century transits of Venus: an Australian and New Zealand overview. *Journal of Astronomical Data, 10,* 219–308.
Orchiston, W. & Bhathal, R. (1991). Illuminating incidents in Antipodean astronomy: Section A of the Royal Society of New South Wales. *Australian Journal of Astronomy, 4,* 49–67.
Orchiston, W. & Brewer, A. (1990). David Ross and the development of amateur astronomy in Victoria. *Journal of the British Astronomical Association, 100,* 173–181.
Orchiston, W. & Buchanan, A. (1993). Illuminating incidents in Antipodean astronomy: Campbell Town, and the 1874 transit of Venus. *Australian Journal of Astronomy, 5,* 11–31.
Ritchie, J. 1882. Letter to John Tebbutt, dated 6 March. In TL.
Russell, H.C. (1873). Letter to John Tebbutt, dated 23 April. In TL.
Russell, H.C. (1881). Letter to John Tebbutt, dated 30 November. In TL.
Russell, H.C. (1892). *Observations of the Transit of Venus, 9 December, 1874; Made at Stations in New South Wales*. Sydney: Government Printer.
Scott, H.H. & Scott, E.O.G. (1935). Not astronomy. *The Examiner,* 20 September.
Sheehan, W. (1995). *The Immortal Fire Within. The Life and Work of Edward Emerson Barnard.* Cambridge: Cambridge University Press.
Smith, D.L. & Jetson, T. (1994). Samuel Tillman and the transit of Venus, Campbell Town, Tasmania, 1874. *Papers and Proceedings of the Tasmanian Historical Research Association, 41,* 141–153.
Stock, A. (1882). Letter to John Tebbutt, dated 2 June. In TL.
Streatfeild, G. (1882). Letter to John Tebbutt, dated 19 September. In TL.
Suttor, N.H. (1882). Letter to John Tebbutt, dated 8 June. In TL.
Swindlehurst, W. (1877a). Letter to John Tebbutt, dated 18 January. In TL.
Swindlehurst, W. (1877b). Letter to John Tebbutt, dated 3 February. In TL.
Swindlehurst, W. (1877c). Letter to John Tebbutt, dated 24 February. In TL.
Swindlehurst, W. (1878). Letter to John Tebbutt, dated 8 May. In TL.
Swindlehurst, W. (1890). Letter to John Tebbutt, dated 13 March. In TL.
Swindlehurst, W. (1891). Letter to John Tebbutt, dated 9 October. In TL.
Tebbutt, J. (1882a). Astronomical notes. *Sydney Morning Herald,* 11 May.
Tebbutt, J. (1882b). Cometary astronomy. *Sydney Morning Herald,* 11 May.
Tebbutt, J. (1882–84). *Journal for Astronomical Observations. The Peninsula, Windsor, N.S.Wales.* MS, Mitchell Library, Sydney (A3750).
Tebbutt, J. (1883). A new comet. *Sydney Morning Herald,* 29 September.

Tebbutt, J. (1885). Astronomical prizes. *Sydney Morning Herald,* 24 March.
Tebbutt, J. (1887). *History and Description of Mr. Tebbutt's Observatory, Windsor, New South Wales*. Sydney: printed for the author.
Tebbutt, J. (1908). *Astronomical Memoirs*. Sydney: printed for the author.
Ward, J.W. (1876). Letter to John Tebbutt, dated 9 September. In TL.
Ward, J.W. (1883a). Letter to John Tebbutt, dated 29 September. In TL.
Ward, J.W. (1883b). Letter to John Tebbutt, dated 8 October. In TL.
Ward, J.W. (1892a). Letter to John Tebbutt, dated 19 November. In TL.
Ward, J.W. (1892b). Letter to John Tebbutt, dated 22 November. In TL.
Waters, B. (1980). *A Reference History of the Astronomical Society of South Australia Inc. Volume 1. The First Decade 1891–1901*. Adelaide: printed for the author.
White, J. (1891). Letter to John Tebbutt, dated 7 June. In TL.
Wiley, D.T. (1879). Letter to John Tebbutt, dated 19 August. In TL.
Willcox & Gibbs (1887). Letter to John Tebbutt, dated 28 April. In TL.
Wright, H. (1882). Letter to John Tebbutt, dated 22 May. In TL.

# Chapter 11
# A Gain in Light Grasp: The Legacy of the Grubb Telescope

## 11.1 Introduction

The 8-in. (20.3-cm) refracting telescope has already been described in Chap. 8 and was manufactured by Grubb of Dublin for Dr. William Bone of Castlemaine, Victoria. Bone (Bio Box 8.3) ordered the telescope in 1881 and began building a substantial observatory for it. In a letter to Tebbutt dated 5 May 1882 he wrote:

> … all my leisure time has been occupied in pushing forward with the building of my new Observatory … I have built my Observatory myself, a 15 feet Dome upon a square room of 20 ft. sq. the Dome being 2 feet higher than a hemisphere of that diameter will I think afford me ample room. (Bone 1882a).

A later letter (Bone 1882b) contains further design details, of the dome shutter and the brick and concrete foundation and base for the telescope mounting.

Although Bone had hoped that the telescope would arrive in time for the 9 December 1882 transit of Venus, various delays prevented its dispatch from Dublin before July 1883, and it only became operational at Castlemaine Observatory in November 1883 (Bone 1883).

The telescope cost Bone £600 (Tebbutt 1908a), which was a very considerable sum in those days. For example, it was more or less equivalent to the annual salary of the Government Astronomer of New South Wales at that time (see Orchiston 1987a, Table 3). It therefore comes as no surprise that such a substantial outlay, when coupled with the cost of the observatory building, strained Bone's financial resources (Bone 1883; Merrifield 1886). Although Bone had plans to use the telescope for a range of observational programmes (see Orchiston 1987b), fate decreed otherwise, for he died of pneumonia less than 2 years later. He was only 48 years of age.

By purchasing the Grubb refractor in 1886, John Tebbutt was able to upgrade Windsor Observatory to a standard comparable to that of many moderately-sized professional observatories. Meanwhile, the increased light grasp of the new telescope and the fact that it had a drive offered him opportunities to substantially


W. Orchiston, *John Tebbutt*, Historical & Cultural Astronomy,
DOI 10.1007/978-3-319-44521-2_11

modify the range of existing observational programmes if he so wished and even to engage in spectroscopic analyses.

This chapter, which represents a greatly-expanded version of Orchiston (1982), and also draws on Orchiston (2000, 2004) begins by discussing the logistics of relocating the telescope from Castlemaine to Windsor, and then proceeds to examine the range of observations that were in fact carried out once it was safely ensconced in Tebbutt's observatory. These observations are placed on a broader chronological canvas, for by combining them with work carried out earlier using the smaller Jones and Cooke refractors we are able to view Tebbutt's overall contribution to nineteenth-century positional astronomy.

## 11.2 The Tyranny of Distance: Tebbutt's Purchase of the Telescope

Following Bone's death a Mr. C. Pritchard of Sydney was contracted to arrange the sale of the telescope, and he made contact with Tebbutt. For some time Tebbutt had been contemplating the purchase of a larger instrument, and so he expressed immediate interest. He then called on his Melbourne Observatory colleague, Robert Ellery (Bio Box 7.3), to provide technical information about the telescope, which he did, at the same time pointing out that it was '… a splendid instrument and fit for any observatory as far as the object glass is concerned, I never looked thro a better one' (Ellery 1886).

After some negotiating, Tebbutt agreed to purchase the telescope for £400, and one of Ellery's staff was charged with the dismantling, packing and freighting, which involved a great deal of organisation, effort, worry and expense (see Orchiston 1986). The way in which Ellery co-operated throughout the evaluation, purchase and transportation phases of the enterprise says something about the wonderful relations that for the most part existed between Australian professional astronomers and serious non-professional astronomers at the time. Of all of the nineteenth-century professional astronomers, Ellery was one of the most supportive in this regard. In 1883 he had assisted Bone when the Grubb telescope arrived in Castlemaine (see Bone 1883), and subsequently, in February 1884, when Bone observed an unexplained oblong reddish object while searching for Comet 12P/Pons-Brooks (Butterfield 1884). At about the same time, Ellery provided Biggs of Launceston (Bio Box 10.1) with invaluable assistance, including the long-term loan of a Melbourne Observatory transit telescope (see Orchiston 1985a). Perhaps Ellery's ready willingness to assist Australian non-professional astronomers was motivated by fond memories of his own days as an independent astronomer before he joined the ranks of the professional astronomers (see Orchiston 2015).

The transportation of the ex-Bone Grubb telescope from Castlemaine to Windsor was a non-trivial exercise involving two rail journeys and one boat trip, and is yet another instance of the 'tyranny of distance' (Blainey 1975) concept at work.

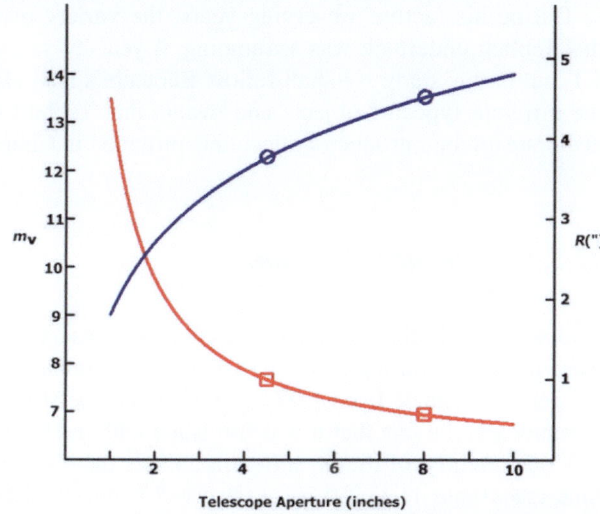


**Fig. 11.1** Plots of the ways in which the limiting visual magnitude ($m_v$), plotted in blue, and the resolving power ($R$), plotted in red, vary with changing telescope aperture. The positions of Tebbutt's 4.5-in and 8-in refractors are shown for comparison (*Plot* Martin George)

Similar problems later were encountered when the late Dr. W.E. McFarlane's 7-in. (17.8-cm) refractor was transferred from Irvinebank in far north Queensland to Riverview Observatory in Sydney in 1922 (see Orchiston 1985b).

As has been described in Chap. 8, although the telescope reached Windsor on 20 May 1886 (just before Tebbutt's 52nd birthday), it was four weeks before it was permanently installed in the small isolated dome located due east of the main observatory building. In 1894 this was replaced by a larger structure (Tebbutt 1895b). Figure 1.12 shows the telescope installed in this observatory.

Initial observations quickly demonstrated that Ellery's glowing assessment of the optics was well founded, and Tebbutt (1887a) concluded that he had indeed acquired a first-class instrument. The ring, square bar and filar micrometers that had been used with the 4.5-in. Cooke refractor were quickly adapted to the new telescope (Tebbutt 1908a).

## 11.3 Exploiting the Increased Light Grasp and a Drive: An Ongoing Commitment to Observational Astronomy

Prior to acquiring the 8-in. Grubb telescope, Tebbutt had pursued an ambitious observing programme, and he continued very much in this tradition with the new instrument, right through to 1904 when he turned 70 years of age. Thereafter, only occasional observations were made. Because of the increase in aperture, the Grubb telescope allowed Tebbutt to reach a significantly fainter limiting magnitude than with the 4.5-in. Cooke refractor and to resolve much closer double stars (see Fig. 11.1).

During his ‘active’ observing years, the variety of observational programmes that Tebbutt undertook was astounding. A year-by-year summary is given in Table 1.1, but in this study we shall follow Baracchi’s lead (1914) and focus, in turn, on the different types of objects and events that Tebbutt observed, and in so doing elaborate on the introductory accounts provided in Chaps. 1, 7 and 8.

### *11.3.1 Cometary Positions*

Undoubtedly, Tebbutt’s greatest passion was reserved for comets. Although they were important throughout his earlier observational ‘career’, they assumed even greater significance from 1886, given the increased light grasp of the 8-in. Grubb refractor and the fact that it was furnished with a reliable drive.

Consequently, of the 50 different comets that Tebbutt recorded during his lifetime (see Table 11.1), 20 were observed between 1853 and 1885 (inclusive), 27 between 1886 and ‘retirement’ at the end of 1903, and just one during both of these intervals (this was periodic comet 2P/Encke). In other words, Tebbutt observed an average of 0.64 comets per year before and 1.67 comets per year after acquiring the Grubb telescope (the rate with the large telescope was 2.6 times greater).

From 1886, Tebbutt recorded or attempted to observe every comet visible from Windsor, and consequently his annual tallies included five different comets in 1886, four in both 1888 and 1892 and three each in 1889, 1894 and 1899. Only in 1895, 1896 and 1900 were no comets detected (notwithstanding searches for three known comets in these last 2 years).

A further reflection of the improved instrumentation is the increased representation in Table 11.1 of periodic comets after 1886, for these were generally fainter than non-periodic comets. Between 1853 and 1885, 19 % of all observed comets were periodic, while the figure for 1886–1903 is 34.5 %.

Another way in which the Grubb refractor allowed Tebbutt to enhance his cometary programme was his ability to follow comets longer, due to the fainter limiting magnitude of the larger telescope. Consequently, longer series of observations became the norm. Some examples are given in Table 11.2, where it can be seen that 1892, 1898 and 1899 were particularly busy years. The only pre-1886 comets for which Tebbutt obtained micrometric data sets exceeding 22 days were C/1864 N1, C/1874 H1, C/1882 F1 and C/1882 R1. With all of these pre- and post-1886 long observational series the object was to obtain as many micrometric positional measurements as possible (e.g. see the second last column in Table 11.1), so that other astronomers could use these when computing the definitive orbital elements of the comets.

The comets that Tebbutt observed are an interesting collection. Among the non-periodic comets, C/1858 L1 (see Fig. 7.8), C/1861 J1 (Tebbutt’s Great Comet—Figs. 1.9; 6.12), C/1874 H1, C/1880 C1 (Fig. 4.1), C/1881 K1 (Tebbutt’s second Great Comet—Fig. 9.4), C/1882 R1 (Fig. 10.6), C/1887 B1 (Fig. 11.2) and C/1901 G1 (Fig. 2.5) were all famous for their impressive tails, while C/1858 L1 (Fig. 1.8), C/1861 J1 (Fig. 6.11), C/1874 H1 (Fig. 11.3; cf. Fig. 8.16), C/1881 K1 (Figs. 9.3 and

**Table 11.1** Different comets observed by Tebbutt (in the order in which he first observed them)[a]

| Year | Comet | Naked eye object | Micrometric observations made | Tebbutt's data used by others |
|---|---|---|---|---|
| 1853 | C/1853 G1 (Schweizer) | × | | |
| | C/1853 L1 (Klinkerfues) | × | | |
| 1858 | C/1858 L1 (Donati)[b] | × | | |
| 1859 | C/1858 L1 (Donati) | × | | |
| 1860 | C/1860 M1 (Great Comet) | × | | |
| 1861 | C/1861 J1 (Tebbutt = Great Comet) | × | | |
| 1862 | 2P/Encke | | | |
| | 109P/Swift-Tuttle | × | × | × |
| 1864 | C/1864 N1 (Tempel) | × | × | × |
| 1865 | C/1865 B1 (Great Southern Comet) | × | × | × |
| | 2P/Encke | | | |
| 1874 | C/1874 H1 (Coggia) | × | × | × |
| 1875 | 2P/Encke | | × | |
| 1878 | 2P/Encke | | × | |
| 1879 | 5D/Brorsen | | × | |
| 1880 | C/1880 C1 (Great Southern Comet) | × | | |
| 1881 | C/1881 K1 (Tebbutt = Great Comet) | × | × | × |
| | C/1881 N1 (Schaeberle) | × | × | × |
| 1882 | C/1882 F1 (Wells) | × | × | |
| | C/1882 R1 (Great Southern Comet)[b] | × | × | × |
| | C/1882 R2 (Barnard) | | × | × |
| 1883 | C/1882 R1 (Great Southern Comet) | × | × | × |
| 1884 | C/1884 A1 (Ross) | | × | × |
| | 12P/Pons-Brooks | | × | × |
| | D/1884 O1 (Barnard) | | × | × |
| 1885 | C/1885 N1 (Barnard) | | × | |
| 1886 | C/1885 X1 (Fabry) | | × | × |
| | C/1885 *X*2 (Barnard) | | × | × |
| | D/1886 K1 (Brooks) | | × | |
| | 7P/Pons-Winnecke | | × | × |
| | 15P/Finlay | | × | × |
| 1887 | C/1887 B1 (Great Southern Comet) | × | | |
| 1888 | C/1888 D1 (Sawerthal) | | × | × |
| | 2P/Encke | | × | |
| | C/1888 R1 (Barnard)[b] | | × | × |
| | 4P/Faye | | | |

(continued)

**Table 11.1** (continued)

| Year | Comet | Naked eye object | Micrometric observations made | Tebbutt's data used by others |
|---|---|---|---|---|
| 1889 | C/1888 R1 (Barnard) | | × | × |
| | C/1889 O1 (Davidson) | × | × | × |
| | 16P/Brooks 2 | | × | × |
| 1890 | C/1890 O2 (Denning) | | × | |
| | 6P/d'Arrest | | × | |
| 1891 | C/1891 F1 (Barnard-Denning) | | × | × |
| | 14P/Wolf | | × | × |
| 1892 | C/1892 E1 (Swift) | × | × | |
| | 7P/Pons-Winnecke | | × | × |
| | 17P/Holmes | | × | × |
| | C/1892 Q1 (Brooks)[b] | | × | |
| 1893 | C/1892 Q1 (Brooks) | | × | |
| | C/1893 N1 (Rordame-Quenisset) | | × | |
| 1894 | C/1894 G1 (Gale) | | × | × |
| | 10P/Tempel 2 | | × | |
| | 2P/Encke | | | |
| 1897 | C/1896 V1 (Perrine) | | × | × |
| 1898 | 2P/Encke | | × | |
| | C/1898 L1 (Coddington-Pauly)[b] | | × | × |
| 1899 | C/1898 L1 (Coddington-Pauly) | | × | × |
| | C/1899 E1 (Swift) | | × | × |
| | 10P/Tempel 2 | | × | |
| 1901 | C/1901 G1 (Great Comet) | × | × | × |
| 1902 | C/1902 R1 (Perrine)[b] | | × | |
| 1903 | C/1902 R1 (Perrine)[b] | | × | |
| | C/1903 H1 (Grigg) | | × | × |
| 1904 | 2P/Encke | | | |
| 1909 | 1P/Halley[b] | | × | |
| 1910 | 1P/Halley | × | × | |
| 1912 | C/1912 R1 (Gale) | | × | |

Naked Eye = This comet reached naked eye magnitude, Micrometric Observations Made = Tebbutt obtained micrometric observations of this comet, Tebbutt's Data Used by Others = Tebbutt's micrometric data were employed by other astronomers for orbital computations

[a]Comet listings after Tebbutt (1908a, 1911, 1912a, 1913). Comet designations after Marsden and Williams (1996)

[b]Observations of this particular comet were continued into the following year

**Table 11.2** Some comets observed for extended periods by Tebbutt, 1886–1901[a]

| Year(s) | Comet | Evenings observed | Number of comparisons |
|---|---|---|---|
| 1886 | 7P/Pons-Winnecke | 32 | 480 |
| | 15P/Finlay | 23 | 289 |
| 1888 | C/1888 D1 (Sawerthal) | 23 | 215 |
| 1888–1889 | C/1888 R1 (Barnard) | 24 | 220 |
| 1892 | 7P/Pons-Winnecke | 24 | 248 |
| 1892–1893 | C/1892 Q1 (Brooks) | 62 | 670 |
| 1894 | C/1894 G1 (Gale) | 27 | 299 |
| 1898–1899 | C/1898 L1 (Coddington-Pauly) | 103 | 764[b] |
| 1899 | C/1899 E1 (Swift) | 28 | 233 |
| | 10P/Tempel 2 | 43 | 413 |
| 1901 | C/1901 G1 (Great Comet) | 32 | 273 |

[a]Data based on Tebbutt (1889b, 1890c, 1893a, 1894b, 1895b, 1899, 1900a, 1902a, 1908a) and selected published papers

[b]This figure is after (Tebbutt 1900a; Tebbutt 1908a, p. 100) gives a figure of 772

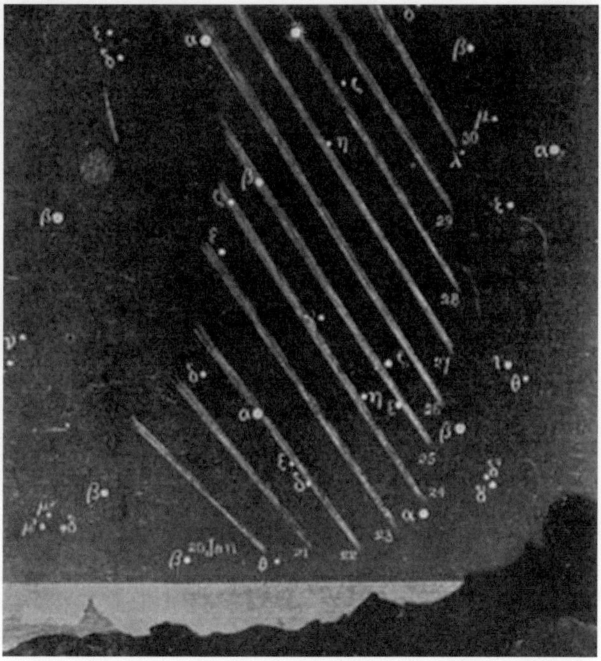


**Fig. 11.2** A sequence of drawings of Comet C/1887 B1 (Great Southern Comet) showing the changing celestial location and length of the tail between 20 and 30 January 1887 (after *Knowledge* 1887)

9.12) and C/1882 R1 all exhibited remarkable changes to the head which collectively contributed to our understanding of the nature of cometary nuclei (e.g. see Chambers 1909).

One of the most interesting of the latter was the Kreutz sun-grazing comet C/1882 R1 which was tidally split at perihelion and was subsequently observed '… to consist of up to six separate components, arranged—like the fragments of

**Fig. 11.3** A drawing of the head of Comet C/1874 H1 (Coggia), by Norman Lockyer on 12 July 1874 (http://longstreet.typepad.com/thesciencebookstore/2015/04/a-beautiful-and-famous-comet-1874.html)

**Fig. 11.4** A drawing showing the various 'stellar points' in the head of Comet C/1882 R1 following nuclear splitting (after Young 1883)

Shoemaker-Levy 9—in a rectilinear train immersed in a sheath of nebulous material' (Sekanina 1997, p. L7; cf. Fig. 11.4). Tebbutt (1882a) mentions this feature in a lengthy newspaper account of the comet, and in one of his research papers (Tebbutt 1883d) specifically refers to the break-up of the nucleus and the problem he had in deciding precisely which part of it he should observe when taking micrometric measurements.

Tebbutt also observed three other non-period comets that suffered nuclear splitting, namely C/1888 D1 (Sawerthal), C/1889 O1 (Davidson) and C/1899 E1 (Swift). Sekanina (1982) has shown that all of these experienced non-tidal (rather than tidal) splitting, and Marsden and Sekanina (1971) postulate that this phenomenon occurs

**Table 11.3** Periodic comets observed by Tebbutt, 1862–1910 (in italics)[a]

| Comet | Period (years) | Perihelion years observed |
|---|---|---|
| 1P/Halley | 76.3 | *1910* |
| 2P/Encke | 3.30 | *1862, 1865*, 1868, 1871, *1875, 1878*, 1881, 1885, *1888*, 1891, *1895, 1898*, 1901, *1905* |
| 4P/Faye | 7.49 | 1873, 1881, *1888*, 1896 |
| 5D/Brorsen | 5.50 | 1873, *1879* |
| 6P/d'Arrest | 6.56 | 1877, *1890*, 1897 |
| 7P/Pons-Winnecke | 5.70 | 1875, *1886, 1892*, 1898 |
| 10P/Tempel 2 | 5.23 | 1873, 1878, *1894, 1899* |
| 12P/Pons-Brooks | 72.2 | *1884* |
| 14P/Wolf | 6.81 | 1884, *1891*, 1898 |
| 15P/Finlay | 6.64 | *1886*, 1893 |
| 16P/Brooks 2 | 7.09 | *1889*, 1896, 1903 |
| 17P/Holmes | 6.89 | *1892*, 1899 |
| 109P/Swift-Tuttle | 132 | *1862* |

Except for Comets 2P/Encke and 109P/Swift-Tuttle, all post-1872 apparitions when each comet was observed (by any astronomer) are listed. On those occasions when Tebbutt observed the comet, the year in question is printed in italics

[a]The periods are given in years, as the mean in each case of all nineteenth-century periods listed for that comet by Marsden and Williams (1996)

as a result of impacts by interplanetary boulders. Tebbutt observed the first two comets after nuclear splitting had occurred (see dates in Sekanina 1982), and in the case of Comet C/1888 D1 (Sawerthal) he specifically mentions this feature in one of his papers. Writing on 12 April 1888, he reports: '… its nucleus, till towards the close of the period of observation, was brilliant and star-like … During the last few days the nucleus was considerably elongated …' (Tebbutt 1888a, p. 376). Tebbutt (1889a, 1890a) also observed Comet C/1889 O1 (Davidson) during the period when others noted two nuclear fragments, but he makes no mention of these in his papers, although he does state that '… the nucleus was small and not brilliant, and was surrounded with an extensive coma' (Tebbutt 1889a). In the case of the third comet, C/1899 E1 (Swift), Tebbutt's timing was unfortunate: his first batch of observations was made before the appearance of fragments A and B, while his second suite of observations post-dated the disappearance of fragments B and C (cf. Sekanina 1982; Tebbutt 1908a).

Let us turn now to periodic comets. In all, Tebbutt observed thirteen different periodic comets between 1862 and 1910, 2P/Encke on eight occasions, 7P/Pons-Winnecke and 10P/Tempel 2 twice, and all others once only. When the numbers of 'missed opportunities' are noted (see Table 11.3), it is apparent that even with the greater light grasp of the Grubb refractor, most of these comets were still beyond its limiting magnitude or else were unsuitably placed for observation from the southern hemisphere.

**Fig. 11.5** Johann Franz Encke (https://en.wikipedia.org)

Of all the periodic comets that Tebbutt observed, the following are of particular interest: 2P/Encke, 109P/Swift-Tuttle (1862), 5D/Brorsen (1879), 16P/Brooks 2 (1889) and 7P/Pons-Winnecke (1892).

After 1P/Halley, Comet 2P/Encke is probably the best researched of all periodic comets (see Marsden and Sekanina 1974), and Tebbutt (1867) was very aware of its special romantic association with New South Wales astronomy. This comet was first observed on 17 January 1786 by the French astronomer, Pierre Méchain (1744–1804; Ten 2014). Two days later it was seen by both Mechain and Jean-Dominique Cassini (1748–1845; Baum 2014; known as Cassini IV) and at the time was '… tolerably large and well defined, and had a bright nucleus but no tail …' (Chambers 1909, p. 59). It was not seen again. In 1795 Caroline Herschel (1750–1848; Ruskin 2014) discovered a small rather unremarkable comet, and although Heinrich Wilhelm Matthias Olbers (1758–1840; Solc 2014) observed it soon after, no accurate positions were obtained so meaningful orbital elements could not be computed. Small, equally unimpressive comets also were discovered in 1805 and 1818, and when the French astronomer, Johann Franz Encke (1791–1865; Fig. 11.5; Meo 2014), investigated the latter (which Pons had followed carefully for 7 weeks) he found that its celestial motion was best fitted by an elliptical orbit with a period of 3.3 years. He then noted that this period accommodated the nondescript comets of 1786, 1795 and 1805. The concept of periodic comets was still a novelty at this time, especially one with so short a period, and the test lay in Encke's hypothesis that the 1818 comet would reappear in southern skies in 1822, and reach perihelion on 24 May.

Southern hemisphere observatories were few and far between in those days, but Australia boasted Parramatta Observatory, and on 2 June one of the astronomers there, James Dunlop detected the comet close to where Encke thought it should be, based on calculations made by Carl Ludwig Christian Rümker (Bio Box 11.1), the other astronomer at Parramatta Observatory (Saunders 2004). Rümker obtained further positional observations every evening from 3 to 13 June, and these were subsequently published in the *Philosophical Transactions of the Royal Society* (Rumker 1829). Because of Encke's accurate prediction and the amount of time and effort that he spent investigating its orbit, astronomers agreed to assign his name to it (Chambers 1909, p. 49). Chambers adds, with some irony, that Encke probably never actually saw 'his' comet.

**Bio Box 11.1: Carl Ludwig Christian Rümker**

Carl Ludwig Christian Rümker (Fig. 11.6) was born in Burg Stargard, Germany, on 18 May 1788. After training and qualifying as a master builder, he decided to teach mathematics in Hamburg (Holland 2014).

In 1809 he moved to England, and after working with the British East India Company and serving in the Royal Navy he met the famous Hungarian astronomer, Baron Franz Xaver von Zach (1754–1832) and decided to study astronomy. Then during 1819–1820 he taught at the School of Navigation in Hamburg.

When the Scottish soldier-astronomer Thomas (later Sir Thomas) Brisbane arrived in Parramatta near Sydney as the new Governor of New South Wales in late 1821 he had already decided to erect an observatory, and he brought two astronomers with him, Carl Rümker and James Dunlop.

As we saw in Chap. 3, Parramatta Observatory was furnished with state-of-the-art instruments (Lomb 2004; Orchiston 1988b), and Rümker used these mainly to observe stars between the zenith and the South Celestial Pole. He also calculated the position of Comet 1P/Encke in 1822, which was recovered by Dunlop, thereby proving this to be a short-period comet. This was only the second periodic comet known, after 1P/Halley.

In 1823 Rümker fell out with Brisbane, left Parramatta Observatory and settled on a tract of land the Government had assigned to him near Sydney. As Holland (2014, p. 1875), recounts: 'Rümker devoted himself to the development of Stargard, as he called the property [after his birth place in Germany], and proved himself an able farmer'. But astronomy continued to beckon, and in 1824 he began making astronomical observations with a small telescope. Over the next 2 years he discovered three comets (e.g. see Rumker 1826).

**Fig. 11.6** A signed drawing of Carl Rümker (https://en.wikipedia.org)

(continued)

(continued)

Then in 1825 Brisbane was recalled to England and in 1826 Rümker was installed as the Director of Parramatta Observatory where he continued his astronomical observations (see Rumker 1829). However, one of his primary responsibilities as New South Wales' first 'Government Astronomer' was to measure an arc of the meridian, as recommended by the Royal Society in England, and in 1829 he travelled to London in order to purchase the necessary instrumentation. There he eventually became involved in astro-politics, and 'As a result of this vitriolic campaign against him, Rümker was formally dismissed from government service on 18 June 1930' (Holland 2014, p. 1875). He then returned to Germany and the School of Navigation in Hamburg, and in 1833 was appointed Director of Hamburg Observatory.

Using the name Christian Karl Ludwig Rümker, he remained at Hamburg Observatory until he retired in 1857. He then moved to Lisbon, Portugal, dying there on 21 December 1862. For details of Rümker's achievements while in Australia see Bergman (1960), Bhathal (2012), Holland (2014) and Saunders (1990, 2004).

**Fig. 11.7** Lewis A. Swift (https://en.wikipedia.org)

Comet 109P/Swift-Tuttle was the first of 13 comets discovered by the American astronomer, Lewis A. Swift (Fig. 11.7; Wlasuk 1996). A conspicuous naked eye object in 1862, it entertained astronomers and the public alike with its active nucleus. Sekanina (1981, p. 1741) has discussed the jets, envelopes and tail bands associated with this comet, and by

> Extrapolating particle trajectories back to the comet's surface, I find that the jets were products of brief bursts of dust from eight discrete emission areas, only one of which was demonstrably active throughout the nearly two months of physical observations.

Tebbutt was one of those who contributed observations, and despite his concerns about their accuracy (e.g. see Tebbutt 1893a) they constituted his first published micrometric cometary positions. Yet his concerns ultimately proved to be fully justified, for in their reanalysis of this comet Marsden et al. (1993) ended up discarding all of Tebbutt's observations.

Comet 5D/Brorsen is a very different story. Tebbutt was one of a number of astronomers who accurately recorded it in 1879, and its reappearance was anticipated 5.5 years later. Despite careful searches then and on several successive calculated return dates it was not found (e.g. see the 1889 and 1890 entries in Tebbutt 1908a), and more recent analyses by Marsden and Sekanina (1971) have indicated that this comet was approaching the end of its active life, and sometime between 1879 and its anticipated next 'appearance' became asteroidal. Tebbutt, therefore was one of those who unknowingly farewelled this comet in 1879.

The only comet apart from Shoemaker-Levy 9 known to have suffered tidal splitting near Jupiter was 16P/Brooks 2 following an approach to within 2.0 Jovian radii from the planet's centre in July 1886. When first observed in early August 1889 there were five different nuclear components (Sekanina 1997), but by late October when the comet was visible from Windsor only two of these remained. Tebbutt found the comet a very faint and difficult object in the Grubb telescope (Tebbutt 1908a, p. 72), and presumably his micrometric observations were of fragment A, the only one which has survived through to the present day.

Comet 7P/Pons-Winnecke was one of two comets that Tebbutt observed on two successive returns. On the first occasion '… it approached so close to star No. 59 [on 25 October 1886] as to be observed with the greatest of difficulty' (Tebbutt 1887b, p. 293). On the next return, Tebbutt was able to improve on this by witnessing a rare event, the occultation of a star by the nucleus of a comet. He reports that on 27 September 1892

> On looking for the comet in the position assigned to it for that night in the ephemeris, I could not find it as a separate object, but a star of the 10th magnitude presented itself fringed with a faint nebulous haze. As this was, doubtless, the comet centrally superimposed on the star, I made four comparisons of this star with my comparison star of the preceding evening. The resulting position agreed to a minute of arc with that given in the ephemeris. Clouds prevented further comparisons, so I was unable to witness the subsequent separation of the comet from the star. (Tebbutt 1908a, pp. 80–81).

One notable feature of Tebbutt's early 'career' as an astronomer is that with just three exceptions (2P/Encke, 5D/Brorsen and C/1882 R2), every single comet which he observed between 1853 and 1882 (inclusive) reached naked eye magnitude (see Table 11.1). Consequently, this was a remarkable period in cometary astronomy, for it featured far more bright comets per year than any 30-year period either before or after. It would appear, therefore, that Tebbutt was weaned into cometary astronomy at a particularly opportune time, and it is interesting to speculate on whether comets would have featured so prominently in his overall observing strategy had there been a deficit of impressive naked eye examples during his early observing years. Perhaps double stars or variable stars would have assumed far greater importance. Nonetheless, it was only from 1882 onwards that he began to systematically observe

**Table 11.4** Comets for which Tebbutt calculated and published orbital elements[a]

| Comet | Publication Venue(s) |
|---|---|
| C/1853 G1 (Schweizer) | *Sydney Morning Herald* newspaper |
| C/1858 L1 (Donati) | *Sydney Morning Herald* newspaper |
| C/1860 M1 (Great Comet) | *Sydney Morning Herald* newspaper |
| C/1861 J1 (Tebbutt = Great Comet) | *Sydney Morning Herald* and *The Empire* newspapers |
| 109P/Swift-Tuttle (1862) | AN, MN |
| C/1864 N1 (Tempel) | AN, MN |
| C/1865 B1 (Great Southern Comet) | AN, MN |
| C/1880 C1 (Great Southern Comet) | AN, Obs |
| C/1881 K1 (Tebbutt = Great Comet) | AN, MN, Obs |
| C/1882 R1 (Great Comet) | *Sydney Morning Herald* newspaper |
| C/1884 A1 (Ross) | AN, MN, Obs |
| C/1894 G1 (Gale) | AN, MN |

[a]AN = *Astronomische Nachrichten*, MN = *Monthly Notices of the Royal Astronomical*, Obs = *The Observatory*

faint comets, and by default—given the relative paucity of bright comets—they became his typical fare with the Grubb telescope.

As Table 11.4 shows, Tebbutt calculated the orbital elements of twelve different comets. Those for the first four comets listed in this table and for Comet C/1882 R1 were published in the Sydney newspapers, while the remainder appeared in three different journals, published in duplicate or (in two instances) in triplicate. In his discussion of Comet C/1882 R1, Tebbutt (1883a) noted that the orbital elements computed by both Kreutz and Frisby differed from his own elements, which were based on a longer arc of the orbit. But he went even further:

> … I have reason to think that an orbit deduced from the whole assemblage of post-perihelion observations will not satisfy the observations taken before perihelion, and that the comet did encounter a very considerable resistance in its passage through the solar atmosphere. (ibid.).

Comet C/1882 R1 was a celebrated Kreutz sun-grazer.

During Tebbutt's lifetime, a large proportion of his micrometric observations of comets were used by others in investigating definitive orbits, as indicated by entries in the last column in Table 11.1 (after Tebbutt 1882a, 1885a, 1908a). In addition, since his death, his observations of Comet C/1892 Q1 (Brooks) and 1P/Halley have been similarly employed (see Van Biesbroeck et al. 1976, and Roser and White 1986, respectively), while Buckley (1979) and Marsden et al. (1978) have also utilised some of his data in their studies.

When it came to cometary publications, Tebbutt was prolific. During his lifetime no fewer than 162 papers appeared in astronomical journals (Table 11.5), but specifically in *Astronomische Nachrichten* (82 papers), *Monthly Notices of the Royal Astronomical Society* (36), *The Observatory* (22), publications of the Royal Society of New South Wales and its predecessor (12), the *Journal of the British Astronomical*

**Table 11.5** Numbers of published papers by Tebbutt dealing with different areas of astronomy, listed by journal[a]

| Object/Event | AJ | AN | AR | AS | BA | MN | NS | RM | Ob | SS | TO |
|---|---|---|---|---|---|---|---|---|---|---|---|
| Comets | 0 | 82 | 1 | 0 | 8 | 36 | 12 | 0 | 22 | 1 | 162 |
| Variable stars | 0 | 6 | 3 | 0 | 2 | 11 | 3 | 0 | 5 | 0 | 30 |
| Double stars | 0 | 0 | 0 | 0 | 3 | 17 | 1 | 0 | 7 | 1 | 29 |
| Minor planets | 1 | 21 | 0 | 0 | 0 | 5 | 0 | 0 | 0 | 0 | 27 |
| Planets | 0 | 6 | 2 | 0 | 3 | 16 | 4 | 0 | 2 | 0 | 33 |
| Solar eclipses | 0 | 4 | 0 | 0 | 0 | 5 | 0 | 0 | 0 | 0 | 9 |
| Lunar eclipses | 0 | 3 | 2 | 0 | 0 | 4 | 0 | 0 | 0 | 0 | 9 |
| Transits of Mercury | 0 | 4 | 0 | 0 | 0 | 1 | 1 | 0 | 0 | 1 | 7 |
| Transits of Venus | 0 | 1 | 1 | 0 | 0 | 1 | 0 | 1 | 1 | 0 | 5 |
| Occultations of planets | 0 | 2 | 0 | 1 | 0 | 5 | 0 | 0 | 0 | 0 | 8 |
| Jovian satellites | 0 | 10 | 0 | 4 | 0 | 18 | 0 | 0 | 1 | 0 | 33 |
| Occultations of stars | 0 | 20 | 1 | 0 | 3 | 4 | 0 | 0 | 4 | 0 | 32 |
| Longitude and latitude | 0 | 1 | 0 | 0 | 0 | 4 | 2 | 0 | 1 | 0 | 8 |
| Other | 0 | 2 | 0 | 1 | 1 | 1 | 4 | 0 | 5 | 0 | 14 |
| Totals | 1 | 162 | 10 | 6 | 20 | 128 | 27 | 1 | 48 | 3 | 406 |

*Journal Key*: AJ = *Astronomical Journal,* AN = *Astronomische Nachrichten,* AR = *Astronomical Register,* AS = *Publications of the Astronomical Society of the Pacific,* BA = *Journal of the British Astronomical Association,* MN = *Monthly Notices of the Royal Astronomical Society,* NS = *Transactions of the Philosophical Society of New South Wales* plus *Journal* and Proceedings of the Royal Society of New South Wales, RM = *Memoirs of the Royal Astronomical Society,* Ob = *The Observatory,* SS = *Southern Science Record*; Totals = Note that this row does not give the total number of *individual* papers published as some papers deal with more than one topic, TO = Total published papers mentioning each topic

[a]The publications are based on entries in Tebbutt (1908a, pp. 118–132), papers in the *Southern Science Record*, and papers published subsequently

*Association* (8), and the *Astronomical Register* and the *Southern Science Record* (1 each). Allowing for duplicated publications, Tebbutt's different comet papers total 118, which is still an amazing output.

Tebbutt's only cometary review paper appeared in the *Southern Science Record* (Tebbutt 1883b), and was mainly about Coggia's Comet (C/1874 H1) and the Great Comet of 1882 (C/1882 R1). Two pages of this paper are devoted to other astronomers' spectroscopic observations of the 1882 comet, and represent the only occasion on which Tebbutt ventured to write on this important emerging area of astrophysics.

Given his penchant for comets, it is only fitting that they were amongst Tebbutt's last astronomical observations of note: in 1909–1910 he recorded Comet 1P/Halley (Fig. 11.8) on 21 evenings (Tebbutt 1911) and in 1912 Comet C/1912 R1 (Gale) was observed on 9 nights (Tebbutt 1912a, 1913). His Comet 1P/Halley observations, although made at the grand old age of 76, were reliable enough to be used during the 1986 apparition of the comet (see Roser and White 1986). Had he been alive in March 1986 surely Tebbutt would have been enthralled when the Giotto space probe

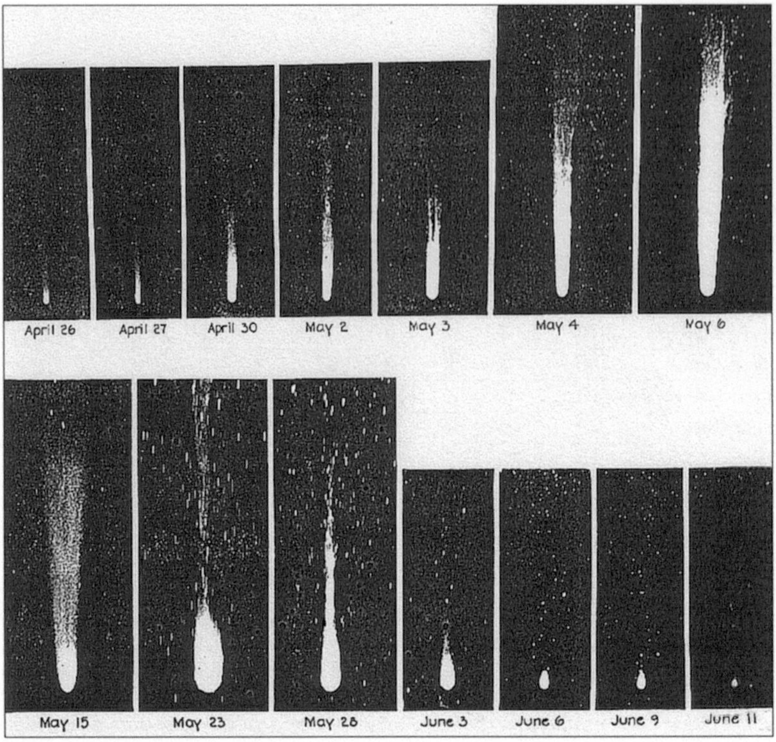


**Fig. 11.8** The length of the tail of Comet 1P/Halley varied enormously in 1910 (https://en.wikipedia.org)

photographed the comet and revealed for the first time the nature of a cometary nucleus, and that gas and dust were emitted from a small number of vents situated in discrete areas of the nucleus (see Fig. 11.9). In other words, only a comparatively small percentage of the surface was active at this time.

In concluding, we should note that in 1914 Pietro Baracchi (Bio Box 11.2), the Government Astronomer of Victoria, described Tebbutt's cometary work as 'remarkable', while his predecessor, Robert Ellery (Bio Box 7.3), had this to say: 'His splendid cometic work, extending over nearly forty years, is in itself a grand contribution to astronomy, independent of the results he secured from observations embracing a wide range of subjects' (Ellery 1901, p. 9).

**Fig. 11.9** This image of Comet 1P/Halley was taken by ESA's Giotto space probe on 14 March 1986. It revealed the presence of a solid but irregularly-shaped (non-spherical) nucleus, and that gas and dust were being emitted from only a small percentage of the surface at the time (*Courtesy* Max-Planck-Gesellschaft)

**Bio Box 11.2: Pietro Paolo Giovanni Ernesto Baracchi**

Pietro Paolo Giovanni Ernesto Baracchi (Fig. 11.10) was born to wealthy parents in Florence, Italy, on 25 February 1851 and studied mathematics and astronomy at school before completing a degree in Civil Engineering. He then served briefly in the Italian Army as an engineer.

**Fig. 11.10** Pietro Baracchi (https://en.wikipedia.org)

(continued)

(continued)

In 1876 Baracchi and two friends emigrated to New Zealand, but soon moved on to Melbourne. For a short time Baracchi worked at Melbourne Observatory, but in early 1877 he was transferred to the Department of Lands and Survey as a draftsman and subsequently trained as a surveyor.

In October 1882 he was transferred back to the Observatory as Third Assistant, and in 1892 was promoted to First Assistant, then when Ellery retired in 1895 he became Acting Director. It was only at the end of 1900 that his formal appointment as Government Astronomer of Victoria was confirmed.

Upon his appointment to the staff of Melbourne Observatory in 1882 Baracchi was involved in a project to investigate the latitudes and longitudes of the Australian colonial observatories, which required some travel (e.g. see Smith 2011). After this, he was in charge of the Great Melbourne Telescope (see Fig. 3.14), which was being used mainly to examine southern galactic objects. He and Ellery also carried out pioneering spectroscopic observations of southern stars (Baracchi 1889; Ellery 1889; cf. Andropoulos and Orchiston 2006).

Towards the end of his Directorship, Baracchi also was involved in a site survey that led—eventually—to the founding of the Commonwealth Solar Observatory on Mt Stromlo, near the Australian capital, Canberra (Bhathal et al. 2013; Frame and Faulkner 2003; Love 1984).

Arguably, one of the most useful papers that Baracchi ever wrote was his 65-page review of 'Astronomy and geodesy in Australia' (Baracchi 1914) which was prepared especially for the *Federal Handbook Prepared in Connection with the Eighty-fourth Meeting of the British Association for the Advancement of Science, Held in Australia, August, 1914*. I should mention that despite the existence of Haynes et al.'s (1996) more recent and wide-ranging *Explorers of the Southern Sky. A History of Australian Astronomy*, I still regularly refer to Barrachi's pioneering paper.

In 1897 Baracchi was awarded an Italian Knighthood, and thereafter was referred to by some as 'Commendatore'. He was described as a man of '… particularly likable disposition, with a genius for making friends' (Perdrix 1979, p. 167), but this did not prevent him from becoming involved in controversy and unpleasant astronomical politics (e.g. see Orchiston 1999).

Already of independent means, Pietro Paolo Giovanni Ernesto Baracchi had married the daughter of a wealthy Melbourne citizen (Merfield 1915), and after retiring in 1915 he lived in luxury until succumbing to cancer on 23 July 1926 (Perdrix 1979).

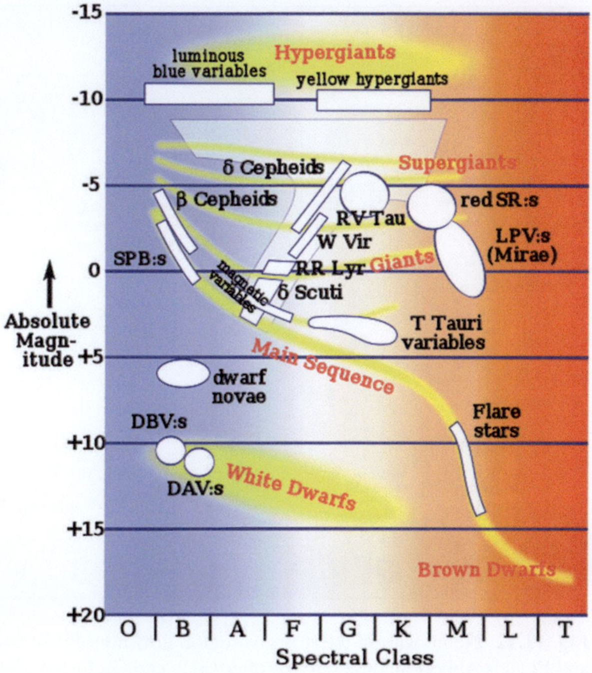


**Fig. 11.11** The two variable stars that Tebbutt mainly observed were η Carinae, a Luminous Blue Variable (*top left* in this diagram) and R Carinae, a Mira-type classical long-period variable—or 'LPV', as shown here (https://en.wikipedia.org)

## *11.3.2 Variable Stars*

Variable stars formed a second major field of observational interest for Tebbutt (see Orchiston 2000) and were the subject of the very first research paper that he prepared. This was titled 'On the desirability of a systematic search for, and observation of variable stars in the Southern Hemisphere' and was presented at a meeting of the Philosophical Society of New South Wales in 1862 (see Tebbutt 1862–1865). Variable star astronomy was still in its infancy at this time (e.g. see Furness 1915; Hogg 1984; Williams and Saladyga 2011).

After acquiring the 8-in. telescope, Tebbutt continued the series of observations of R Carinae and η Argus (or η Carinae, as it is now known) begun in 1880 and 1854, respectively, but he used the smaller Jones refractor as neither star required a larger aperture.

Eta Carinae is a Luminous Blue Variable, a rare group of extremely luminous hot stars way above the Main Sequence (see Fig. 11.11). It remains one of the most intriguing of all known variable stars (Humphreys and Davidson 1994).

In January 1845, when young Tebbutt was just 10 years old, η Argus outshone almost all other stars in the southern sky at a magnitude of –1 (Smith and Frew 2011). By 1850 it had begun to fade, slowly at first but more quickly from 1860, and by 1864 it was close to the limiting magnitude visible with the naked eye (see Frew 2004; Smith and Frew 2011, Fig. 2).

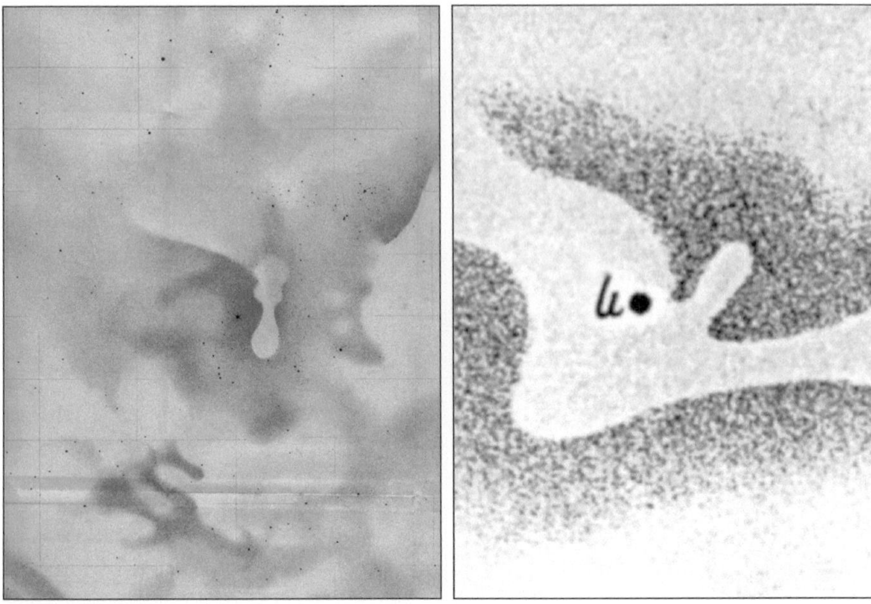

**Fig. 11.12** Nebulosity in the η Argus region drawn by Sir John Herschel in the mid-1830s (*left*) and Francis Abbott in 1863 (*right*) (after Frew and Orchiston 2003)

Tebbutt made one magnitude estimate of η Argus in 1854, but it was only after he discussed this star with William Scott (Bio Box 6.1), the founding Director of Sydney Observatory, that it was added to his regular observing list. In response to a letter from Tebbutt, on 18 April 1860 Scott wrote: 'This is no doubt a subject worthy of further consideration, and I would suggest a frequent comparison between η Argus and some star of nearly the same brilliancy …' (Scott 1860). Tebbutt began his regular monitoring of this star on 4 May 1860, and the timing turned out to be fortuitous for the nebula associated with η Argus was soon to become a bone of contention between the Hobart astronomer Francis Abbott (Bio Box 6.4) and a number of distinguished northern hemisphere astronomers when Abbott suggested that notable changes had occurred since the mid-1830s (Fig. 11.12).

Between 1854 and 1898, Tebbutt made 241 magnitude estimates of η Argus in 35 different years (Table 11.6), and he wrote seven different papers on this work. These were published in *Monthly Notices of the Royal Astronomical Society* (4) and *The Observatory* (3). Two of the *Monthly Notices …* papers also appeared in *Astronomische Nachrichten.*

**Table 11.6** Tebbutt's observations of η Argus, 1854–1898[a]

| Year | Obsns | Year | Obsns | Year | Obsns | Year | Obsns |
|---|---|---|---|---|---|---|---|
| 1854 | 1 | 1869 | 14 | 1878 | 6 | 1889 | 12 |
| 1860 | 2 | 1870 | 21 | 1879 | 2 | 1890 | 5 |
| 1862 | 19 | 1871 | 18 | 1880 | 9 | 1891 | 4 |
| 1863 | 3 | 1872 | 10 | 1881 | 4 | 1892 | 1 |
| 1864 | 6 | 1873 | 12 | 1883 | 1 | 1894 | 2 |
| 1865 | 4 | 1874 | 12 | 1885 | 1 | 1895 | 2 |
| 1866 | 9 | 1875 | 8 | 1886 | 1 | 1896 | 1 |
| 1867 | 8 | 1876 | 9 | 1887 | 1 | 1898 | 1 |
| 1868 | 15 | 1877 | 6 | 1888 | 11 | | |

[a]Annual totals based on Tebbutt (1868, 1869, 1871, 1877b, 1886b, 1888b, 1898a, 1900b)

**Fig. 11.13** A photograph of Elias Loomis taken in the 1880s (https://en.wikipedia.org)

Back in 1863, the Zurich Observatory Director Professor Rudolf Wolf (1816–1893; Charbonneau 2014) had proposed a 46-year period for the star's variability (see Loomis 1869; Wolf 1863), and Tebbutt was interested to see if this would prove to be realistic. Thus, in May 1868 he wrote Sir John Herschel:

> ... I have carefully watched the variations in the light of this star during the past fifteen years. As it is not improbable that the star may soon begin to increase in lustre, I have watched it for the past twelve months, and am still watching it, with more than ordinary interest, in order to secure, if possible, the time and magnitude of the minimum. (Tebbutt 1869).

By this time it was obvious that Wolf's period was wrong (Tebbutt 1868), and after reviewing all of the nineteenth-century observations and earlier estimates by Halley and Lacaille (in the seventeenth and eighteenth centuries, respectively), the Professor of Natural Philosophy at Yale University, astronomer Elias Loomis (1811–1889; Fig. 11.13; Newton 1890), suggested a period of 67 years. In his analysis, Loomis (1869) depended entirely on Tebbutt's magnitude estimates from 1854 and 1862–1867 in deriving the light curve.

Subsequently, Tebbutt (1871) published a list of deduced magnitudes of η Argus, based on his observations from 1854 to 1870, inclusive, and this was supplemented with a follow-up list for 1871–1876 (Tebbutt 1877b). Thereafter, he made further observations between 1877 and 1885 (inclusive), but despite an intention to publish these (Tebbutt 1886b), he never did so. However, he did note the magnitude at about 7.0 on 17 March 1886, which indicated that the gradual decline of the 1860s and 1870s had continued.

A notable change to this pattern occurred the following year. In a paper written in May 1887 and published in *The Observatory*, Tebbutt (1888b) reports:

> After making some telescopic comparisons of R Carinae on the evening of the 19th instant I turned the instrument on η Argus for the first time since 1887, April 23, and was immediately struck with the clear and sparkling character of its rays as compared with the dull light emitted by it for some years past. I proceeded at once to compare the star with its neighbours, and repeated these comparisons under favourable conditions on the following two evenings. Careful comparisons … showed that the variable had increased fully half a magnitude since my last observations …

Tebbutt wondered if this increase portended '… another magnificent outburst …' (ibid.), and he decided to keep a close watch on the star. His observing journals reveal that further estimates were indeed made—as indicated in Table 11.6—but these were never published. In 1900 he wrote in *The Observatory* that he would discuss *all* of his η Argus observations just as soon as he could secure the necessary leisure from other more pressing commitments (Tebbutt 1900b), but this never occurred.

David Frew has taken η Carinae magnitude estimates made by Tebbutt and other observers between 1835 and 1915, corrected them using currently-accepted values for the magnitudes of the comparison stars and plotted out the light curve (see Fig. 11.14). Of particular note are the small-scale fluctuations of $m_v$ >0.2 that should be interpreted as real rather than observer 'noise', and the minor increase in magnitude that Tebbutt noted during the late 1880s. This increase also was reported by Henry Russell (1888). Another feature of Fig. 11.14, as already noted by Frew (2004), is that it does not support the revised light curve postulated by Polcaro and Viotti (1993) on the basis of previously-unnoticed magnitude estimates reportedly made during 1860–1865 by a French astronomer based in New Caledonia.

The other variable star that Tebbutt carried out a long-term study of was the Mira-type long period variable, R Carinae (Orchiston 2000). He reports:

> My attention was first directed to this star by the *Uranometria Argentina* … in the autumn of 1880. Perceiving from the remarks of Dr. Gould that it must prove one of the most remarkable variable stars south of the equator, I at once placed it in my list of objects for regular and systematic observation. (Tebbutt 1881c).

The 'Dr. Gould' who inspired Tebbutt was Benjamin Apthorp Gould (Fig. 11.15; Comstock 1924; Hodge 1971), the American-born founding Director of the Argentine National Observatory, and the *Uranometria Argentina* was a 1879 publication of his listing 'Gould designations' for 7756 bright stars located within 100° of the South Celestial Pole.

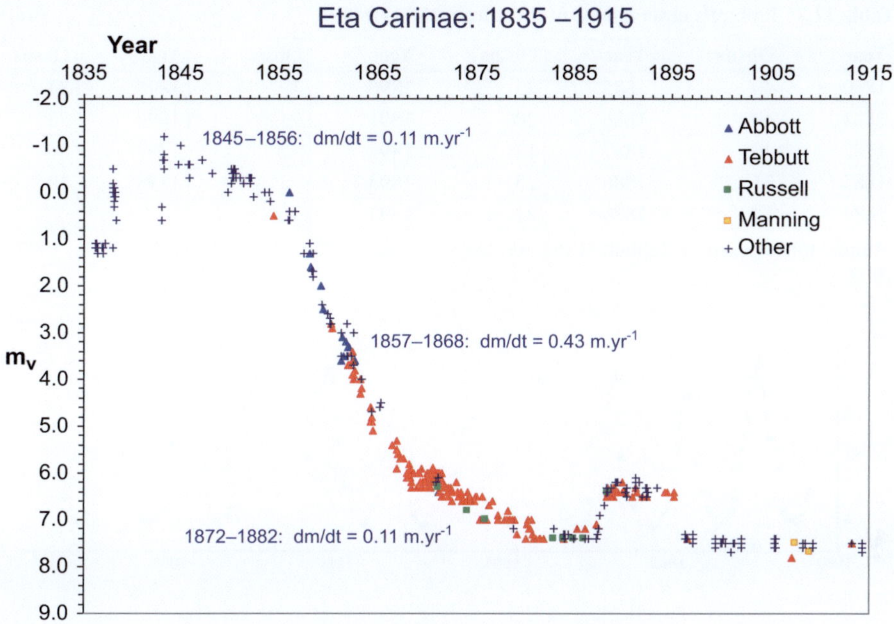


**Fig. 11.14** An historic light curve of η Carinae, with Tebbutt's visual observations shown as *red triangles* (after Frew and Orchiston 2003); note that the 1838–1850 section of this light curve has been superseded by the light curve presented in Smith and Frew (2011)

**Fig. 11.15** Dr. Gould (after *Harper's Encyclopedia* ...1905, p. 99)

**Table 11.7** Tebbutt's observations of R Carinae, 1880–1898[a]

| Year | Obsns | Year | Obsns | Year | Obsns | Year | Obsns |
|---|---|---|---|---|---|---|---|
| 1880 | 40 | 1885 | 23 | 1890 | 17 | 1895 | 33 |
| 1881 | 49 | 1886 | 20 | 1891 | 21 | 1896 | 33 |
| 1882 | 40 | 1887 | 13 | 1892 | 12 | 1897 | 30 |
| 1883 | 31 | 1888 | 23 | 1893 | 1 | 1898 | 18 |
| 1884 | 14 | 1889 | 21 | 1894 | 10 | | |

[a]Annual totals based on Tebbutt (1862–65, 1883c, 1886a, 1890b, 1896b, 1896–98, 1897e, 1897f, 1908a)

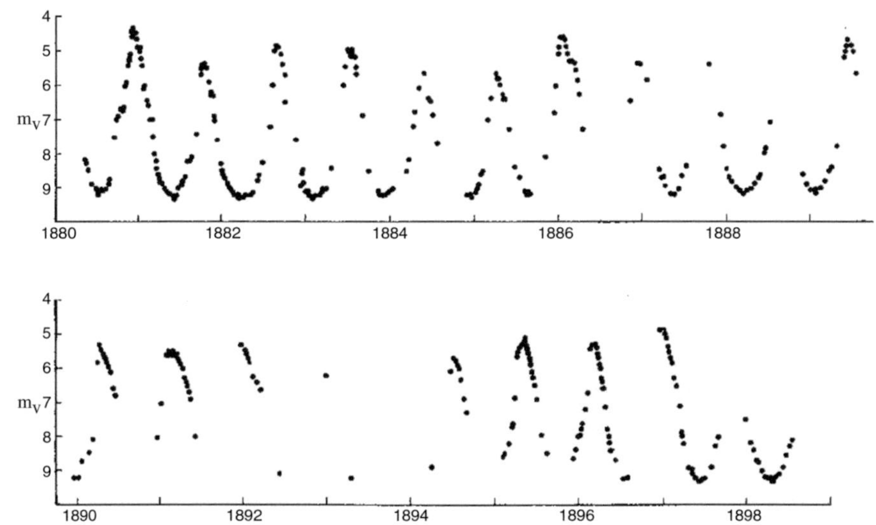


**Fig. 11.16** A light curve of R Carinae based on John Tebbutt's observations 1880–1898 (*Plot* Wayne Orchiston)

Tebbutt proceeded to observe R Carinae from 1880 to 1898, and in all made 450 individual magnitude estimates (see Table 11.7). When the star was within naked eye range he used binoculars to observe it, but had to resort to the 3.25-in. (8.3-cm) Jones refractor when it was fainter.

Between 1880 and 1900 Tebbutt published seven different papers on his R Carinae observations, in *Monthly Notices of the Royal Astronomical Society* (4), *Journal of the British Astronomical Association* (2) and *Journal and Proceedings of the Royal Society of New South Wales* (1), and these reported his observations up to and including August 1897. His later observations were never published. Tebbutt (1890b) derived a period of 312 days based on his observations from 1886 to 1890, and revised this to 311 days when he published his 1890–1895 observations (Tebbutt 1896b).

A light curve for R Carinae based on Tebbutt's published observations and his unpublished observations of 1897–1898 is shown in Fig. 11.16, where 22 different minima are represented. In *The Observations of Variable Stars*, Bateson (1958,

**Table 11.8** Suspected variable stars observed by Tebbutt

| Star | Dates observed | Magnitude ($m_v$) |
|---|---|---|
| Argelander No. 2765 | 1875: June 23 and August 24 | 9 |
| Lalande 27095 | 1876: August 24 and 25 | ~9 |
| μ Doradus | 1877: February 26 and March 14 | ~8.5 |
| Olber's variable | 1876: August 1 and 16 | 7 |

p. 23) writes that R Carinae '... has a mean range of magnitude from 4.5 to 9.6, though maxima may be as bright as 3.9 or no brighter than 5.3'. Fig. 11.16 reveals that a number of the maxima observed by Tebbutt were marginally fainter than 5.3.

When Tebbutt began monitoring R Carinae in 1880 he was the only astronomer to do so, until joined in 1891 by the South African independent astronomer, Alexander William Roberts (1857–1938; Snedegar 2015; Williams 1988). In 1901, Roberts published an investigation into variations in the period of R Carinae, based on his own observations and those of Tebbutt and Gould. He found that although the mean period was 309.3 days, the period actually varied in a sinusoidal fashion between a minimum of 305.8 days and a maximum of 312.8 days. More recently, Bateson listed the period as 308.58 days in his 1973 observing chart of R Carinae. Roberts (1901, p. 556) concluded his R Carinae paper with the following comments:

> One is tempted to wander into speculation as to the cause, or causes, of this anomaly; but until we know more about the conditions of motion, or about the chemical changes, or about both combined, that produce the type of long period variation that we find in stars of the same class as R Carinae, such excursions are unprofitable.

In 1901, stellar astrophysics was still in its infancy.

Apart from η Argus and R Carinae, the only other known variable stars which Tebbutt made a conscious effort to observe—albeit briefly—were Lalande 14599, Lacaille 4189, Star 157 in the *Uranometria Argentina*, Mira, S Carinae and T Centauri. The variability of the first of these was raised by the Irish astronomer John Birmingham (1816–1884; Mohr 2004) in February 1875, and was confirmed by Tebbutt in June of that year (see Tebbutt 1875a). Tebbutt (1879–81) was led to Lacaille 4189 by the *Uranometria Argentina* (Gould 1879), and although he made a number of observations of it and of Star 157 in 1880, and solitary observations of Algol and Mira (o Ceti) in 1877 (Tebbutt 1873–1879), he did not publish any of these. He also made a few magnitude estimates of S Carinae in April–May 1863 (Tebbutt 1863) and in May 1881 (Tebbutt 1879–81), and he made comparisons of T Centauri on four nights in May–June 1896 (Tebbutt 1894–1896, 1897c).

During the 1870s, he also examined four different suspected variable stars (see Table 11.8). Astronomers were alerted to three of these via the scientific journal, *Nature*, and Tebbutt was one of those who responded. Although he concluded that all were non-variable during the short intervals over which he observed them, he merely recorded these findings in manuscript form (see Tebbutt (1873–1879) and never bothered to publish them.

Very early in his astronomical career Tebbutt (1862–1865) saw systematic variable star searches as a particularly worthwhile pursuit for the non-professional astronomer, and he reinforced this viewpoint in 1880:

> I believe many of the stars in Lalande's catalogue will ultimately be found to be variable. A thorough comparison of the old astronomer's records with the present state of the heavens would be fruitful in the discovery of variable stars. (Tebbutt 1880c)

However, he never decided 'to practise what he preached', but in the course of conducting other observations he did carry out fortuitous magnitude estimates of five different variable stars, and he brought these to the attention of other astronomers. One was B.A.C. 2472, and he suspected its variability when attempting to observe its occultation by the Moon on 22 April 1874 (Tebbutt 1880e). At the same time he recorded his strong suspicion that the nearby star Lalande 14571 also was variable (ibid.). In November 1882 while observing Comet C/1882 R1 he found evidence for the variability of two adjacent stars, Lalande 19034 and 19093 (Tebbutt 1883e), and in February 1883, while observing this same comet, he used 19 Leporis as a reference star and on checking his catalogues immediately realised that it, too, was variable (Tebbutt 1883f).

Finally, in any review of Tebbutt's contribution to international variable star astronomy we must not forget his remarkable discovery of Nova V728 Scorpii 1862, which has already been discussed in Chap. 8 (see, also, Tebbutt 1877a).

### *11.3.3 Double Stars*

A third observational field in which Tebbutt was particularly active with the 8-in. telescope was double stars. By the start of the nineteenth century, astronomers were beginning to accept that most double (and multiple) stars represented dynamic gravitationally-bound systems and not coincidental line-of-sight juxtapositions of stars (see Clerke 1893). The challenge, then, was to systematically search for and catalogue new double stars and to establish their orbital parameters (Aitken 1918). This could be done by making precise measurements of changes in the position angle and separation of the components over time. Tebbutt was not interested in searching for new double stars—he left that to his Sydney Observatory colleagues (see Haynes et al. 1996)—but he was eager to monitor known doubles.

On 28 February 1880, Tebbutt (1879–81) wrote in his journal: 'I made my first serious attempts to measure double stars by means of the position micrometer on the 4½ inch equatorial last evening and this morning. The first star selected last evening was α Crucis'. He then proceeded to measure the position angle and separation of α Centauri, using as a reference the book *Lectures on Practical Astronomy* (ibid.) by James Challis (1879). Challis (1803–1882; Sheehan 2014) was the Plumian Professor of Astronomy at Cambridge University and Director of the University Observatory. For Tebbutt, these observations marked the start of a long-term commitment to double stars.

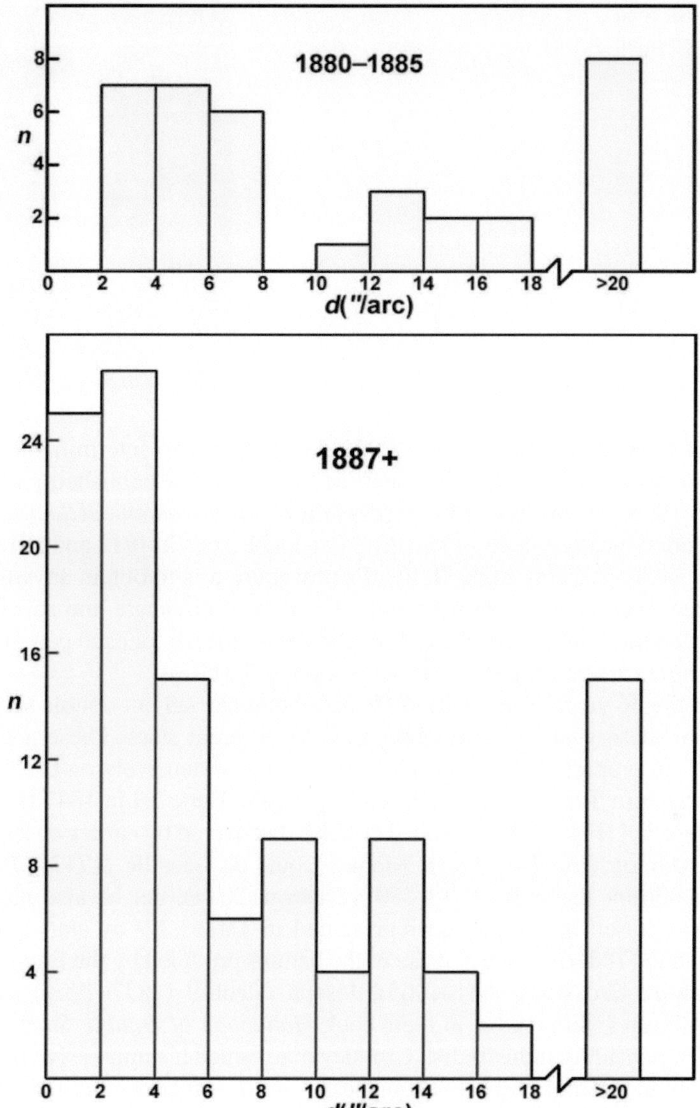


**Fig. 11.17** Histograms showing the separations of double stars observed by Tebbutt before (upper) and after (lower) the purchase of the 8-in. Grubb telescope (*Plots* Martin George)

The arrival of the Grubb telescope (with its drive, and increased light grasp) allowed Tebbutt to greatly expand this programme and include doubles with smaller separations (see Fig. 11.17). Between 1886 and 1902 inclusive, he observed double stars every single year, except in 1897 and 1898. The most intensive activity occurred in 1887 and 1888, immediately after acquiring the new telescope, and

**Fig. 11.18** Sir John Herschel (https://en.wikipedia.org)

during the three-year period 1900–1902. He continued to intermittently observe double stars after his official retirement, and his very last published paper, in the November 1915 issue of *Monthly Notices of the Royal Astronomical Society* (Tebbutt 1915), reported measurements of two different double stars in 1913 and 1915.

On a typical observing night, Tebbutt's procedure was to obtain anywhere up to ten separate micrometric measurements of both position angle and separation for each star, and he would then reduce these and derive means for each parameter, with position angle expressed to 0.1° and separation to 0.01″.

During the 36 years from 1880 to 1915, Tebbutt carried out double star work in all but 9 of these years, and he observed 133 different stars. These are listed in Table 11.9. In preparing his observing lists, he drew extensively on the 1847 catalogue of southern double stars published by Sir John Herschel in 1847 (see Tebbutt 1880f). Herschel (Fig. 11.18) included doubles discovered by earlier observers such as James Dunlop (Bio Box 11.3), Nicolas Louis de Lacaille (1713–1762; Glass 2013) and Jérôme Lalande (1732–1807; Dumont 2014), but he also made many discoveries himself and these are represented in Table 11.9 by stars with an 'h' prefix. Tebbutt (1881d) also made use of the listings provided by the English astronomers Edward Crossley (1841–1905), Joseph Gledhill (1837–1906) and James Maurice Wilson (1836–1931) in their book *Handbook of Double Stars* (Crossley et al. 1879). A wealthy industrialist, Crossley maintained his impressive Bermerside Observatory in Halifax (between Manchester and Leeds), where Gledhill was his

**Bio Box 11.3: James Dunlop**

James Dunlop (Fig. 11.19) was born in Dalby, Scotland, on 31 October 1793. After very little formal schooling he worked in a factory and showed a mechanical bent, which included the manufacture of telescopes (Service 1890).

When Thomas Brisbane was preparing to move to Australia and take up the post of Governor of New South Wales, he decided Dunlop's mechanical skills would be useful and so he offered him a position as astronomer, responsible for maintaining the scientific instruments at Parramatta Observatory. Thus, both Dunlop and Rümker accompanied Brisbane to Sydney.

(continued)

(continued)

**Fig. 11.19** An oil painting of James Dunlop by Joseph Backler dating to about 1843 (https://en.wikipedia.org)

Once the Observatory was operational, Dunlop not only looked after the instruments but also shared in the systematic observation of stars in the southern sky—especially after Rümker left the Observatory (Rutledge 2009)—that eventually resulted in the publication of *Catalogue of 7385 Stars, Chiefly in the Southern Hemisphere, from Observations … in … 1822–26 at … Paramatta [sic]* (Richardson 1835). As Bhathal (2012) has noted, though, later investigations showed that the positions of many of the stars in the catalogue were in error.

Rümker was installed as Director of Parramatta Observatory in 1826 following Brisbane's return to England, and Dunlop retreated to his residence in Parramatta where he used a home-made 9-in. (22.9-cm) reflector to conduct a survey of clusters and galaxies (Dunlop 1828; c.f. Cozens and White 2001; Cozens et al. 2010). Although there also were some problems with his catalogue (Bhathal 2012), with the benefit of hindsight one of his most significant discoveries was the extragalactic nebula now known as NGC 5128, which much later was correlated with the discrete radio source Centaurus A (see Robertson et al. 2010).

In 1827 Dunlop left New South Wales and went to work at Brisbane's new observatory at Makerstoun in Scotland (Rutledge 2009), then when Rümker decided to accept the Hamburg Observatory post in 1830 Dunlop returned to Parramatta Observatory as Superintendent. He held this post from 1831 until 1847, when the Observatory officially closed (Orchiston 1988b), but apart from discovering comets in 1833 and 1834 he contributed comparatively little to astronomy during this interval (Saunders 2004), and the Observatory became run down. This was because Dunlop '… did not have the theoretical or intellectual expertise to run the Observatory. He had always been a follower rather than a leader and had prospered under the guidance of Brisbane and Rumker. He did not have the standing of either Brisbane and [*sic*] Rumker in the political or scientific circles in London and Sydney to obtain the resources to run the Observatory effectively … This was compounded by Dunlop's ill health and from about 1837 he found it a chore to look after the Observatory and carry out observations of the night sky' (Bhathal 2012, p. 122).

After leaving Parramatta Observatory, James Dunlop retired to his land at Brisbane Waters, near Sydney, and he died there on 22 September 1848.

**Table 11.9** Double stars observed by Tebbutt, 1880–1915

| Star name | 1880 | 1 | 2 | 4 | 5 | 6 | 7 | 8 | 9 | 1890 | 1 | 2 | 3 | 4 | 5 | 6 | 9 | 1900 | 1 | 2 | 5 | 6 | 7 | 9 | 1911 | 3 | 5 |
|---|---|---|---|---|---|---|---|---|---|---|---|---|---|---|---|---|---|---|---|---|---|---|---|---|---|---|---|
| δ Argus | | | | | | | | | | | | | | | × | | | | | × | | | | | | | |
| γ Argus A.B. | | × | | | | | × | × | | | | | | | | | | × | | | | | | | | | |
| γ Argus A.C. | | × | | | | | | × | | | | | | | | | | × | | | | | | | | | |
| γ Argus A.D. | | | | | | | | × | | | | | | | | | | | | | | | | | | | |
| μ Argus | | | | | | | | | | | | | | | | | | | | × | | | | | | | |
| ν Argus | × | | | | | | | | | | | | | | | | | | | × | | | | | | | |
| ε Camelopardalis | | | | | | | | | | | | | | | | | | | × | | | | | | | | |
| α Centauri | × | × | × | × | × | × | × | × | | × | × | × | × | × | × | × | × | × | × | × | × | | × | × | × | × | × |
| γ Centauri | | | × | | | | × | × | | × | | × | | | × | × | × | × | | × | × | | × | × | × | | |
| k Centauri | | | | | | | | | | | | | | | | | | | | × | | | | | | | |
| y Centauri | | | | | | | | | | | | | | | | | | | | × | | | | | | | |
| α Circini | | | | | × | | | | | | | | | | | | | | | × | | | | | | | |
| γ Circini | | | | | | | | | | | | | | | | | | | | × | | | × | × | | | |
| δ Corvi | | | | | | | | × | | | | | | | | | | | | | | | | | | | |
| γ C. Australis | | | | | | | × | × | × | × | × | × | | × | × | | | × | | × | × | | × | × | × | × | |
| k C. Australis | | × | | | | | × | | | | × | | | | | | | | | | | | | | | | |
| α Crucis | × | | | | | | × | × | | | | × | | | × | | | × | | × | | | | | | | |
| f Eridani | | | × | | | | | | | | | | | | | | | × | | × | | | | | | | |
| p Eridani | × | | × | | × | | × | × | × | × | × | × | | × | × | × | | × | | × | | × | × | | | | |
| θ Eridani | | | × | | | | | | | | | | | | | | | × | | × | | | | | | | |
| 12 Eridani | | | × | | | | | | | | | | | | | | | | | × | | | | | | | |
| θ Gruis | | | | | | | × | | | | | | | | × | | | | × | | | | | | | | |
| β Hydrae | | | | | | | | | | | | | | | | | | | × | × | | | | | | | |
| m Hyrae | | | | | | | | | | | | | | | | | | | | × | | | | | | | |
| N Hydrae | | | | | | | | | | | | | | | | | | | × | | | | | | | | |

| | | | | | | | | | | | | | | | | | | | | | | | | | | | |
|---|---|---|---|---|---|---|---|---|---|---|---|---|---|---|---|---|---|---|---|---|---|---|---|---|---|---|---|
| 59 Hydrae | | | | | | | | | | | | | | | | | | | | × | | | | | | | |
| θ Indi | | | | | × | | × | | | | | | | | × | | | | | | | | | | | | |
| d Lupi | | | | | | | | | | | | | | | | | | | | × | | | × | | | | |
| ε Lupi | | | | | | | | | | | | | | | | | | | | × | | | | | | | |
| ε Lupi/Dunlop 182 | | | | | | | | | | | | | | | | | | | | × | | | | | | | |
| K Lupi | | | | | | | | | | | | | | | | | | | | × | | | | | | | |
| μ Lupi | | | × | | | | | | | | | | | | | | | | | × | | | × | | | | |
| ε Lupi | | × | | | × | | × | | | | | | | | | | | | | × | | | | | | | |
| π Lupi | | | | | | | | × | | | | × | | | × | | | × | | × | | | × | × | | | |
| ξ Lupi | | × | | | | | × | | | | | | | | | | | | | | | | | | | | |
| β Muscae | | | | | | | | × | | | | × | | | × | × | | × | × | × | × | | × | × | × | | |
| θ Muscae | | | | | | | | | | | | | | | | | | | | × | | | | | | | |
| A Ophiuchi | | | | | | | | | | | | | | | | | | | | × | | | | | | | |
| ρ Ophiuchi | | | | | | | | | | | | × | | | × | | | | | × | | | | | | | |
| 36 Ophiuchi | | × | | | | | × | | | | | × | | | × | | | | | | | | | | | | |
| 39 Ophiuchi | | | | | | | | | | | × | | | | | | | | | | | | | | | | |
| β Phoenicis | | | | | | | | | | | | | | × | | | | | × | × | | | × | | | | |
| ζ Phoenicis | | | | | | | | | | | | | | × | | | | × | | | | | | | | | |
| ι Pictoris | | | × | | | | | | | | | | | | | | | × | | | | | | | | | |
| γ P. Austrini | | | | | | | | × | | | | | | | | × | | | × | × | | | | | | | |
| V Puppis | | × | | | | | | × | | | | | | | | × | | | | | | | | | | | |
| θ Reticuli | | | × | | | | | | | | | | | | | | | × | | | | | | | | | |
| A Scorpii | | | | | | | | | | | | | | | | × | | | | × | | | | | | | |
| α Scorpii | | | | | | | | | | | | | | | | | | × | | | | | | | | | |
| β Scorpii | | | | | | | × | | | | | | | | | | | | | | | | | | | | |
| ν Scorpii | | | | | | | | | | | | | | | | | | | | × | | | | | | | |

(continued)

**Table 11.9** (continued)

| Star name | 1880 | 1 | 2 | 4 | 5 | 6 | 7 | 8 | 9 | 1890 | 1 | 2 | 3 | 4 | 5 | 6 | 9 | 1900 | 1 | 2 | 5 | 6 | 7 | 9 | 1911 | 3 | 5 |
|---|---|---|---|---|---|---|---|---|---|---|---|---|---|---|---|---|---|---|---|---|---|---|---|---|---|---|---|
| β Tucanae | | | × | | | | | | | | | | | | | | | | | | | | | | | | |
| k Tucanae | | | | | | | | | | | | | | × | | | | | × | | | | | | | | |
| λ Tucanae | × | | | | | | | | | | | | | | | | | × | | | | | | | | | |
| H Velorum | | | | | | | | | | | | | | | | | | | | × | | | | | | | |
| s Velorum | | × | × | | × | | | | | | | | | | | | | | | | | | | | | | |
| γ Virginis | × | | | | | | × | × | | | | × | | | × | | | × | | × | | | | | | | |
| Arg. G. Cat 19385 | | | | | | | | | | | | | | | | | | × | | | | | | | | | |
| β 416 | | | | | | | | | | | | | | | | | | × | | × | × | | × | × | × | | |
| Brisbane 1201 | | | | | × | | | | | | | | | | | | | | | | | | | | | | |
| Brisbane 3574 | | | | | | | | | | | | | | | | | | | × | | | | | | | | |
| Brisbane 5784 | | | | | | | | × | | × | | | | | | | | | | | | | | | | | |
| Brisbane 6556 | | | | | | | × | × | × | | × | × | | | × | | | | | | | | | | | | |
| Δ 1500 | | | | | | | × | | | | | | | | | | | | | | | | | | | | |
| Dunlop 109 | | | | | | | | | | | | | | | | | | | × | | | | | | | | |
| Dunlop 180 | | | | | | | | | | | | | | | | | | | | × | | | | | | | |
| h 2167 | | | | | | | × | | | | | | | | | | | | | | | | | | | | |
| h 3447 | | | | | | | | | | | | | | × | | | | | | | | | | | | | |
| h 3556 | | | | | | | | | | | | | | | | | | × | | | | | | | | | |
| h $3586_2$ | | | | | | | | | | | | | | | | | | × | | | | | | | | | |
| h $3586_3$ | | | | | | | | | | | | | | | | | | × | | | | | | | | | |
| h 3592 | | | | | | | | | | | | | | × | | | | | | | | | | | | | |
| h 3752 | | | | | | | | | | | | | | | | | | | | × | | | | | | | |
| h 3784 | | | | | | | | | | | | | | | | | | | × | | | | | | | | |
| h 3789 | | | | | | | | | | | | | | | | | | | × | | | | | | | | |

| | | | | | | | | | | | | | | | | | | | | | | | | | | | |
|---|---|---|---|---|---|---|---|---|---|---|---|---|---|---|---|---|---|---|---|---|---|---|---|---|---|---|---|
| h 3823 | | | | | | | | | | | | | | | | | | × | × | | | | | | | | |
| h 3834 | | | | | | | | | | | | | | | | | | | × | | | | | | | | |
| h 3891 | | | | | | | | | | | | | | | | | | | × | × | | | | | | | |
| h 3960 | | | | | | | | | | | | | | | | | | | × | | | | | | | | |
| h 4087 | | | | | | | × | | | | | | | | | | | | | × | | | | | | | |
| h4220 | | | | | | | | | | | | | | | | | | | | × | | | | | | | |
| h 4252 | | | | | | | | | | | | | | | | | | | | × | | | | | | | |
| h 4306 | | | | | | | × | | | | | | | | | | | | | | | | | | | | |
| h 4949 | | | | | | | | | | | | | | | | | | | | × | | | | | | | |
| h 5014 | | | | | | | | | | | | | | | | | | | | × | | | × | | | | |
| h 5027 | | | | | | | × | | | | | | | | | | | | | | | | | | | | |
| h 5075 | | | | | | | | | | | × | | | | | | | | | | | | | | | | |
| h5092 | | | | | × | | | | | | | | | | | | | | | | | | | | | | |
| h5292 2/3 | | | | | | | | | | | | | | × | | | | | | | | | | | | | |
| Jacob 9 | | | | | | | | | | | | | | | | | | | | × | | | | | | | |
| Lacaille 1181 | | | × | | | | | | | | | | | | | | | | | | | | | | | | |
| Lacaille 1475 | | | × | | | | | | | | | | | | | | | | | | | | | | | | |
| Lacaille 1889 | | | | | | | | | | | | | | | | | | | × | | | | | | | | |
| Lacaille 2145 | | × | × | | × | | × | | | | | | × | × | × | | | × | × | × | | | × | | | | |
| Lacaille 2224 | | | | | | | | | | | | | | | | | | | × | | | | | | | | |
| Lacaille 2275 | | | | | | | | | | | | | | | | | | | × | | | | | | | | |
| Lacaille 2402 | | | | | | | | | | | | | | | | | | | × | | | | | | | | |
| Lacaille 2640 | | × | × | | | | × | × | | | | | | | | | | | | | | | | | | | |
| Lacaille 2685 | | | | | × | | | | | | | | | | | | | | | | | | | | | | |
| Lacaille 2779 | | | | | | | | | | | | | | | | | | | × | | | | | | | | |
| Lacaille 3366 | | | | | | | × | | | | | | | | | | | | | | | | | | | | |

(continued)

**Table 11.9** (continued)

| Star name | 1880 | 1 | 2 | 4 | 5 | 6 | 7 | 8 | 9 | 1890 | 1 | 2 | 3 | 4 | 5 | 6 | 9 | 1900 | 1 | 2 | 5 | 6 | 7 | 9 | 1911 | 3 | 5 |
|---|---|---|---|---|---|---|---|---|---|---|---|---|---|---|---|---|---|---|---|---|---|---|---|---|---|---|---|
| Lacaille 4031 | | × | | | | | | | | | | | | | | | | | | | | | | | | | |
| Lacaille 4102 | | | | | | | | | | | | | | | | | | | | × | | | | | | | |
| Lacaille 4268 | | | | | × | | | | | | | | | | | | | | | | | | | | | | |
| Lacaille 4350 | | | × | | | | | | | | | | | | | | | | | | | | | | | | |
| Lacaille 5147 | | | | | | | | × | | | | | | | | | | | | | | | | | | | |
| Lacaille 5632 | | | | | | | | × | | | | | | | | | | | | | | | | | | | |
| Lacaille 5750 | | | | | | | | | | | | | | | | | | | | × | | | | | | | |
| Lacaille 5893 | | | | | | | | | | | | | | | | | | | | × | | | | | | | |
| Lacaille 6477 | | | | | × | | × | × | | | | × | | | | | | | | × | | | | | | | |
| Lacaille 7194 | | | | | | | | | | | | | | | | | | | | × | | | | | | | |
| Lacaille 7267 | | | | | × | | × | | | | | | | | | | | | | | | | | | | | |
| Lacaille 7924 | | | | | × | | × | | | | | | | | × | | | | | | | | | | | | |
| Lacaille 8443 | | | | | × | | × | | | | | | | | | | | | | | | | | | | | |
| Lacaille 8550 | | | | | × | | × | | | | | | | × | | | | | | | | | | | | | |
| Lacaille 8687 | | | | | × | | × | | | | | | | | | | | | | | | | | | | | |
| Lalande 4219 | | | | | | | | | | × | × | | | | | | | | | | | | | | | | |
| P.I. 127 | | | | | | | | | | | | | | | | | | × | | | | | | | | | |
| Stone 4019 | | | | | | | | | | | | | | | × | | | | | | | | | | | | |
| Sydney $_2$ 213 | | | | | | | | × | | | | | | | | | | | | | | | | | | | |
| Sydney $_2$ 287 | | | | | | | | × | | | | | | | | | | | | | | | | | | | |
| Sydney $_1$ 411 | | | | | | | | × | | | | | | | | | | | | | | | | | | | |

*Note*: Based on Tebbutt (1885b, 1889c, 1891b, 1892c, 1893b, 1894c, 1895c, 1896d, 1897d, 1901b, 1902b, 1903b, 1906, 1908b, 1910, 1912b, 1915)

**Fig. 11.20** A photograph of J.E. Gore taken in 1875 (https://en.wikipedia.org)

**Fig. 11.21** Dr. T.J.J. See (https://en.wikipedia.org)

Astronomical Assistant, while Wilson was associated with Temple Observatory in Rugby (see Orchiston 2002, 2016, Chap. 13).

Tebbutt also was prompted to observe certain doubles when data were requested by other astronomers. Thus, in 1889 and 1895 he measured γ Coronae Australis for the well-known Irish astronomer John Ellard Gore (1845–1910; Fig. 11.20; Elliott 2014; FitzGerald 1966) and for Dr. Thomas Jefferson Jackson See (1866–1962; Fig. 11.21; Ashbrook 1984; Peterson 2004), a controversial astronomer from the University of Chicago. In 1893, See (see Ashbrook 1984) asked for measurements of α Centauri, and in 1895 Tebbutt sent See and Danish-born Dr. August William Doberck (1852–1941; MacKeown 2007), the Director of Hong Kong Observatory, observations of this same star; See also received measurements of γ Centauri and γ Coronae Australis (see Tebbutt 1908a). Such requests, however, related to less than 3 % of the double stars that Tebbutt observed.

In 1902, Tebbutt's routine changed following the Cape Observatory's publication of the monumental *Reference Catalogue of Southern Double Stars* prepared by the

**Table 11.10** Double stars most commonly observed by Tebbutt

| Star | Number of years observed | Date range | Total Number of nights observed |
|---|---|---|---|
| α Centauri | 25 | 1880–1915 | 207 |
| p Eridani | 16 | 1880–1907 | 64 |
| γ Coronae Australis | 15 | 1887–1913 | 77 |
| γ Centauri | 14 | 1882–1911 | 70 |
| β Muscae | 11 | 1888–1911 | 56 |
| Lacaille 2145 | 11 | 1881–1907 | 43 |
| α Crucis | 7 | 1880–1902 | 15 |
| π Lupi | 7 | 1888–1909 | 29 |
| γ Virginis | 7 | 1880–1902 | 15 |
| β (=Burnham) 416 | 6 | 1900–1911 | 27 |
| Brisbane 6556 | 6 | 1887–1895 | 11 |

one-time Sydney-based independent astronomer, R.T.A. Innes (1899). Innes (Bio Box 8.1) presented a thorough treatment of earlier Australian efforts in double star astronomy, and Tebbutt was duly impressed (Roseby 1901) and determined to make 'Repeated measurements … of fifty-four pairs of the most interesting double stars …' in the catalogue (Tebbutt 1908a, p. 106). In all, he assigned 82 different evenings to this work.

The 11 most frequently-observed stars represented in Table 11.9 are listed in Table 11.10, led by α Centauri (which was measured in 25 different years between 1880 and 1915). Since Table 11.9 does not indicate the total number of individual nights on which measures were made of any particular star, these statistics have been included in Table 11.10 for the most frequently-observed stars. Stars that Tebbutt only occasionally monitored were generally observed on just a single occasion, or at most on a few different nights, in any one 'observing year'.

Between 1885 and 1915, Tebbutt systematically published his reduced double star observations in 17 successive papers in *Monthly Notices of the Royal Astronomical Society* (and it is the data in these which were drawn on in preparing Tables 11.9 and 11.10 here). In addition, he published 12 other papers on double stars, in *The Observatory* (7), *Journal of the British Astronomical Association* (3) and *Journal and Proceedings of the Royal Society of New South Wales* and the *Southern Science Record* (1 each). However, all of the data in six of the seven papers published in *The Observatory* were included in the series of papers that appeared in *Monthly Notices of the Royal Astronomical Society*. Furthermore, the paper published by the Royal Society of New South Wales was a duplicate of one that appeared in the *Southern Science Record*, and data in these papers were included in a paper published in *The Observatory*.

Some of Tebbutt's distance and position angle measurements were used by other astronomers during his lifetime. By way of example, his observations of α Centauri were utilised by Biggs (1887), Downing (see Tebbutt 1884), Powell (1886, 1892) and See (1893, 1896); of γ Centauri by Gore (1892a) and See (1896); of γ Coronae

Australis by Gore (1892b), Powell (1890) and Sellors (1892); of γ Virginis by See (1896); of p Eridani by Gore (1887) and of β 416 by Voute (1908).

Tebbutt's connection with Biggs's double star work is interesting. Launceston's Alfred Barrett Biggs (Bio Box 10.1) was a remarkable non-professional scientist and was particularly active in astronomy during the 1880s as Tebbutt's protégé. Between 1878 and 1888 he carried out an extensive series of micrometric observations of α Centauri, and he subsequently combined these with measures made earlier by Lacaille, Herschel, Maclear, Powell, Ellery, Suliger and Tebbutt in order to calculate the orbital elements and a period of 83.7 years (cf. See's later figure of 81.1 years). He also computed the mean separation of the two stars as 20.043 Astronomical Units (Biggs 1887). Elsewhere, I have suggested that

> This interesting paper, undoubtedly Biggs's best, and his sole assault on mathematical astronomy, owes much to Biggs's friend and mentor, John Tebbutt. It was Tebbutt who provided him with the measurements of the other observers, and it was Tebbutt who introduced an unsure Biggs to the intricacies of orbital computations and guided him safely through the calculations. The paper could, by rights, have been co-authored. (Orchiston 1985a, p. 17).

It is clear that Tebbutt was capable of carrying out double star orbital reductions—as he virtually did for Biggs in the case of α Centauri. Given his wealth of double star measures secured over so long a time span, Tebbutt must at one time or another have been tempted to reduce some of these and publish orbital elements and periods for his more frequently observed stars. We must assume that it was pressure of other astronomical work that prevented him from adding this extra burden to an already over-extended workload. But even without doing this, Tebbutt made an important contribution to southern hemisphere double star astronomy (See 1896), and his role as a pioneer in this field deserves to be more widely recognised.

### *11.3.4 Minor Planet Positions*

Another of Tebbutt's astronomical interests was minor planets. The existence of these small Solar System bodies was only established between 1 January 1801 and 29 March 1807 when four different minor planets were discovered (see Cunningham, 2014, 2016; Cunningham et al. 2009, 2011). The next discovery occurred in 1845, and by April 1877, when Tebbutt first observed 4 Vesta, 172 of these objects had been discovered (see Pilcher 1979).

Minor planets, or asteroids as they also are called, had a special place in Solar System astronomy during the second half of the nineteenth century. According to Clerke (1893, p. 95),

> Theoretical and practical astronomy both derived profit from the admission of these apparently insignificant strangers to the rights of citizenship of the solar system. The disturbance of their motions by their giant neighbour afforded a more accurate knowledge of the Jovian mass ... the anomalous character of their orbits presented geometers with highly stimulating problems in the theory of perturbations; while ... the sure prospect of further detections

> [i.e. discoveries] powerfully incited to the exploration of the skies; observers became more numerous and more zealous in view of the prizes held out to them …

Tebbutt was not interested in these 'prizes' and took no part in the search for new minor planets. What interested him were their anomalous orbits, and he hoped that his observations would be used to improve scientific understanding of these (see Tebbutt 1880a).

The first minor planet observations that Tebbutt made were almost fortuitous. After searching unsuccessfully for Comet 6P/d'Arrest on the morning of 13 April 1877, he records in his journal that 'I also availed myself of the opportunity to observe the planet Vesta which was fortunately situated nearly on the same parallel as B.A.C. 7781 a star of the 6th magnitude …' (Tebbutt 1873–1879). This star was used for the micrometric comparisons. Tebbutt subsequently also made a few observations of 1 Ceres and 2 Pallas with the 4.5-in. telescope, but minor planets only featured prominently in his observing schedule after the arrival of the larger equatorial. As Table 11.11 indicates, they were observed every single year between 1887 and 1904 (inclusive), except in 1892 and 1894.

Although the greatest number of different minor planets followed in any 1 year was 8, over the 28 years under discussion 23 different bodies were observed micrometrically. Most popular were 1 Ceres and 6 Hebe which were researched in 8 and 7 different years respectively, followed by 2 Pallas, 4 Vesta and 11 Parthenope (5 years), 3 Juno (4 years) and 8 Flora (3 years). The annual breakdown of Tebbutt's minor planet observations is given in Table 11.11. In 1896 he observed 22 Kalliope and 140 Siwa at the request of Professor Julius Bauschinger (1860–1934) the newly-appointed Director of the Astronomische Reichen-Institut in Berlin, and 6 Hebe and 11 Parthenope at the request of Dr. Karl Theodor Robert Luther (1822–1900) from the Bilk Observatory in Dusseldorf; while his numerous observations of 7 Iris in 1899 were for an investigation being conducted by Dr. Johannes Karl Richard Riem also from the Astronomische Reichen-Institut (Tebbutt 1908a).

Table 11.11 reveals Tebbutt's *modus operandi*. His aim was to obtain as large a number of micrometric positional observations as possible for each minor planet, over a comparatively short period. The longest sequences of such observations were for 7 Iris (30 nights), 42 Isis (25), 324 Bamberga and 1 Ceres (both 23), though 8 Flora, 2 Pallas, 11 Parthenope and 17 Thetis also featured prominently. Perhaps the most notable of these was the 324 Bamberga series: since this minor planet was only discovered in 1892 (Table 11.12), just 11 years before Tebbutt made his measurements, its orbital parameters would not have been well known.

1 Ceres was observed most of all minor planets, on 80 evenings between 1882 and 1904, followed by 6 Hebe (63 evenings, 1889–1904), 11 Parthenope (47 evenings, 1893–1901), 7 Iris (46 evenings, in 2 years only, 1898 and 1899), 2 Pallas (41 evenings, 1879–1902) and 4 Vesta (36 evenings, 1877–1902). However, all of the minor planets on Tebbutt's observing list were characterised by comparatively bright opposition magnitudes (see Table 11.12). Other parameters of interest are listed in this table, including diameters and rotation periods. In the days when

**Table 11.11** Minor planets observed by Tebbutt, 1877–1904[a]

| *Minor Planet* | *1877* | *1879* | *1880* | *1882* | *1885* | *1887* | *1888* | *1889* | *1890* | *1891* | *1893* | *1895* | *1895* |
|---|---|---|---|---|---|---|---|---|---|---|---|---|---|
| 29 Amphitrite | | | | | | | | | | | | | |
| 43 Ariadne | | | | | | | | | 3/30 | | | | |
| 324 Bamberga | | | | | | | | | | | | | |
| 1 Ceres | | | | 10/159 | 5/59 | | | | 10/76 | 11/106 | | 14/189 | 14/189 |
| 61 Danaë | | | | | | | | | | | | | |
| 78 Diana | | | | | | | | | | | | | |
| 164 Eva | | | | | | | | | | | | | |
| 8 Flora | | | | | | | | 8/146 | | | | | |
| 6 Hebe | | | | | | | | 5/57 | | | 6/75 | 11/143 | 11/143 |
| 7 Iris | | | | | | | | | | | | | |
| 42 Isis | | | | | | | | | | | | | |
| 3 Juno | | | | | | 3/54 | | 11/168 | 4/34 | | | | |
| 22 Kalliope | | | | | | | | | | | | | |
| 18 Melpomene | | | | | | | | | | | | | |
| 71 Niobe | | | | | | | | | | | | | |
| 2 Pallas | | 5/58 | 3/42 | | 5/53 | | 17/266 | | | | | | |
| 11 Parthenope | | | | | | | | | | | 7/106 | | |
| 194 Prokne | | | | | | | | | | | | | |
| 80 Sappho | | | | | | | | | | | | | |
| 140 Siwa | | | | | | | | | | | | | |
| 17 Thetis | | | | | | | | | | | | | |
| 92 Undina | | | | | | | | | 3/38 | | | | |
| 4 Vesta | 1/– | | | | | 9/93 | | | | | | | |
| Number Obs | 1 | 1 | 1 | 1 | 2 | 2 | 1 | 3 | 4 | 1 | 2 | 2 | 2 |
| *Minor planet* | *1896* | *1897* | *1898* | *1899* | *1900* | *1901* | *1902* | *1903* | *1904* | *Total number of evenings* | | | |

**Table 11.11** (continued)

| | | | | | | | | | | | | | |
|---|---|---|---|---|---|---|---|---|---|---|---|---|---|
| 29 Amphitrite | | | | | | | | 4/45 | | 4 | | | |
| 43 Ariadne | | | | | | | | | | 3 | | | |
| 324 Bamberga | | | | | | | | 23/236 | | 23 | | | |
| 1 Ceres | | | | | 23/317 | 6/60 | | | 1/10 | 80 | | | |
| 61 Danaë | 7/75 | | | | | | | | | 7 | | | |
| 78 Diana | | | | | | 4/47 | | | | 4 | | | |
| 164 Eva | | | | | | | 7/65 | | | 7 | | | |
| 8 Flora | 9/113 | | | | | | 17/176 | | | 34 | | | |
| 6 Hebe | 11/162 | 4/58 | | | 13/189 | | | | 13/178 | 63 | | | |
| 7 Iris | | | 16/211 | 30/380 | | | | | | 46 | | | |
| 42 Isis | | | 8/107 | | | | 25/270 | | | 33 | | | |
| 3 Juno | | | | | 8/96 | | | | | 26 | | | |
| 22 Kalliope | 8/84 | | | | | | | | | 8 | | | |
| 18 Melpomene | | | | | | 10/136 | | | | 10 | | | |
| 71 Niobe | | 7/122 | | | | | | | | 7 | | | |
| 2 Pallas | | | | | | | 11/134 | | | 41 | | | |
| 11 Parthenope | 18/196 | 13/125 | | | | 9/114 | | | | 47 | | | |
| 194 Prokne | | 14/170 | | | | | | | | 14 | | | |
| 80 Sappho | 3/34 | | | | | | | | | 3 | | | |
| 140 Siwa | 10/91 | | | | | | | | | 10 | | | |
| 17 Thetis | | | | 16/204 | | | | 5/77 | | 21 | | | |
| 92 Undina | 4/55 | | | | | | | | | 7 | | | |
| 4 Vesta | | | 6/73 | 7/81 | | | 13/124 | | | 36 | | | |
| Number Obs | 8 | 4 | 3 | 3 | 3 | 4 | 5 | 3 | 2 | | | | |

[a]Note: Based on Tebbutt (1887a, 1889b, 1890c, 1891c, 1892b, 1893a, 1894b, 1895b, 1896c, 1897c, 1898b, 1899, 1900a, 1901a, 1902a, 1903a, 1904, 1908a, and published papers)
Key: 5/58 = 58 positional measurements, obtained on 5 evenings
-- = observed, but detailed information not available
Numner Obs = Total numbner of minor planets observed

**Table 11.12** Some features of the minor planets observed by Tebbutt[a]

| Number and name | Discovery date | Opposition magnitude ($m_v$) | Diameter (km) | Period (h) |
|---|---|---|---|---|
| 1 Ceres | 1801 | 7.0 | 1025 | 9.1 |
| 2 Pallas | 1802 | 6.3 | 583 | 7.9 |
| 3 Juno | 1804 | 6.9 | 249 | 7.2 |
| 4 Vesta | 1807 | 6.0 | 555 | 5.3 |
| 6 Hebe | 1847 | 7.1 | 206 | 7.3 |
| 7 Iris | 1847 | 6.7 | 222 | 7.1 |
| 8 Flora | 1847 | 7.8 | 160 | 13.6 |
| 11 Parthenope | 1850 | 8.7 | 155 | 10.7 |
| 17 Thetis | 1852 | 9.3 | 97 | 12.3 |
| 18 Melpomene | 1852 | 7.7 | 164 | 11.6 |
| 22 Kalliope | 1852 | 9.2 | 175 | 4.1 |
| 29 Amphitrite | 1854 | 8.6 | 199 | 5.4 |
| 42 Isis | 1856 | 8.8 | 104 | 13.6 |
| 43 Ariadne | 1857 | 8.8 | 78 | 5.8 |
| 61 Danaë | 1860 | | 88 | 11.5 |
| 71 Niobe | 1861 | | 106 | 11.2 |
| 78 Diana | 1863 | 9.2 | 144 | 8? |
| 80 Sappho | 1864 | 9.2 | 84 | >20 |
| 92 Undina | 1867 | | 184 | 15.9 |
| 140 Siwa | 1874 | | 105 | |
| 164 Eva | 1876 | 9.3 | 111 | 27.3 |
| 194 Prokne | 1879 | 8.9 | 195 | 15.7 |
| 324 Bamberga | 1892 | 7.4 | 256 | 8? |

Opposition magnitudes are after Roth (1962, p. 122)
Approximate diameters are after Bowell et al. (1979)
Rotation periods are after Tedesco (1979)
[a]Discovery dates are after Pilcher (1979)

Tebbutt was observing, before the advent of asteroidal photoelectric photometry, the rotation of minor planets was hardly thought of.

Between 1880 and 1905, Tebbutt published 27 different papers on his micrometric minor planet work, which appeared in *Astronomische Nachrichten* (21 papers), *Monthly Notices of the Royal Astronomical Society* (5) and the *Astronomical Journal* (1).

One of the few individuals to make use of Tebbutt's published data while he was still alive was the German astronomer, Dr. Johannes Riem. In 1902 he published an account of 7 Iris which drew on Tebbutt's observations made in 1898 and 1899 (Tebbutt 1908a).

In the century since Tebbutt's demise Solar System astronomy has changed in ways that he could never have imagined. First, there has been a total redefinition of what is deemed to constitute a minor planet—and indeed a planet (see Dick 2013). Thus, Pluto is now a dwarf planet, not a planet, and Ceres (by far the largest body in the Asteroid Belt) also is classified as a 'dwarf planet'. I was at those infamous August 2006 meetings of the International Astronomical Union where this occurred, and I voted *against* any change to the *status quo*. However, in my opinion sanity did

**Fig. 11.22** NASA's Dawn spacecraft orbited minor planet 4 Vesta from July 2011 to September 2012 and this is a synthesised view of some of the best images that were obtained during this period (*Image* NASA/JPL-Caltech/UCAL/MPS/DLR/IDA; https://en.wikipedia.org)

not prevail on this occasion, and now we have one dwarf planet and hundreds of thousands of minor planets (or asteroids) inhabiting the Asteroid Belt!

The second major change that has occurred is thanks to the advent of space mission fly-bys. NASA's Dawn Mission has provided detailed close-up images of two of the minor planets that Tebbutt observed, 4 Vesta (see Fig. 11.22) and 1 Ceres (see Fig. 11.23). These images revealed that Ceres is spherical in shape while Vesta is an oblate spheroid. Both bodies are pock-marked with craters. The three conspicuous adjacent craters in Fig. 11.22 are known colloquially as the 'Snowman Craters'.

Finally, we now know that some minor planets have satellites. Thus, 22 Kalliope has a satellite named Linus. This was discovered in August 2001, is ~28 km in diameter and orbits its parent minor planet at ~13.2 Kalliope radii (Descamps et al. 2008). Photometric data also suggest that 61 Danaë—another minor planet observed by Tebbutt—may have a satellite (see Cellino et al. 1985), but this has yet to be confirmed.

**Fig. 11.23** Dwarf planet Ceres, imaged from a distance of 13,642 km by NASA's Dawn spacecraft on 2 May 2015; the two bright spots (centre and centre right) are associated with the craters Oxo and Haulani, respectively (*Image* NASA/JPL-Caltech/UCLA/MPS/DLR/ IDA; https://en.wikipedia.org)

### *11.3.5 Planetary Positions*

A fifth observational field in which Tebbutt was particularly active was positional observations of planets (Table 11.13), and between 1869 and 1905 he conducted micrometric observations of 26 different conjunctions of stars and planets. He also recorded three mutual conjunctions of two planets (in 1879, 1882 and 1891); one occultation of a star by a Jovian satellite (in 1878); and one occultation of a star by a planet (in 1888). After the introduction of photoelectric photometry, such occultations assumed considerable importance as a means of establishing the existence of

**Table 11.13** Planetary phenomena observed by Tebbutt, 1869–1905[a]

| Year | Date | Event | Bodies Involved |
|---|---|---|---|
| 1869 | February 19–25 | C | Uranus and δ and γ Geminorum |
| 1878 | October 05 | O | Jupiter's Satellite Io and a star |
| 1879 | June 30 | C | Mars and Saturn |
| 1880 | November 20 | C | Jupiter and Star 363 Washington General Catalogue |
| 1881 | March 08–15 | C | Uranus and χ Leonis |
| 1882 | August 02 | C | Venus and Mars |
| 1885 | February 10 | C | Uranus and η Virginis |
| 1886 | January 10 | C | Saturn and μ Geminorum |
| 1887 | Jan 30–Feb 14 | C | Saturn and δ Geminorum |
| | April 21 | C | Jupiter and Lalande 25797 |
| 1888 | May 14–25 | C | Jupiter and β1 Scorpii |
| | June 10 | O | Jupiter and 47 Librae |
| 1891 | April 07 | C | Venus and Jupiter |
| 1892 | November 01 | C | Mars and ι Aquarii |
| 1893 | April 03–11 | C | Saturn and γ Virginis |
| 1894 | October 05 | C | Uranus and α Librae |
| 1895 | January 05, 10 | C | Jupiter and ι Geminorum |
| | May 06–16 | C | Saturn and k Virginis |
| | May 19, 20 | C | Venus and ε Geminorum |
| | August 04–12 | C | Saturn and 96 Virginis |
| 1897 | September 05–13 | C | Uranus and 41 Librae |
| | December 22–28 | C | Neptune and 114 Tauri |
| 1898 | April 02–09 | C | Jupiter and η Virginis |
| | April 18–May 16 | C | Uranus and ω1 and ω2 Scorpii |
| 1899 | February 22–28 | C | Neptune and 114 Tauri |
| 1900 | July 12–16 | C | Jupiter and β1 Scorpii |
| 1901 | November 21–23 | C | Venus and ξ Sagittarii |
| | November 25 | C | Venus and h2 Sagittarii |
| | November 25 | C | Venus and Star 5212 in Radcliffe Catalogue (1890) |
| 1904 | July 18–Aug 05 | C | Uranus and Star 4648 in Radcliffe Catalogue |
| 1905 | July 15–23 | C | Uranus and I Sagittarii |

C = conjunction; O = occultation

[a]Based on Tebbutt (1864–69b, 1882–84, 1884–86, 1887a, 1889b, 1892b, 1893a, 1894b, 1895b, 1896c, 1898b, 1899, 1900a, 1901a, 1902a, 1908a) and published papers

planetary or sub-planetary atmospheres (e.g. see Millis et al. 1993) and planetary ring systems (e.g. Elliot et al. 1977; Millis et al. 1977), but in Tebbutt's day visual observing did not offer such precision.

Although Tebbutt carried out his first planetary positional observations in 1869 (Tebbutt 1864–69b), he only began this programme in earnest in 1880. However, his attempts to make useful observations during the intervening decade were thwarted. Thus, in 1875 he used the *Sydney Morning Herald* newspaper to alert local people to the probable occultation of the star 3 Sagittarii by Mars on 30 June (Tebbutt 1875c), but inclement weather prevented observation of this rare phenomenon. He had similar bad luck on 14 September 1879 when an occultation of 64 Aquarii by Jupiter was predicted (Tebbutt 1879c), and heavy clouds prevented him from observing what would have proved a spectacular event (Tebbutt 1879d).

While he usually was clinical in recording these spectacular events, from time to time the majesty of the situation found its way into Tebbutt's journals and published accounts. Here is his description of the 30 June 1879 conjunction of Mars and Saturn:

> At the time of closest approach of centres the planets appeared to the naked eye as one large and brilliant star. In the telescope they were, of course, widely separated … Mars appeared to be the usual fiery yellow-red colour, but Saturn was … a light sap green colour. … (Conjunction of Mars and Saturn 1879; cf. Tebbutt 1879a).

Three years later, during his observation of the 1882 conjunction of Venus and Mars, Tebbutt was moved to write: 'The contrast between the colours of the planets as seen in the comet eyepiece was very marked and beautiful, Mars was fiery red while Venus was beautifully white. The latter was gibbous' (Tebbutt 1882–84).

Tebbutt's observations of conjunctions of planets and stars became more frequent following his acquisition of the Grubb telescope, and between 1887 and 1905 (inclusive) he spent about 100 evenings obtaining micrometric positions of all of the then-known planets, with the exception of Mercury (see Table 11.14). As the entries in this table indicate, Jupiter, Saturn and Uranus were the most prolifically observed, but it was the observations of Uranus and Neptune that were of greatest value. In the case of Uranus, it was irregularities in its orbit that led to the discovery of Neptune (see Grosser 1979), while anomalies in Neptune's orbit ultimately enticed Percival Lowell (1855–1916; Strauss 2014) to search for Pluto (Tombaugh and Moore 1980). Tebbutt's precise micrometric observations revealed that the actual positions of both planets differed significantly from those listed in the *Nautical Almanac*. His observations and those of others helped planetary scientists monitor the positions of these wayward outer planets.

Between 1878 and 1906, Tebbutt published 23 different papers on the aforementioned micrometric observations of planetary conjunctions and occultations. These appeared in *Monthly Notices of the Royal Astronomical Society* (15), *Astronomische Nachrichten* (6) and *The Observatory* and *Journal of the British Astronomical Association* (1 each).

Before closing this review of Tebbutt's observations, several other planetary events should be mentioned. On 5 October 1878, Tebbutt witnessed a very rare event, the apparent occultation of a ninth magnitude star by Jupiter's first satellite (Io). He watched the two bodies come together until they could no longer be separated in the telescope, and it was only seven minutes later that he detected indications of separation. However, he was not convinced that an occultation actually occurred, and voiced this concern in a report published in the *Sydney Morning Herald*:

**Table 11.14** Positional observations of planets and stars by Tebbutt 1869–1905[a]

| Year | Venus | Mars | Jupiter | Saturn | Uranus | Neptune |
|---|---|---|---|---|---|---|
| 1869 | | | | | 4/- | |
| 1880 | | | 1/- | | | |
| 1881 | | | | | 6/- | |
| 1885 | | | | | 1/- | |
| 1886 | | | | 1/20 | | |
| 1887 | | | 1/20 | 5/72 | | |
| 1888 | | | 12/230 | | | |
| 1892 | | 1/20 | | | | |
| 1893 | | | | 7/137 | | |
| 1894 | | | | | 1/20 | |
| 1895 | 2/- | | 2/44 | 12/229 | | |
| 1897 | | | | | 8/213 | 6/120 |
| 1898 | | | 7/91 | | 10/264 | |
| 1899 | | | | | | 6/80 |
| 1900 | | | 5/92 | | | |
| 1901 | 4/45 | | | | | |
| 1904 | | | | | 8/95 | |
| 1905 | | | | | 6/96 | |

12/229 = 229 positional measurements, obtained on 12 evenings, – = observed, but detailed information is not available

[a]Data based on Tebbutt (1887a, 1889b, 1893a, 1894b, 1895b, 1896c, 1898b, 1899, 1900a, 1901a, 1902a, 1908a) and published papers

> The occultation, if it actually amounted to one, was certainly not central; it appeared to be a graze by the satellite's northern limb. It seems probable that the star was occulted by the planet itself some time previously. (Tebbutt 1878f; cf. Tebbutt 1878d).

Henry Russell subsequently confirmed this, as the higher magnification used with the 11.5-in. Schroeder refractor at Sydney Observatory revealed that an occultation did not actually take place (see Tebbutt 1878a).

On 6 February 1892 there was a conjunction of Venus and Jupiter which promised to be an attractive sight, and the general populace was accordingly forewarned through the newspapers. While dense cloud prevented Tebbutt from making any observations, people living elsewhere in New South Wales were more fortunate, and Tebbutt (1892a) reported their observations in the *Journal of the British Astronomical Association.*

The 1897 conjunction of 41 Librae and Uranus already listed in Table 11.13 also deserves further comment. Soon after beginning his series of observations of this event, Tebbutt became convinced that an occultation would very likely occur on 8 September, a night when he would be absent from Windsor (Tebbutt 1897a, b). He accordingly asked two Sydney colleagues, Charles James Merfield (Bio Box 11.4) and Hugh Wright (Bio Box 2.1) to observe this rare phenomenon. Both of these astronomers, and T.H. Close, saw the planet within 5 arc seconds of the star at nightfall (Merfield 1897; Wright 1897), and as a result of his subsequent analysis Merfield concluded that an occultation had earlier taken place. Tebbutt reported Merfield's observations in both of his papers, and in the one published by the British Astronomical Association (Tebbutt 1897b) he paid homage to his Sydney colleague by reporting his conclusion that the occultation would have occurred at 16 h 05 m 11 s local time.

Finally we should note that, as with comets and minor planets, our knowledge of the six planets observed by Tebbutt has changed enormously since his era, thanks largely to a succession of space probes, dedicated landers and the Hubble Space Telescope.

**Bio Box 11.4: Charles James Merfield**

Charles James Merfield (Fig. 11.24) was born in Ararat, Victoria, on 28 April 1866, and trained in mathematics, surveying and civil engineering. For a number of years he was employed by the Government supervising '... the construction of new railway lines through the then virgin territory of Victoria' (Obituaries 1932). He then moved to Sydney and was appointed a draftsman with the New South Wales Railways in 1890 where his mathematical skills were greatly appreciated (ibid.).

Merfield had a long-standing interest in astronomy and a particular passion for mathematical astronomy, but this only came to the fore in Sydney, initially through the influence of his friend and neighbour, the Reverend Dr. Thomas Roseby (see Bio Box 12.4). However, he soon established a close friendship with Tebbutt and drew inspiration from his many achievements: 'Your kindly influence upon me the last few years, through the medium of your letters and occasionally in person, has tended towards the cultivation of strict habits ...

**Fig. 11.24** C.J. Merfield (*Courtesy* Peter Hingley)

(continued)

(continued)

[and] it has been mainly through your advice and encouragement; that I have attained a certain knowledge for which I shall ever feel grateful' (Merfield 1901).

In 1894, Merfield began calculating cometary orbits and stellar proper motions, and over the next 10 years he published a number of research papers in *Astronomische Nachrichten* and *Monthly Notices of the Royal Astronomical Society* which served to establish his international reputation. Three of these, on the definitive orbital elements of C/1898 L1 (Coddington-Pauly), C/1899 E1 (Swift) and the Great Comet of 1901 (C/1901 G1) drew praise from professional colleagues, and after the appearance of the first of these Dr. Roseby was moved to write: '... the definitive orbital elements which Mr. Merfield had computed were a long way in advance of all that has ever been done in Australia before ... [and] may be said to be the coping-stone to the mathematical astronomy of Australia up to the close of the century' (Reports of the Branches 1900). Apart from these computations, from 1894 Merfield also immersed himself in observational astronomy (mainly of comets, lunar occultations and star colours) with a 7.5-in. (19.1-cm) reflector.

Like his great friend and mentor, Tebbutt, Merfield also contributed to the popularisation of astronomy by writing frequently for the newspapers. He also was a stalwart of the NSW Branch of the British Astronomical Association from the time this group was formed at the start of 1895, serving for two sessions as President and as founding Director of the Branch's first observing section, which was devoted to Star Colours (Orchiston 1988a).

Mimicking R.T.A. Innes (Bio Box 8.1), Merfield also made the transition to professional ranks, eventually joining the staff of Sydney Observatory in 1904, where he was finally able to indulge his interest in minor planet orbits (Orchiston 2015). Later (in 1908) he moved to Melbourne Observatory, where he eventually became Deputy Director and Australia's foremost authority on solar eclipses. He also was founding President of the Astronomical Society of Victoria (Perdrix 1972).

Merfield died unexpectedly on 23 January 1931 as a result of a motor accident (Crommelin 1932; Mr. C.J. Merfield 1932; Obituaries 1932). In addition to astronomy, he also was known internationally for his avocational achievements in photography.

### 11.3.6 Other Observations of Jupiter and Uranus

In addition to his focus on planetary positions, Tebbutt carried out two short-term studies of Jovian surface features and a brief investigation of the brightness of Uranus.

The Jovian belts have been known since the time of Galileo, and the Great Red Spot (GRS) was first definitely observed in 1665 by Giovanni Domenico Cassini (1625–1712; Moore 1995). However,

> Although it was not a new feature, it was in 1878 that attention was first directed to the Red Spot, as it was then rising to a prominence that had not previously been recorded; but it was from 1879 to 1882 that it dominated the surface of the planet and achieved immortal fame, not only on account of its being the most conspicuous individual spot ever seen in the matter of size and darkness but also because of its strikingly red colour. (Peek 1958, p. 130).

More recently, Hockey (1992, 1999) has shown that the dramatic 'reddening' of the GRS actually began in 1869. Long known to be a major atmospheric disturbance, the Pioneer and Voyager space probes have provided much information on this enormous 'anticyclonic vortex' whose surface area is sometimes greater than that of the Earth (Moore 1995).

Following its 1869 reddening, the GRS became an object of intense interest for astronomers, and for a few brief months in 1880 Tebbutt was one of these. It would seem that he was inspired, at least in part, by the Jovian studies conducted by his Sydney colleague, G.D. Hirst (Bio Box 10.3), who was known for his visual acuity (see Hirst 1876, 1878, 1879–80, 1880). While Tebbutt (1879–81) commented on the colour of the GRS in his journal, he was more interested in its rotation period, and made micrometric timings of its transit across the central meridian of the planet on 12 different evenings between 31 October and 30 December 1880, inclusive (Tebbutt 1881b). What Tebbutt did not do when he published his transit times was to use these to compute the rotation period of the GSR, which Peek (1958, p.144) lists as 09 h 55 m 34 s for 1879–80 and 09 h 55 m 36 s for 1880–1881. However, the Director of Athens Observatory, Dr. Julius Schmidt (Fig. 11.25), subsequently incorporated Tebbutt's data into just such an analysis (Schmidt 1882).

**Fig. 11.25** Julius Schmidt in 1884 (https://en.wikipedia.org)

The other Jovian feature which attracted Tebbutt's attention was '... a black and nearly circular spot on the disk ...' which was conspicuous in the 4.5-in. Cooke telescope on 23 January 1896 (Tebbutt 1896a). On 27 January the larger Grubb telescope revealed that there were in fact two different dark spots, separated in longitude, but both lying within the North Tropical Zone (the bright zone immediately north of the North Equatorial Belt). Peek (1958, p. 83) reports that such spots are characteristic features of the southern part of this zone but are rarely observed in the northern half of the zone.

Meanwhile, Tebbutt proceeded to monitor these two spots, which he termed A and B. The 'original' spot, B, was recorded on 17 different occasions between 23 January and 25 May (when it had almost faded from view), while the new spot, A, was observed on nine occasions between 27 January and 4 April (by which time it had become excessively faint). Tebbutt published these observations in the *Journal of the British Astronomical Association* in the hope that they

> ... will prove of some service in the determination of the rotation-period of the planet. When I can get an opportunity I shall probably myself attempt to deduce a period from all the published observations of the spots. (Tebbutt 1896a).

Regrettably, the 'opportunity' never arose.

Tebbutt also was interested in the 'observability' of Uranus for users of modest-sized telescopes. In a paper published in Sydney in 1878 (Tebbutt 1878b) he drew attention to the fact that Uranus would be in the most favourable part of its orbit for Earth-based observation between that date and 1882. In a follow-up paper, also published in the *Journal and Proceedings of the Royal Society of New South Wales* (Tebbutt 1880d), he estimated the apparent visual magnitude of Uranus at opposition in 1880. Alexander (1965) has assigned 2 pages of his textbook on Uranus to Tebbutt's 2 short papers.

### *11.3.7 Solar and Lunar Eclipses*

In addition to the foregoing five distinct long-term programs of observations, and his occasional observations of Jovian surface features, Tebbutt undertook a number of programmes associated with what he termed 'phenomena of an occasional character'.

Perhaps the most visually appealing of these were eclipses, and between 1857 and 1891 he observed 15 lunar and 9 solar eclipses (Table 11.15). Unfortunately, all of the latter were partial eclipses, and thus they did not offer the research possibilities typical of total solar eclipses (such as studies of the corona, the chromosphere and prominences). During the nineteenth century, astronomers also used data derived from both solar and lunar eclipses (along with lunar occultations) to investigate the motion of the Moon. Mitchell (1932, p. 69) describes how 'After making allowance for the gravitational attraction of every conceivable form of disturbance, it is unmistakable that the moon departs from her theoretical place in a very irregular manner'.

**Table 11.15** Solar and lunar eclipses observed by Tebbutt, 1857–1891[a]

| Year | Date | Solar (S) or Lunar (L) | Total (T) or Partial (P) |
|---|---|---|---|
| 1857 | September 18 | S | P |
| 1861 | January 11 | S | P |
| 1866 | September 24 | L | T |
| 1867 | March 20 | L | T |
| 1868 | August 18 | S | P |
| 1869 | July 23 | L | P |
| 1870 | January 17 | L | T |
| | June 28 | S | P |
| 1871 | December 12 | S | P |
| 1873 | May 12 | L | T |
| | November 04/05 | L | T |
| 1874 | May 01 | L | P |
| 1876 | September 17 | S | P |
| 1877 | February 28 | L | T |
| 1878 | February 02 | S | P |
| 1879 | December 29 | L | P |
| 1880 | June 22/23 | L | T |
| | December 16 | L | T |
| 1881 | December 06 | L | P |
| 1882 | November 10 | S | P |
| 1885 | September 24 | L | P |
| 1887 | February 08 | L | T? |
| 1890 | December 12 | S | P |
| 1891 | May 24 | L | T |

[a]Based on Tebbutt (1864–69a, 1873–79, 1879–81, 1884–86, 1886–87, 1887a, 1891c, 1892b, 1908a), newspaper articles and published papers

During lunar eclipses, studies also were made of colour graduations at totality, and timings were recorded of the passage of the Earth's shadow across distinctive craters and other lunar features. In more recent times, such timings have been used with great effect to investigate changes in the oblateness of the Earth's upper atmosphere (e.g. see Soulsby 1992).

When he observed lunar eclipses, Tebbutt usually attempted to record the times of first, second, third and fourth contacts, but because of the nebulous demarcation of the shadow boundary he was frustrated by the resulting imprecise nature of the results, noting that '… no class of observations is more unsatisfactory than that which relates to the determination of the contacts in a lunar eclipse …' (Tebbutt 1873–1879, pp. 40–42). Despite this, he usually also tried to record the times when the shadow intersected specific lunar features. The crater Tycho (see Fig. 11.26) almost always was one of these.

**Fig. 11.26** A photograph of a lunar eclipse showing the location of the distinctive 'ray crater' Tycho—which is indicated by the black arrow (*Photograph courtesy* John Drummond)

When it came to recording colour-related aspects of lunar eclipses Tebbutt was on firmer ground, and he was particularly interested in variations apparent at totality. For example, during the 16 December 1880 eclipse he recorded

> Observations were particularly directed to the variety and depth of the colour of the shadow. Its periphery was dark steel blue, a little further in it was of a light straw colour, and still nearer the centre of a dull red or copper colour. (Tebbutt 1881a).

In contrast, during the 24 May 1891 eclipse, he noted that at totality the northern portion of the Moon was a light coppery colour while the southern sector was dark grey (Tebbutt 1890–1892).

Nor was Tebbutt oblivious to the beauty of a total lunar eclipse. In the case of the 22/23 June 1880 event he conducted observations with the naked eye, his marine telescope, the transit telescope and the 4.5-in. refractor, and noted that 'The moon's eclipse was very interesting owing to her being at the time in the eastern branch of the Milky Way' (Tebbutt 1879–81). At totality, the Moon was a coppery colour.

With the acquisition of the Grubb telescope, Tebbutt felt there were more important calls on his time and he largely turned his back on eclipses, although he did use the total lunar eclipse of 23 June 1899 as an opportunity to observe lunar occultations (Tebbutt 1908a). A further reason for his declining interest in lunar eclipses is noted in his Journal during 1896: 'There is [now] no satisfaction in observing eclipses of the moon, the earth's shadow being so ill-defined' (Tebbutt 1896–1898). As we have seen, Tebbutt strove to make precise astronomical observations, and he felt that lunar eclipses did not offer this option.

Tebbutt published nine papers on his solar eclipse observations, in *Monthly Notices of the Royal Astronomical Society* (5) and *Astronomische Nachrichten* (4); his lunar eclipse observations appeared in eight short research papers, in *Monthly Notices of the Royal Astronomical Society* (4), *Astronomische Nachrichten* (3) and *The Astronomical Register* (1). Since seven of these papers were duplicates, he in fact only wrote ten different papers on his eclipse work. About half of all eclipses that he observed were not reported in academic journals at all; some were discussed in newspaper articles, while data on others were confined to his observing journals. In 1924, the Harvard College Observatory astronomer, Willard James Fisher (1867–1934; Hockey and MacPherson 2014) used some of Tebbutt's published data in his analysis of the brightness of lunar eclipses that occurred between 1860 and 1922 (Fisher 1924).

### *11.3.8 Transits of Mercury*

Although they were not of the same importance as transits of Venus (see Chap. 8), transits of Mercury had their own visual charm and therefore were popular with astronomers during the nineteenth century. One of the research objectives of astronomers was to determine the extent of Mercury's atmosphere. A halo around the planet suggested the existence of an atmosphere (Clerke 1893), so the presence or absence of a halo was a matter of considerable interest, as was the search for a Mercurian satellite. Transits of Mercury also assumed an altogether different importance when Newcomb (1903) used ingress and egress timings as a test for minute changes in the rotation period of the Earth.

Tebbutt successfully observed four transits of Mercury, on 6 May 1878, 8 November 1881, 9 May 1891 and 10 November 1894, and on each occasion where possible he noted the times of the ingress and egress contacts, looked out for a halo and any evidence of a 'black drop effect', and searched for any sign of a satellite of the planet.

The 1878 transit, which had started before sunrise, was observed with both the 3.25-in. and 4.5-in. telescopes (Tebbutt 1878c). Through the former, Tebbutt thought he could distinguish a faint whitish spot on the disk of Mercury, but this was not visible shortly afterwards in the larger refractor. As the time of egress approached, Tebbutt observed an interesting phenomenon:

> At 20 h. 48 m. 19.3 s. observatory mean solar time, when I estimated the thread of light between the limbs of the sun and planet to be about a second of arc in breadth, the planet suddenly threw out a narrow ligament and immediately became somewhat pear-shaped, as represented in Fig. 1 … The ligament observed by me was nearly as black as the planet itself, and rendered it very difficult to estimate the time of the geometrical or apparent internal contact of limbs … At 20 h. 49 m. 0.3 s. Windsor mean time, I noted the appearance of the planet to be that shown in figure 2. … (ibid.)

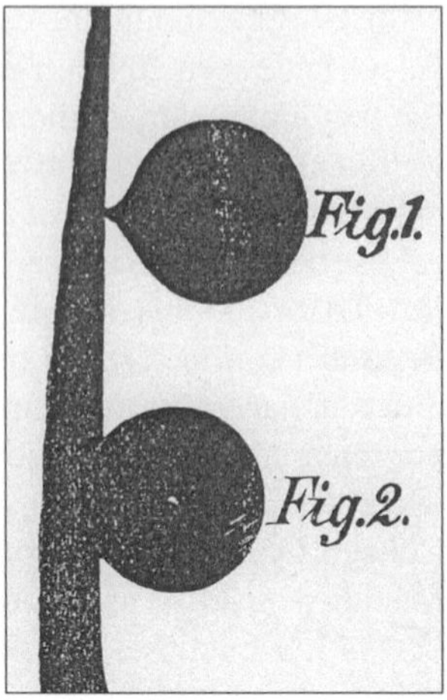


**Fig. 11.27** Two of Tebbutt's drawings of the 6 May 1878 transit of Mercury (after Tebbutt 1878e)

Tebbutt's Figures 1 and 2 are reproduced here in Fig. 11.27. In the course of the transit, Tebbutt (1878e) noted the absence of any halo or ring round the planet, or anything resembling a satellite.

The entire 7/8 November 1881 transit was observed from Windsor, and Tebbutt was quick to point out the uneventful nature of the first and second contacts:

> There was no distortion of the planet; there was no black drop, no triangular black ligament like that experienced by me in May, 1878, nor was there even the shadowy connection observed at the contacts of *Venus* in 1874. (Tebbutt 1882b, p. 103).

The egress contacts were equally uneventful, but in the course of this transit Tebbutt on a number of occasions noted the same faint whitish spot in the centre of the planet's disk as seen in 1878. Meanwhile, he looked unsuccessfully for a halo or ring around the planet and for a satellite.

Although Tebbutt was greeted by clear skies for the 1891 transit, the ingress contacts were not accurately recorded because '... at the critical moment the vibrations of the limbs of the sun and planet were very great' (Tebbutt 1891a, p. 25). Consequently,

> ... just at the time when I expected geometrical internal contact to take place the planet became somewhat pear-shaped, its limb being connected with that of the sun by a triangular black ligament, whose base was on the planet and its apex on the solar limb. The ligament suddenly broke at the same time, and for fully 14 seconds the vibrations were so great that the triangular ligament was repeatedly seen separated both from the disc of the planet and the limb of the sun. (Tebbutt 1891a, pp. 25–26).

Towards the end of the transit, the all too familiar white spot made its appearance, '... dancing about on the planet's disc' (Tebbutt 1891a, p. 27). By the time of the egress the seeing had improved, and Tebbutt obtained accurate contact timings. This time, there was no sign of the black ligament. Nor could he detect any evidence of a halo round the planet during the transit, or a satellite.

**Fig. 11.28** Simon Newcomb (https://en.wikipedia.org)

The black drop was not observed during the 1894 transit, and on this occasion Tebbutt (1895a) also noted the absence of a halo or a satellite.

Tebbutt's observations of the four transits must have confirmed, to his satisfaction, that Mercury lacked an atmosphere and a satellite, and with the benefit of hindsight we now know that he was correct on both counts. The faint whitish spots which he saw during three of the transits were noted by other observers, and although these are now thought to be due to an optical phenomenon (Antoniadi 1974, p. 23) they still have not been adequately explained.

In 1882, the noted Canadian-born American astronomer Simon Newcomb (1835–1909; Fig. 11.28; Archibald 1924) used data from transits of Mercury observed between 1677 and 1881 to investigate possible changes in the rotation period of the Earth (Newcomb 1882), and he included Tebbutt's 1881 ingress and egress times in this study (see Tebbutt 1882b). In a later study, he concluded that although the contact time discrepancies noted for transits between 1690 and 1894 '... give colour to the hypothesis of variability in the measure of time ... they are not large enough to prove it' (Newcomb 1903, pp. 317–318). More recently, Morrison and Ward (1975) used Tebbutt's observations in their analysis of transits of Mercury that occurred between 1677 and 1973.

### *11.3.9 Jovian Satellite Phenomena*

Another major field of observational activity undertaken by Tebbutt was Jovian satellite phenomena, in order to assist in an understanding of the precise orbits of the various satellites. During Tebbutt's own lifetime he was to see the number of known satellites rise from four (the so-called Galilean satellites) to nine (see Table 11.16). However, his focus was solely on the four Galilean satellites.

**Table 11.16** Jovian satellite discoveries to 1916 (after Moore 1995, p. 87)

| No. | Name | Discoverer | Year | Mean distance from Jupiter (km) |
|---|---|---|---|---|
| I | Io | Galileo/Marius | 1610 | 421,600 |
| II | Europa | Galileo/Marius | 1610 | 670,900 |
| III | Ganymede | Galileo/Marius | 1610 | 1,070,000 |
| IV | Callisto | Galileo/Marius | 1610 | 1,880,000 |
| V | Amalthea | Barnard | 1892 | 181,300 |
| VI | Himalia | Perrine | 1904 | 11,480,000 |
| VII | Elara | Perrine | 1905 | 11,737,000 |
| VIII | Pasiphae | Melotte | 1908 | 23,500,000 |
| IX | Sinope | Nicholson | 1914 | 23,700,000 |

**Fig. 11.29** D.P. Todd (https://en.wikipedia.org)

Tebbutt carried out his first 'familiarisation' observations on ten different evenings in April–May 1864 (Tebbutt 1864–69a), and he must have enjoyed the challenge for he began serious Jovian satellite observing in 1866. From that year until 1902 he recorded Jovian satellite phenomena every single year except in 1871, 1872, 1883, 1884, 1893, 1894 and 1901. His earliest observations, with the small Jones refractor, were solely of eclipses, but following the arrival of the larger Cooke telescope in 1872 he began observing a few transits. In 1877, he received some welcome encouragement from Dr. David Peck Todd (Fig. 11.29; Giclas 2014) of the U.S. Naval Observatory, who was conducting an analysis of Satellite IV:

> In collecting recently all the observations of eclipses of this satellite since 1840, I was much struck by their deficiency. I was very glad, however, to find myself putting down so large a number to the credit of your little observatory in New South Wales. I trust that you continue to observe the eclipses of the satellites. (Todd 1877).

The acquisition of the 8-in. Grubb telescope in 1886 allowed Tebbutt to regularly observe eclipses, transits *and* occultations, and his first view of Jupiter with this instrument made a deep impression:

> I pointed the telescope on Jupiter and Mars. The definition was excellent. Jupiter was a magnificent sight … I could observe the occultation and reappearance from eclipse of Jupiter's first satellite. I accordingly observed these two phenomena with much greater accuracy than I could possibly obtain with the 4½ inch equatorial. (Tebbutt 1884–86, 22 May 1886 entry).

Tebbutt's Jovian satellite observations from 1866 to 1902 are summarised in Table 11.17, but to the tallies listed there should be added four shadow transits which were timed between 1896 and 1898 (inclusive). Tebbutt's busiest years were 1888, 1887, 1896 and 1890. From 1887, approximately equal numbers of transits (152) and eclipses (157) were observed, at the expense of occultations (88). Similar numbers of transit ingresses and egresses were observed, but eclipse disappearances were much harder to record than reappearances. Conversely, Tebbutt observed far more occultation disappearances than reappearances.

Regardless of the type of phenomenon, between 1887 and 1902 inclusive, just over half (52 %) of Tebbutt's observations were of Satellite 1 (203 of the 389 total), with 88 observations of Satellite II (23 %), 83 of Satellite III (21 %) and only 15 of Satellite IV (4 %).

In all, Tebbutt published 33 papers on his Jovian satellite work, in *Monthly Notices of the Royal Astronomical Society* (18), *Astronomische Nachrichten* (10), *Publications of the Astronomical Society of the Pacific* (4) and *The Observatory* (1). Three papers were published in both *Monthly Notices* and *Astronomische Nachrichten*, which means that Tebbutt wrote 30 different papers on this work.

Most of these were clinical, descriptive papers, simply reporting on the timings themselves, but occasionally Tebbutt would be drawn by the majesty of scene to describe the planet itself. On 22 September 1879 he observed an eclipse of Satellite IV, and was moved to write:

> The physical appearance of the planet was exceedingly beautiful at the time of the reappearance. In addition to two long dull red equatorial belts there was a large oval pink patch south of the equator; it was edged with a narrow band of white of a brighter colour than the general ground of the disc … The sky was beautifully clear, and the definition unusually good. … (Tebbutt 1880b, pp. 37–38)

Tebbutt did not carry out any analyses of his Jovian satellite observations, although he did remark in 1879 that "It appears from the observations [made in 1878] that the errors of the theoretical eclipses of the fourth satellite were very large." (Tebbutt 1879b). Meanwhile, his transit observations of Satellite III carried out between 1866 and 1897 later were used by the British astronomer and Director of Durham Observatory, Professor Ralph Allen Sampson (1866–1939; Elliott 2014; Sampson 1910), while four years earlier the Utrecht University astronomer, Professor Jean Abraham Chrétien Oudemans (1827–1906; Fig. 11.30; Orchiston et al. 2017), had used Tebbutt's eclipse data from 1894 to 1899 in his *Investigation on the Mutual Occultations and Eclipses of the Satellites of Jupiter* (see Tebbutt 1908a).

### *11.3.10 Lunar Occultations of Stars*

Another of Tebbutt's observational interests was occultations of stars by the Moon. These events were used to establish accurate longitudes for the world's leading observatories. Along with lunar occultations of planets, they also were employed by

**Table 11.17** Jovian satellite phases observed by Tebbutt, 1866–1903[a]

| Year | Satellite | Satellite Phenomena | | | | Annual total |
|---|---|---|---|---|---|---|
| | | Transit | Occultation | Eclipse | Total | |
| 1866 | | | | | | 18 |
| | I | | | 7 | 7 | |
| | II | | | 6 | 6 | |
| | III | | | 5 | 5 | |
| 1867 | | | | | | 21 |
| | I | | | 9 | 9 | |
| | II | | | 4 | 4 | |
| | III | | | 4 | 4 | |
| | IV | | | 4 | 4 | |
| 1868 | | | | | | 5 |
| | I | | | 2 | 2 | |
| | II | | | 2 | 2 | |
| | III | | | 1 | 1 | |
| 1869 | | | | | | 2 |
| | I | | | 1 | 1 | |
| | II | | | 1 | 1 | |
| 1870 | | | | | | 3 |
| | I | | | 2 | 2 | |
| | III | | | 1 | 1 | |
| 1873 | | 1 | | | | 7 |
| | I | 1 | | | 1 | |
| | II | | | 1 | 1 | |
| | III | 3 | | | 3 | |
| | IV | | | 2 | 2 | |
| 1874 | | | | | | 11 |
| | I | | | 6 | 6 | |
| | II | | | 1 | 1 | |
| | III | 4 | | 4 | 4 | |
| 1875 | | | | | | 14 |
| | I | | | 5 | 5 | |
| | II | | | 4 | 4 | |
| | III | | | 5 | 5 | |
| 1876 | | | | | | 17 |
| | I | | | 6 | 6 | |
| | II | | | 5 | 5 | |
| | III | | | 6 | 6 | |

(continued)

**Table 11.17** (continued)

| Year | Satellite | Satellite Phenomena | | | | Annual total |
|---|---|---|---|---|---|---|
| | | Transit | Occultation | Eclipse | Total | |
| 1877 | | | | | | 8 |
| | I | | | 1 | 1 | |
| | II | | | 1 | 1 | |
| | III | | | 5 | 5 | |
| | IV | | | 1 | 1 | |
| 1878 | | | | | | 27 |
| | I | | | 13 | 13 | |
| | II | | | 4 | 4 | |
| | III | | | 6 | 6 | |
| | IV | | | 4 | 4 | |
| 1879 | | | | | | 19 |
| | I | | | 7 | 7 | |
| | II | | | 3 | 3 | |
| | III | | | 6 | 6 | |
| | IV | | | 3 | 3 | |
| 1880 | | | | | | 14 |
| | I | | | 6 | 6 | |
| | II | | | 3 | 3 | |
| | III | | | 4 | 4 | |
| | IV | | | 1 | 1 | |
| 1881 | | | | | | 3 |
| | I | | | 2 | 2 | |
| | II | | | 1 | 1 | |
| 1882 | | | | | | 4 |
| | I | | | 2 | 2 | |
| | III | | | 2 | 2 | |
| 1885 | | | | | | 15 |
| | I | | | 6 | 6 | |
| | II | | | 6 | 6 | |
| | III | | | 2 | 2 | |
| | IV | | | 1 | 1 | |
| 1886 | | | | | | 27 |
| | I | 1 | 1 | 9 | 11 | |
| | II | 1 | 1 | 5 | 7 | |
| | III | 1 | 1 | 6 | 8 | |
| | IV | | | 1 | 1 | |
| 1887 | | | | | | 46 |
| | I | 10 | 2 | 9 | 21 | |
| | II | 6 | | 6 | 12 | |
| | III | 2 | 5 | 6 | 13 | |

(continued)

**Table 11.17** (continued)

| Year | Satellite | Satellite Phenomena | | | | Annual total |
|---|---|---|---|---|---|---|
| | | Transit | Occultation | Eclipse | Total | |
| 1888 | | | | | | 75 |
| | I | 14 | 9 | 12 | 35 | |
| | II | 11 | 7 | 6 | 24 | |
| | III | 4 | 1 | 11 | 16 | |
| 1889 | | | | | | 26 |
| | I | 5 | 4 | 4 | 13 | |
| | II | 3 | 1 | 1 | 5 | |
| | III | 3 | | 3 | 6 | |
| | IV | 2 | | | 2 | |
| 1890 | | | | | | 41 |
| | I | 8 | 5 | 8 | 21 | |
| | II | 2 | 2 | 4 | 8 | |
| | III | 2 | | 5 | 7 | |
| | IV | 2 | 1 | 2 | 5 | |
| 1891 | | | | | | 31 |
| | I | 7 | 4 | 4 | 15 | |
| | II | 5 | 1 | 4 | 10 | |
| | III | 4 | 1 | | 5 | |
| | IV | | | 1 | 1 | |
| 1892 | | | | | | 14 |
| | I | 3 | | 3 | 6 | |
| | II | 1 | 1 | 2 | 4 | |
| | III | 2 | 2 | | 4 | |
| 1895 | | | | | | 24 |
| | I | 5 | 2 | 7 | 14 | |
| | II | | 1 | 2 | 3 | |
| | III | 1 | 1 | 3 | 5 | |
| | IV | | 1 | 1 | 2 | |
| 1896 | | | | | | 43 |
| | I | 9 | 7 | 8 | 24 | |
| | II | 3 | 2 | 3 | 8 | |
| | III | 2 | 4 | 3 | 9 | |
| | IV | | | 2 | 2 | |
| 1897 | | | | | | 32 |
| | I | 7 | 5 | 6 | 18 | |
| | II | 2 | 3 | 2 | 7 | |
| | III | 1 | 2 | 2 | 5 | |
| | IV | | 1 | 1 | 2 | |

(continued)

**Table 11.17** (continued)

| Year | Satellite | Satellite Phenomena | | | | Annual total |
|---|---|---|---|---|---|---|
| | | Transit | Occultation | Eclipse | Total | |
| 1898 | | | | | | 11 |
| | I | 2 | 1 | 2 | 5 | |
| | II | 3 | | | 3 | |
| | III | 1 | | 2 | 3 | |
| 1899 | | | | | | 23 |
| | I | 3 | 5 | 4 | 12 | |
| | II | 2 | 2 | 3 | 7 | |
| | III | 4 | | | 4 | |
| 1900 | | | | | | 8 |
| | I | 1 | 1 | 3 | 5 | |
| | II | 1 | | 1 | 2 | |
| | III | 1 | | | 1 | |
| 1902 | | | | | | 23 |
| | I | 4 | 3 | 4 | 11 | |
| | II | 4 | | 3 | 7 | |
| | III | 1 | | 3 | 4 | |
| | IV | | 1 | | 1 | |
| Totals | | 163 | 91 | 358 | | 612 |

[a]Data based on published papers

Tra = Transits (ingress + egress), Occ = Occultations (disappearances + reappearances), Ecl = Eclipses (disappearances + reappearances), Ann Total = Annual totals

**Fig. 11.30** J.A.C. Oudemans (https://en.wikipedia.org)

**Table 11.18** Lunar occultation phases observed by Tebbutt, 1864–1902[a]

| Year | Phases | Year | Phases | Year | Phases | Year | Phases |
|---|---|---|---|---|---|---|---|
| 1864 | 2 | 1876 | 14 | 1888 | 56 | 1897 | 134 |
| 1866 | 8 | 1877 | 9 | 1889 | 34 | 1898 | 36 |
| 1867 | 2 | 1878 | 5 | 1890 | 27 | 1899 | 78 |
| 1868 | 32 | 1880 | 8 | 1891 | 56 | 1900 | 26 |
| 1869 | 12 | 1881 | 5 | 1892 | 45 | 1901 | 3 |
| 1870 | 22 | 1882 | 10 | 1893 | 45 | 1902 | 2 |
| 1873 | 38 | 1885 | 54 | 1894 | 48 | | |
| 1874 | 17 | 1886 | 43 | 1895 | 65 | | |
| 1875 | 12 | 1887 | 24 | 1896 | 161 | | |

[a]Based on data in Tebbutt (1908a) and published papers

professional astronomers to monitor irregularities in the Moon's orbit. In this regard, Simon Newcomb, wrote:

> Should the question be asked, What is to-day the most important unresolved problem growing out of celestial motions? a survey of the field could, it seems to me, lead to but one reply. It is that of the discrepancies between the observed mean motion of the Moon and the most exhaustive investigations of the theory of that motion. Twenty-five years have now passed since the publication of my "Researches on the Motion of the Moon," and during that time little of a comprehensive and definitive character has been added to the solution of the problem. (Newcomb 1903, p. 316).

In 1863 August, Tebbutt observed three different lunar occultations (Tebbutt 1863), and after the completion of his first observatory building these events became a regular part of his observing programme (see Table 11.18). Between 1864 and 1903 (inclusive), he observed lunar occultations every single year except in 1865, 1871, 1872, 1879, 1883, 1884 and 1903, initially with the Jones and Cooke refractors, and from 1886 generally with the Grubb refractor.

Tebbutt referred to each individual disappearance or reappearance of a star as a 'phase', and remarked in his 1900 *Annual Report* (Tebbutt 1901a, p. 10) that a total of 1,127 phases had been observed at Windsor Observatory since systematic lunar occultation work began in 1864. From the figures in Tebbutt's published papers it is apparent that about 75 % of all lunar occultations were observed with the Grubb refractor. As can be seen from Table 11.18 the two successive years 1896 and 1897 stand out in particular, with total annual phases of 161 and 134 respectively. The highest annual totals in the preceding period, 1864–1895, are only 65 (in 1895), 56 (in 1891 and 1888) and 54 (in 1885). After a 'high' of 78 phases in 1899 the totals fall off very rapidly, and by 1902 Tebbutt was recording only "a few" lunar occultations. (Tebbutt 1903a, p. 11)

Tebbutt published 32 papers documenting his lunar occultation measurements. Most of his results appeared in a succession of predominantly annual papers in *Astronomische Nachrichten* (20 papers), but other papers were in *Monthly Notices of the Royal Astronomical Society* and *The Observatory* (4 each), *Journal of the*

*British Astronomical Association* (3) and the *Astronomical Register* (1). Only 3 of these (all in *Monthly Notices*) were duplicates.

Tebbutt was pleased to see his lunar occultation observations put to good use by two German scholars during his lifetime. In 1884, Professor Auwers (Fig. 8.26) from Berlin investigated a new fundamental meridian for Australia and in the process used Tebbutt's 1873–1876 observations to derive an improved longitude for Windsor Observatory. In this paper, Auwers praised Tebbutt's work:

> These observations ... prove the great care with which Mr. Tebbutt has worked. I cannot omit to point out that in my experience in this branch of research there is no second series of observations approaching at all in extent to his... (cited in Tebbutt 1908a, pp. 42–43).

Later, in 1896, Dr. Hugo Clemens from the University of Gottingen employed Tebbutt's 1864–1870 occultation data as the basis for his doctoral thesis on the longitude of Windsor Observatory. Other attempts to refine the value of the longitude by means of time-signal exchanges and triangulation (surveying) are discussed in Chap. 8.

### *11.3.11 Lunar Occultations of Planets*

In addition to 'conventional' lunar occultations of stars (see earlier), Tebbutt also observed eight different lunar occultations of planets between 1875 and 1906, the great majority after his acquisition of the Grubb telescope. As Table 11.19 indicates, only four planets (Venus, Mars, Jupiter and Saturn) were involved. Where possible, Tebbutt noted the disappearance and reappearance times of the planets themselves, and in the case of Jupiter and Saturn the principal satellites.

In a note published on the 1875 occultation of Mars, which was observed in full daylight, Tebbutt (1875b) specifically mentions the unique nature of this particular observation: 'I may mention that, although the records of the Observatory for the past ten years contain a long list of lunar occultations of fixed stars, this is the only instance of an observed occultation of a planet'.

**Table 11.19** Occultations of planets by the Moon observed by Tebbutt, 1875–1906[a]

| Year | Date | Planet |
|---|---|---|
| 1875 | August 12 | Mars |
| 1886 | April 16 | Jupiter |
| 1888 | March 9 | Venus |
| | June 13 | Saturn |
| 1893 | May 25 | Saturn |
| 1894 | May 1 | Venus |
| 1900 | September 29 | Jupiter |
| 1906 | October 27 | Saturn |

[a]Based on Tebbutt (1887a, 1889b, 1894b, 1895b, 1901a, 1908a) and published papers

Sometimes Tebbutt observed unusual optical effects associated with these occultations. While observing the 1893 event he noted that

> … for a brief space of time, somewhat less than a second, after the total disappearance … of the ring of Saturn … a faint copper-coloured streak of light was visible on the moon's dark limb: It was perpendicular to the major axis of the ring and lay along the moon's limb, which, however, was not itself visible in the telescope, unless so far as this faint luminosity rendered it visible. The streak appeared to be about the semidiameter of the ball [disk of the planet] in length. (Tebbutt 1892–1894).

Tebbutt (1894a) attributed this streak of light to '… the existence of a low-lying lunar atmosphere, or of a faint luminous atmosphere round the planet itself'. The following year, he noted a similar phenomenon during the occultation of Venus: 'After the brilliant light of the planet had vanished, a faint silvery streak was left on the limb which rapidly shortened till the final speck of it disappeared…' (Tebbutt 1892–1894)

Among the most captivating of all astronomical events, these occultations of planets were documented by Tebbutt in eight different papers which were published in *Monthly Notices of the Royal Astronomical Society* (5), *Astronomische Nachrichten* (2) and *Publications of the Astronomical Society of the Pacific* (1).

## 11.4 Concluding Remarks

There is little doubt that Tebbutt made extensive use of the 8-in. telescope, but what is equally apparent is that this instrument caused little change to the basic thrust of his observational strategy. Table 1.1 presents the types of observations that he carried out between 1864 and 1905 (inclusive), and there was no wholesale change following acquisition of the Grubb telescope.

Rather, there is a shift of emphasis. The access to a drive, the fainter limiting magnitude and greater resolving power of the new telescope meant that if he wished to do so Tebbutt could now observe fainter comets, variable stars and minor planets; more lunar occultations and Jovian satellite phenomena; and closer double stars. In fact, he chose not to alter his lunar occultation programme (there were adequate stars for his purpose, without resorting to fainter magnitudes) or to expand his variable star repertoire. Instead he decided to expand his comet, double star, Jovian satellite phenomena and minor planet programmes. But even with just these, the time required for the reduction and writing up of observations became an issue, and in his 1892 *Annual Report* he alludes to this: 'Many suggestions have been received from eminent astronomers as to work desirable to be done, but the author regrets that he is unable to comply with them' (Tebbutt 1893a, p. 21). As we shall see in Chap. 14, by this time overwork and illness were already beginning to take their toll.

Finally, it is interesting to note that after acquiring the Grubb telescope Tebbutt did not embark on any spectroscopic studies or on any systematic search programmes. At the time, the quest for new southern double stars and variable stars was still in its infancy and would have proved lucrative, but Tebbutt was captivated by celestial mechanics. The only search programme that he ever engaged in dates to 1869 and the small Jones telescope. On 22 July he noted in his journal: 'I commenced a search after red stars in the Southern Hemisphere … [for] the formation of a Catalogue of such objects' (Tebbutt 1864–69a). That evening he found 4 stars, followed by 8 more on 26 July and 26 on the evening of 3/4 August. Apparently this was the last date on which he searched for red stars (most of which are variable), and what could have proved a valuable contribution to southern variable star astronomy came to an abrupt and premature end.

**Acknowledgements** I am grateful to Dr. David Frew (University of Hong Kong), Martin George (National Astronomical Research Institute of Thailand and University of Southern Queensland), the late Peter Hingley (former Royal Astronomical Society Librarian), Professor Nick Lomb (University of Southern Queensland, Toowoomba), the late Dr. Brian Marsden (formerly Minor Planet Center, Smithsonian Astrophysical Observatory), the late John Perdrix (Perth) and staff at the Mitchell Library (Sydney) for their assistance in various ways. I also wish to thank John Drummond (Patutahi, New Zealand), Martin George, the late Peter Hingley and the Max-Planck-Gesellschaft (Germany) for kindly supplying Figs. 11.1, 11.9, 11.17, 11.24 and 11.26.

## References[1]

Aitken, R.B. (1918). *The Binary Stars*. New York: McMurtrie.

Alexander, A.F.O'D. (1965). *The Planet Uranus. A History of Observation, Theory and Discovery*. London: Faber & Faber.

Andropoulos, J. & Orchiston, W. (2006). Melbourne Observatory and the genesis of astrophysics in Australia. *Anglo-Australian Observatory Newsletter, 109,* 18–20.

Antoniadi, E.M., 1974. *The Planet Mercury*. Shaldon: Reid.

Archibald, R.C. (1924). Simon Newcomb 1839–1909. Biography of his life and work. *Memoirs of the National Academy of Sciences, 17,* 19–69.

Ashbrook, J. (1984). The sage of Mare Island [T.J.J. See]. In J. Ashbrook, *The Astronomical Scrapbook. Skywatchers, Pioneers, and Seekers in Astronomy* (pp. 111–115). Cambridge (Mass.): Sky Publishing Corporation.

Baracchi, P. (1889). Spectra of southern stars observed at the Melbourne Observatory with the McLean Direct-vision Spectroscope attached to the South Equatorial. *Monthly Notices of the Royal Astronomical Society, 50,* 66–71.

Baracchi, P. (1914). Astronomy and geodesy in Australia. In G.H. Knibbs (Ed.), *Federal Handbook Prepared in Connection with the Eighty-fourth Meeting of the British Association for the Advancement of Science, Held in Australia, August, 1914* (pp. 326–390). Melbourne: Government Printer.

Bateson, F.M. (1958). *The Observation of Variable Stars*. Printed for the author.

Baum, R. (2014). Cassini, Jean-Dominique. In Hockey et al., 381–382.

Bergman. G.F.J. (1960). Christian Carl Ludwig Rümker (1788–1862), Australia's first Government Astronomer. *Royal Australian Historical Society Journal and Proceedings, 46,* 547–589.

Bhathal, R. (2012). Some scientific aspects of Parramatta Observatory. *Journal and Proceedings of the Royal Society of New South Wales, 145,* 111–127.

[1]The following abbreviation is used: TL = Letters to J. Tebbutt, bound manuscript letters in the Mitchell Library, Sydney.

Bhathal, R., Sutherland. R. & Butcher, H. (2013). *Mt Stromlo Observatory: From Bush Observatory to the Nobel Prize*. Melbourne: CSIRO Publishing.

Biggs, A.B. (1887). α Centauri, with a graphic projection of its orbit from its apparent curve. *Papers and Proceedings of the Royal Society of Tasmania,* 79–82.

Blainey, G. (1975). *The Tyranny of Distance. How Distance Shaped Australia's History*. Melbourne: Macmillan

Bone, W. (1882a). Letter to John Tebbutt, dated 5 May. In TL.

Bone, W. (1882b). Letter to John Tebbutt, dated 13 June. In TL

Bone, W. (1883). Letter to John Tebbutt, dated 14 November. In TL.

Bowell, E., Gehrels, T. & Zellner, B. (1979). Magnitudes, colours, types and adopted diameters of the asteroids. In Gehrels, 1108–1129.

Buckley, R.J. (1979). New orbits for some nineteenth century comets. *Journal of the British Astronomical Association, 89,* 256–264.

Butterfield, G. (1884). Letter to John Tebbutt, dated 21 February. In TL.

Cellino, A., Pannunzio, R., Zappalà, V., Farinella, P. & and Paolicchi, P. (1985). Do we observe light curves of binary asteroids? *Astronomy and Astrophysics, 144,* 355–362.

Challis, J. (1879). *Lectures on Practical Astronomy and Astronomical Instruments*. Cambridge: Deighton, Bell and Co.

Chambers, G.F. (1909). *The Story of Comets Simply Told for General Readers*. Oxford: Clarendon Press.

Charbonneau, P. (2014). Wolf, Johann Rudolf. In Hockey et al., 2366–2368.

Clerke, A. (1893). *A Popular History of Astronomy during the Nineteenth Century*. London: Adam and Charles Black.

Comstock, G.C. (1924). Benjamin Apthorp Gould. *Memoirs of the National Academy of Sciences, 17,* 155–180.

Conjunction of Mars and Saturn. *Sydney Morning Herald,* 2 July 1879.

Cozens, G. & White, G.L. (2001). James Dunlop: Messier of the southern sky. *Sky and Telescope, June,* 112–116.

Cozens, G., Walsh, A. & Orchiston, W. (2010). James Dunlop's historical catalogue of southern nebulae and clusters. *Journal of Astronomical History and Heritage, 13,* 59–73.

Crommelin, A.C.D., 1932. Comet notes. *Journal of the British Astronomical Association, 42,* 155–156.

Crossley, J., Gledhill, J. & Wilson, J.M. (1879). *Handbook of Double Stars*. London: Macmillan.

Cunningham, C.J. (2014). *The First Four Asteroids: A History of their Impact on English Astronomy in the Nineteenth Century*. Ph.D. Thesis, University of Southern Queensland, Toowoomba, Australia.

Cunningham, C.J. (2016). *Discovery of the First Asteroid, Ceres. Historical Studies in Asteroid Research*. New York: Springer.

Cunningham, C.J., Marsden, B.G. & Orchiston, W. (2009). How the first dwarf planet became the asteroid Ceres. *Journal of Astronomical History and Heritage, 12,* 240–248.

Cunningham, C.J., Marsden, B.G. & Orchiston, W. (2011). Giuseppe Piazzi: the controversial discovery and loss of Ceres in 1801. *Journal for the History of Astronomy, 42,* 283–306.

Descamps, P., et al. (2008). New determination of the size and bulk density of the binary asteroid 22 Kalliope from observations of mutual eclipses. *Icarus, 196*, 578–600.

Dick, S.J. (2013). *Discovery and Classification in Astronomy: Controversy and Consensus*. New York: Cambridge University Press.

Dumont, S. (2014). Lalande, Joseph-Jérôme. In Hockey et al., 1264–1266.

Dunlop, J. (1828). A catalogue of nebulae and clusters of stars in the southern hemisphere. *Philosophical Transactions of the Royal Society, 118,* 113–151.

Ellery, R.J. (1886). Letter to John Tebbutt, dated 25 March. In TL.

Ellery, R.L.J. (1889). Preliminary spectroscopic survey of southern stars, made at Melbourne Observatory with a Maclean direct-vision spectroscope on the 8-inch equatorial. *Monthly Notices of the Royal Astronomical Society, 49,* 439–445.

Ellery, R.J. (1901). A brief history of the beginnings and growth of astronomy in Australasia. In *Report of the Eighth Meeting of the Australasian Association for the Advancement of Science* (pp. 1–17). Melbourne: AAAS.

Elliott, I. (2014). Gore, John Ellard. In Hockey et al., 830–832.

Elliot, J.L., Dunham, E. & Mink, D. (1977). The rings of Uranus. *Nature, 267,* 328–330.

Fisher, W.J. (1924). The brightness of lunar eclipses 1860–1922. *Smithsonian Miscellaneous Collections, 76* (9), 1–61.

FitzGerald, A.P. (1966). John Ellard Gore (1845–1910). *Irish Astronomical Journal, 7,* 213–219.

Frame, T. & Faulkner, D. (2003). *Stromlo: An Australian Observatory*. Sydney: Allen & Unwin.

Frew, D. (2004). The historical record of η Carinae. I. The visual light curve 1595–2000. *Journal of Astronomical Data, 10,* 6–76.

Frew, D. & Orchiston, W. (2003). Nineteenth century Australian observations of η Carinae: international controversy and astrophysical implications. Paper presented at the IAU General Assembly, Sydney.

Furness, C.E. (1915). *An Introduction to the Study of Variable Stars*. Boston: Houghton Mifflin.

Gehrels, T. (Ed.) (1979). *Asteroids*. Tucson: University of Arizona Press.

Giclas, H.L. (2014). Todd, David Peck. In Hockey et al, 2163.

Glass, I. (2013). *Nicolas-Louis de La Caille, Astronomer and Geodesist*. Oxford: Oxford University Press.

Gore, J.E. (1887). On the orbit of p Eridani. *Monthly Notices of the Royal Astronomical Society, 48,* 26–27.

Gore, J.E. (1892a). On the orbit of γ Centauri. *Monthly Notices of the Royal Astronomical Society, 52,* 505–506.

Gore, J.E. (1892b). On the orbit of γ Coronae Australis. *Monthly Notices of the Royal Astronomical Society, 52,* 503–504.

Gould, B.A. (1879). *Resultados del Observatorio Nacional Argentino en Cordoba. Volume 1: Uranometria Argentina.* Buenos Aires: Argentine National Observatory.

Grosser, M. (1979). *The Discovery of Neptune*. New York: Dover Publications.

*Harper's Encyclopedia of United States History, Volume IV*. New York: Harper & Brothers (1905).

Haynes, R., Haynes, R., Malin, D. & McGee, R. (1996). *Explorers of the Southern Sky. A History of Australian Astronomy*. Cambridge: Cambridge University Press.

Herschel, Sir John F.W. (1847). *Results of Astronomical Observations Made During the Years 1834, 5, 6, 7, 8 at the Cape of Good Hope*. London: Smith, Elder.

Hirst, G.D. (1876). Some notes on Jupiter during his opposition of 1876. *Journal and Proceedings of the Royal Society of New South Wales, 10,* 83–98.

Hirst, G.D. (1878). Notes on Jupiter during his opposition, 1878. *Journal and Proceedings of the Royal Society of New South Wales, 12,* 238–240.

Hirst, G.D. (1879–80). An evening with Jupiter. *The Australian, 3,* 69–77.

Hirst, G.D. (1880). Remarks on the colours of Jupiter's belts, and some changes observed thereon during the opposition of 1880. *Journal and Proceedings of the Royal Society of New South Wales, 14,* 77–79.

Hockey, T.A. (1992). Seeing red: observations of colours in Jupiter's Equatorial Zone on the eve of the modern discovery of the Great Red Spot. *Journal for the History of Astronomy, 23,* 93–105.

Hockey, T.A. (1999). *Galileo's Planet. Observing Jupiter before Photography*. Bristol: Institute of Physics.

Hockey, T. & MacPherson, H. (2014). Fisher, Willard James. In Hockey et al., 721.

Hockey, T. et al. (Eds.) 2014. *Biographical Encyclopedia of Astronomers. 2nd Edition*. New York: Springer.

Hodge, J.E. (1971). Benjamin Apthorp Gould and the founding of the Argentine National Observatory. *The Americas, 28,* 152–175.

Hogg, H.S. (1984). Variable stars. In O. Gingerich (Ed.), *The General History of Astronomy. Volume 4. Astrophysics and Twentieth Century Astronomy to 1950: Part A* (pp.73–89). Cambridge: Cambridge University Press.

Humphreys, R.M. & Davidson, K. (1994). The Luminous Blue Variables: astrophysical geysers. *Publications of the Astronomical Society of the Pacific, 106,* 1025–1051.

Innes, R.T.A. (1899). Reference catalogue of southern double stars. *Annals of the Cape Observatory, 2* (2).

*Knowledge*, 1 November 1887.

Lomb, N. (2004). The instruments from the Parramatta Observatory. *Historical Records of Australian Science, 15,* 211–222.

Loomis, E. (1869). On the period of η Argus. *Monthly Notices of the Royal Astronomical Society, 29,* 298–299.

Love, R. (1984). Science and government in Australia 1905–15: Geoffrey Duffield and the foundation of the Commonwealth Solar Observatory. *Historical Records of Australian Science, 6,* 171–188.

MacKeown, P.K. (2007). William Doberck – double star astronomer. *Journal of Astronomical History and Heritage, 10,* 49–64.

Marsden, B.G. & Sekanina, Z. (1971). Comets and nongravitational forces. IV. *The Astronomical Journal, 76,* 1135–1151.

Marsden, B.G. & Sekanina, Z. (1974). Comets and nongravitational forces. VI. Periodic Comet Encke 1786–1971. *The Astronomical Journal, 79,* 413–419.

Marsden, B.G. & Williams, G.V. (1996). *Catalogue of Cometary Orbits 1996*. Cambridge (Mass.): Minor Planet Center.

Marsden, B.G., Sekanina, Z. & Everhart, E. (1978). New osculating orbits for 110 comets and analysis of original orbits for 200 comets. *The Astronomical Journal, 83,* 64–71.

Marsden, B.G., Williams, G.V., Kronk, G.W. & Waddington, W. (1993). Update on Comet Swift-Tuttle. *Icarus, 105,* 420–426.

Meo, M. (2014). Encke, Johann Franz. In Hockey, et al., 661–662.

Merfield, C.J. (1897). Uranus. Conjunction with the star 41 Librae. *Journal of the British Astronomical Association, 8,* 87–88.

Merfield, C.J. (1901). Letter to John Tebbutt, dated 18 May. In TL.

Merfield, C.J. (1915). Letter to John Tebbutt, dated 8 October. In TL.

Merrifield, G. (1886). Letter to John Tebbutt, dated 5 June. In TL.

Millis, R.L., Wasserman, L.H. & Birch, P.V. (1977). Detection of rings around Uranus. *Nature, 267,* 330–331.

Millis, R.L., et al. (1993). Pluto's radius and atmosphere: results from the entire 9 June 1988 occultation data set. *Icarus, 105,* 282–297.

Mitchell, S.A. (1932). *Eclipses of the Sun*. New York: Columbia University Press.

Mohr, P. (2004). A star in the western sky: John Birmingham, astronomer and poet. *The Antiquarian Astronomer, 1,* 24–33.

Moore, P. (1995). *The Guinness Book of Astronomy*. Enfield: Guinness.

Morrison, L.V. & Ward, C.G. (1975). An analysis of the transits of Mercury: 1677–1973. *Monthly Notices of the Royal Astronomical Society, 173,* 183–206.

Mr. C.J. Merfield. *Nature, 129,* 228 (1932).

Newcomb, S. (1882). Discussion and results of observations on transits of Mercury, from 1677 to 1881. *Astronomical Papers Prepared for the Use of the American Ephemeris and Nautical Almanac, 1,* 363–487.

Newcomb, S. (1903). On the desirableness of a re-investigation of the problems growing out of the mean motion of the Moon. *Monthly Notices of the Royal Astronomical Society, 63,* 316–324.

Newton, H.A. (1890). *Memoir of Elias Loomis 1811–1889*. Washington: National Academy of Sciences (pp. 213–252).

Obituaries. Charles James Merfield. *Monthly Notices of the Royal Astronomical Society, 92,* 258–260 (1932).

Orchiston, W. (1982). John Tebbutt and the Whakatane Eight Inch Refractor: a review of "The Australian Connection". *Southern Stars, 29,* 215–244.

Orchiston, W. (1985a). The contribution of Alfred Barrett Biggs to Australian and Tasmanian astronomy. *Records of the Queen Victoria Museum, 89,* 1–28.

Orchiston, W. (1985b). Life-history of a telescope: the 17.8-cm refractor at the Riverview Observatory in Sydney. *Australian Journal of Astronomy, 1,* 57–80.

Orchiston, W. (1986). *Observatories of Victoria. 1: Dr William Bone and the Castlemaine Observatory*. Melbourne, Astronomical Society of Victoria (Historical Notes No.3).

Orchiston, W. (1987a). Contribution of the Lands Department to the development of astronomy in New South Wales during the nineteenth century. *Australian Journal of Astronomy, 2,* 65–74.

Orchiston, W. (1987b). Dr William Bone, and the role of the amateur observatory in Australian astronomy. *Southern Stars, 32,* 111–128.

Orchiston, W. (1988a). Illuminating incidents in antipodean astronomy: formation of the New South Wales Branch of the British Astronomical Association. *Journal of the British Astronomical Association, 98,* 75–84.

Orchiston, W. (1988b). Illuminating incidents in Antipodean astronomy: Phillip Parker King and the founding of Sydney Observatory. *Vistas in Astronomy, 32,* 285–301.

Orchiston, W. (1999). Comets and communication: amateur-professional tension in Australian astronomy. *Publications of the Astronomical Society of Australia*, 16, 212–221.

Orchiston, W. (2000). John Tebbutt of Windsor, New South Wales: a pioneer southern hemisphere variable star observer. *Irish Astronomical Journal, 27,* 47–54.

Orchiston, W. (2002). From Crossley to Carter: the life and times of an historic Cooke refractor. *Journal of the antique Telescope Society, 24,* 9–24.

Orchiston, W. (2004). John Tebbutt and observational astronomy at Windsor Observatory. *Journal of the British Astronomical Association, 114,* 141–154.

Orchiston, W. (2015). The amateur-turned-professional (ATP) syndrome: two Australian case studies. In W. Orchiston, D. Green & R. Strom (Eds.), *New Insights from Recent Studies in Historical Astronomy: Following in the Footsteps of F. Richard Stephenson* (pp. 259–350). Cham: Springer.

Orchiston, W. (2016*). Exploring the History of New Zealand Astronomy: Trials, Tribulations, Telescopes and Transits*. Cham: Springer.

Orchiston, W., Steinicke, W. & Mumpuni, I. (2017). J.A.C. Oudemans and nineteenth century astronomy in the Dutch East Indies. In Orchiston, W. & Vahia, M. (eds.). *Researching the History of Astronomy in Southeast Asia. A Review of Current Projects and Future Prospects and Possibilities*. Cham: Springer.

Peek, B.M. (1958). *The Planet Jupiter*. London: Faber & Faber.

Perdrix, J.L. (1972). The Astronomical Society of Victoria: the beginning. *Journal of the Astronomical Society of Victoria, 25,* 30–36.

Perdrix, J.L. (1979). Baracchi, Pietro Paolo Giovanni Ernesto (1851–1926), astronomer. In B. Nairn & G. Serle (Eds.), *Australian Dictionary of Biography, Volume 7* (pp. 166–167). Melbourne: Melbourne University Press.

Peterson, C.J. (2004). The education of an astronomical maverick: T.J.J. See and the University of Missouri. *Journal for the History of Astronomy*, 35, 293–304.

Pilcher, F. (1979). Circumstances of minor planet discovery. In Gehrels, 1130–1154.

Polcaro, V.F. & Viotti, R. (1993). A forgotten episode of the η Carinae light curve in 1860–1865. *Astronomy and Astrophysics, 274,* 807–810.

Powell, E.B. (1886). On the orbit of α Centauri. *Monthly Notices of the Royal Astronomical Society, 46,* 336–340.

Powell, E.B. (1890). γ Coronae Australis. *Monthly Notices of the Royal Astronomical Society, 50,* 299–301.

Powell, E.B. (1892). Note on the orbit of α Centauri. *Monthly Notices of the Royal Astronomical Society, 52,* 506–508

Report of the Branches. New South Wales Branch. *Journal of the British Astronomical Association, 11,* 20 (1900).

Roberts, A.W. (1901). Secular variation in the period of R Carinae. *Monthly Notices of the Royal Astronomical Society, 61,* 552–556.

Robertson, P., Cozens, G., Orchiston, W. & Slee, B. (2010). Early Australian optical and radio observations of Centaurus A. *Publications of the Astronomical Society of Australia, 27,* 402–430.

Roseby, T. (1901). Letter to John Tebbutt, dated 3 May. In TL.

Roser, S. & White, G.L. (1986). John Tebbutt's observations of Halley's Comet in 1910. *Proceedings of the Astronomical Society of Australia, 6,* 354–358.

Roth, G.D. (1962). *The System of Minor Planets*. London: Faber & Faber.

Rumker, C. (1826). Observations made at Stargard, New South Wales, of the comet, which appeared in the constellation Leo, in July and August 1824. *Memoirs of the Royal Astronomical Society, 2,* 284.

Rumker, C. (1829). Astronomical observations made at the Observatory at Paramatta [*sic*]. *Philosophical Transactions of the Royal Society of London, 119,* i–vi, 1–152.

Ruskin, S. (2014). Herschel, Caroline Lucretia. In Hockey et al., 947–949.

Russell, H.C. (1888). On the increasing magnitude of Eta Argus. *Journal and Proceedings of the Royal Society of New South Wales, 22,* 76–78.

Rutledge, S. (2009). For further consideration: Mr James Dunlop, Esq. *Journal and Proceedings of the Royal Society of New South Wales, 142,* 17–28.

Sampson, R.A. (1910). On the old observations of the eclipses of Jupiter's satellites. *Memoirs of the Royal Astronomical Society, 59,* 199–256.

Saunders, S.D. (1990*). Astronomy in Colonials New South Wales: 1788 to 1858*. Ph.D. thesis, Department of General Philosophy, University of Sydney, Sydney, Australia.

Saunders, S. (2004). Sir Thomas Brisbane's legacy to colonial science: colonial astronomy at Parramatta Observatory, 1822–1848. *Historical Records of Australian Science, 15,* 177–209.

Schmidt, J.F. (1882). Jupiter. *Astronomische Nachrichten, 101,* 147–160.

Scott, W. (1860). Letter to John Tebbutt, dated 18 April. In TL.

See, T.J.J. (1893). On the orbit of α Centauri. *Monthly Notices of the Royal Astronomical Society, 54,* 102–117.

See, T.J.J. (1896). *Researches on the Evolution of Stellar Systems. Volume 1*. Lynn: Nichols Press.

Sekanina, Z. (1981). Distribution and activity of discrete emission areas on the nucleus of periodic comet Swift-Tuttle. *The Astronomical Journal, 86,* 1741–1773.

Sekanina, Z. (1982). The problem of split comets in review. In L.L. Wilkening (Ed.), *Comets* (pp. 251–287). Tucson: University of Arizona Press.

Sekanina, Z. (1997). The problem of split comets revisited. *Astronomy and Astrophysics, 318,* L5–L8.

Sellors, R.P. (1892). The orbit of γ Coronae Australis. *Monthly Notices of the Royal Astronomical Society, 53,* 45–47.

Service, J. (1890). *Thir Notandums, Being the Literary Recreations of Laird Canticarl of Mongrynen (of Kittle Memory) to which is Appended a Biographical Sketch of James Dunlop* (pp. 129–222). Sydney: Angus & Robertson.

Sheehan, W. (2014). Challis, James. In Hockey et al., 397–398.

Smith, N. & and Frew, D (2011). A revised historical light curve of Eta Carinae and the timing of close periastron encounters. *Monthly Notices of the Royal Astronomical Society, 415,* 2009–2019.

Smith, R. (2011). Pietro Baracchi and his sojourn in Port Darwin. *Journal of Northern Territory History, 22,* 89–96.

Snedegar, K. (2015). *Lost in the Stars. AW Roberts at the Crossroads of Mission, Science, and Race in South Africa 1883–1938*. Lanhan: Lexington Books.

Solc, M. (2014). Olbers, Heinrich Wilhelm Matthias. In Hockey et al., 1606–1608.

Soulsby, B. (1992). An improved oblateness for the Earth's thermosphere from lunar eclipse observations. *Proceedings of the Astronomical Society of Australia, 10,* 131–133.

Strauss, D. (2014). Lowell, Percival. In Hockey et al., 1351–1354.

Tebbutt, J. (1862–1865). On the desirability of a systematic search for, and observation of variable stars in the Southern Hemisphere. *Transactions of the Philosophical Society of New South Wales,* 126–139.

Tebbutt, J. (1863). Diary. MS, Mitchell Library, Sydney (A3786).

Tebbutt, J. (1864–69a). *Journal. Observatory, Peninsula, Windsor, New South Wales.* MS, Mitchell Library, Sydney (A3747).

Tebbutt, J. (1864–69b). Untitled Journal. MS, Mitchell Library, Sydney (A3784).

Tebbutt, J. (1867). On the progress and present state of astronomical science in New South Wales. In *Catalogue of the Natural and Industrial Products of New South Wales. Forwarded to the Paris Universal Exhibition of 1867, by the New South Wales Exhibition Commissioners* (pp. 55–64). Sydney: Government Printer, Appendix.

Tebbutt, J. (1868). On the variability of η Argus. *Monthly Notices of the Royal Astronomical Society, 28,* 266–268.

Tebbutt, J. (1869). On the period of η Argus. *Monthly Notices of the Royal Astronomical Society, 29,* 296–298.

Tebbutt, J. (1871). On η Argus. *Monthly Notices of the Royal Astronomical Society, 31,* 210–211.

Tebbutt, J. (1873–79). Untitled journal. MS, Mitchell Library, Sydney (A3783).

Tebbutt, J. (1875a). Mr. Birmingham's suspected new star. *Astronomical Register, 13,* 223.

Tebbutt, J. (1875b). Occultation of Mars, observed at Windsor, N.S.W. *Monthly Notices of the Royal Astronomical Society, 36,* 43.

Tebbutt, J. (1875c). A rare phenomenon. *Sydney Morning Herald,* 30 June.

Tebbutt, J. (1877a). Discovery of a remarkable variable star. *Sydney Morning Herald,* 19 November.

Tebbutt, J. (1877b). On the magnitude of η Argus. *Monthly Notices of the Royal Astronomical Society, 37,* 280–281.

Tebbutt, J. (1878a). Astronomical. *Sydney Morning Herald,* 11 October.

Tebbutt, J. (1878b). Note on the planet Uranus. *Journal and Proceedings of the Royal Society of New South Wales, 12,* 220–221.

Tebbutt, J. (1878c). Note on the transit of Mercury, on May 6th, 1878. *Journal and Proceedings of the Royal Society of New South Wales, 12,* 226–227.

Tebbutt, J. (1878d). Occultation of a fixed star by Jupiters [*sic.*] first satellite. *Astronomische Nachrichten, 93,* 379–380.

Tebbutt, J. (1878e). On the transit of Mercury and the longitude of the Windsor Observatory. *Astronomische Nachrichten, 93,* 61–62.

Tebbutt, J. (1878f). A rare phenomenon. *Sydney Morning Herald,* 8 October.

Tebbutt, J. (1879a). Conjunction of Mars and Saturn, June 30th, 1879. *The Observatory, 3,* 26.

Tebbutt, J. (1879b). Eclipses of Jupiter's satellites observed at Windsor, N.S. Wales in 1878. *Astronomische Nachrichten, 95,* 119–122.

Tebbutt, J. (1897c). Occultation of 64 Aquarii by Jupiter, Sep. 14th, 1879. *Journal and Proceedings of the Royal Society of New South Wales, 13,* 165–166.

Tebbutt, J. (1879d). To the Editor of the Herald. *Sydney Morning Herald,* 19 September.

Tebbutt, J. (1879–81). *Journal for Astronomical Observations. The Peninsula, Windsor, N.S. Wales.* MS, Mitchell Library, Sydney (A3749).

Tebbutt, J. (1880a). Astronomical and meteorological notes. *Sydney Morning Herald,* 11 December.

Tebbutt, J. (1880b). Eclipses of Jupiter's satellites. *Astronomische Nachrichten, 97,* 37–40.

Tebbutt, J. (1880c). The new variable star in Gemini. *Sydney Morning Herald,* 5 February.

Tebbutt, J. (1880d). Note on the opposition magnitudes of Uranus and Jupiter. *Journal and Proceedings of the Royal Society of New South Wales, 14,* 23.

Tebbutt, J. (1880e). On the variability of B.A.C. 2472. *Monthly Notices of the Royal Astronomical Society, 40,* 437.

Tebbutt, J. (1880f). Southern double stars. *The Observatory, 3,* 514–515.

Tebbutt, J. (1881a). Astronomical and meteorological notes. *Sydney Morning Herald,* 10 January.

Tebbutt, J. (1881b). Observations of the Red Spot on Jupiter. *Monthly Notices of the Royal Astronomical Society, 41,* 331–332.

Tebbutt, J. (1881c). On the variable star R. Carinae. *Journal and Proceedings of the Royal Society of New South Wales, 15,* 380–385.
Tebbutt, J. (1881d). Southern double stars. *The Observatory, 4,* 211–212.
Tebbutt, J. (1882a). Notes on comets. *Sydney Morning Herald,* 21 December.
Tebbutt, J. (1882b). Transit of Mercury, 1881, November 7–8. *Monthly Notices of the Royal Astronomical Society, 42,* 103–104.
Tebbutt, J. (1882–84). *Journal for Astronomical Observations. The Peninsula, Windsor, N.S.Wales.* MS, Mitchell Library, Sydney (A3750).
Tebbutt, J. (1883a). The Great Comet. *Sydney Morning Herald*, 7 April.
Tebbutt, J. (1883b). Notes on comets. *The Southern Science Record, 3,* 42–49.
Tebbutt, J. (1883c). Observations of the variable star R Carinae from May 1880 to Sept. 1883. *Monthly Notices of the Royal Astronomical Society, 44,* 14–17.
Tebbutt, J. (1883d). Post-perihelion observations of the Great Comet (b) 1882. *Monthly Notices of the Royal Astronomical Society, 43,* 383–394.
Tebbutt, J. (1883e). The stars Lalande 19034 and 19093. *Astronomical Register, 21,* 39.
Tebbutt, J. (1883f). 19 Leporis. *The Observatory, 6,* 154.
Tebbutt, J. (1884). Astronomical notes. *Sydney Morning Herald,* 6 December.
Tebbutt, J. (1884–86). *Journal for Astronomical Observations. The Peninsula. Windsor, N.S.Wales.* MS, Mitchell Library, Sydney (A3751).
Tebbutt, J. (1885a). Astronomical notes. *Sydney Morning Herald,* 6 November.
Tebbutt, J. (1885b). Observations of double stars at Windsor, New South Wales. *Monthly Notices of the Royal Astronomical Society, 46,* 50–55.
Tebbutt, J. (1886a). Observations of the variable star R Carinae from Sept. 1883, to April 1886. *Monthly Notices of the Royal Astronomical Society, 46,* 487–488.
Tebbutt, J. (1886b). The variable star η Argus. *The Observatory, 9,* 194–195.
Tebbutt, J. (1886–87). *Journal for Astronomical Observations. The Peninsula. Windsor, N.S.Wales.* MS, Mitchell Library, Sydney (A3752).
Tebbutt, J. (1887a). *History and Description of Mr. Tebbutt's Observatory, Windsor, New South Wales*. Sydney: printed for the author.
Tebbutt, J. (1887b). Observations of Winnecke's Comet, 1886, made at Windsor, New South Wales. *Monthly Notices of the Royal Astronomical Society, 47,* 293–301.
Tebbutt, J. (1888a). Observations of Comet a 1888 (Sawerthal), made at Windsor, N.S. Wales. *Monthly Notices of the Royal Astronomical Society, 48,* 376–379.
Tebbutt, J. (1888b). η Argus and W.B. XIII. 595. *The Observatory, 11,* 299–300.
Tebbutt, J. (1889a). Observations of Comet 1889 … (Davidson). *Astronomische Nachrichten, 122,* 349–350.
Tebbutt, J. (1889b). *Report of Mr. Tebbutt's Observatory, The Peninsula, Windsor, New South Wales, For the Year 1888*. Sydney: printed for the author.
Tebbutt, J. (1889c). Results of double-star measures at Windsor, New South Wales, during the years 1886, 1887, and 1888. *Monthly Notices of the Royal Astronomical Society, 50,* 23–31.
Tebbutt, J. (1890a). Observations of comets at Windsor N.S. Wales. *Astronomische Nachrichten, 123,* 409–410.
Tebbutt, J. (1890b). Observations of the variable star R Carinae from November, 1886, to June, 1890. *Monthly Notices of the Royal Astronomical Society, 51,* 46–47.
Tebbutt, J. (1890c). *Report of Mr. Tebbutt's Observatory, The Peninsula, Windsor, New South Wales, For the Year 1889*. Sydney: printed for the author.
Tebbutt, J. (1890–1892). *Journal for Astronomical Observations. The Peninsula, Windsor, New South Wales*. MS, Mitchell Library, Sydney (A3754).
Tebbutt, J. (1891a). Beobachtungen des Mercurdurchgangs 1891 Mai 9. *Astronomische Nachritchten, 128,* 25–28.
Tebbutt, J. (1891b). Double star measures at Windsor, New South Wales, 1889 and 1890. *Monthly Notices of the Royal Astronomical Society, 51,* 331–333.
Tebbutt, J. (1891c). *Report of Mr. Tebbutt's Observatory, The Peninsula, Windsor, New South Wales, For the Year 1890*. Sydney: printed for the author.

Tebbutt, J. (1892a). The conjunction of Venus and Jupiter, February 6, 1892. *Journal of the British Astronomical Association, 2,* 325–327.

Tebbutt, J. (1892b). *Report of Mr Tebbutt's Observatory, The Peninsula, Windsor, New South Wales, For the Year 1891*. Sydney: printed for the author.

Tebbutt, J. (1892c). Results of double-star measures with the 8-inch equatorial at Windsor, N.S. Wales, in 1891. *Monthly Notices of the Royal Astronomical Society, 52,* 386–388.

Tebbutt, J. (1892–1894). *Journal; for Astronomical Observations. The Peninsula, Windsor, New South Wales*. MS, Mitchell Library, Sydney (A3755).

Tebbutt, J. (1893a). *Report of Mr. Tebbutt's Observatory, The Peninsula, Windsor, New South Wales, For the Year 1892*. Sydney: printed for the author.

Tebbutt, J. (1893b). Results of double-star measures with the 8-inch equatorial at Windsor, N.S. Wales, in 1892. *Monthly Notices of the Royal Astronomical Society, 53,* 369–372.

Tebbutt, J. (1894a). The occultation of Saturn, May 25, 1893. *Publications of the Astronomical Society of the Pacific, 6,* 23–24.

Tebbutt, J. (1894b). *Report of Mr. Tebbutt's Observatory, The Peninsula, Windsor, New South Wales, For the Year 1893*. Sydney: printed for the author.

Tebbutt, J. (1894c). Windsor measures of α Centauri in 1893. *Monthly Notices of the Royal Astronomical Society, 54,* 123.

Tebbutt, J. (1894–1896). Untitled journal. MS, Mitchell Library, Sydney (A3756).

Tebbutt, J. (1895a). Observation of the transit of Mercury, 1894 November 10. *Astronomische Nachrichten, 137,* 93–94.

Tebbutt, J. (1895b). *Report of Mr. Tebbutt's Observatory, The Peninsula, Windsor, New South Wales, For the Year 1894*. Sydney: printed for the author.

Tebbutt, J. (1895c). Results of double-star measures with the 8-inch equatorial at Windsor, New South Wales, in 1894. *Monthly Notices of the Royal Astronomical Society, 55,* 308–311.

Tebbutt, J. (1896a). Observations of dark spots on the planet Jupiter, January to May, 1896. *Journal of the British Astronomical Association, 7,* 59–62.

Tebbutt, J. (1896b). Observations of the variable star R Carinae from 1890 December to 1895 August. *Monthly Notices of the Royal Astronomical Society, 56,* 351–352.

Tebbutt, J. (1896c). *Report of Mr. Tebbutt's Observatory, The Peninsula, Windsor, New South Wales, For the Year 1895*. Sydney: printed for the author.

Tebbutt, J. (1896d). Results of double-star measures with the 8-inch equatorial at Windsor, New South Wales, in 1895. *Monthly Notices of the Royal Astronomical Society, 56,* 353–357.

Tebbutt, J. (1896–98). Untilted journal. MS, Mitchell Library, Sydney (A3757).

Tebbutt, J. (1897a). Equatorial comparisons of Uranus with 41 Librae, and a probable occultation of the star by the planet. *Monthly Notices of the Royal Astronomical Society, 58,* 21–22.

Tebbutt, J. (1897b). On the occultation of 41 Librae by Uranus on 1897 September 8. *Journal of the British Astronomical Association, 8,* 86–87.

Tebbutt, J. (1897c). *Report of Mr. Tebbutt's Observatory, The Peninsula, Windsor, New South Wales, For the Year 1896*. Sydney: printed for the author.

Tebbutt, J. (1897d). Results of double-star measures with the 8-inch equatorial at Windsor, New South Wales, in 1896. *Monthly Notices of the Royal Astronomical Society, 57,* 584–585.

Tebbutt, J. (1897e). The variable star R Carinae. *Journal of the British Astronomical Association, 7,* 138–138.

Tebbutt, J. (1897f). The variable star R Carinae. *Journal of the British Astronomical Association, 8,* 85.

Tebbutt, J. (1898a). Eta Argus Compared With its Neighbours. MS (inserted between pages 227 and 228 in Tebbutt, J., 1896–1898. Journal), Mitchell Library, Sydney (A3757).

Tebbutt, J. (1898b). *Report of Mr. Tebbutt's Observatory, The Peninsula, Windsor, New South Wales, For the Year 1897*. Sydney: printed for the author.

Tebbutt, J. (1899). *Report of Mr Tebbutt's Observatory, The Peninsula, Windsor, New South Wales, For the Year 1898*. Sydney: printed for the author.

Tebbutt, J. (1900a). *Report of Mr Tebbutt's Observatory, The Peninsula, Windsor, New South Wales, For the Year 1899*. Sydney: printed for the author.

Tebbutt, J. (1900b). The variable star η Argus. *The Observatory, 23,* 256.

Tebbutt, J. (1901a). *Report of Mr Tebbutt's Observatory, The Peninsula, Windsor, New South Wales, For the Year 1900*. Sydney: printed for the author.

Tebbutt, J. (1901b). Results of double-star measures with the 8-inch equatorial at Windsor, New South Wales, in the years 1899 and 1900. *Monthly Notices of the Royal Astronomical Society, 61,* 501–505.

Tebbutt, J. (1902a). *Report of Mr. Tebbutt's Observatory, The Peninsula, Windsor, New South Wales, For the Year 1901*. Sydney: printed for the author.

Tebbutt, J. (1902b). Results of double-star measures with the 8-inch equatorial at Windsor, New South Wales, in 1901. *Monthly Notices of the Royal Astronomical Society, 62,* 485–488.

Tebbutt, J. (1903a). *Report of Mr. Tebbutt's Observatory, The Peninsula, Windsor, New South Wales for the Year 1902*. Sydney: printed for the author.

Tebbutt, J. (1903b). Results of double-star measures with the 8-inch equatorial at Windsor, New South Wales, in 1902. *Monthly Notices of the Royal Astronomical Society, 64,* 58–69.

Tebbutt, J. (1904). *Report of Mr. Tebbutt's Observatory, The Peninsula, Windsor, New South Wales, For the Year 1903*. Sydney: printed for the author.

Tebbutt, J. (1906). Measures of southern binary stars. *Monthly Notices of the Royal Astronomical Society, 66,* 547.

Tebbutt, J. (1908a). *Astronomical Memoirs*. Sydney: printed for the author.

Tebbutt, J. (1908b). Measures of southern binary stars in 1907. *Monthly Notices of the Royal Astronomical Society, 68,* 396.

Tebbutt, J. (1910). Measures of southern binary stars in 1909. *Monthly Notices of the Royal Astronomical Society, 70,* 655.

Tebbutt, J. (1911). Observations of Halley's Comet at Windsor, N.S. Wales. *Monthly Notices of the Royal Astronomical Society, 71,* 224–226.

Tebbutt, J. (1912a). Gale's Comet. *Journal of the British Astronomical Association, 23,* 139–140.

Tebbutt, J. (1912b). Measures of southern binary stars in 1911. *Monthly Notices of the Royal Astronomical Society, 72,* 563.

Tebbutt, J. (1913). Komet 1912a (Gale). Observations at Windsor, N.S. Wales. *Astronomische Nachrichten, 193,* 313–314.

Tebbutt, J. (1915). Measures of southern binary stars. *Monthly Notices of the Royal Astronomical Society, 76,* 36.

Tedesco, E.F. (1979). Lightcurve parameters of asteroids. In Gehrels, 1098–1107.

Ten, A.E. (2014). Méchain, Pierre-François-André. In Hockey et al., 1446–1447.

Todd, D.P. (1877). Letter to John Tebbutt, dated 27 August. In TL

Tombaugh, C. & Moore, P. (1980). *Out of the Darkness. The Planet Pluto*. Harrisburg: Stackpole Books.

van Biesbroeck, G. Vesely, C.D. & Marsden, B.G. (1976). Orbits of comets 1892 VI and 1911 V. *The Astronomical Journal, 81,* 125–126.

Voute, J. (1908). On the orbit of β 416. *Monthly Notices of the Royal Astronomical Society, 68,* 563–565.

Williams, T.R. (1988). Roberts of Lovedale and eclipsing binary stars. In S. Dunlop & M. Gerbaldi (Eds.), *Stargazers: The Contribution of Amateurs to Astronomy* (pp. 48–49). Berlin: Springer-Verlag.

Williams, T.R. & Saladyga, M (2011). *Advancing Variable Star Astronomy. The Centennial History of the American Association of Variable Star Observers*. Cambridge: Cambridge University Press.

Wlasuk, P.T., 1996. 'So much for fame!': the story of Lewis Swift. *Quarterly Journal of the Royal Astronomical Society, 37,* 683–707.

Wolf, R. (1863). Schreiben des Herrn Prof. Wolf … *Astronomische Nachrichten, 60,* 59–62.

Wright, H. (1897). Note. *Journal of the British Astronomical Association, 8,* 87.

Young, C.A. (1883). The Great Comet of 1882. *Popular Science Monthly, 22 (January),* 289–299.

# Chapter 12
# Power, Politics and Prestige: The Russell–Tebbutt Feud

## 12.1 Introduction

Henry Chamberlain Russell (Fig. 12.1) was born at West Maitland, New South Wales, on 17 March 1836 (nearly 2 years after Tebbutt), and came from a distinguished family. His father was the Honourable Bourn Russell, a member of the Legislative Council from 1858 to 1880. After completing his B.A. degree at the University of Sydney in 1859 H.C. Russell was appointed a computer at Sydney Observatory under founding Director, William Scott (Orchiston 1998b; Wood 1958). In 1870 he became Government Astronomer of New South Wales and rose to prominence in both astronomy and meteorology. He also played a leading role in the Royal Society of New South Wales and the Australasian Association for the Advancement of Science, and was a member of the Senate of his *alma mater* for more than 32 years (see Bhathal 1991; Walsh 1976). A Fellow of the Royal Society and a Companion of the Order of St. Michael and St. George, H.C. Russell died in 1907.

During the second half of the nineteenth century, Russell and Tebbutt (Fig. 12.2) were the most prominent astronomers in New South Wales and initially were friends, but during the 1880s there was a gradual breakdown in relations and by 1891 they had become bitter enemies. In the late 1880s and throughout the 1890s, the Russell–Tebbutt feud permeated New South Wales astronomy, forcing all of the leading astronomers to 'take sides', and in doing so a number of Russell's own staff were placed in a precarious position. The feud also erupted on the pages of *The Observatory*, thereby becoming an international issue.

This chapter examines the Russell–Tebbutt feud and its impact on New South Wales astronomy, and draws freely on material already published in Orchiston (2000b, 2002).



W. Orchiston, *John Tebbutt*, Historical & Cultural Astronomy,
DOI 10.1007/978-3-319-44521-2_12

**Fig. 12.1** Henry Chamberlain Russell (Orchiston collection)

**Fig. 12.2** John Tebbutt (after The Windsor (N.S.W.) Observatory 1889)

## 12.2 From Friend to Foe: Narrative of a Deteriorating Relationship

George Robarts Smalley died on 11 July 1870 and Russell succeeded him as Government Astronomer of New South Wales and Director of Sydney Observatory (Wood 1958). Despite this official ranking, Tebbutt felt free to publicly correct him, which he did on 29 September 1870 in reference to a report which appeared in the *Sydney Morning Herald* 2 days earlier where Russell mentioned that four different aurorae had been seen in 1870. Tebbutt begged to differ: ‘Permit me to state that, on looking through my meteorological journal, I find that no fewer than *seven* [aurorae] have been recorded’ (Tebbutt 1870). He then proceeded to list these. Aurorae were rarely seen as far north as Sydney (at latitude 34° S), or even under the somewhat darker skies of nearby Windsor, so 1870 was an exceptional year.

Despite this article, Russell and Tebbutt were at first close friends. This is reflected in the popular article, ‘On the progress and present state of astronomical science in New South Wales …’, where Tebbutt makes several complimentary comments about Russell, including the following:

> It is matter for congratulation to astronomers that such good use is now being made of the fine equatorial of the Sydney Observatory [i.e. the 7.25-in Merz], and the present director is entitled to the cordial sympathy and co-operation of all lovers of the science. (Tebbutt 1871, p. 12).

It is equally apparent that Tebbutt’s sentiments were reciprocated, as the following letter from Russell demonstrates:

> Many thanks for the kind way you have spoken to me, manifesting a feeling which I fully reciprocate and which I hope will increase as the years roll over as to our mutual benefit and the advantage of the science to which we have devoted our lives. (Russell 1870d).

This mutual respect and admiration continued throughout the 1870s, both at a personal and a professional level. Each expressed concern about the health of the other and their respective families (see Russell 1871c, f, 1872a, h, j, 1876b, 1877), and Tebbutt was invited to join the Russell family for 1 pm dinner when on daytime visits to Sydney (e.g. see Russell 1872h, 1874b).

When Russell’s life was threatened during the notorious ‘letter bomb incident’ of 1877 (see Wood 1958), Tebbutt was quick to offer his heartfelt sympathy. Russell’s response says something of his personality and his basic *modus operandi* in interpersonal relations:

> The stories in the paper about my being quite unnerved were all newspaper nonesense [*sic*] I have felt and indeed still feel that it is a terrible thing to be obliged to recognise the existence of such a diabolical enemy. the more so that I did not before know I had one. *It has always been my rule not to offend any one’s feelings unless stern duty demanded it*, and I thought that I had so far passed quietly through life. I feel now that the kind hand which has so recently preserved me will continue to do so if there be yet work for me to do. (Russell 1877; my italics).

As we shall see, 10 years later Russell was only too happy to bend these rules—about not offending anyone’s feeling—when it came to John Tebbutt, but the situation was different in the 1870s.

When it came to astronomy and meteorology, Russell (1872i, 1873b, 1874b) welcomed and appreciated Tebbutt’s observations, and offered to publish any auroral observations he should make (Russell 1871d). He invited Tebbutt to use the telescopes and the library at Sydney Observatory (Fig. 12.3), and on at least one occasion offered him a bed for the night (Russell 1870a, 1871d, 1874b).

**Fig. 12.3** Sydney Observatory made Darunee Lingling Orchiston welcome in 2015 and John Tebbutt equally welcome back in the 1870s (*Photograph* Wayne Orchiston)

For his part, Russell (1871b) was eager to solicit suggestions from Tebbutt of possible new meteorological observers in western New South Wales, and he also sought information about historic Parramatta Observatory (Russell 1871a), the ruins of which (see Fig. 12.4) Tebbutt had first visited in January 1868 (Tebbutt 1864–69).

Founded by Governor Sir Thomas Brisbane (1773–1860; Fig. 12.5; Morrison-Low 2004) in 1821, Parramatta Observatory closed in 1847 (see Goodin 1947; Orchiston 1988e; Saunders 2004), but not before its astronomers had discovered a number of new comets, double stars, star clusters and nebulae, and accumulated observations for the famous Parramatta Catalogue (e.g. see Cozens et al. 2010; Richardson 1835; Saunders 1990). After it was closed, instruments from Parramatta Observatory went into storage and subsequently were transferred to Sydney Observatory when this institution opened in 1858 (see Lomb 2004; Orchiston 1988e). Tebbutt lobbied successfully for an obelisk to be erected in Parramatta Park to mark the site of Parramatta Observatory (see Fig. 12.6). Given this historical link with Sydney Observatory it is understandable that Russell had a special interest in Parramatta Observatory.

In 1870 Tebbutt was busy preparing a paper on the history and current state of astronomy in New South Wales, and Russell (1870b, d) was happy to provide him with details about the work done at Sydney Observatory. On the other hand,

**Fig. 12.4** Remains of the pillars that once supported the Parramatta Observatory transit telescope. Examining them is the British science historian, the late Anita McConnell (*Courtesy* Julian Holland)

**Fig. 12.5** Sir Thomas Brisbane, who founded Parramatta Observatory (https://en.wikipedia.org)

Russell (1870c) sought specific meteorological data from Tebbutt, and the following year invited him to input to a paper outlining the history of meteorology in New South Wales: 'If you would like to make additions to what I have said

**Fig. 12.6** The obelisk that was erected in 1880, following Tebbutt's lobbying, to mark the site of Parramatta Observatory (after McGuinness 2014)

about your own metr work and observatory I will gladly insert them I did not attempt a list of your instruments because I did not know what you have' (Russell 1871a). This request may have mystified Tebbutt since a copy of his first meteorological monograph (Tebbutt 1868) was already in the Sydney Observatory library, and it contained full documentation on the meteorological instruments at Windsor Observatory.

In 1871 Russell (1871e) also was encouraging Tebbutt to observe η Argus and to prepare a paper for the Royal Society of New South Wales, and later in the year he invited Tebbutt to join the expedition that was going to Queensland to observe the total solar eclipse (Russell 1871g; cf. Lomb 2016—see Fig. 12.7). Founding Director of Sydney Observatory, the Reverend William Scott (Bio Box 6.1) was also a member of this expedition, and he wrote Tebbutt an even more importunate letter:

**Fig. 12.7** A watercolour painting by Silvester Diggles of the *S.S. Governor Blackall* at Eclipse Island in the Claremont Group, North Queensland, in 1871 (*Courtesy* State Library of NSW, Australian Eclipse Expedition 1871, SV*/Ecl/1)

> Now cannot you be induced to go with us? You could take your own Telescope, and follow out your own views ... Remember you will never have another opportunity of seeing a Total Eclipse. Moreover your reputation as a zealous Astronomer would suffer if you declined so good an opportunity of furthering the cause. (Scott 1871).

Although the cost was only £25 (ibid.), a modest sum for Tebbutt, illness in the family forced him to decline the kind invitation. Needless to say, his reputation remained intact—despite Scott's concerns—but true to the prediction, Tebbutt was to die without seeing a total solar eclipse. He missed the Tasmanian eclipse of 1910 (see Baracchi 1914) and died six years before the next eclipse that was visible from eastern Australia (see Haynes et al. 1993; The Solar Eclipse 1922).

Despite this setback, Russell (1872g) invited Tebbutt to join him in discussing a research strategy for the 1874 transit of Venus, and looked forward to co-operating where possible (Russell 1874a). Both astronomers ended up with useful results (Orchiston 2004b; Russell 1874c).

As with the eclipse invitation, Russell was unsuccessful in getting Tebbutt to support him in forming a local astronomical group in 1872. On 24 April he first broached the subject:

> Will you join me in an attempt to form a little Astronomical Society as a branch or otherwise if thought desirable of the Royal Society [of New South Wales]; there are a number of persons who would like to join and we could then do what you suggested long ago publish monthly notices and so have a means of recording anything of Interest. I would like to hear from you on the subject or I will come and see you if you prefer it. (Russell 1872a).

Tebbutt obviously was not encouraging, judging from Russell's next letter: 'I think you are quite right about the few here who know anything of astronomy but my idea is that in the way suggested we should be cultivating a knowledge of the science …' (Russell 1872b). In fact, Russell persisted with the idea and in 1876 succeeded in forming an Astronomy Section of the Royal Society of New South Wales (see Russell 1876c), in which Tebbutt was an active member (Orchiston and Bhathal 1991). In the interim, Tebbutt was elected a Fellow of the Royal Astronomical Society (in 1873), and Russell (1873a) was quick to congratulate him on this honour.

When Tebbutt began seriously entertaining the purchase of a larger telescope to replace his 3.25-in. refractor, it was Russell more than anyone else who helped in the search. On 1 May 1872 he advised Tebbutt (Russell 1872b) that there was an 8.5-in. (21-cm) Browning-With reflector available in Sydney, and 1 week later wrote that a Mr F.L. Edwards was about to leave for England and had an equatorially-mounted 5-in. (12.7-cm) Troughton & Simms refractor for sale at £120 (Russell 1872c). On 13 May he wrote that he had been inquiring about telescope prices from various Sydney opticians, and had acquired Merz's catalogue. He also discussed the relative merits of Merz and Schroeder objectives, preferring the latter (Russell 1872d).

As we know, after reviewing the various alternatives, Tebbutt opted for the 4.5-in. Cooke refractor owned by Anthony Hordern and imported new just 2 years earlier (Hordern 1870). On 23 May, Tebbutt arranged to visit Sydney and test the telescope (Hordern 1872), and the following day Russell (1872e) wrote describing the mounting and the quality of the optics (which gave a pleasing performance). Russell compared the Hordern and Edwards refractors, preferring the former. But he still had one reservation: 'I am sure the glass is worthy of a good trial, but I wish you could be tempted to get a larger one. 4½ inches cannot be final it will make you wish for more' (ibid.). Russell was basing this assessment on his experience with the 7-in. Merz refractor (Fig. 6.5) at Sydney Observatory. In the long run he proved to be right, but it was not until 1886 that Tebbutt acquired his 8-in. Grubb refractor (see Orchiston 1982, 1986, 1987b). At any rate, Tebbutt went ahead and purchased the Cooke telescope from Hordern, and from all accounts was pleased with it (see Russell 1872f).

Nor did Russell's guidance on astronomical instrumentation end there, for when about to leave for Europe in 1875 he made the following offer: '… if I can do anything for you in the way of Instruments or otherwise I shall be very glad' (Russell 1875a). The following year he offered to provide Tebbutt with data on the relative merits of ring and filar micrometers (Russell 1876a), and later that same year provided this response when Tebbutt wrote about acquiring a larger transit telescope:

> I do not think you would hear of such a transit instrument as you want in Sydney. I have never seen or heard of one that size here except in the Observatory, but you would get one at home [England] quickly, but if you send for it great care would have to be taken that you did not get one that was useless as there are plenty of that sort in the Old Country for sale. (Russell 1876d).

A reliable transit telescope was a prerequisite for any astronomer wishing to carry out systematic positional astronomy (see Stott and Hughes 1987), and the instrument Tebbutt wished to replace had an aperture of only 2 inches (5.1 cm). By comparison, at that time there was a 3.75-in. (9.5-cm) instrument at Sydney Observatory (see Wood 1958). When Tebbutt did obtain a larger instrument, 4 years later, it was a Cooke with a 3-in. (7.6-cm) objective (but by that time Sydney Observatory had acquired a 6-in. (15.2-cm) Troughton & Simms transit instrument—see Orchiston 1987a).

Russell (1876a) also came to Tebbutt's assistance in 1876 when he was seeking a publisher for his third meteorological monograph (Tebbutt 1877), and obtained two different quotes for him (though in the end neither of these was accepted).

Despite these amiable relations, there were occasional tension points during the 1870s. Walter Gale (Bio Box 12.1) knew Tebbutt personally for more than 30 years and told the late Dr. Harley Weston Wood (1911–1984; Fig. 12.8; Robertson 1985), a former Government Astronomer of New South Wales, that Tebbutt was a 'prickly character' (pers. comm., 1981). He was proud of his achievements in astronomy and meteorology, and took offence easily if he felt at all slighted.

**Fig. 12.8** Dr. Harley Wood in 1959 (Orchiston collection)

**Bio Box 12.1: Walter Frederick Gale**

Walter Frederick Gale (Fig. 12.9) successfully combined observational astronomy and popular astronomy. He was born in Sydney on 17 November 1865, and after completing his schooling worked for 5 years in the insurance and commercial fields before joining the Savings Bank of New South Wales in 1888. He remained with the Bank until 1925, rising to the position of Manager and Chief Inspector at Head Office (Wood 1981). He then worked as Manager of the Hoskins Investment Company, which was founded by George Hoskins (see Orchiston and Bembrick 1997), a wealthy industrialist and fellow-member of the New South Wales Branch of the British Astronomical Association.

Gale inherited an early interest in astronomy from his father, but the Great Comet of 1882 (C/1882 R1) made a major impact (see Finds Venus Unexciting 1943). Gale also acknowledged that it was Tebbutt's newspaper articles more than anything else which enticed him to seriously take up astronomy as a hobby (Gale 1886; c.f. Swindlehurst 1890). In 1884 he made a 7-in. (17.8-cm) reflecting telescope (Gale 1928), which was destined to be the first of many (Wood 1981). The largest had an aperture of 12 inches (30.5 cm) (Obituary 1945). In 1928, Gale was to reminisce about '… how I had to strive for ten

**Fig. 12.9** A cartoon showing Walter Gale and his interest in the planet Mars (Orchiston collection)

(continued)

(continued)

years of the best part of my young manhood to equip myself with a decent telescope, after "messing about" with mirrors of my own grinding mounted on "bits of board"'. (Gale 1928). During the twentieth century, he was one of Australia's leading non-professional 'telescope-brokers' and reportedly '… knew the history and characteristics of every astronomical instrument in Australia, and could tell many anecdotes relating to them' (Obituary 1945). Over the years a large number of instruments passed through his hands, and these ranged in aperture up to the ex-Tebbutt 8-in. (20.3-cm) refractor (Orchiston 1982, 1997b), two 18-in. (45.7-cm) reflectors (see Orchiston and Bembrick 1995) and the 20-in. (50.8-cm) 'Catts Telescope' (Orchiston 2010).

Unlike some non-professional telescope-makers, Gale used his instruments for observational astronomy. He observed the planets, particularly Mars, Jupiter and Saturn (Wood 1981), and believed that the so-called canals of Mars were genuine naturally-occurring surface features and that the planet '… may be inhabited by a race of sentient beings, perhaps not cast in the same mould as we are, but of a type suited to the conditions of the planet …' (Gale 1921). He also discovered a number of double stars and a planetary nebula (Wood 1981), experimented successfully with astronomical photography (Obituary 1945), and participated in a number of solar eclipse expeditions (ibid.).

Like Tebbutt, Gale was addicted to comet-searching and he independently discovered seven different comets, three of which (C/1894 G1, C/1912 R1 and 34P/Gale) now bear his name (Wood 1946). He and R.T.A. Innes tried to form the Australian Astronomical Society in 1892 (Orchiston and Bhathal 1984), and in 1895 they succeeded in founding the New South Wales Branch of the British Astronomical Association. Gale served as its first Secretary (Orchiston 1988c). In 1935 he emulated Tebbutt by receiving the Jackson-Gwilt Medal and Gift from the Royal Astronomical Society, for '… his discoveries of comets and his work for astronomy in New South Wales' (Wood 1981). For 28 years he was on the Board of Visitors of Sydney Observatory (Wood 1946).

Gale also believed in using the Sydney newspapers to bring astronomical discoveries and upcoming events to the attention of the public (Orchiston 1997a). For example, between October 1891 and February 1896 he contributed articles to the *Australian Star*, *Daily Telegraph*, *Evening News*, and the *Sydney Morning Herald* on four different comets (including C/1894 G1 which he discovered); a conjunction of Venus and Jupiter; the planets Mars, Jupiter and Uranus; a transit of Mercury and lunar occultations of Antares and Spica. Like Innes and Tebbutt, he also wrote about the astronomical record of Sydney Observatory (Gale 1894b).

Walter Gale died on 1 June 1945, and was remembered by his many friends for his '… personal qualities of helpfulness, enthusiasm, kindness, tolerance and understanding …' (Obituary 1945). In New South Wales, he was regarded by some as the State's leading independent astronomer after Tebbutt.

The December 1874 transit of Venus—which both Russell and Tebbutt successfully observed (see Figs. 12.10 and 12.11; cf. Fig. 8.18; Lomb, 2011; Orchiston 2004b; Russell 1883, 1892c; Tebbutt 1875a, 1883)—provided a case in point where Tebbutt felt 'hard done by', and with hindsight this event can be seen to mark the start of a systematic deterioration in the previously amiable relations between the two astronomers.

In January 1875, just 1 month after the grand event, Russell presented a paper on the transit at a meeting of the Royal Society of New South Wales, and an account appeared shortly thereafter in the *Sydney Morning Herald*. Upon reading this

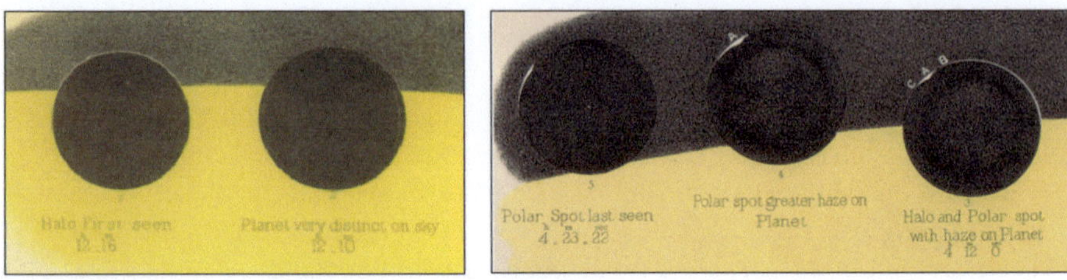


**Fig. 12.10** Some of Russell's drawings of the ingress (*left*) and egress (*right*) phases of the 1874 transit (after Russell 1892c, Plates XXVII and XXV, respectively)

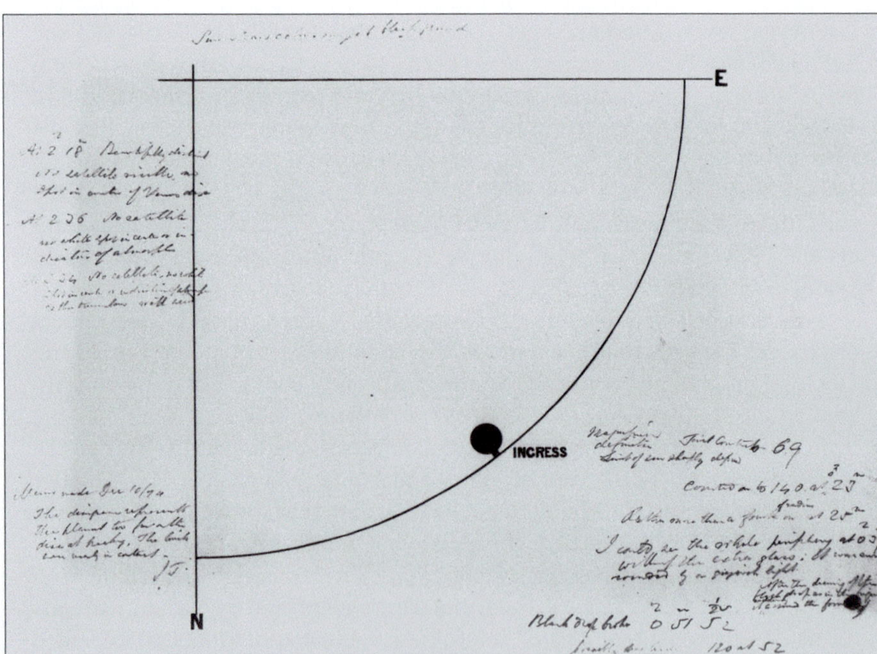


**Fig. 12.11** One of Tebbutt's original drawings of the ingress, showing the 'black drop effect' (*Courtesy* The Mitchell Library, Sydney)

Tebbutt (1875d) was peeved that Russell omitted to mention his observations, but Russell (1875b) was quick to explain:

> I am sorry Mr. Tebbutt was not at the meeting … had he been he would have known that … I was anxious to use them [his observations] as part of the results obtained in New South Wales, but this of course could only be done with his sanction, which I had not time to ask.

Upon seeing Russell's explanation, the noted Australian clergyman-botanist, Dr. William Woolls (1814–1893; Fig. 12.12; Cable 1976; Gilbert 1985; Thompson 1986), wrote Tebbutt: 'I am glad to see by today's Herald that Mr. Russell did not intend to act discourteously towards you. You have done so much for the promotion of science, that I think you are worthy of all honour' (Woolls 1875; his underlining).

Russell's letter seemed to placate Tebbutt who thanked him for his kind explanation, and went on:

> My sanction would have been quite unnecessary for the employment of my *published* observations. Indeed, I shall always be glad to see them discussed in the colony, more particularly by the author of the able paper read before the Royal Society. It appeared strange to me that my friend, Mr. Russell, should obtain data from Armidale and Raymond Terrace, and even condescend to the "back yard of a house in Hunter-Street," to the *apparent* neglect of the only private observatory in the Southern Hemisphere recognised by the British Admiralty in the pages of the Nautical Almanac. This was the reason for sending my note. I trust I may say, without being subjected to any imputation of boasting, that results proceeding from the observatory in question have been discussed and adopted by distinguished astronomers, both in Europe and America. (Tebbutt 1875e; his italics).

It is interesting to see that while professing respect for his 'friend' Russell, Tebbutt used this 'transit incident' to promote himself and the international status of Windsor Observatory. This brought forth an immediate charge by an anonymous critic that Tebbutt was trying to disparage Sydney Observatory, which Tebbutt was quick to refute:

> I am not at all conscious of having "tried to take the wind out of the sails of the legitimate establishment under the superintendence of the Government Astronomer," but I am quite determined that it should not be taken out of my own sails. The class of observations to

**Fig. 12.12** The Reverend Dr. William Woolls (https://en.wikipedia.org)

> which I usually confine my observations is one seldom regarded by the "legitimate establishment," consequently I ought to be exonerated from the charge brought against me by "P.G.K."
>
> There is, I am sorry to say, a class of very great men in this corner of the world who think that an amateur should confine himself to the art of star-gazing simply, and the said amateur is a very good fellow, and is patted on the back while he is content to keep within the limits assigned to him. The moment, however, he oversteps those limits and marks out an independent course for himself he is assailed as an individual who interferes with the "legitimate establishment," and if he cannot be frowned down he is quietly ignored. According to the views of these great men there should be only one authority in the colony on astronomical matters, and that authority the "Government" or "legitimate" one. This is the spirit against which I protest, and one which New South Wales must not adopt if she intends, like our generous "American cousins," to make astronomy flourish within her. It is a spirit which I do not remember to have encountered outside of the colonies. (Tebbutt 1875f).

Once again, Tebbutt had exploited the situation to expand the 'transit incident' far beyond its original bounds and raise the interesting spectre of professional–non-professional competition in nineteenth-century science. Although he does not name the 'very great men', he was probably referring to Henry Russell, given the basically amicable relations that existed between Windsor and Melbourne Observatories.

Tebbutt then used the 'transit incident' as an excuse to attack the publications record of Sydney Observatory under the directorship of Russell's predecessor, G.R. Smalley, and concluded:

> I know that the remarks I have made will draw upon me the displeasure of a certain party in the colony; but I am quite content to become the victim provided New South Wales, by adopting the course I have pointed out, asserts her proper place among the scientific countries of the world. She has done well in the matter of the Transit of Venus, only let the publication of these results be supplemented by that of our astronomical labours since the retirement of our first able director. (Tebbutt 1875b).

One of the independent astronomers who participated in the Sydney Observatory transit programme was Dr. Horatio Wright (Bio Box 10.5), and he leapt into print to defend Russell's '… arduous labours … and untiring exertions' (which Tebbutt did not dispute anyway). Wright (1875) wrote the editor that he was dismayed to see '… a certain amount of personal ill-feeling displayed by some of your correspondents …' This all but ended what had developed into a rather bitter media campaign that dragged on for just over three months. In the process, readers of the *Sydney Morning Herald* witnessed a change in the way in which Tebbutt viewed Russell and Sydney Observatory.

The entire 'transit incident' does raise the thorny question of the relative status of Russell and Tebbutt—both in their own eyes and in those of their scientific colleagues, national and international. Tebbutt's superior ranking in New South Wales astronomy was apparent as early as 1862 when he rather than Russell was offered the post of Government Astronomer of New South Wales upon Scott's resignation (Orchiston 1988d, 1998b)—something that Russell was well aware of. Then, in 1866, it was Tebbutt who was asked to prepare a report on New South Wales astronomy for the Paris International Exhibition (see Tebbutt 1867), but when this was published (in revised form) by the New South Wales Government in 1871 (see Tebbutt 1871), Russell (1871h) contributed a chapter on meteorology for the very same tome. So Tebbutt and Russell had already begun to establish their respective 'territories' in astronomy and meteorology as early as 1870. By the end of 1874, Tebbutt had further consolidated his position. He was now a Fellow of the Royal Astronomical Society and continued to contribute with almost indecent frequency to international astronomical journals and to the Sydney and Hawkesbury district newspapers.

Foibles that Tebbutt could not tolerate were unreliability or incompetence, and because of this his relationship with Russell was destined to deteriorate further during the late 1870s. In October 1878, he arranged to see Russell during a visit to Sydney but this did not eventuate because Russell (1878a) mistook the time. When Tebbutt objected, Russell (1878b) explained:

> Your note about the time you would call was quite right; but I read it rather hurriedly and as it was usual for you to come in the morning so understood it; I am sorry I did so, but amid the multitude of things requiring my attention here I sometimes read letters in great haste.

Then in the following year, Russell waited until 12 December to answer a letter written by Tebbutt on 10 October (and even then he only did so after being prompted by Tebbutt). Russell (1879) was a little put out by this 'encouragement', explaining that with nearly two hundred correspondents to attend to it was inevitable that letters would sometimes be overlooked. Obviously Tebbutt expected that his should not be.

Tebbutt's next gripe was in 1880 when the 4-year-old Astronomy Section of the Royal Society of New South Wales lapsed (Orchiston and Bhathal 1991). The then Secretary of the Section was W.J. Macdonnell (Bio Box 9.2), and he had clear views on the failure, as the following letter to Tebbutt indicates:

> Like yourself I very much regret that Section A has fallen through as it certainly ought to have been the leading one. I think the Council has a great deal to do with the suppressing of the sectional meetings, especially Liversidge, who I know thinks they interfere with the main meetings …
>
> I am afraid there is too much cliqueism in the Society just now, and a little fresh blood on the Council would do a great deal of good—Russell, Liversidge and their party have had it all their own way for many years and cannot brook the slightest opposition …
>
> I think the coolness shown by our official leaders will account for the discontinuance of the Astro. Section. (Macdonnell 1880).

Not only was Russell one of the 'official leaders' referred to, but he also was the founder and Chairman of the 'Astro. Section' (see Orchiston and Bhathal 1991). Meanwhile, Russell's friend, London-born Archibald Liversidge (1847–1927; Fig. 12.13) was by this time the Professor of Geology and Mineralogy at Sydney University and the founding Dean of the Faculty of Science. A shy retiring man who never married, Liversidge was a stalwart of the Royal Society of New South Wales, serving as Honorary Secretary from 1874 to 1884 and Editor of the *Journal and Proceedings* for many years. Later he would serve three sessions as President, in 1886, 1890 and 1901 (Mellow 1974).

Despite a cooling in their relations, Russell personally made a point of trying to keep Tebbutt in touch with significant developments at Sydney Observatory. Thus,

**Fig. 12.13** Professor Archibald Liversidge (adapted from Russell 1892c)

on 20 September 1880 he wrote about a new telescope that was nearing completion:

> I am just putting up a cheap observatory for the large reflector which I hope to shew you in position next time you come. The observatory will be round, 26 feet in diameter and 18 feet in height, and with the one exception of the great reflector at Melbourne it is I believe the largest telescope in the Southern Hemisphere. (Russell 1880a).

This unfortunate reference to the 48-in. 'Great Melbourne Telescope' conjures up images of a large 'Great Sydney Telescope' but in reality it was no more than a 15-in. (38-cm) reflector. Its principal claim to fame, though, was that it featured a modified English equatorial mounting reminiscent of the one later used for the Palomar Telescope (Fig. 12.14; see Orchiston 2000a, 2001a).

Russell, who ground the mirror himself, inveigled the government into committing funds for an observatory on the basis that the new telescope should reveal the newly-discovered satellites of Mars (see Ashbrook 1984; Gingerich 1970). Despite a careful search, Russell (1877) had been unable to sight these through the largest telescope previously at the Observatory—the fine 11.5-in. (29.2-cm) Schroeder refractor.

Although Russell's new reflector was operational by mid-October 1880 (Russell 1880b), there is no evidence that it was ever used for any serious observational programmes. What is more, we do not even know if this telescope ever gave Russell those long-awaited glimpses of Phobos and Deimos.

What we do know, however, is that the 26-ft diameter observatory that Russell had purpose-built for his new reflector would later do fine service at Sydney Observatory as home to the Sydney astrograph once Russell made a commitment to the international Astrographic Catalogue and Carte du Ciel programme. This observatory featured a conical dome reminiscent of those favoured by Tebbutt at his Windsor Observatory, but in this instance the whole structure—and not just the dome—rotated. Fig. 12.15 shows this observatory along with the Sydney astrograph.

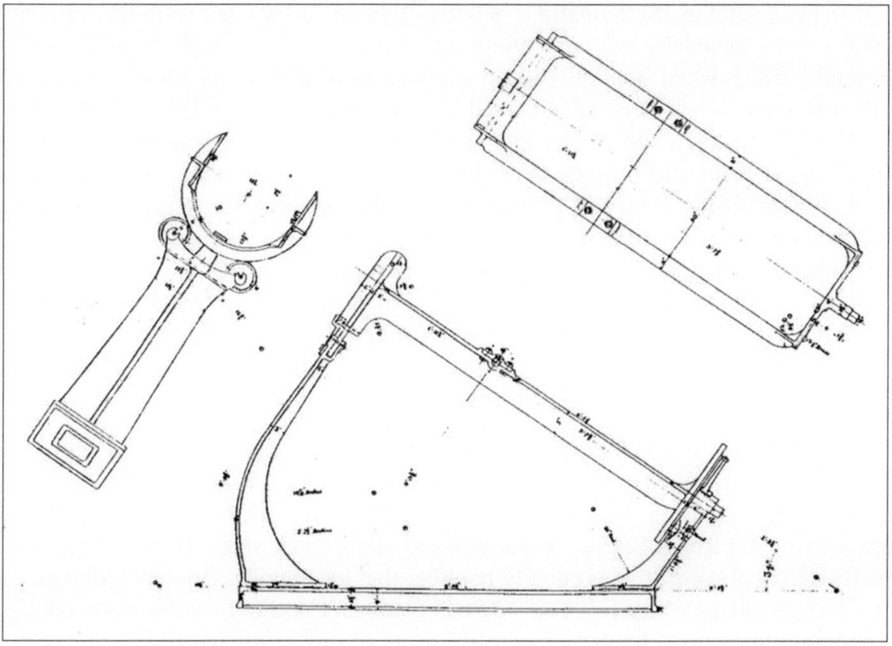

**Fig. 12.14** Russell's original drawings of the 15-in. telescope mounting (after Orchiston 2001a)

**Fig. 12.15** The 26-ft diameter observatory and the 13-in. (33-cm) astrograph with its 7.25-in. (18.4-cm) Merz guidescope (after Russell 1892a)

So, by the end of 1880, relations between Tebbutt and Russell were still amiable, even if punctuated by ongoing minor irritations. This was still the situation in February 1881 when Tebbutt had an accident and injured his foot, and Russell (1881b) immediately wrote and offered his sympathy.

This all changed the following month, when a paper or papers by Russell appeared in print and omitted to mention assistance that Tebbutt had provided. Without recourse to copies of Tebbutt's outward letters, it is impossible to be absolutely certain about which of Russell's publications was involved, but Tebbutt was very likely referring to one or other of the following papers, both of which were read at the 1 December 1880 meeting of the Royal Society of New South Wales and appeared shortly afterwards in the Society's *Journal and Proceedings*:

"Thunder and hail storms in New South Wales". (see Russell 1880d)
"Recent changes in the surface of Jupiter." (see Russell 1880c).

On 18 December 1880, Tebbutt had published his own account of changing Jovian surface features in the *Sydney Morning Herald* (see Tebbutt 1880a), and it is quite possible that this resulted from a compilation that he had earlier forwarded Russell for his Royal Society paper. On the other hand, since Tebbutt kept daily meteorological records, it is equally possible that he supplied Russell with data on storms at Windsor for the other Royal Society paper. Certainly, storms were of long-standing interest to Tebbutt and were the topic of one of the earliest papers of his published by the predecessor of the Royal Society of New South Wales (see Tebbutt 1863). Whatever the facts of the matter, Tebbutt was furious, even though he received reassurances from friends like Commercial Bank Manager at nearby Richmond and astronomy enthusiast Thomas Henry Furnival Griffin (1851–1938):

> I sympathise with you in the matter of the oversight of your assistance rendered to Mr. Russel [*sic*] which that gentleman has omitted to acknowledge … however, as you well point out, the approval which all of your work meets with, & the acknowledgement of its value rendered by the scientific world outside NSW is a far greater recompense than you could receive near home. (Griffin 1881).

This matter of peer recognition and perceived status as an international astronomer was something that was to preoccupy Tebbutt for the remainder of his life. As an obvious corollary, he also was concerned about Russell's astronomical output, and the balance between astronomy and meteorology at Sydney Observatory. Tebbutt believed that as a publicly-funded institution, the Observatory should contribute equally to both sciences, and he was becoming more and more concerned about the increasing emphasis which Russell placed on meteorology at the expense of astronomy. This disquiet was echoed by others, including the surveyor A.O. Powys (1881a): 'It seems—it is—a pity better use is not made with the Sydney instruments. Perhaps if the Government were to appoint a Government meteorologist we might have some chance of an astronomer' [his underlining]. Meanwhile, the astronomy populariser George Butterfield (Bio Box 9.1) goes a little further in his letter to Tebbutt dated 5 May 1881:

> I suppose we may congratulate ourselves on having an Astronomer [as the new President of the Royal Society of New South Wales], or one who does his best in that line … I have an

> idea however that a "full private" whose place of abode has a Royal name stands pretty high "One straw differeth from another in glory". (Butterfield 1881a).

Next came an important event that would test the level of Russell's regard for Tebbutt without any possible ambiguity, and this was Tebbutt's discovery of Comet C/1881 K1 (Orchiston 1999). As we saw in Chaps. 6 and 9, during the nineteenth century comet discoveries were regarded as notable achievements, whether made by professional or non-professional astronomers. Tebbutt immediately advised Russell of his find, but the reply he received three days later expressed reservations rather than offering congratulations:

> For one reason I am rather sorry it has come now, and that is, because it imparts a seeming confirmation to one of those irresponsible and unsigned predictions that have been published in England of late, apparently with the object of frightening the multitude. (Russell 1881c).

This was hardly the response that Tebbutt anticipated, and it would not have been appreciated (Orchiston 1999). And when the discovery was subsequently applauded in the 8 September 1881 issue of *Nature*, Butterfield (1881b) was moved to remark: 'How cold & indifferent we are here compared with the interest it caused in England'. Meanwhile, when Russell did actually succeed in securing positional and spectroscopic observations of the comet, he initially reported on these in one of the local newspapers rather than communicating with Tebbutt directly (Russell 1881a).

A little later, Powys (1881b) wrote to Tebbutt discussing Russell's lack of involvement in astronomy and concluded with the following comment: 'People naturally ask, How is it that he does not discover these comets?' This attitude was one that Scott had first encountered soon after the founding of Sydney Observatory (see Orchiston 1998b), and merely reflected public ignorance of the valid research role of a professional observatory.

From this critical date in 1881, communication between Russell and Tebbutt became less frequent (see Table 12.1) and the relationship more strained, although Russell did try to maintain a friendly disposition in most of the letters he wrote Tebbutt in 1882 and 1883. In one (Russell 1882a), he even thanked Tebbutt for coming to his aid in the newspaper. Later in the year, Russell was again in difficulty in the media, and this evoked the following response from Griffin (1882) via a letter to Tebbutt:

**Table 12.1** Letters and other communications from Russell to Tebbutt, 1859–1904

| Period | Number of letters etc. |
|---|---|
| 1860–1864 | 9 |
| 1865–1869 | 0 |
| 1870–1874 | 37 |
| 1875–1879 | 20 |
| 1880–1884 | 34 |
| 1885–1889 | 30 |
| 1890–1894 | 11 |
| 1895–1899 | 14 |
| 1900–1904 | 5 |
| Total: | 160 |

> I missed that curious communication of Russell's to which you refer. It is amazing to me that a professional astronomer should commit himself so frequently as he does, and I fear his bretheren [*sic*] in science on the whole put little faith in him from all that I hear.

This was a clear vote of no confidence in Russell the astronomer, and when he produced a 16-page booklet titled *The Sydney Observatory: History and Progress* (Russell 1882b), Tebbutt just had to be critical. This evoked the following reply from Russell (1884): 'As to the History and progress of the Observatory it is too much condensed to serve its proper purpose, but as it was written at the request of the Government and had to be confined within a few pages it is as it is'. One has to wonder whether Tebbutt would have been more tolerant had he been privy to all of the facts.

So by the mid-1880s Russell's letters to Tebbutt had become formal and to the point: once they commenced with 'My Dear Mr Tebbutt', but now it was a terse 'Dear Sir' (e.g. see Russell 1886, 1887a). Moreover, this deplorable breakdown in relations and Tebbutt's concerns about his perceived status in the eyes of the public were becoming known to overseas astronomers. On 3 May 1887 the Astronomer Royal for Scotland, Charles Piazzi Smyth (Fig. 12.16), wrote Tebbutt a melodramatic letter which ended with an intriguing suggestion:

> But alas the public of N.S.W. are strangely unaware, even after a quarter of a century, that they have in you the rare example of peculiar astronomical genius, which is almost entirely wanting in Anglo-Saxons of the Mother Country, as well as N.S.W., excepting as I said in your instance. If they go on much longer thinking lightly of your orbital calculations—I wonder whether it would be within the range of possibilities that you should come home & do for Great Britain and her Cometic Astronomy what you have been doing so long for that of Australia. … (Smyth 1887).

Fate was to decree that Tebbutt would never leave his beloved Australia—even for a brief holiday—and that he personally would never meet Smyth.

As the relationship between Tebbutt and Russell deteriorated, there were brief moments of reprieve. For instance, Tebbutt discovered that it was Russell who pointed C. Pritchard of Sydney in his direction when the late Dr. Bone's 8-in. Grubb refractor came on the market in 1886

**Fig. 12.16** A painting of Charles Piazzi Smyth by the Scottish artist John Faed (https://www.wikipedia.org)

**Fig. 12.17** William Harkness (after *Popular Science Monthly* 1903, p. 86)

> … and [was] placed in my hands for disposal. M[r] Russell informed me he did not require such an instrument for the Observatory but indicated that it would be well to bring it to your knowledge as if you did not require such an instrument yourself you might possibly know of some gentleman who would be convinced to buy it … (Pritchard 1886).

As we know, Tebbutt proceeded to purchase this telescope, and the fact that it did excellent service for Australian astronomy over the next twenty years or so can at least in part be attributed to Russell's initial influence!

There was a notable warming of relations between the Sydney and Windsor Observatories for much of 1887 when Russell was overseas, and this was welcomed by both parties. In March, William Harkness (1837–1903; Fig. 12.17) from the U.S. Naval Observatory wrote that he was working up the 1874 transit of Venus observations from Melbourne, Sydney and Windsor Observatories, and asked that the longitude of Tebbutt's observatory be reinvestigated (Harkness, 1887). As Acting Government Astronomer in Sydney, H.A. Lenehan (Bio Box 7.2) was only too happy to cooperate, and arranged for the telegraphic exchange of time signals between the two observatories (Lenehan 1887a, b). Because of the international importance of the outcome, Russell would have had to oblige had he been in Sydney at the time, but he may not have been quite so encouraging or cooperative.

Meanwhile, others at Sydney Observatory took advantage of Russell's absence to not only learn more about Tebbutt first hand, but also to offer him whatever assistance they could. Thus, the technician, William Isaac Masters fabricated an adapter so that one of Tebbutt's micrometers could be used with the 8-in. telescope (see Masters 1887; Pollock 1887a), while one of the Assistant Astronomers, James Arthur Pollock (Bio Box 12.2), wrote on 2 May thanking Tebbutt for information he had supplied and the trouble he had gone to. Further, he indicated that if he struck trouble with his investigation he would again seek Tebbutt's help (Pollock 1887a).

**Bio Box 12.2: James Arthur Pollock**

James Arthur Pollock (Fig. 12.18) was born at Douglas, Ireland, on 17 November 1865. After completing a Bachelor of Engineering degree at the Royal University of Ireland he and other members of the Pollock family emigrated to Sydney in January 1885 (McCaughan 1988).

**Fig. 12.18** A photograph of Professor Pollock taken during WWI (*Courtesy* University of Sydney Archives, G14/12)

Initially Pollock worked as an Astronomical Assistant at Sydney Observatory, but soon left in order to study at Sydney University, graduating with a B.Sc. and the University Medal in Physics in 1889. The following year he was appointed a Demonstrator in Physics at the University, and in 1899 became the Professor of Physics (Prof. J.A. Pollock … 1922).

Professor Pollock carried out wide-ranging research, was awarded a D.Sc. in 1906 and was elected a Fellow of the Royal Society (London) in 1916 (McCaughan 1988). In addition, he was '… an important factor in all University affairs' (R.T. 1923). Professor Pollock died on 24 May 1922 at the age of only 57 after a short illness (Prof. J.A. Pollock … 1922).

It is clear from the warm tone of the foregoing letters that with the exception of Russell, staff at Sydney Observatory had great regard for Tebbutt, and this is also evidenced by their responses when he distributed copies of his booklet *History and Description of Mr. Tebbutt's Observatory, Windsor, New South Wales* (Tebbutt 1887) towards the end of that year. For instance, Pollock (1887b) had this to say:

> On looking over your book one cannot help being struck with the zeal & energy with which you have worked single handed for so many years in the cause of science. For myself I can only say my admiration of yourself & your work is unbounded.

Lenehan's letter (1887c) was in a similar vein, and even Russell (1887b) was moved to comment that 'There are few, if any private observatories in the world which can shew such a record as you'. Coming from him, this was a considerable compliment. The following year, Pollock resigned from Sydney Observatory (Lenehan 1888), and in a letter dated 19 June 1888 he offered to help Tebbutt out in any way possible, deeming this a great honour (Pollock 1888). As we have seen from Bio Box 12.2, later Pollock went on to become Professor of Physics at the University of Sydney.

The year 1888 also brought with it an incident that was to dash any prospect of a rapprochement between Russell and Tebbutt and remove once and for all any semblance of civility in their future letters. This was the appearance of Russell's classic historical paper, 'Astronomical and meteorological workers in New South Wales, 1788–1860' (Russell 1888), which discussed others who were active during the 1850s but omitted any mention whatsoever of Tebbutt. Edward White (Bio Box 6.3), from Melbourne Observatory, was quick to point out that Russell's paper was incomplete without the inclusion of Tebbutt's labours (White 1889), while Tebbutt (1890) made the same point in a note that was published in *The Observatory*.

This omission must have been a calculated one, for Russell was just two years Tebbutt's junior and would have seen Tebbutt's numerous astronomical articles in the Sydney newspapers long before he joined the Sydney Observatory staff on 1 January 1859. The 1850s were a critical period in Australian astronomy, falling as they did between the closure of Parramatta Observatory and the full flowering of the various colonial observatories, and only two Australian astronomers were prominent at this time: Hobart's Francis Abbott (Bio Box 6.4; Orchiston 1992) and Windsor's John Tebbutt (Orchiston 2004a). As we saw in Chap. 7, between 1853 and 1860, Tebbutt carried out wide-ranging astronomical observations, of the Sun, aurorae, meteors, the zodiacal light, lunar occultations, lunar eclipses, Jupiter and its satellites, comets, double stars and variable stars (see Table 7.1). If anyone did, he certainly deserved a place in Russell's paper, and to Tebbutt's way of thinking by not giving credit where it was due Russell was abusing his power and authority. This deliberate snub was later seen by some (e.g. Lenehan 1907c) as the starting point of what was to degenerate into a bitter feud between the two former allies.

As a result of this incident, Tebbutt also reacted negatively towards Russell's friends and colleagues. In 1888 Russell and a number of his associates in the Royal Society of New South Wales were instrumental in forming the Australasian Association for the Advancement of Science, a major development for science in Australia and New Zealand (MacLeod 1988). One of those involved was British-born Richard (later, Sir Richard) Threlfall (1861–1932; Fig. 12.19), the Foundation Professor of Physics at the University of Sydney in 1886 (Home 1986), and when he invited Tebbutt to join the new society and contribute to Section A (Astronomy) he received a rebuff (Threlfall 1888a, b).

As news of the growing tension between Tebbutt and Russell began to circulate throughout the Australian astronomical community some looked to take advantage of the situation. On 12 January 1889 the Brisbane astronomer and geographer James Park Thomson (Bio Box 12.3) wrote Tebbutt on behalf of the Royal Australian

**Fig. 12.19** An undated photograph of Richard Threlfall (https://www.wikipedia.org)

Geographical Society demanding (not politely requesting) that he write an article on Australian rainfall for them (Thomson 1889a). When Tebbutt refused, Thomson (1889b) responded by threatening to ask Russell to do the honours. Tebbutt was unmoved.

**Bio Box 12.3: James Park Thomson**

James Park Thomson (1854–1941; Fig. 12.20) was born at Ulst, Shetland Islands, Scotland on 20 June 1854 and after spending a couple of years in the United States and South America returned to Scotland and learnt the basics of marine engineering in Glasgow. In 1876 he went to New Zealand and from 1877 spent 2 years learning the rudiments of surveying. From 1880 to 1884 Thomson worked as a surveyor in Fiji (Kitson 1990).

**Fig. 12.20** The Thomson Medal, showing the likeness of James Park Thomson (*Courtesy* Museum Victoria)

(continued)

**Bio Box 12.3:** (continued)

In 1884 Thomson settled in Brisbane and was employed as a draftsman in the Survey Department. In 1885 he formed the Queensland Branch of the Royal Geographical Society and actively solicited for new members. He also served as President of the Australasian Association for the Advancement of Science. Thomson wrote prolifically on scientific subjects. 'Always a keen traveller with wide connections, Thomson lectured in such towns as Charleville, Roma, Longreach, Blackall and on Thursday Island. He raised locals' awareness of their specific environments, while giving them a sense of union with a wider world' (ibid.).

In 1900 the Queensland Branch of the Royal Geographical Society named its foundation medal after Thomson and he was its first recipient. He also received the Peak Award from the Royal Geographical Society in London in 1902, and the following year an honorary LL.D. from Queen's University in Kingston, Ontario.

Thomson had an interest in observational astronomy and in the popularisation of the discipline (see Orchiston 1997a). In 1882 he observed the transit of Venus while living in Fiji, and in 1894 he recorded a transit of Mercury and published an account in *Monthly Notices of the Royal Astronomical Society* (Thomson 1895), but it was only upon his retirement in 1922 that he erected a private observatory in Brisbane (Eglinton Family Papers n.d.).

In 1895 a group of 76 non-professional astronomers and prominent Brisbane citizens came together and formed the Brisbane Astronomical Society, ostensibly to keep the 6-in. (15.2-cm) Grubb refractor owned by the highly-regarded Scottish-born Brisbane architect Francis Drummond Greville Stanley (1839–1897) in Queensland and prevent it from being sold to an interstate astronomer (Orchiston 1997b). Three Trustees were appointed, and one of these was J.P. Thomson, who as to play a leading role in the growth and eventual demise of the young Society (Orchiston 1998a). An anonymous newspaper article penned in 1896 forecast a rosy future for the Society: '... although the society has barely got out of its swaddling clothes, so to speak, the active interest taken in star-gazing by many of the members ... augers well for the promise and ultimate performance of this youngest of our scientific guilds' (The Brisbane Astronomical Society 1896). In fact, this was not to be for, like Thomson, British-born fellow Trustee Dudley Eglinton F.R.A.S. (1850–1937; Hicks and Frankland 1998) was strong willed, and the two men had clashing personalities, not an ideal situation for a fledgling Society. Furthermore, the two conflicting functions—public astronomy vs. observational astronomy—tore at the heart of the new Society. Regular meetings were held, and although Meteor, Lunar, Solar, Planetary and Double Star Sections were mooted, '... it would appear that membership response to these proposals was shockingly poor and eventually little work in any one of these fields

(continued)

**Bio Box 12.3:** (continued)
was accomplished' (Page 1959, p. 21). A year after its formation, Thomson lamented this situation, and opined that 'There is no room for such an institution here. Fifty years hence will no doubt be time enough for such a Society' (Thomson 1897).

Two other destabilising factors emerged early in the history of the Society. In the first place, there was dissension over the siting of the telescope (once its removal from Stanley's observatory became necessary), and second, there was debate over whether or not new members should be accepted into the Society (and if so what their status would be *vis-à-vis* the foundation members, and what their rights of access to the telescope would be). After Stanley died in 1897 Thomson arranged for title of the telescope to be transferred to the Queensland Branch of the Royal Geographical Society, but Eglinton who was absent from the meeting had the motion rescinded at a subsequent meeting!

By 1900 membership and enthusiasm had dwindled, and Eglinton tried to generate popular interest through a series of public lectures. These provided a short-lived reprieve, but the Society held its final meeting in 1901. This effectively marked the end of the Brisbane Astronomical Society (although a special meeting was held in 1911 to discuss the fate of the telescope, and this instrument was used briefly in 1916 for public demonstrations).

One can only speculate on the fate of the Society had there been a strong professional astronomy presence in Queensland at the time to counterbalance the competing influences of Thomson and Eglinton. Unfortunately, such was not the case (Haynes et al. 1993).

James Park Thomson died at Kilcoy, Queensland, on 10 May 1941 (Kitson 1990) a little over 1 month short of his 87th birthday.

Others were eager to express their own concerns about Russell. Amongst these was Charles Egeson, one of Russell's Assistant Meteorologists at Sydney Observatory, who on 29 March 1890 wrote Tebbutt a carefully-worded letter:

> It surprises me to think of the amount of work that you have done by private effort. In this Observatory which is unworthy of the name we have telescopes and observers in abundance but there is no result …
>
> Our aspirations are like a metal track that guides the intellectual engine and its physical train over the undulating fields of life. To know when to apply the brake and when to steam up is the problem which the driver of the engine has to solve in order to get to the terminus of human destiny. I have known some … also known to you, who have wasted life's fuel in simply sounding the steam whistle, that others might hear them, and have in consequence stuck at the foot of the first big hill that presented itself, for want of steam. They are now raking up the last embers in the fire hole in vain endeavour to proceed.

Four and a half months earlier, Egeson (1889b) had commented on yet another of Russell's meteorological 'blunders' that had appeared in a local newspaper. In this same year Egeson (1889a) produced a little book titled *Egeson's Weather System of Sun-Spot Causality: Being Original Researchers in Solar and Terrestrial Meteorology* which ran counter to Russell's meteorological theories and hardly strengthened the relationship between the two men. In 1890, Russell was only too happy to see Egeson released from his Sydney Observatory post, reputedly for insubordination and neglect of duty.

## 12.3 A Fully Frontal Attack: The *Sydney Morning Herald* and the Pamphlet

One of the reasons for Tebbutt's growing animosity towards Russell was the Observatory's increasing devotion to meteorology, at the expense of astronomy. Matters came to a head in 1891. On 9 September the Minister of Public Instruction tabled in Parliament documentation on the number of staff employed at Sydney Observatory in the years 1880 and 1890, total salaries paid in those 2 years and the overall cost of running the Observatory for the 10 years ended 1890. The total Government allocation over this decade was £41,103, and £24,597 of this (60 %) was expended on the Observatory's meteorological functions.

A response to this report appeared in the *Sydney Morning Herald* on 15 September in the form of a letter by an anonymous writer using the *nom de plume* 'TRIA'. It drew attention to the comparatively small outlay allocated to astronomy and attributed this to the absence of a Board of Visitors at the Observatory to which the Government Astronomer had to report. Furthermore,

> The observations at our Observatory are published irregularly and in divers places. It behoves the Government to alter this. It should request the Astronomer to present his observations once a year and should charge itself with their printing, so that the world at large may benefit by the outlay we make on the science. As a matter of justice the Government Astronomer should claim this for his protection. As a matter of right the people of this colony interested in the science should know what is being done at the Observatory … (TRIA 1891).

This letter looks suspiciously like the work of R.T.A. Innes (Bio Box 8.1), who was already beginning to make his mark in New South Wales after emigrating from Scotland (Orchiston 2001b, 2003, 2015). First, 'TRIA' devotes a long paragraph to similar problems which originally existed at the Royal Observatory, Greenwich, and second, the *nom de plume* is an interesting play on Innes's initials. By September 1891, Innes (Fig. 12.21) was already well aware of Russell's recent history of apathy towards astronomy (this is alluded to in Innes 1891a), and in the years that immediately followed he openly used the press to air his views on local astronomical issues and he was consistently critical of Russell.

Tebbutt waited a few days to see if any discussion would arise from TRIA's letter but nothing eventuated, so on 21 September he prepared a letter and sent this off to the same newspaper. He felt that

**Fig. 12.21** Robert Innes in the 1890s (*Courtesy* South African Astronomical Observatory)

> … as a contributor to the revenue out of which the Observatory is maintained, and a gratuitous promoter of the science of astronomy for upwards of thirty years, I had an undoubted claim to discuss the question raised by the anonymous correspondent referred to, and in my discussion of it I do not think I have in any way transgressed the limits of fair and honest criticism. (Tebbutt 1891b).

Obviously the Editor did not agree, for the letter never appeared in print, and Tebbutt's request for its return (he even sent the necessary postage) fell on deaf ears. But the Editor was one of Russell's acquaintances, so he should have known that this would not be the end of the matter, for when Tebbutt had a point to make—and in this case some political point-scoring to indulge in—he would not be silenced! To use Tebbutt's own words: '… I am determined not to be thus quietly snuffed out' (ibid.).

His solution was the obvious one for a gentleman of means: he was in the habit of publishing and distributing his own *Annual Reports* and meteorological monographs, so he would publish the 'missing letter' as a pamphlet, after first adding further critical comments, and then would distribute it privately.

Late in September 1891, *The Sydney Observatory and the 'Sydney Morning Herald'. A Plea for Astronomy in New South Wales* (Tebbutt 1891b) rolled off the printing press in Sydney (see Fig. 12.22). The 'missing letter' occupied three and a half pages of this little 8-page document, which suggests that quite apart from the political implications, its excessive length and detail may have been a factor (but certainly not the deciding factor) in the original decision for the newspaper to suppress it. There is no logic in reproducing the entire letter here, but the salient points that it makes deserve to be listed:

- The Board of Visitors that was formed in Scott's day was allowed to lapse under Russell's directorship.
- For many years the Observatory has not produced an Annual Report, '… hence I am ignorant of what the Observatory has really done for astronomical science'.
- The published astronomical work of the Observatory over the past 20 years is '… quite unworthy of such a highly-equipped institution'.

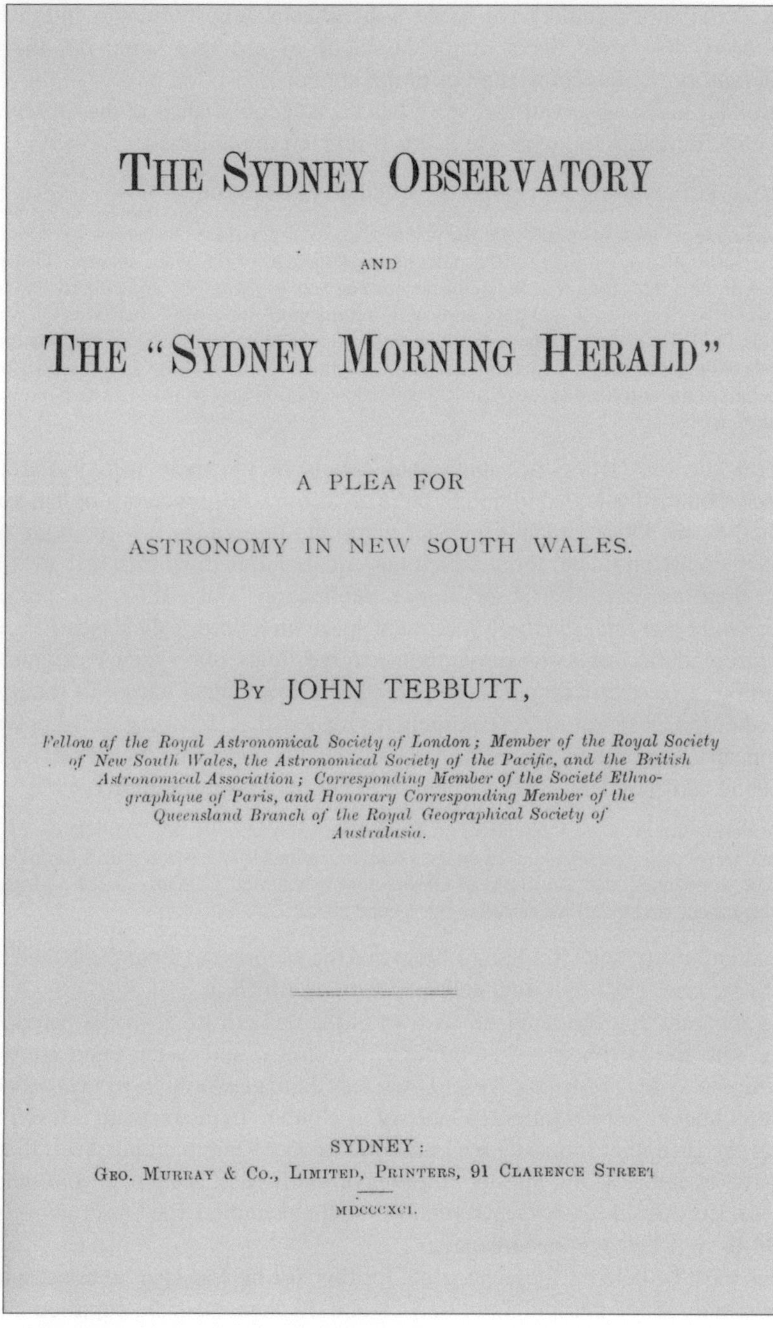

THE SYDNEY OBSERVATORY

AND

THE "SYDNEY MORNING HERALD"

A PLEA FOR

ASTRONOMY IN NEW SOUTH WALES.

BY JOHN TEBBUTT,

*Fellow of the Royal Astronomical Society of London; Member of the Royal Society of New South Wales, the Astronomical Society of the Pacific, and the British Astronomical Association; Corresponding Member of the Societé Ethnographique of Paris, and Honorary Corresponding Member of the Queensland Branch of the Royal Geographical Society of Australasia.*

SYDNEY:
GEO. MURRAY & CO., LIMITED, PRINTERS, 91 CLARENCE STREET

MDCCCXCI.

**Fig. 12.22** The cover of Tebbutt's booklet about Sydney Observatory and the letter suppressed by the *Sydney Morning Herald* (Orchiston collection)

- The Observatory could have made a significant contribution to international astronomy by producing a major catalogue of southern stars, but the Cape Observatory '… has forestalled us of the honour …'
- Doubtless much observational work has been accomplished at the Observatory since its founding, but what use is this if it is left unpublished.

Tebbutt's concluding paragraph is particularly interesting:

> In conclusion, I may say that by far the greater force of the Sydney Observatory is devoted to the cause of meteorology, to the obvious disadvantage of the sister science. That the efforts of the establishment on behalf of meteorology are praiseworthy and effective no one knows better than myself, and the results will become more and more valuable as the years roll on; but I would suggest the high importance of publishing all that has been done there for astronomy, and also of separating the two departments, so that while the present efforts on behalf of meteorology are continued, a better contribution may be made for the improvement of astronomy.

Supporting the 'letter' are additional details of the meteorology–astronomy imbalance, on the basis that although there are those whose meteorological thirst is quenched by the Observatory's weather maps and rain tables, '… there are others who have an astronomical thirst which has equal claims to be satisfied …' (ibid.). Tebbutt then discusses the Observatory's publications since 1862, and points out that '… while one linear inch of horizontal space on a library shelf is sufficient for the accommodation of astronomy, about twelve times that space is required for meteorology'. He then compares the published astronomical output of the Sydney and Melbourne Observatories, reminding his readers that 'It is very well-known that in European astronomical circles the Melbourne Observatory has long been regarded as the chief astronomical establishment in Australia …' Furthermore,

> Considering its profuse publication of meteorological work, its non-publication of its annual astronomical results, and its sparse communications to the astronomical journals, is it to be wondered at that the [Sydney] Observatory is regarded as chiefly a meteorological establishment, and therefore placed in the second class?

As added 'insurance', Tebbutt (1891a) had the suppressed letter published in full on 26 September 1891 in a rival newspaper, the *Australian*.

It is apparent that the letter on its own is far less critical than the pamphlet in dealing with the astronomy–meteorology dichotomy, and had it been allowed to appear in the *Sydney Morning Herald* then less damage would have been done and the matter may have died quickly—at least as a public issue. Instead, it festered for years as the pamphlet found its way onto the desks of innumerable Australian and overseas astronomers, and into the libraries of politicians, academics and others of influence. Between the newspaper version and the pamphlet, the Observatory's perceived 'ills' reached a wide audience.

Innes must be held partly responsible for this, for he was keen to assist with the dissemination of the pamphlet, and on 28 and 30 September he supplied Tebbutt with names of those people whom he thought should go on the mailing list (Innes 1891b, c). One month later, he suggested that the pamphlet could be sold cheaply through the principal booksellers in Sydney (Innes 1891d), while his friend, Walter

**Fig. 12.23** Walter Gale (*Courtesy* British Astronomical Association)

Gale (Fig. 12.23), had already offered to distribute copies in the scientific circles in which he moved (Gale 1891).

Tebbutt had to feel gratified that many individuals penned letters supporting his stand. Among these were well-known Australian and overseas astronomers such as Joseph Brooks (1891; Bio Box 8.5), Agnes Clerke (1891), J.U.C. Colyer (1891), R.L.J. Ellery (1891; Bio Box 7.3), Walter Gale (1891; Bio Box 12.1), R.T.A. Innes (1891b; Bio Box 8.2), the Reverend Dr. Thomas Roseby (1892), the Reverend W. Swindlehurst (1891; Bio Box 10.9) and Melbourne Observatory's E.J. White (1893; Bio Box 6.3).

The Government Astronomer of Victoria, Robert Ellery (Fig. 12.24), felt that the pamphlet

> … will stir up our friend to do a little more standard astrl work and print it. At the same time I am afraid these strictures worry him a bit.
>
> I have a very strong opinion about meteorological work as done at present, and the amount of time and money spent on printing volume upon volume of base observations in various parts of the world is appalling, and the amount of space already required to store these books is equally alarming. (Ellery 1891; his underlining).

Joseph Brooks, the 'Field Astronomer' in the New South Wales Lands Department (Orchiston 1987a), particularly regretted the little astronomy that had been done at the Observatory,

> … and this the more as I know that the Transit Circle was purchased by the Lands Department for the Trigonometrical Survey and when it arrived here it was transferred to the Observatory for the express purpose of determining the positions of any stars which might be required for that Survey. (Brooks 1891).

Russell's non-cooperation led eventually to strained relations between the two men, and Brooks informed Tebbutt that '… we have ceased to be on speaking terms even' (ibid.). So Tebbutt was not the only one Russell had difficulties with.

Meanwhile, the Reverend Dr. Thomas Roseby (Bio Box 12.4) informed Tebbutt that at the University of Sydney Professor Threlfall (Fig. 12.19) had been speaking about the relative merits of Sydney and Windsor Observatories and stated that it was only because of the work carried out by Tebbutt that the longitude of Sydney

**Fig. 12.24** Robert Ellery (*Courtesy* MuseumVictoria NM 59450)

Observatory was known (Roseby 1892). In this regard, back in 1878, Tebbutt (1878) had published a paper proposing a correction to the accepted longitude of that Observatory (cf. Russell 1878c), and this was followed by a further paper two years later (Tebbutt 1880b; cf. Tebbutt 1888).

**Bio Box 12.4: The Reverend Dr. Thomas Roseby**

The Reverend Dr. Thomas Roseby (Fig. 12.25) was born in Sydney on 8 April 1844, and was a brilliant scholar, graduating from the University of Sydney with M.A., LL.B. and LL.D. degrees and a Gold Medal (Phillips 1976). In 1872 he went to Dunedin, New Zealand, as a Congregational minister, and then spent from 1885 to 1889 in Ballarat before returning to Sydney (Dun 1919; Phillips 1976). He played a leading role in the development of the Church through to his retirement in 1911 and beyond (W.F.G. 1920).

**Fig. 12.25** The Reverend Dr. Thomas Roseby (*Courtesy* The Mitchell Library, Sydney)

(continued)

**Bio Box 12.4:** (continued)

Although active in astronomy while in Dunedin (Roseby 1882–83), Roseby came into his element with the founding of the New South Wales Branch of the British Astronomical Association in 1895. From the start he was a stalwart of the Branch, and for many years he was on the Council and served several sessions as President (see Orchiston 1988c). He also gave numerous popular presentations at Branch meetings, and '… by means of monthly notanda kept the members informed of the latest progress' (W.F.G. 1920).

Roseby maintained a private observatory, but 'His observational work was limited by the many calls upon his time, and was principally devoted to such double stars as were within the range of his 4-inch refractor' (ibid.). He also was interested in comets, and published a short paper on the orbital elements of Comet C/1894 G1 (Gale) in *Monthly Notices of the Royal Astronomical Society* (Roseby 1896), having been elected a Fellow of the Society the preceding year. However, his forté was the popularisation of astronomy (Leavitt, 1887; Orchiston 1997a), and Phillips (1976, p. 59) relates that he '… often gave educational evenings to church groups and students'. An original and creative thinker, Roseby '… saw no basic conflict between religion and science …' (ibid.).

The Reverend Dr. Thomas Roseby died of pneumonia on 16 December 1918 (W.F.G. 1920).

The Sydney independent astronomer J.W. Ward (Bio Box 10.7), proud owner of a 6-in. (15.2-cm) refractor, responded to Tebbutt's pamphlet by writing about the poor treatment that he had received from Russell over the years. To him the reason was simple: 'But Sir I am but a Bookmaker and cannot place half the alphabit [*sic*] after my name' (Ward 1892). Nor, incidentally, could Tebbutt!

One of Russell's Assistant Astronomers, Richard Pickering Sellors (Bio Box 12.5), wrote a tactful and carefully-worded letter which carried veiled support for Tebbutt (see Fig. 12.26), but his superior (and Russell's deputy), Henry Alfred Lenehan (Bio Box 7.2), was totally non-committal. He acknowledged receipt of the copy of *The Australian* with Tebbutt's severe criticism of the Observatory, and advised that Russell would write him in due course (Lenehan 1891). In fact no such communication is to be found in the 'Tebbutt Letters' in the Mitchell Library.

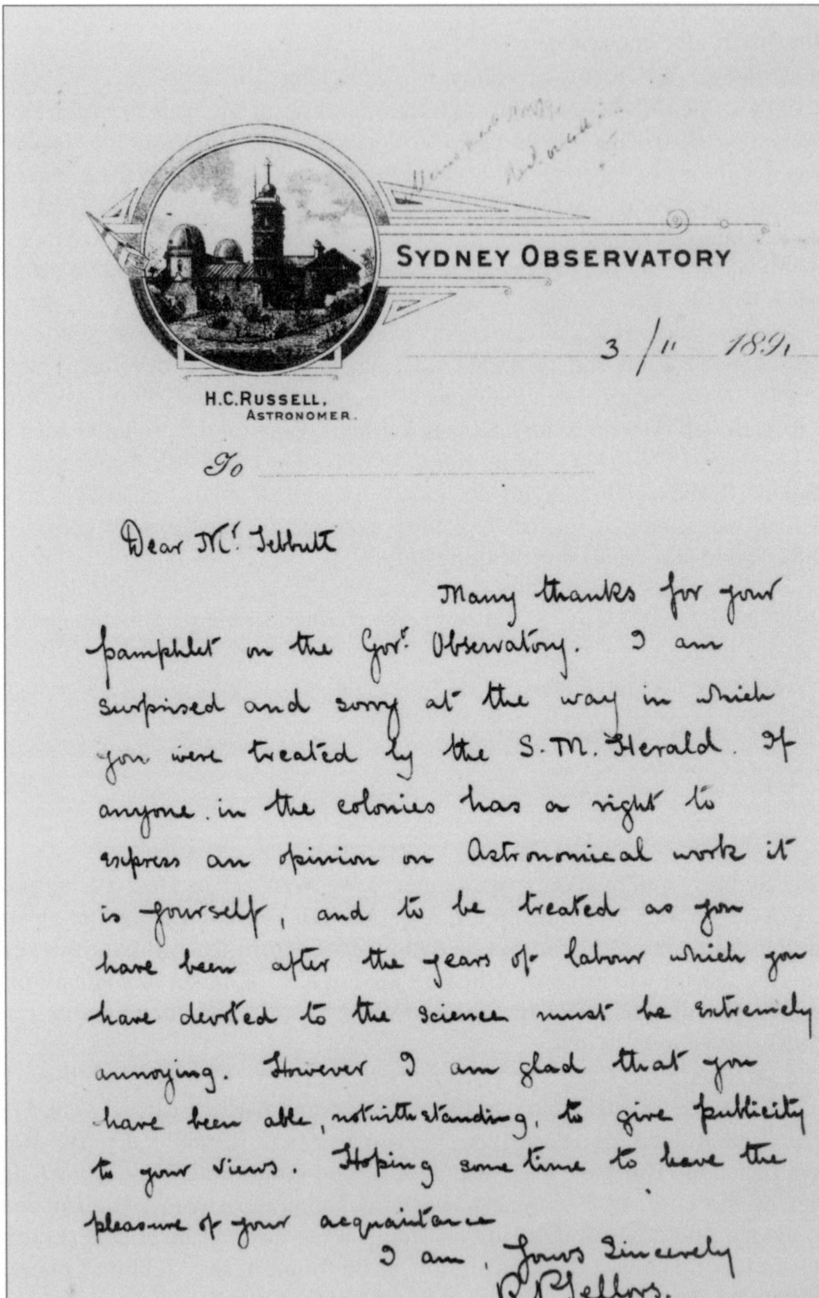
SYDNEY OBSERVATORY

3/11 1891

H.C. RUSSELL,
ASTRONOMER.

To

Dear M^r^ Tebbutt

Many thanks for your pamphlet on the Gov^t^ Observatory. I am surprised and sorry at the way in which you were treated by the S.M. Herald. If anyone in the colonies has a right to express an opinion on Astronomical work it is yourself, and to be treated as you have been after the years of labour which you have devoted to the science must be extremely annoying. However I am glad that you have been able, notwithstanding, to give publicity to your views. Hoping some time to have the pleasure of your acquaintance

I am, Yours Sincerely
R.P. Sellors.

**Fig. 12.26** R.P. Sellors' Letter to John Tebbutt (*Courtesy* The Mitchell Library, Sydney)

**Bio Box 12.5: Richard Pickering Sellors**

Richard Pickering Sellors was born in 1869 or 1870, and after completing a B.A. degree at the University of Sydney with a Gold Medal in Mathematics was appointed an Assistant Astronomer at Sydney Observatory in 1890 (Mr. R.P. Sellors … 1924).

For 10 years he worked at the Observatory, mainly using the 11.5-in. refractor to research double stars (Sellors 1892), but Russell later refused to allow him to publish his work (e.g. see Sellors 1894a), which caused some resentment.

In 1890, soon after its formation, Sellors joined the Astronomical Society of the Pacific, and the following year was elected a Fellow of the Royal Astronomical Society (Mr. R.P. Sellors … 1924). In March 1895 he joined the NSW Branch of the British Astronomical Association, but rarely participated in Branch activities (e.g. see Orchiston 1988c).

After leaving Sydney Observatory in 1900, Sellors joined the Lands Department and worked there until his retirement in 1924. For many years he was the Chief Computer in the Trigonometrical Survey, and Secretary of the Surveyors' Board (Mr. R.P. Sellors … 1924).

Richard Pickering Sellors died on 11 February 1937.

One of the non-astronomers to comment was H. Daly of Sydney, who mentioned Russell's '… excessive vanity. He likes to be mentioned in connection with everything & have kudos on all hands …' (Daly 1891). Another to write Tebbutt was his close friend T.H.F. Griffin (1891), who advocated a formal visit to the Government Minister in charge of the Observatory.

The *Daily Telegraph* newspaper acknowledged receipt of its copy of the Tebbutt pamphlet on 31 October, and mentioned that through this little publication 'Mr. Tebbutt repeats an old complaint of his that while the Sydney Observatory does much valuable meteorological work, the sister science of astronomy is greatly neglected' (Publications received 1891).

This whole affair also served to highlight the achievements of Windsor Observatory by comparison, and one whose eyes were opened was the Sydney Observatory technician, W. Masters:

> I confess astonishment as to the extent of your observatory, I was given the idea all the instruments were very diminutive … If there are any little trifles in the way of repair I could easily arrange for them. Have always several days leave in reserve … which can be taken … at short notice. (Masters 1892).

Given these comments, and others, do Tebbutt's claims have any substance? In order to provide some numerical quantification, I have prepared Table 12.2. This lists publications by Russell and Tebbutt in astronomy, meteorology and other topics at ten-yearly intervals from 1861 to 1900, where Russell's increased focus on meteorology during the 1880s and his accelerated astronomical output during the

**Table 12.2** Publications by Russell (R) and Tebbutt (T), 1861–1900 (cf. Orchiston 1989)

| | 1861–1870 | | 1871–1880 | | 1881–1890 | | 1891–1900 | | Totals | |
|---|---|---|---|---|---|---|---|---|---|---|
| Discipline | R | T | R | T | R | T | R | T | R | T |
| Astronomy | 0 | 36 | 33 | 75 | 13 | 134 | 24 | 110 | 70 | 355 |
| Meteorology | 3 | 1 | 18 | 1 | 22 | 0 | 18 | 1 | 61 | 3 |
| 'Other' | 0 | 0 | 3 | 0 | 7 | 1 | 3 | 1 | 13 | 2 |
| Totals | 3 | 37 | 54 | 76 | 42 | 135 | 45 | 112 | 144 | 360 |

1890s are obvious. However, it should be borne in mind that a number of astronomical publications listed actually report the work of subordinate staff even though they appeared in Russell's name (e.g. Russell 1893, 1894f, 1896, 1897), a practice which caused considerable resentment and may have contributed to the decisions by both Pollock and Sellors to quit the Observatory (e.g. see Pollock 1891).

Upon examining this table further, what is even more illuminating is to compare Russell's and Tebbutt's overall record of publication *across all disciplines* during this 40-year period, but as I have stressed elsewhere, '… it is important to remember that Tebbutt's output was the envy of many a professional observatory' (Orchiston 2000b, p. 10). So Tebbutt's claim that the published astronomical output of Sydney Observatory was '… quite unworthy of such a highly-equipped institution …' needs to be seen in context.

But what Tebbutt forgot, or conveniently ignored, was that Sydney Observatory was under great pressure from politicians, farmers and even the general public to provide meteorological rather than astronomical data. In many cases rural livelihoods depended upon accurate weather forecasts. South Australia's Sir Charles Todd (Bio Box 12.6) elaborates:

> That meteorology should have been taken up so energetically [in Australia] and been so liberally supported by the several Colonial Governments, on whose purse, in building up a new nation, there are many claims, is not, however, without a sufficient cause. To successfully occupy and establish industries in new countries, a knowledge of climate and meteorological conditions under which we are to labour is essential to success. … (Todd 1893).

In this context, Russell's bias towards meteorology was understandable. But it was ironic, given his excellent intentions when he took on the post of Government Astronomer in 1870. In one of his first letters to Tebbutt, he lamented the inordinate amount of time and effort that Smalley put into meteorology and the trigonometric survey of New South Wales, pledging that '… I should not like to have the credit of doing as little for Astronomy as he did' (Russell 1870a).

**Bio Box 12.6: Sir Charles Todd**

Charles Todd (Fig. 12.27) was born in London on 7 July 1826, and in December 1841 commenced working at the Royal Observatory Greenwich under George Biddell Airy (Symes 1976).

(continued)

**Bio Box 12.6:** (continued)

**Fig. 12.27** Sir Charles Todd (*Courtesy* The Mitchell Library, Sydney)

From 1848 to 1854 Todd worked as Assistant Astronomer at Cambridge University Observatory under Professor Challis, before returning briefly to Greenwich '… as superintendent of the galvanic apparatus for the transmission of time signals … [In the process] Todd became fascinated with telecommunications' (ibid.) which would serve him well in his next appointment.

This happened all too soon: Todd married, and on 5 November 1855 he and 18-yr-old Alice Todd arrived in Adelaide, South Australia, after he had accepted the position of Astronomical and Meteorological Observer and Head of the Electric Telegraph Department. Upon his arrival Todd found that his Electric Telegraph Department was almost non-existent, without a single telegraph line, but he quickly rectified this in March 1856 with a line established between Adelaide and the city's port. Nearly two and a half years later the telegraph systems of South Australia and Victoria were linked near Mount Gambier, and in 1868 Todd connected Adelaide and Sydney by telegraph. But his most ambitious scheme was to telegraphically link Port Augusta in South Australia and Darwin on the northern coast of Australia in the Northern Territory. In 1872 this remarkable scheme (referred to as the 'Overland Telegraph') was completed, and subsequently Australia was linked telegraphically to England via the undersea cable (Clune 1955). In 1870, while the Overland Telegraph was being constructed, the South Australian Government combined the Post Office and the Telegraph Department and Todd became the colony's new Postmaster General.

Todd also was a pioneering Australian meteorologist (Gibbs 1975) and used Australian and overseas telegraph networks to gather data. Thus, 'In the early 1880s Todd and his staff at the West Terrace Observatory in Adelaide were drawing inter-continental weather charts that had greater geographical reach than any other jurisdiction in the world. Todd's ability to pull together the individual threads of technology, weather science, and a widely dispersed group of weather observers put him in the forefront of the profession' (Charles Todd (pioneer) n.d.).

(continued)

**Bio Box 12.4:** (continued)

Despite heavy commitments demanded by his telegraphic schemes and meteorological role, Todd also found time to devote to astronomy, surveying and the introduction of electricity in Adelaide. He used astronomical observations to determine the position of Adelaide Observatory and created a time service for the colony, eventually having a time ball installed on a tower at Semaphore, Port Adelaide (Abell and Kinns 2010). He also carried out successful observations of the 1874 and 1882 transits of Venus (see Todd 1883; cf. Edwards 2004).

At an international level, Charles Todd was a Fellow of the Royal Astronomical Society, the Royal Meteorological Society and the Society of Electrical Engineers. In 1886 he was awarded an honorary M.A. by Cambridge University and in 1889 he achieved an even higher honour when elected a Fellow of the Royal Society. 'These two distinctions gave him great personal satisfaction …' (Symes 1876). At a local level Todd was a stalwart of the Royal Society of South Australia and was partially responsible for the formation of its Astronomy Section (see Orchiston 1998a).

In June 1893 Todd was knighted for his long and multifarious services to South Australia. Sir Charles was a popular employer and generally was happy and good humoured, but 'His besetting "weakness" was his constant punning of which, as a connoisseur of tea from his early days, the hackneyed example was "I'd be odd without my T"' (Symes 1976).

Sir Charles Todd died in Adelaide on 29 January 1910. For further overall details, see Cryle (2011) and for more information about his astronomical achievements, see Edwards (1993).

Tebbutt also was well aware that in the nineteenth century most so-called astronomical observatories were involved in trigonometrical surveys, and in a range of other scientific disciplines, such as meteorology, seismology, geomagnetism and tidal studies (see Orchiston 1988a). While he focussed his attack on astronomy and meteorology, Tebbutt forgot to mention that staff at Sydney Observatory also were engaged in geomagnetic and tidal studies. At least Russell had been wise enough to jettison the trigonometrical survey of the colony when he became Government Astronomer of New South Wales (Orchiston 1987a).

## 12.4 Paying a Heavy Price: The Consequences of Confrontation

While Tebbutt's brash, frontal attack certainly did not enthuse Russell, it did at least sting him into action on both the observational and publication fronts, thereby achieving the desired effect. Thus, on 24 June 1892, Russell was pleased to be able to advise

Tebbutt that the Observatory's meteorological results which '… had been waiting the printer's pleasure …' for so many years were about to be published (Russell 1892b; his underlining), as was astronomical work that had been waiting since 1880.

This same year saw the somewhat belated appearance of an attractively-bound and illustrated book about observations of the 1874 transit of Venus made in New South Wales (Russell 1892c; see Fig. 12.28). This volume contained more than 70 pages of text and 40 pages of plates, but not one word about Windsor Observatory,

**Fig. 12.28** The colourful cover of Russell's 1874 transit of Venus book, which only was published in 1892 (Orchiston collection)

even though Tebbutt not only recorded the event but also published accounts in *Astronomische Nachrichten* and the *Memoirs of the Royal Astronomical Society* (Tebbutt 1875a, c). All this was well known to Russell, who had even congratulated Tebbutt on his achievement at the time (see Russell 1874c)! Tebbutt's friend, Griffin (1893), was appalled by Russell's action.

Tebbutt (1893b) responded by submitting a long letter of complaint to the *Australian Star*, and this matter soon became international news when it entered the pages of *The Observatory* astronomical journal. In the October 1893 issue, mention is made of Tebbutt's newspaper letter, and the Editors (British astronomers) then make their position clear:

> We shall hope that this matter may be cleared up satisfactorily, but, as it stands at present, Mr. Tebbutt's position is, that after 40 years' astronomical work in New South Wales he is denied the right "to be styled a New South Wales observer." (Observatories … 1893).

When this happened, Russell (1894e) was quick to reply and express no surprise that '… Mr. Tebbutt should make this complaint after my experience of others which have gone before it …' Russell pointed out that the book only purported to report on those observing stations under his personal direction, '… in which it would obviously have been impossible to include Mr. Tebbutt's work, for he acted quite independently of me both in observing and publication' (cf. Russell 1894d). The Editors of *The Observatory* were unmoved, and at the end of Russell's letter presented their viewpoint: '… we feel bound to say that after reading his [Russell's] letter, and looking again carefully at the volume in question, we sympathize with Mr. Tebbutt'.

Tebbutt (1894b) also was dissatisfied with Russell's explanation, and after reading the relevant issues of *The Observatory*, Edward White from Melbourne Observatory wrote to Tebbutt about the 'shabby treatment' he had received from Russell, but reassured him:

> … now the subject has been pointedly referred to, you need not worry yourself as to the result. From the experience I have gained during my travels in Europe & America I should infer that Tebbutt was a better known name in the astronomical world than Russell, and I frequently hear surprise that such a quantity of first class work should be done by a private individual in such an out of the way corner of the globe.
>
> It is a pity that Russell is so narrow (in mind I mean, not body) he shows it in many petty ways. (White 1893).

Innes (1894a) wrote in a similar vein: 'I hope the lesson may do him good, he cannot injure your reputation but his pettiness is pitiful'. White (1893) and Wright (1896) later attributed this pettiness to jealousy.

In yet another snub, Russell purposely omitted any mention of Tebbutt in his 'Presidential Report' to the Royal Society of New South Wales in 1892. Although four and a half pages of the published version were devoted to astronomical activities in New South Wales (Russell 1892d), to all intents and purposes Windsor Observatory did not exist!

In order to revitalise his flagging astronomical interests, Russell became immersed in the Astrographic Catalogue and Carte du Ciel project, an international collaboration aimed at photographing the entire sky down to magnitude 14 (see Gingerich 1992; Turner 1912). Back in 1887 he had committed both Melbourne and Sydney Observatories to this Project (White 1988), and in September 1890 the 13-in. (33-cm) objective was received from Howard Grubb (Wood 1958). The 13-in. Sydney Astrograph (with its 7.25-in. Merz guidescope) subsequently was constructed, and as we have seen (Fig. 12.15) it was installed in the old observatory that Russell had erected in 1880 to house his 15-in. (38.1-cm) reflector.

Russell (1892a) then published a popular book about the 'Sydney Star Camera' (as he called it) in order to promote this project at a local level, but some in the local astronomical community were appalled that public taxes were used to produce this—and the transit of Venus volume—when there was such a backlog of astronomical observations awaiting publication. The South African astronomer, Alexander William Roberts (1893), facetiously referred to '… men who issue "picture books" and "Reminiscences" instead of scientific reports …', while *The Observatory* also referred to the transit of Venus book as '… a delightful picture book … which seems to be a little late in coming' (Notes 1893).

The Astrographic Catalogue and Carte du Ciel project came at a critical time in international astronomy as positional astronomy gave way to astrophysics (Clerke 1903; Langley 1889; Meadows 1984a, b), ushered in by the application of spectroscopy and photography to astronomical issues. It was within this environment of international change that Melbourne and Sydney Observatories ventured into astronomical photography, but they did so only in order to apply this new technique to positional astronomy. Consequently, this was a methodological rather than a philosophical change of direction for the observatories, but it was one that was greeted with suspicion by many of the nation's astronomers—Tebbutt included.

Russell somewhat exacerbated this by forecasting that '… photography will, in the course of a few years, take the place of the astronomer in recording astronomical events, just as in the same way it had taken the place of the artist in portraiture' (Tebbutt 1893a). As might be expected from a dedicated visual observer, Tebbutt (ibid.) took exception to this report. In reality, visual observations would remain the norm in Australian astronomy for many years, but as the potential of astrophotography became apparent, a number of Australian non-professional astronomers also began experimenting for themselves, with some success (e.g. see Reports of the Branches … 1895). Perhaps the most prominent of these was Melbourne's David Ross (Bio Box 9.3).

During 1893 Walter Gale visited England, and on 10 November he addressed a meeting of the Royal Astronomical Society, in the process referring to Sydney Observatory's earlier preoccupation with meteorology. This was subsequently reported in the Sydney press (see Astronomy in New South Wales 1894), and brought forth letters of support and elaboration from both Innes (1894d) and Tebbutt (1894a). Russell (1894c) also responded, and objected vehemently to some of what Gale had purportedly said. Gale (1894b) published the last word on the controversy on 6 February 1894, after his return to Sydney.

Further friction between Russell and Tebbutt surfaced over the discovery of a comet (C/1894 G1) by Walter Gale in April 1894, and this expanded to involve others in the local astronomical community as well. Gale (1894a) was more than a little taken aback when *The Age* newspaper in Melbourne reported that 'Mr Russell … has succeeded in securing observations of *a new comet* …', the clear inference being that it was Russell who made the actual discovery. Ellery (1894) subsequently clarified the matter: that it was a paragraph in the *Sydney Morning Herald* which '… led me to believe that Russell had first obsd. the Comet'. Roseby was not at all amused by this incident, and on 21 April 1894 he wrote Tebbutt:

> I am sure it will be pleasant to you to find that the credit of the colony is being in some measure sustained in spite of the indiffce of our official observatory. I think the failure to make some graceful references to Mr Gale's discovery on the part of our Govt: Astronomer is not creditable to his taste or good feeling. But, in fact, it was more than good fortune. The discovery could only have been made by one who knew the heavens well, & so was able to detect the stranger. (Roseby 1894; his underlining).

Innes (1894b) summarised the situation perfectly: 'Mr Russell gives no credit to anybody outside the observatory for any work over the Comet'.

The comet also created problems for other staff at Sydney Observatory. For example, on 1 May 1894 R.P. Sellors wrote to Tebbutt, after reading his orbital elements in the local newspaper:

> … I think you may care to see those which I have determined from my observations with the 11½ inch, and which I enclose.
>
> As I have not been permitted to compute them during office hours, but had to work on them during what little spare time I can call my own (often late at night after my evening's observations) I trust it will be some excuse for my being so late in the field with them.
>
> I am now refused permission to publish my results in any way, so I must ask you to please treat this communication as private. (Sellors 1894a; his underlining).

Meanwhile, Russell continued to issue more incorrect press statements and to further alienate himself from the strong astronomical contingent in Sydney (e.g. see Butterfield 1894; Fielder 1896; Lenehan 1903; Merfield 1896, 1903; Tebbutt 1897a, b; Thomson 1896; Wright 1903). One consequence of this was that he was passed over as President of the New South Wales Branch of the British Astronomical Association when this group was formed in Sydney at the end of 1894, and Tebbutt was offered—and accepted—the post (Orchiston 1988c). This topic is discussed in the following chapter.

At about the same time, Russell instructed his staff that no information whatsoever was to be supplied to public inquirers without his permission (Sellors 1894b). When Sellors sought approval to provide some data that Tebbutt had requested, Russell told him that he would attend to it personally (ibid.), and just one day later he wrote Tebbutt: '… it is usual to write to the head of the department for any information from records. I have much pleasure in stating that we have not yet received the numbers [journals] you refer to …' (Russell 1894a). When Tebbutt objected, Russell (1894b) retorted with: 'Yes the rule of the Government service is strict, but … I have to obey the rule'. In the past, either 'the rule' had not existed, or it conveniently had been ignored! Tebbutt also received a letter from Innes (1894c) at this time indicating that he, too, had been sent 'stiff letters' by Russell.

In 1895, Russell (1895a) wrote Tebbutt a terse letter complaining that two reprints that he had sent were omitted from the Windsor Observatory *Annual Report*, but he had to recant just two days later (Russell 1895b) when he discovered that his secretary had omitted to send them!

Meanwhile, Innes took Tebbutt's stand of 1891 one step further in June 1895, by writing to the Minister of Public Instruction about the implementation of Sydney Observatory annual reports and the introduction of a Board of Visitors. On 24 June he advised Tebbutt that if he did not receive a reply shortly he would write again, '… asking for a deputation to be received or say that I shall circularize the M.P.s so that when the estimates come on an explanation may be forthcoming' (Innes 1895a). This proved to be a fruitful approach, for 3 months later he was able to report (Innes 1895b) that steps were being taken to publish an annual report and that the matter of a Board was under consideration (although nearly 15 years were to elapse before this actually was set up).

In 1894 and 1895 a severe drought in New South Wales was preoccupying the lives and minds of people, and the efficacy of meteorological forecasts became a topic of public debate, as did prayers for rain. Understandably, it did not take long for Russell and Sydney Observatory to be drawn into the fray, as evidenced by an article which appeared in the *Windsor and Richmond Gazette* of 1 September 1894, which reads in part:

> Weather-wise Wragge is really a splendid weather prophet; and though our own Russell don't speak very plainly, it is clear that he is mighty jealous of the Seer of Banana-land. Russell is a jealous person, anyhow; he feared—and therefore hated—poor Egeson—small pumpkins as he was compared to Wragge—and now he sneers at the latter's deductions. The fact is our man is all behind the times—even in his own particular department—that is meteorology. As an astronomer he is infinitely inferior to … Mr John Tebbutt, of Windsor, who can run rings round the Fort Street man … Anyhow, Russell has had a pretty long innings; and as he is certainly not up to date, either in astronomy or meteorology, it is about time we paid our money to a first-class man. (Mr. Astronomer Tebbutt 1894).

At that time, British-born Clement Lindley Wragge (1852–1922; Fig. 12.29; Gibbs 1975; Wilson 1990) was the Government Meteorologist in Queensland, and

**Fig. 12.29** A photograph of Clement Wragge in about 1901 (https://en.wikipedia.org)

as the newspaper article indicates he was a popular public figure. However, some of his meteorological ideas and practices were questionable, and he was not popular with the nation's Government Astronomers (see Home and Livingston 1994).

Be that as it may, a similar article to the earlier one appeared some months later in the *Redfern Chronicle*, in Sydney, and was subsequently reprinted elsewhere. This also compared and contrasted Russell, Tebbutt, and Wragge, and includes the following perceptive comments:

> … Russell had made it a study—in fact, he was more a meteorologist than an astronomer. And now we have that wretched Wragge saying he isn't. So that our Government Astronomer seems just now to be neither one thing nor t'other. He is in a state of suspension … Whatever the man is he holds fast to his billet and regularly draws his six or seven hundred a year for doing something or other upon Observatory Hill—among other things being his garden parties to other distinguished Civil Servants and pensioners… No one here attempts to refute Mr. Tebbutt, while even the amateurs we have with us worry Russell's calculations. Of course he wiped out Egeson; but it will require a stronger man to wipe out either Wragge or Tebbutt. (Tebbutt, Wragge, and Russell 1895).

Both letters indicate the degree of public feeling against Russell in some camps, but they also illustrate the power of the press at a time when libel laws were obviously somewhat more lax than they are today. Russell could hardly have relished 'trial by media', but by then he must have become rather used to it, and he soon had R.T.A. Innes to contend with once again. Just before his departure for South Africa in early 1896, Innes penned a long letter to the Editor of the *Sydney Morning Herald* which basically reiterated the charges that Tebbutt had laid in his 1891 pamphlet (see Innes 1896).

The Sydney independent astronomer, Hugh Wright (Bio Box 2.1), also referred to the matter of the relative astronomical outputs of the Sydney and Windsor Observatories after reading Tebbutt's 1897 *Annual Report* (Tebbutt 1898):

> Without your labours this Colony would almost be unknown in the astronomical world. Whatever is being done at Sydney Observatory remains a mystery, or nothing is done. I believe the retiring president of the Royal Society of N.S.W. asked Mr. Russell to give him a summary of astronomy in the Colony for the past year, and the reply received was to the effect that Mr. Russell had been too busy during the period with routine work to give the science much attention, and consequently all other workers were ignored. (Wright 1898; his underlining).

The following year, after reading Tebbutt's 1898 *Annual Report*, Wright (1899) was moved to write: 'What a pity it is that you are not in charge of the Government Observatory! Then work of other than a red-tape character would be done …'

Charles Merfield (Bio Box 11.4) rejoined the fray in 1902, with the following comments:

> It is hopeless to expect any encouragement from our state observatories as in most cases the gentlemen in charge are more interested in collecting their salaries than in astronomy. Indeed I have often said that if you eliminate your work for astronomy in these climes, then there is very little left to say any thing about. (Merfield 1902).

Russell took ill with Bright's disease in October 1903, and went on leave, finally retiring on 28 February 1905. His successor, as Acting Government Astronomer, was Henry Lenehan (Bio Box 7.2), who inherited the controversy over the Observatory's output, but took a very different stance to that adopted by his predecessor. In writing to John Tebbutt, he said: '… I sincerely trust that if I can be of any service to you or others in observing or doing any thing in my power to serve the ends of scientific men I will have the chance of doing it …' (Lenehan 1904a). As a further indication of his attitude, in another letter written this same year, Lenehan (1904b) marks Tebbutt's retirement by congratulating him on a lifetime of achievement in astronomy and meteorology, and concludes: 'I am personally pleased that our intercourse has been so pleasant and in my humble way have ever born testimony to your work as a scientist and friend—a friendship I hope will be continued for many years'. Lenehan signs the letter, '… your sincere friend and admirer'.

Nor did Russell's departure from the Observatory protect him from continued criticism. Thus, on 30 January 1906 Merfield wrote:

> Really I think that one of our recent Govt. astronomers should have been paying into the Govt. funds instead of receiving thereof for the honour of the title he used, as I dont consider that he has advanced the science of astronomy in this state one bit; he has left nothing of value behind him so far as I can see. (Merfield 1906).

By at this stage Merfield was in an excellent position to 'see' first hand, for he was employed at Sydney Observatory (Orchiston 2015), and later he was to describe the equipment there as

> … a scrap heap of old iron and brass, for the most part. One good instrument remains, the meridian circle…
>
> The late Govt. Astronomer Mr. Russell had some hazy idea that he could improve on the construction of astronomical instruments from such firms as Cooke, Grubb etc. etc. … (Merfield 1909).

As a result, he remarked that the 11.5-in. Schroeder refractor (Fig. 8.8) is a '… useless piece of mechanism …' suitable only for public night demonstrations. Lenehan (1907a) had earlier mentioned that another of the telescopes was nonfunctional because Russell had dismembered it for a double pendulum of his own design, which did not work. To make matters worse, it was soon discovered that there were problems with work done to that date for the Sydney Zone of the Astrographic Catalogue project in that the settings of more than half of the plates were in error (Merfield 1908).

Russell died on 22 February 1907, and these adverse opinions of him soon reached the ears of his relatives. On 30 September 1907, Lenehan wrote Tebbutt that he had recently met the son, Mr. T. Russell, who '… was boiling at some remarks that he had heard of the unscientific work of his late esteemed father …' (Lenehan 1907b). Lenehan went on to point out that '… the family had a very exalted opinion of his attainments which his work left behind does not justify'. This is a useful assessment, coming as it did from one who knew and worked with Russell for a great many years.

Less than 2 weeks later Lenehan wrote that Russell's widow

> … is imbued with the false notion that all the animus that existed between yourself & HCR emanated from you. I immediately told her that the trouble was caused in the early stages

> by HCR ignoring you in his publication on the astronomers & workers in Australia. *Unfortunately they look on the goings & writings of the departed through spectacles tinted with great prejudice*. (Lenehan 1907c; my italics).

In reality, the breakdown of amiable relations between Tebbutt and Russell was a long drawn out affair, which commenced well before the appearance of Russell's 1888 paper (although this certainly served to accelerate the process), and lasted right up to his death.

It is equally clear that this relationship deeply upset and irritated Tebbutt, who in 1874 summarised his philosophy on 'fair play' in the *Sydney Morning Herald*: 'I am not unmindful of praise more than I am regardless of censure; but if I am to have either the one or the other with any degree of weight, it must come from those who are competent to give it' (Tebbutt 1874). While Russell was certainly qualified to offer either, Tebbutt felt that he did not give credit where it was due, and that he abused his position as Government Astronomer. Within a year of Russell's death, Tebbutt wrote in his *Astronomical Memoirs*:

> Since 1861 I have scarcely ever been directly attacked on astronomical matters; the main opposition locally to my efforts has been by the method of ignoring, a mode of opposition which is difficult to meet. (Tebbutt 1908, p. 28).

After Russell's departure, one of Tebbutt's aims—the separation of astronomy and meteorology at Sydney Observatory—was eventually achieved. In 1908 the Federal Government took over responsibility for the meteorological functions of the state observatories, but this turned out to be a two-edge sword for with the loss of funding that had been tied to meteorology the financial survival of these observatories would sooner or later become an issue.

Thus, when the Public Service Board carried out an investigation 'Into the Working of the Sydney Observatory' in 1909, the research output of that institution was still an issue, and comparisons were still being made with Windsor Observatory (see Public Service Board 1909), even though Tebbutt had officially retired as an active astronomer several years earlier!

## 12.5 Concluding Remarks

Underlying the Russell–Tebbutt feud was competition for local power and prestige, and because of the environment in which this occurred Russell was always going to be the victor at a local level.

From a political perspective, he held the upper hand. As a senior government employee he knew the vagaries of the public service and had the ear of Parliament. This he used to great effect over the years to obtain funding for the 1874 and 1882 transits of Venus programmes; purchase the Schroeder refractor, a new 6-in. (15.2-cm) transit telescope and the 13-in. (33-cm) astrographic telescope; add a new wing and dome to the observatory building; and expand the rural meteorological network (see Bhathal 1991; Orchiston 1988b; Wood 1958).

He also was able to exert considerable influence through the Royal Society of New South Wales and the Australasian Association for the Advancement of Science, and it is of interest that with the passage of time Tebbutt took less and less interest in the affairs of the former, while he was never active in the latter (even when pressed to give papers at the conferences, and serve on the Astronomy Section Committee).

Finally, through his membership of the Senate of the University of Sydney, and ultimately as Vice-Chancellor for a while, Russell was able to influence policy there and at the same time maintain a watching brief over staff with serious astronomical interests (such as J. Cook, G.H. Knibbs and J.A. Pollock). Had anyone wished to propose Tebbutt for an honorary degree then Russell would have been the obvious referee—with the power of veto.

Clearly, there was room for only one 'king' of New South Wales astronomy, and as Joseph Brooks (1897) so aptly put it, '… Mr Russell does not like anyone to be "near the throne"'. It was only when it came to a wider realm—the world stage—that Tebbutt was able to gain the ascendancy as Australia's leading astronomer. At the international level, commitment and passion triumphed over parochial power and politics, and Tebbutt reigned supreme.

**Acknowledgements** I wish to thank Dr. Richard Dodd (formerly Carter Observatory, Wellington, New Zealand), Dr Richard Gillespie (Museum Victoria), Professor Nick Lomb (University of Southern Queensland, Toowoomba), Nyree Morrison (University of Sydney Archives), the late Dr. Harley Wood (formerly Sydney Observatory) and staff at the Mitchell Library (Sydney) for their assistance in various ways, and the Julian Holland (Sydney), Mitchell Library (Sydney), MuseumVictoria (Melbourne), South African Astronomical Observatory, State Library of New South Wales and Sydney University Archives for kindly providing Figs. 12.4, 12.7, 12.11, 12.18, 12.20, 12.21 and 12.24–12.27.

## References[1]

Abell, L. & Kinns, R. (2010). 'Telegraph' Todd and the Semaphore time ball. Journal of the Historical Society of South Australia, 38, 42–57.

Ashbrook, J. (1984). John Tebbutt, his observatory, and a probable nova. In J. Ashbrook, *The Astronomical Scrapbook. Skywatchers, Pioneers and Seekers in Astronomy* (pp. 66–71). Cambridge (Mass.): Sky Publishing Corporation.

Astronomy in New South Wales. *Daily Telegraph*, 11 January 1894.

Baracchi, P. (1914). Astronomy and geodesy in Australia. In G.H. Knibbs (Ed.), *Federal Handbook Prepared in Connection with the Eighty-fourth Meeting of the British Association for the Advancement of Science, Held in Australia, August, 1914* (pp. 326–390). Melbourne: Government Printer.

Bhathal, R. (1991). Henry Chamberlain Russell—astronomer, meteorologist and scientific entrepreneur. *Journal and Proceedings of the Royal Society of New South Wales, 124,* 1–21.

Brooks, J. (1891). Letter to John Tebbutt, dated 8 October. In TL.

Brooks, J. (1897). Letter to John Tebbutt, dated 28 April. In TL.

[1] The following abbreviation is used: TL = Letters to J. Tebbutt, bound manuscript letters in the Mitchell Library, Sydney.

Butterfield, G. (1881a). Letter to John Tebbutt, dated 5 May. In TL.
Butterfield, G. (1881b). Letter to John Tebbutt, dated 14 November. In TL.
Butterfield, G. (1894). Letter to John Tebbutt, dated 1 May. In TL.
Cable, K.J. (1976). Woolls, William (1814–1893). In G. Serle & R. Ward (Eds.), *Australian Dictionary of Biography. Volume 6: 1851–1890. R-Z* (pp. 437–438). Melbourne: Melbourne University Press.
Charles Todd (pioneer) n.d. Web site: https://en.wikipedia.org/wiki/Charles_Todd_(pioneer)
Clerke, A. (1891). Letter to John Tebbutt, dated 6 December. In TL.
Clerke, A. (1903). *Problems in Astrophysics*. London: Adam and Charles Black.
Clune, F. (1955). *Overland Telegraph*. Sydney: Angus & Robertson.
Colyer, J.U.C. (1891). Letter to John Tebbutt, dated 3 October. In TL.
Cozens, G., Walsh, A. & Orchiston, W. (2010). James Dunlop's historical catalogue of Southern nebulae and clusters. *Journal of Astronomical History and Heritage, 13,* 59–73.
Cryle, D. (2011). From outback icon to imperial time lord: "Reinventing" Sir Charles Todd (1910–2010). *Journal of Australian Studies, 35,* 65–82.
Daly, H. (1891). Letter to John Tebbutt, dated 29 October. In TL.
Dun, W.S. (1919). Presidential address. *Journal and Proceedings of the Royal Society of New South Wales, 53,* 1–28.
Edwards, P.G. (1993). Charles Todd and the Adelaide Observatory. *Proceedings of the Astronomical Society of Australia, 10,* 349–354.
Edwards, P.G. (2004). Charles Todd's observations of the transits of Venus. *Journal of Astronomical History and Heritage, 7,* 1–7.
Egeson, C. (1889a). *Egeson's Weather System of Sun-Spot Causality: Being Original Researchers in Solar and Terrestrial Meteorology.* Sydney: Turner & Henderson.
Egeson, C. (1889b). Letter to John Tebbutt, dated 16 November. In TL.
Egeson, C. (1890). Letter to John Tebbutt, dated 29 March. In TL.
Eglinton Family Papers. Oxley Library, Brisbane.
Ellery, R.J. (1891). Letter to John Tebbutt, dated 29 October. In TL.
Ellery, R.J. (1894). Letter to John Tebbutt, dated 11 April. In TL.
Fielder, A. (1896). The lunar eclipse. *Sunday Times,* 30 August.
Finds Venus Unexciting. Newsclipping, 1943, Riverview Archives.
Gale, W. (1886). Letter to John Tebbutt, dated 2 May. In TL.
Gale, W. (1891). Letter to John Tebbutt, dated 31 October. In TL.
Gale, W. (1894a). Letter to John Tebbutt, dated 10 April. In TL.
Gale, W. (1894b). Mr. Gale and the Sydney Observatory. *Daily Telegraph,* 6 February.
Gale, W. (1921). Astronomical facts and fictions. *Transactions of the Sydney Lodge of Research, 8,* 1–14.
Gale, W. (1928). Letter to Mr Longbottom, dated 8 May. In BAA Letter Archives.
Gibbs, W.J. (1975). *The Origins of Australian Meteorology*. Canberra: Australian Government Publishing Service (Department of Science Bureau of Meteorology, Historical Notes No. 1).
Gilbert, L. (1985). *William Woolls, 1814–1893: "A Most Useful Colonist"*. Canberra: Mulini Press.
Gingerich, O. (1970). The satellites of Mars: prediction and discovery. *Journal for the History of Astronomy, 1,* 109–115.
Gingerich, O. (Ed.) (1984). *The General History of Astronomy. Volume 4. Astrophysics and Twentieth Century Astronomy to 1950: Part A*. Cambridge, Cambridge University Press.
Gingerich, O. (1992). The Great Comet and the "Carte". In *The Great Copernicus Chase and Other Adventures in Astronomical History* (pp. 189–194). Cambridge (Mass.): Sky Publishing Corporation.
Goodin, V.W.E. (1947). Parramatta Observatory. The story of an absurdity. *Journal and Proceedings of the Royal Australian Historical Society, 33,* 173–187.
Griffin, T.H.F. (1881). Letter to John Tebbutt, dated 5 April. In TL.
Griffin, T.H.F. (1882). Letter to John Tebbutt, dated 5 October. In TL.

Griffin, T.H.F. (1891). Letter to John Tebbutt, dated 29 September. In TL.
Griffin, T.H.F. (1893). Letter to John Tebbutt, dated 24 June. In TL.
Harkness, W. (1887). Letter to John Tebbutt, dated 31 March. In TL.
Haynes, R.D., Haynes, R.A. & Kitson, W.S. (1993). The history of astronomy in Queensland. *Vistas in Astronomy, 36,* 231–252.
Hicks, S. & Frankland, K (1998). Amateur scientists: finding and preserving their contributions to the growth of scientific research in Australia. Web site: http://www.asap.unimelb.edu.au/confs/recovering/hicks.htm
Home, R. (1986). First physicist of Australia: Richard Threlfall at the University of Sydney, 1886–1898. *Historical Records of Australian Science, 6,* 333–357.
Home, R. & Livingston, K.T. (1994). Science and technology in the story of Australian federation: the case of meteorology, 1876–1908. *Historical Records of Australian Science, 10,* 109–127.
Hordern, A. (1870). Letter to John Tebbutt, dated 6 July. In TL.
Hordern, A. (1872). Letter to John Tebbutt, dated 23 May. In TL.
Innes, R.T.A. (1891a). Letter to John Tebbutt, dated 26 August. In TL.
Innes, R.T.A. (1891b). Letter to John Tebbutt, dated 28 September. In TL.
Innes, R.T.A. (1891c). Letter to John Tebbutt, dated 30 September. In TL.
Innes, R.T.A. (1891d). Letter to John Tebbutt, dated 24 October. In TL.
Innes, R.T.A. (1894a). Letter to John Tebbutt, dated 13 March. In TL.
Innes, R.T.A. (1894b). Letter to John Tebbutt, dated 26 April. In TL.
Innes, R.T.A. (1894c). Letter to John Tebbutt, dated 14 November. In TL.
Innes, R.T.A. (1894d). Mr. Gale and the Sydney Observatory. *Daily Telegraph,* 16 January.
Innes, R.T.A. (1895a). Letter to John Tebbutt, dated 24 June. In TL.
Innes, R.T.A. (1895b). Letter to John Tebbutt, dated 27 September. In TL.
Innes, R.T.A. (1896). Weather forecasting at the Observatory. *Sydney Morning Herald,* 8 January.
Kitson, W.S., 1990. Thomson, James Park (1854–1941). *Australian Dictionary of Biography, Volume 12*. Melbourne: Melbourne University Press (http://adb.anu.edu.au/biography/thomson-james-park-8797/text15427, accessed online 10 April 2016).
Langley, S.P. (1889). *The New Astronomy*. Boston: Houghton, Mifflin and Company.
Leavitt, T.W.H. (1887*). Australian Representative Men*. Melbourne: Wells and Leavitt.
Lenehan, H.A. (1887a). Letter to John Tebbutt, dated 7 July. In TL.
Lenehan, H.A. (1887b). Letter to John Tebbutt, dated 8 July. In TL.
Lenehan, H.A. (1887c). Letter to John Tebbutt, dated 28 December. In TL.
Lenehan, H.A. (1888). Letter to John Tebbutt, dated 4 December. In TL.
Lenehan, H.A. (1891). Letter to John Tebbutt, dated 28 September. In TL.
Lenehan, H.A. (1903). Letter to John Tebbutt, dated 24 April. In TL.
Lenehan, H.A. (1904a). Letter to John Tebbutt, dated 19 January. In TL.
Lenehan, H.A. (1904b). Letter to John Tebbutt, dated 13 September. In TL.
Lenehan, H.A. (1907a). Letter to John Tebbutt, dated 15 September. In TL.
Lenehan, H.A. (1907b). Letter to John Tebbutt, dated 30 September. In TL.
Lenehan, H.A. (1907c). Letter to John Tebbutt, dated 11 October. In TL.
Lomb, N. (2004). The instruments from the Parramatta Observatory. *Historical Records of Australian Science, 15,* 211–222.
Lomb, N. (2011). *Transit of Venus: 1631 to the Present.* Sydney: NewSouth & Powerhouse Museum.
Lomb, N. (2016). Australian solar eclipse expeditions: the voyage to Cape York in 1871. *Journal of Astronomical History and Heritage, 19,* 79–95.
McCaughan, J.B.T. (1988). Pollock, James Arthur (1865–1922). In G. Serle (Ed.), *Australian Dictionary of Biography. Volume 11: 1891–1939. Nes-Smi* (pp. 253–254). Melbourne: Melbourne University Press.
Macdonnell, W.J. (1880). Letter to John Tebbutt, dated 5 May. In TL.
McGuinness, M. (2014). Under the Southern Skies—A Rich History of Astronomy in the Colony of New South Wales. History Services Blog, at: http://historyservicesnswblog.blogspot.com/2014/01/under-southern-skies-rich-history-of.html

MacLeod, R. (ed.)(1988). *The Commonwealth of Science: ANZAAS and the Scientific Enterprise in Australasia, 1888–1988*. Melbourne: Oxford University Press.

Masters, W. (1887). Letter to John Tebbutt, dated 9 May. In TL.

Masters, W. (1892). Letter to John Tebbutt, dated 3 April. In TL

Meadows, J. (1984a). The new astronomy. In Gingerich, 59–72.

Meadows, J. (1984b). The origins of astrophysics. In Gingerich, 3–15.

Mellow, D.P., 1974. Liversidge, Archibald (1846–1927). In *Australian Dictionary of Biography, Volume 5*. Melbourne: Melbourne University Press [http://adb.anu.edu.au/biography/liversidge-archibald-4027/text6394, accessed online 9 April 2016].

Merfield, C.J. (1896). Letter to John Tebbutt, dated 28 February. In TL.

Merfield, C.J. (1902). Letter to John Tebbutt, dated 20 January. In TL.

Merfield, C.J. (1903). Letter to John Tebbutt, dated 21 April. In TL.

Merfield, C.J. (1906). Letter to John Tebbutt, dated 30 January. In TL.

Merfield, C.J. (1908). Letter to John Tebbutt, dated 5 October. In TL.

Merfield, C.J. (1909) Letter to John Tebbutt, dated 9 January. In TL.

Morrison-Low, A.D. (2004). The soldier-astronomer in Scotland: Thomas Makdougall Brisbane's scientific work in the northern hemisphere. *Historical Records of Australian Science, 15,* 151–176.

Mr. Astronomer Tebbutt. *Windsor and Richmond Gazette,* 1 September 1894.

Mr. R.P. Sellors public servant retires. *Sydney Morning Herald*, 25 July 1924 (page 8).

Notes. *The Observatory, 16,* 367 (1893).

Obituary (1945). Obituary. Walter F. Gale, F.R.A.S. *Journal of the British Asronomical Association, 56,* 18–19.

Observatories. Mr. Tebbutt's, N.S. Wales. *The Observatory, 16,* 361–362 (1893).

Orchiston, W. (1982). John Tebbutt and the Whakatane Eight Inch Refractor: a review of "The Australian Connection". *Southern Stars, 29,* 215–244.

Orchiston, W. (1986). *Observatories of Victoria. 1: Dr William Bone and the Castlemaine Observatory*. Melbourne: Astronomical Society of Victoria (Historical Notes No.3).

Orchiston, W. (1987a). Contribution of the Lands Department to the development of astronomy in New South Wales during the nineteenth century. *Australian Journal of Astronomy, 2,* 65–74.

Orchiston, W. (1987b). Dr William Bone, and the role of the amateur observatory in Australian astronomy. *Southern Stars*, *32,* 111–128.

Orchiston, W. (1988a). The emergence of professional astronomy in nineteenth century Australia. *Search, 19,* 76–81.

Orchiston, W. (1988b). From research to recreation: the rise and fall of Sydney Observatory. *Vistas in Astronomy, 32,* 49–63.

Orchiston, W. (1988c). Illuminating incidents in Antipodean astronomy: formation of the New South Wales Branch of the British Astronomical Association. *Journal of the British Astronomical Association, 98,* 75–84.

Orchiston, W. (1988d). Illuminating incidents in Antipodean astronomy: John Tebbutt and the Sydney Observatory directorship of 1862. *Australian Journal of Astronomy, 2,* 149–158.

Orchiston, W. (1988e). Illuminating incidents in Antipodean astronomy: Phillip Parker King and the founding of Sydney Observatory. *Vistas in Astronomy, 32,* 285–301.

Orchiston, W. (1989). The role of the amateur in early Australian astronomy. *Search, 20,* 16–23.

Orchiston, W. (1992). The contribution of Francis Abbott to Tasmanian and Australian astronomy. *Vistas in Astronomy, 35,* 315–344.

Orchiston, W. (1997a). The role of the amateur in popularising astronomy: an Australian case study. *Australian Journal of Astronomy, 7,* 33–66.

Orchiston, W. (1997b). The role of the large refracting telescope in Australian amateur astronomy: an historical perspective. *Australian Journal of Astronomy, 7,* 89–114.

Orchiston, W. (1998a). Amateur-professional collaboration in Australian science: the earliest astronomical groups and societies. *Historical Records of Australian Science, 12,* 163–182.

Orchiston, W. (1998b). Mission impossible: William Scott and the first Sydney Observatory directorship. *Journal of Astronomical History and Heritage, 1,* 21–43.

Orchiston, W. (1999). C/1881 K1: a forgotten "Great Comet" of the nineteenth century. *Irish Astronomical Journal, 26,* 33–44.

Orchiston, W. (2000a). Illuminating incidents in Antipodean astronomy: H.C. Russell and the origin of the Palomar-type mounting. *Journal of the Antique Telescope Society, 19,* 13–15.

Orchiston, W. (2000b). Politics and personalities in Australian astronomy: the Russell–Tebbutt feud. *AAO Newsletter, 95,* 8–11.

Orchiston, W. (2001a). Australian innovation in telescope design: origin of the horseshoe mounting. *AAO Newsletter, 97,* 12–13.

Orchiston, W. (2001b). From amateur astronomer to observatory director: the curious case of R.T.A. Innes. *Publications of the Astronomical Society of Australia, 18,* 317–327.

Orchiston, W. (2002). Tebbutt vs Russell: passion, power and politics in nineteenth century Australian astronomy. In S.M.R. Ansari (Ed.), *History of Oriental Astronomy* (pp. 169–201). Dordrecht: Kluwer (IAU Joint Discussion 17).

Orchiston, W. (2003). Passion and patronage: the Afro-Australian astronomy of Robert Thorburn Ayton Innes. *Monthly Notices of the Astronomical Society of Southern Africa, 62,* 4–13.

Orchiston, W. (2004a). John Tebbutt and observational astronomy at Windsor Observatory. *Journal of the British Astronomical Association, 114,* 141–154.

Orchiston, W. (2004b). The nineteenth century transits of Venus: an Australian and New Zealand overview. *Journal of Astronomical Data, 10,* 219–308.

Orchiston, W., 2010. The changing role of the 'Catts Telescope': The life and times of a nineteenth century 20-inch Grubb reflector. *Journal of Astronomical History and Heritage*, *13,* 235–254.

Orchiston, W. (2015). The amateur-turned-professional syndrome: two Australian case studies. In W. Orchiston, D.A. Green & R. Strom (Eds.), *New Insights from Recent Studies in Historical Astronomy: Following in the Footsteps of F. Richard Stephenson* (pp. 259–350). Cham: Springer.

Orchiston, W. & Bembrick, C. (1995). The role of the large reflecting telescope in amateur astronomy: an Australian case study. *Australian Journal of Astronomy, 6,* 53–74.

Orchiston, W. & Bembrick, C. (1997). The potency of patronage: George Hoskins and the New South Wales Branch of the British Astronomical Association. *Australian Journal of Astronomy, 7,* 1–15.

Orchiston, W. & Bhathal, R. (1984). Illuminating incidents in Antipodean astronomy: attempted formation of the Australian Astronomical Society in 1892. *Proceedings of the Astronomical Society of Australia, 5,* 422–425.

Orchiston, W. & Bhathal, R. (1991). Illuminating incidents in Antipodean astronomy: Section A of the Royal Society of New South Wales. *Australian Journal of Astronomy, 4,* 49–67.

Page, A. (1959). Astronomy in Queensland 1849–1959. *Proceedings of the Astronomical Society of Queensland,* 19–24.

Phillips, W. (1976). Roseby, Thomas (1844–1918). *Australian Dictionary of Biography. Volume 6.* Melbourne: Melbourne University Press [http://adb.anu.edu.au/biography/roseby-thomas-4505/text7367, accessed online 11 April 2016].

Pollock, J.A. (1887a.) Letter to John Tebbutt, dated 2 May. In TL.

Pollock, J.A. (1887b). Letter to John Tebbutt, dated 28 December. In TL.

Pollock, J.A. (1888). Letter to John Tebbutt, dated 19 June. In TL.

Pollock, J.A. (1891). Letter to John Tebbutt, dated 22 April. In TL.

*Popular Science Monthly, 63,* 86 (1903).

Powys, A.O. (1881a). Letter to John Tebbutt, dated 17 May. In TL.

Powys, A.O. (1881b). Letter to John Tebbutt, dated 21 September. In TL.

Pritchard, C. (1886). Letter to John Tebbutt, dated 21 January. In TL.

Prof. J.A. Pollock, F.R.S. *Nature, 110,* 359–361 (1922).

Public Service Board (1909). Investigation Under Section 9 of the Public Service Act Into the Working of the Sydney Observatory. MS, Archives Office of New South Wales, Sydney (8/400.2).

Publications received. *Daily Telegraph,* 31 October 1891.

Reports of the Branches. New South Wales (Sydney). *Journal of the British Astronomical Association, 5,* 344–345 (1895).
Richardson, W. (1835). *Catalogue of 7385 Stars, Chiefly in the Southern Hemisphere, Prepared from Observations Made in the Years 1822, 1823, 1824, 1825, and 1826, at the Observatory at Paramatta [sic], New South Wales …* London: His Majesty's Stationery Office.
Roberts, A.W. (1893). Letter to John Tebbutt, dated 28 November. In TL.
Robertson, B. (1985). Obituaries. Harley Wood. *Proceedings of the Astronomical Society of Australia, 6,* 111–112.
Roseby, T. (1882–83). A plea for the stars. *New Zealand Journal of Science and Technology, 1,* 63–68.
Roseby, T. (1892). Letter to John Tebbutt, dated 25 June. In TL.
Roseby, T. (1894). Letter to John Tebbutt, dated 21 April. In TL.
Roseby, T. (1896). Elliptical orbit elements of Comet *b* 1894 (Gale). *Monthly Notices of the Royal Astronomical Society, 56,* 329.
R.T. (1923). James Arthur Pollock, 1865–1922. *Proceedings of the Royal Society, A104,* xv–xxii.
Russell, H.C. (1870a). Letter to John Tebbutt, dated 27 October. In TL.
Russell, H.C. (1870b). Letter to John Tebbutt, dated 23 November. In TL.
Russell, H.C. (1870c). Letter to John Tebbutt, dated 30 November. In TL.
Russell, H.C. (1870d). Letter to John Tebbutt, dated 3 December. In TL.
Russell, H.C. (1871a). Letter to John Tebbutt, dated 5 January. In TL.
Russell, H.C. (1871b). Letter to John Tebbutt, dated 23 February. In TL.
Russell, H.C. (1871c). Letter to John Tebbutt, dated 17 March. In TL.
Russell, H.C. (1871d). Letter to John Tebbutt, dated 28 March. In TL.
Russell, H.C. (1871e). Letter to John Tebbutt, dated 26 June. In TL.
Russell, H.C. (1871f). Letter to John Tebbutt, dated 4 November. In TL.
Russell, H.C. (1871g) Letter to John Tebbutt, dated 6 November. In TL.
Russell, H.C. (1871h). Meteorology in New South Wales. In *The Industrial Progress of New South Wales: Being A Report of the Intercolonial Exhibition of 1870, at Sydney; Together With a Variety of Papers Illustrative of the Industrial Resources of the Colony* (pp. 579–615). Sydney: Government Printer.
Russell, H.C. (1872a). Letter to John Tebbutt, dated 24 April. In TL.
Russell, H.C. (1872b). Letter to John Tebbutt, dated 1 May. In TL.
Russell, H.C. (1872c). Letter to John Tebbutt, dated 8 May. In TL.
Russell, H.C. (1872d). Letter to John Tebbutt, dated 13 May. In TL.
Russell, H.C. (1872e). Letter to John Tebbutt, dated 24 May. In TL.
Russell, H.C. (1872f). Letter to John Tebbutt, dated 10 June. In TL.
Russell, H.C. (1872g). Letter to John Tebbutt, dated 18 June. In TL.
Russell, H.C. (1872h). Letter to John Tebbutt, dated 18 October. In TL.
Russell, H.C. (1872i). Letter to John Tebbutt, dated 11 November. In TL.
Russell, H.C. (1872j). Letter to John Tebbutt, dated 14 November. In TL.
Russell, H.C. (1873a). Letter to John Tebbutt, dated 23 April. In TL.
Russell, H.C. (1873b). Letter to John Tebbutt, dated 6 June. In TL.
Russell, H.C. (1874a). Letter to John Tebbutt, dated 7 March. In TL.
Russell, H.C. (1874b). Letter to John Tebbutt, dated 14 May. In TL.
Russell, H.C. (1874c). Letter to John Tebbutt, dated 15 December. In TL.
Russell, H.C. (1875a). Letter to John Tebbutt, dated 10 February. In TL.
Russell, H.C. (1875b). The transit of Venus. *Sydney Morning Herald*, 18 January.
Russell, H.C. (1876a). Letter to John Tebbutt, dated 10 April. In TL.
Russell, H.C. (1876b). Letter to John Tebbutt, dated 25 May. In TL.
Russell, H.C. (1876c). Letter to John Tebbutt, dated 26 May. In TL.
Russell, H.C. (1876d). Letter to John Tebbutt, dated 2 October. In TL.
Russell, H.C. (1877). Letter to John Tebbutt, dated 26 September. In TL.
Russell, H.C. (1878a). Letter to John Tebbutt, dated 7 October. In TL.

Russell, H.C. (1878b). Letter to John Tebbutt, dated 14 October. In TL.
Russell, H.C. (1878c). On the longitude of Sydney Observatory. *Journal and Proceedings of the Royal Society of New South Wales, 12,* 222–225.
Russell, H.C. (1879). Letter to John Tebbutt, dated 12 December. In TL.
Russell, H.C. (1880a). Letter to John Tebbutt, dated 20 September. In TL.
Russell, H.C. (1880b). Letter to John Tebbutt, dated 7 October. In TL.
Russell, H.C. (1880c). Recent changes in the surface of Jupiter. *Journal and Proceedings of the Royal Society of New South Wales, 14,* 63–76.
Russell, H.C. (1880d). Thunder and hail storms in New South Wales. *Journal and Proceedings of the Royal Society of New South Wales, 14,* 51–61.
Russell, H.C. (1881a). The Comet. *Sydney Morning Herald,* 26 May.
Russell, H.C. (1881b). Letter to John Tebbutt, dated 8 February. In TL.
Russell, H.C. (1881c). Letter to John Tebbutt, dated 25 May. In TL.
Russell, H.C. (1882a). Letter to John Tebbutt, dated 22 February. In TL.
Russell, H.C. (1882b). *The Sydney Observatory. History and Progress*. Sydney: Government Printer.
Russell, H.C. (1883). Observations of the transit of Venus, 1874, December 8–9, made at stations in New South Wales. *Memoirs of the Royal Astronomical Society, 47,* 49–88.
Russell, H.C. (1884). Letter to John Tebbutt, dated 30 September. In TL.
Russell, H.C. (1886). Letter to John Tebbutt, dated 20 April. In TL.
Russell, H.C. (1887a). Letter to John Tebbutt, dated 15 January. In TL.
Russell, H.C. (1887b). Letter to John Tebbutt, dated 29 December. In TL.
Russell, H.C. (1888). Astronomical and meteorologial workers in New South Wales, 1788 to 1860. In A. Liversidge & R. Etheridge, R. (Eds.), *Report of the First Meeting of the Australasian Association for the Advancement of Science* (pp. 45–94). Sydney: Australasian Association for the Advancement of Science.
Russell, H.C. (1892a). *Description of the Star Camera, at Sydney Observatory*. Sydney: Government Printer.
Russell, H.C. (1892b). Letter to John Tebbutt, dated 24 June. In TL.
Russell, H.C. (1892c). *Observations of the Transit of Venus, 9 December, 1874; Made at Stations in New South Wales*. Sydney: Government Printer.
Russell, H.C. (1892d). *Presidential Address*. Sydney: Royal Society of New South Wales.
Russell, H.C. (1893). Measures of double stars. *Astronomische Nachrichten, 132,* 145–154.
Russell, H.C. (1894a). Letter to John Tebbutt, dated 6 November. In TL.
Russell, H.C. (1894b). Letter to John Tebbutt, dated 12 November. In TL.
Russell, H.C. (1894c). Mr. Gale and the Sydney Observatory. *Daily Telegraph,* 15 January.
Russell, H.C. (1894d). Mr. Tebbutt and the transit of Venus, 1874. *Australian Star,* 22 March.
Russell, H.C. (1894e). Mr. Tebbutt and the transit of Venus, 1874. *The Observatory, 17,* 81–82.
Russell, H.C. (1894f). Recent measures of double stars made at Sydney. *Astronomische Nachrichten, 135,* 417–424.
Russell, H.C. (1895a). Letter to John Tebbutt, dated 6 May. In TL.
Russell, H.C. (1895b). Letter to John Tebbutt, dated 8 May. In TL.
Russell, H.C. (1896). Measures of double stars made at Sydney. *Astronomische Nachrichten, 141,* 137–148.
Russell, H.C. (1897). Measures of double stars made at Sydney 1896. *Astronomische Nachrichten, 143,* 233–242.
Saunders, S.D. (1990). *Astronomy in Colonial New South Wales: 1788 to 1858*. Ph.D. Thesis, Department of General Philosophy, University of Sydney, Sydney, Australia.
Saunders, S.D. (2004). Sir Thomas Brisbane's legacy to colonial science: colonial astronomy at the Parramatta Observatory, 1822–1848. *Historical Records of Australian Science, 15,* 177–209.
Scott, W. (1871). Letter to John Tebbutt, dated 31 October. In TL.
Sellors, R.P. (1892). The orbit of $\gamma$ Coronae Australia. *Monthly Notices of the Royal Astronomical Society, 53,* 45–47.

Sellors, R.P. (1894a). Letter to John Tebbutt, dated 1 May. In TL.
Sellors, R.P. (1894b). Letter to John Tebbutt, dated 5 November. In TL.
Smyth, P. (1887). Letter to John Tebbutt, dated 3 May. In TL.
Stott, C., and Hughes, D.W. (1987). The amateur's small transit telescope in the nineteenth century. *Quarterly Journal of the Royal Astronomical Society, 28,* 30–42.
Swindlehurst, W. (1890). Letter to John Tebbutt, dated 15 April. In TL.
Swindlehurst, W. (1891). Letter to John Tebbutt, dated 9 October. In TL.
*Sydney Morning Herald,* 10 September 1891.
Symes, G.W. (1976). Todd, Sir Charles (1826–1910), astronomer, meteorologist and electrical engineer. In Nairn, B. (Ed.). *Australian Dictionary of Biography. Volume 6* (pp. 280–282). Melbourne: Melbourne University Press.
Tebbutt, J. (1863). On Australian storms. *Transactions of the Philosophical Society of New South Wales,* 153–164.
Tebbutt, J. (1864–69). *Journal. Observatory, Peninsula, Windsor, New South Wales*. MS, Mitchell Library, Sydney (A3747).
Tebbutt, J. (1867). On the progress and present state of astronomical science in New South Wales. In *Catalogue of the Natural and Industrial Products of New South Wales. Forwarded to the Paris Universal Exhibition of 1867, by the New South Wales Exhibition Commissioners* (pp. 55–64). Sydney: Government Printer. Appendix.
Tebbutt, J. (1868). *Meteorological Observation Made at the Private Observatory of John Tebbutt, Junr., The Peninsula, Windsor, New South Wales, in the Years 1863, 1864, 1865, and 1866.* Sydney: printed for the author.
Tebbutt, J. (1870). The aurora australis. *Sydney Morning Herald,* 29 September.
Tebbutt, J. (1871). On the progress and present state of astronomical science in New South Wales. In *The Industrial Progress of New South Wales: Being a Report of the Intercolonial Exhibition of 1870, at Sydney; Together With a Variety of Papers Illustrative of the Industrial Resources of the Colony* (pp. 617–632). Sydney: Government Printer.
Tebbutt, J. (1874). The Melbourne Evening Herald v. science. *Sydney Morning Herald,* 15 June.
Tebbutt, J. (1875a). Beobachtung des Venusdurchganges in Windsor (N.-S.-Wales). *Astronomische Nachrichten, 85,* 173–176.
Tebbutt, J. (1875b). New South Wales and the transit of Venus. *Sydney Morning Herald,* 21 April.
Tebbutt, J. (1875c). Observations of the transit of Venus, 1874, December 8–9, made at Windsor, N.S. Wales. *Memoirs of the Royal Astronomical Society, 47,* 89–92.
Tebbutt, J. (1875d). The transit of Venus. *Sydney Morning Herald,* 16 January.
Tebbutt, J. (1875e). The transit of Venus. *Sydney Morning Herald,* 20 January.
Tebbutt, J. (1875f). The transit of Venus. *Sydney Morning Herald,* 5 February.
Tebbutt, J. (1877). *Results of Meteorological Observations Made at the Private Observatory of John Tebbutt, The Peninsula, Windsor, New South Wales, in the Years 1871, 1872, 1873, 1874, 1875 & 1875*. Sydney: printed for the author.
Tebbutt, J. (1878). Proposed correction to the assumed longitude of the Sydney Observatory. *Journal and Proceedings of the Royal Society of New South Wales, 12,* 63–69.
Tebbutt, J. (1880a). Jupiter. *Sydney Morning Herald,* 18 December.
Tebbutt, J. (1880b). On the longitude of Sydney Observatory. *Journal and Proceedings of the Royal Society of New South Wales, 14,* 19–21.
Tebbutt, J. (1883). Observations of the transit of Venus, 1874, December 8–9, made at Windsor, N.S.W. *Memoirs of the Royal Astronomical Society, 47,* 89–92.
Tebbutt, J. (1887). *History and Description of Mr. Tebbutt's Observatory, Windsor, New South Wales*. Sydney: printed for the author.
Tebbutt, J. (1888). On the difference of longitude between Mr. Tebbutt's Observatory, Windsor, New South Wales, and the Government Observatories at Sydney and Melbourne. *Monthly Notices of the Royal Astronomical Society, 48,* 313–316.
Tebbutt, J. (1890). Astronomical and meteorological workers in N.S. Wales, 1778 to 1860. *The Observatory, 13,* 211–212.

Tebbutt, J. (1891a). Our Observatory. *Australian,* 26 September.
Tebbutt, J. (1891b). *The Sydney Observatory and the "Sydney Morning Herald"*. Sydney: printed for the author.
Tebbutt, J. (1893a). Astronomical photography. *Australian Star,* 3 November.
Tebbutt, J. (1893b). The Government Astronomer and the transit of Venus. *Australian Star,* 20 June.
Tebbutt, J. (1894a). Mr. Gale and the Sydney Observatory. *Daily Telegraph,* 20 January.
Tebbutt, J. (1894b). Mr. Tebbutt and the transit of Venus, 1874. *Australian Star*, 16 March. Reprinted in *The Observatory, 17,* 179–180.
Tebbutt, J. (1897a). Flood records in New South Wales. *Australian Star,* 14 April.
Tebbutt, J. (1897b). The longest record of floods in New South Wales. *Windsor and Richmond Gazette,* 17 April.
Tebbutt, J. (1898). *Report of Mr. Tebbutt's Observatory, The Peninsula, Windsor, New South Wales, For the Year 1897*. Sydney: printed for the author.
Tebbutt, J. (1908). *Astronomical Memoirs*. Sydney: printed for the author.
Tebbutt, Wragge, and Russell. *Windsor and Richmond Gazette,* 21 September 1895.
The Brisbane Astronomical Society (1896). Newspaper clipping bound in TL.
The Solar Eclipse. *The Sydney Mail,* 27 September 1922 (pp. 14–27).
The Windsor (N.S.W.) Observatory. *Town and Country Journal,* 2 November 1889.
Thompson, M.M.H. (1986). *William Woolls: A Man of Parramatta*. Sydney: Hale & Iremonger.
Thomson, A. (1896). Mr. Russell and the lunar eclipse. *Sunday Times,* 30 August.
Thomson, J.P. (1889a). Letter to John Tebbutt, dated 12 January. In TL.
Thomson, J.P. (1889b). Letter to John Tebbutt, dated 2 March. In TL.
Thomson, J.P. (1895). The transit of Mercury, 1894 November 11; egress. *Monthly Notices of the Royal Astronomical Society, 55,* 316–319.
Thomson, J.P. (1897). Letter to John Tebbutt, dated 6 June. In TL.
Threlfall, R. (1888a). Letter to John Tebbutt, dated 23 May. In TL.
Threlfall, R. (1888b). Letter to John Tebbutt, dated 19 June. In TL.
Todd, C. (1883). Observations of the transit of Venus, 1874, December 8–9, at Adelaide, South Australia. *Memoirs of the Royal Astronomical Society, 47,* 93–96.
Todd, Sir Charles (1893). Meteorological work in Australia: a review. In R.Tate, E.H. Rennie & W.H. Bragg (Eds.), *Report of the Fifth Meeting of the Australasian Association for the Advancement of Science, Held at Adelaide, South Australia, September, 1893* (pp. 246–270). Sydney: AAAS.
TRIA (1891). Our Observatory. *Sydney Morning Herald,* 15 September.
Turner H.H. (1912). *The Great Star Map. Being a General Account of the International Project Known as the Astrographic Chart*. London: Murray.
Walsh, G.P. (1976). Russell, Henry Chamberlain (1836–1907). In Serle, G., and Ward, R. (Eds.). *Australian Dictionary of Biography. Volume 6: 1851–1890. R-Z* (pp. 74–75). Melbourne: Melbourne University Press.
Ward, J.W. (1892). Letter to John Tebbutt, dated 22 November. In TL.
W.F.G. [Walter F. Gale], 1920. Obituaries. Thomas Roseby. *Monthly Notices of the Royal Astronomical Society, 80,* 354.
White, E.J. (1889). Letter to John Tebbutt, dated 20 November. In TL.
White, E.J. (1893). Letter to John Tebbutt, dated 21 November. In TL.
White, G. (1988). The Carte du Ciel—the Australian connection. In S. Débarbat, J.A. Eddy, H.K. Eichhorn & A.R. Upgren (Eds.), *Mapping the Sky. Past Heritage and Future Directions* (pp. 45–51). Dordrecht: Kluwer.
Wilson, P.D. (1990). Wragge, Clement Lindley (1852–1922). In J. Ritchie (Ed.), *Australian Dictionary of Biography. Volume 12: 1891–1939. Smy-Z* (pp. 576–577). Melbourne: Melbourne University Press.
Wood, H. (1946). Obituaries. Walter Frederick Gale. *Monthly Notices of the Royal Astronomical Society, 104,* 29–30.

Wood, H. (1958). *Sydney Observatory 1858 to 1958*. Sydney: Government Printer (Sydney Observatory Papers No. 31).

Wood, H. (1981). Gale, Walter Frederick (1865–1945). In B. Nairn & G. Serle G. (Eds.), *Australian Dictionary of Biography. Volume 8: 1891–1939. Cl-Gib* (p. 609). Melbourne: Melbourne University Press.

Woolls, W. (1875). Letter to John Tebbutt, dated 18 January. In TL.

Wright, H. (1875). The transit of Venus. *Sydney Morning Herald*, 23 April.

Wright, H. (1896). Letter to John Tebbutt, dated 13 August. In TL.

Wright, H. (1898). Letter to John Tebbutt, dated 8 May. In TL.

Wright, H. (1899). Letter to John Tebbutt, dated 6 March. In TL.

Wright, H. (1903). Letter to John Tebbutt, dated 27 April. In TL.

# Chapter 13
# Astronomers in Isolation: A New Branch of the British Astronomical Association

## 13.1 Introduction

Societies are a major phenomenon in the evolution of any discipline (see Berman 1975; Brasch 1931; Daniels 1967; Fox 1980; Oleson and Brown 1976; Rothenberg 1981; Schofield 1963), and this was no more apparent than in the nineteenth century Australia, with its vast geographical distances, mere handful of active scientists (by world standards), and physical isolation from the main centres of scientific activity in the northern hemisphere (Branagan 1972; Hoare 1974, 1976; Inkster and Todd 1988; MacLeod 1988; Maiden 1918; Pescott 1961; Piesse 1913).

In Australia, the earliest astronomical groups were Sections A of the Royal Societies of New South Wales and Victoria, founded in 1876 and 1879, respectively, and the Australian Comet Corps, which Tebbutt formed in 1882 (Orchiston 1982, 1998; Orchiston and Bhathal 1991). However, none of these survived for any appreciable period of time (see Table 4.4).

It was not until the 1890s that successful local astronomical groups were established in Brisbane, Sydney, Melbourne and Adelaide, but only the Adelaide and Sydney groups have survived in one form or another through to the present day. The former was founded in 1892 as the Astronomy Branch of the Royal Society of South Australia (Waters 1980), but later gained its independence, becoming the Astronomical Society of South Australia (Waters 1981). The Sydney group, the New South Wales (henceforth NSW) Branch of the British Astronomical Association, was established in 1895, just 5 years after the formation of the parent body in England (Kelly 1948), and in 2004 was reconstituted as the Sydney City Skywatchers.

From the start, the British Astronomical Association (henceforth BAA) had a policy of encouraging branches and, by 1930, seven had been formed (ibid.). However, only two of these, the West of Scotland Branch (later renamed the Scottish Branch) and the NSW Branch, displayed any degree of longevity. Other Australian branches, in Melbourne and Perth, quickly died (Orchiston and Perdrix 1990, 2002). The NSW Branch therefore is an enigma.



W. Orchiston, *John Tebbutt*, Historical & Cultural Astronomy,
DOI 10.1007/978-3-319-44521-2_13

In this chapter, we explore the sequence of events that led to the formation of the NSW Branch, and survey the vital first 5 years of the new body when—like any newborn society—its very survival was in the balance. In so doing, we examine the critical role that John Tebbutt played in the founding and early evolution of the group. This chapter has been developed from a paper that was published in anticipation of the centennial of the Branch (Orchiston 1988b), but it also draws on material included in Orchiston and Perdrix (1990, 2002).

## 13.2 Astronomical Agitation in Sydney: The Mooted Australian Astronomical Society

At the same time that Sir Charles Todd (Bio Box 12.6; Fig. 13.1) and his professional and non-professional colleagues were discussing the formation of a local astronomical society in Adelaide, two other astronomers were doing likewise in Sydney (see Orchiston and Bhathal 1984). They were Walter Gale (Bio Box 12.1) and Robert Innes (Bio Box 8.2), both well-known names in the nineteenth century Australian astronomy.

In the decade following Tebbutt's formation of the Australian Comet Corps, the nature of astronomy changed somewhat in the Sydney region. Because of his focus on meteorology at Sydney Observatory, H.C. Russell had estranged himself from many in the strong non-professional astronomical fraternity and from some of his professional colleagues (Orchiston 2002), yet public interest in astronomy had continued to grow. This was in part because of the impact of the 1874 and 1882 transits of Venus, and the succession of impressive naked eye comets that graced Australian skies between 1874 and 1887 inclusive (see Haynes et al. 1996). Consequently, not only were there many more active independent astronomers in the region, but there were now highly-qualified men with strong astronomical interests and considerable observing experience working in such places as the New South Wales Lands

**Fig. 13.1** Sir Charles Todd, Director of Adelaide Observatory (https://community.history.sa.gov.au)

Department (Orchiston 1987) and the University of Sydney. One other significant factor was the presence of a number of key individuals with advanced mathematical skills and interests, including Joseph Brooks (Bio Box 8.5), Thomas Furber (Bio Box 8.4), Robert Innes, George Knibbs (Bio Box 13.1), C.J. Merfield (Bio Box 11.4) and John Tebbutt. By 1891, there were between 40 and 50 people in the general vicinity of Sydney actively involved in astronomy in one capacity or another (see Fig. 4.3), and it seemed that the time was ripe for the formation of a local astronomical group of some kind.

**Bio Box 13.1: George Handley Knibbs**

George Handley Knibbs (Fig. 13.2) was born in Sydney on 13 June 1858, and trained as a surveyor (Bambrick 1983; Obituary 1930; Orchiston 1987).

**Fig. 13.2** A photograph of G.H. Knibbs taken in about 1898 when he was very active in the newly-formed New South Wales Branch of the British Astronomical Association (*Courtesy* Mitchell Library 524)

In 1881, he obtained a post as a surveyor in the New South Wales Lands Department and in 1890 accepted a Lectureship in Surveying at Sydney University, eventually becoming Acting Professor of Physics. Knibbs was appointed to the senior government post of Commonwealth Statistician in 1906, and in 1921 became Director of the Commonwealth Institute of Science and Technology (a forerunner to the Commonwealth Scientific and Industrial Research Organisation).

Given his surveying background, Knibbs (1900) had a strong interest in astronomy, and in 1900 he published a detailed paper on 'The Sun's motion in space' which received glowing accolades from the Director of Melbourne Observatory (Baracchi 1914). Knibbs was elected an FRAS in 1896. He served as both President and Secretary of the Royal Society of New South Wales, and for one session was President of the Australasian Association for the Advancement of Science. In addition, Knibbs was a Vice-President and later President of the NSW Branch of the British Astronomical Association.

(continued)

**Bio Box 13.1** (continued)

Knibbs was made a C.M.G. in 1911, and in 1923 he was knighted for his services to Australia. Sir George Handley Knibbs died in Melbourne on 30 March 1929 from coronary disease following a long illness.

However, Gale and Innes drew their inspiration for a new group not so much from their local colleagues as from the examples recently set by the founding of the Astronomical Society of the Pacific in 1889 (Bracher 1989) and the BAA in 1890 (McKim 1990). Tebbutt was an early member of both societies and in March 1891 he loaned the first publications of the American society to Innes, who immediately responded: 'I feel Australian enough to wish we had a "live" Astronomical Society here like they have in California, where the population is not much greater' (Innes 1891a). Similar sentiments were expressed in a later letter (dated 26 August), when Innes mentions that, although the local need is there, the time may not be right. He goes on: 'I do not like to mention our chief difficulty; you may guess it, but it is not L.s.d' (Innes 1891b; 'L.s.d.' refers to pounds, shillings and pence, the Australian currency of the day). He was specifically referring to Russell, whose hostility towards some of the local astronomers was already public knowledge (see Chap. 12).

These problems notwithstanding, Gale, Innes and Tebbutt proceeded to discuss the concept of a new society, and within a week Innes (1891c) was referring to it as '… our proposed local or Australian Astronomical Society'. He had already decided on the name, and that it should be an autonomous society rather than a resuscitated section of the local Royal Society or the NSW Branch of the BAA. These were important decisions, for there was some pressure—both from London and within Australia—to establish a BAA Branch. As early as January 1891, the Secretary of the BAA advised Tebbutt that branches (in general) were anticipated and that the required rules were being prepared, and in the same letter (Maunder 1891a) and in a subsequent one (Maunder 1891b) he asked Tebbutt to publicize the BAA within Australia, particularly through the media.

Meanwhile, there must have been some thought given to local branches (or at least a branch) by the increasing band of Australian-based BAA members for the Reverend W. Swindlehurst of Newcastle (Bio Box 10.9) could not have been alone in October 1891 when he expressed the following view in a letter to Tebbutt: 'I trust it is not hoping too much to expect before long to be able to set up an Australian Branch of the BAA' (Swindlehurst 1891). At this time, there were already 17 members living in Australia, representing just 2.9 % of the total membership of 583 (Report of the Council 1891). Of these, seven were in NSW, six in Victoria, three in South Australia and one in Queensland (based on List of Members 1891; and New Members listed in Volumes 1 and 2 of the *Journal of the British Astronomical Association*). Tebbutt was one of these, having been admitted to membership on 24 June 1891 (Notices of the Association 1891). Although he was not the first from Australia to join the Association (there were even four foundation members), he was the first to publish in the *Journal*. His paper, on the colour of r Velorum (Tebbutt 1891), appeared before he became a member.

Gale (1894b) recognised the urgent need for some sort of local astronomical group or society and was keen to work with Innes, but he was not fully convinced of the wisdom of founding a totally new society. Tebbutt also doubted that a new society was the best alternative:

> … considering the few who were really interested in astronomical work he [i.e. Tebbutt] did not think an attempt to form an independent society would succeed, and he therefore suggested that an effort should be made to resuscitate the astronomical branch of the Royal Society [of New South Wales] … or to form a branch of the British Astronomical Association … (Reports of the Branches 1895a, p. 244).

There was immediate agreement between Gale and Innes that a renascent Section A of the local Royal Society was out of the question, given 'The high fees … and "clique" rule …' (Gale 1895d), and it was Innes (1891d) who insisted on forming a separate society.

In a letter to Tebbutt dated 17 January 1892, he advised that 'Mr Gale and I are now going to move actively in setting up an Astronomical Society' (Innes 1892a), and he discussed in some detail whether Russell would have any active involvement and, depending on the answer to that question, whether Tebbutt should be the founding President or Vice-President. Accompanying the letter was a 2-page document in Innes's handwriting titled 'Provisional Memorandum of Association to be sent to all those asked to attend the preliminary meeting', and Innes asked for Tebbutt's comments on this. The document gave the name of the group ('The Australian Astronomical Society'), outlined its aims and objectives, listed the entrance fee and annual subscription (one guinea, in each case) and discussed the management structure. Membership would be open to ladies *and* gentlemen and, although national in scope, the Society would be based in Sydney.

Tebbutt still insisted that a new society was not the preferred option and Gale began having doubts, but Innes (1892b) only decided to abandon this approach when some other keen astronomers that he contacted indicated that they would not join the new society. This left the gate open to form a local branch of the BAA.

## 13.3 Organising Astronomy in Sydney: Formation of the BAA Branch

Gale was the one who took up the new initiative. After leaving the matter of a local astronomical society in abeyance for a year, he had discussions with the BAA Council while in London in 1893 and returned to Sydney convinced that the time was right to inaugurate a local society or a branch of the BAA (Gale 1894a). The first alternative had already been considered and rejected so that left him with just one choice, and on 11 July 1894 he announced:

> I am interesting myself just now in the formation of a branch of the British Astronomical Association, which I think will succeed. When in London, I chatted with the Council concerning the matter and received much encouragement. The parent society will undertake to

> send out, in time for each meeting, a sufficient number of new contributions to fill the evening. With this reserve force it should be possible to make our branch a live institution from its inception.
>
> The arrangement will certainly be better than attempting the formation of a new local society. Already thirty-two have promised to join and eight of them will be of use as contributors. I trust you will see your way to actively support such an undertaking. Personally, I am anxious for its success, partly on account of the official sneer at the idea of organization. (Gale 1894c).

The 'official sneer' referred to Russell.

Tebbutt had a much more positive attitude and strongly supported the initiative, but strangely—given his earlier views—was having some second thoughts. These were voiced in a letter to Gale where he expressed the hope that at some point in the future it might be possible to form a strictly Australian Society. Gale (1894d) agreed but, in the short term, was confident that the BAA Branch would be a success:

> It is to be hoped, as you suggest, that in the future we may have a strictly Australian Astronomical Society.
>
> I found such an undertaking as founding one at present would be certain to end in failure. With the fostering aid of the B.A.A. I am sanguine of success.

By 23 July his potential membership list had risen to 46, and he anticipated that '… more than fifty percent [of these] will take up the study with more or less earnestness' (ibid.). Meanwhile, Knibbs (1894) was delighted to hear that '… a branch of the British Royal Astronomical Society [*sic*] …' was to be formed, and pledged his support and that of a number of his University colleagues.

After canvassing and receiving further support for the mooted Branch, in October 1894 Gale teamed with Innes, Tebbutt and ten other 'local' BAA members (Table 13.1) to petition the parent body for the formation of the NSW Branch (Gale

**Table 13.1** Members of the BAA who petitioned for the formation of the NSW Branch in 1894

| Name[a] | Position(s) subsequently held on the Branch Council, 1895–1899 |
|---|---|
| J.W. Askew | None |
| G. Butterfield | None |
| C.C. Cater | None |
| T.W. Craven | Treasurer (1895, 1896), Councillor (1899) |
| F.D. Edmonds | Councillor (1895, 1896) |
| W.F. Gale | Secretary (1895, 1896, 1897), Councillor (1898, 1899) |
| R.T.A. Innes | Vice-President (1895) |
| G.H. Knibbs | Vice-President (1895, 1896), President (1897, 1898), Councillor (1899) |
| R.D. Lewers | Councillor (1895) |
| C. Matthews | None |
| C.J. Merfield | Councillor (1897), Vice-President (1899) |
| J. Tebbutt | President (1895, 1896), Councillor (1897, 1898) |
| H. Wright | Councillor (1895, 1896), Secretary (1897, 1898, 1899) |

[a]The 'Names' are after Baracchi (1914, p. 359)

1894e; Wright 1908). He was full of confidence, and shortly afterwards wrote to W.H. Wesley at the Royal Astronomical Society:

> You will be glad to hear that the success of the N.S.W. Branch of the British Astronomical Assn. seems beyond doubt, so far as mere membership is concerned. I anticipate the number will exceed fifty in six months. It should be possible to get some good work out of those who can observe. (Gale 1894f).

On 28 November, Council issued its Warrant of approval (Reports of the Branches 1923), and on 14 January 1895 Gale (1895c) was able to announce the good news in a letter to Tebbutt. He also advised that the inaugural meeting of the Branch would be held on 30 January, in the rooms of the Royal Geographical Society, and mentioned that '… it is unanimously desired that you will accept the office of President and, after the business of the evening, deliver an address on any astronomical subject' (ibid.).

Tebbutt accepted the Presidency (Gale 1895d), but not without reservations, given his non-Sydney domicile. Innes (1895a) was delighted with this decision and offered his congratulations in a letter dated 25 January 1895: 'I am glad you are going to be our President (B.A.A.) as it is your proper place as the leading amateur astronomer of the continent, I use the word amateur of course in its proper sense'. Similar letters were sent by others, including the Reverend Dr. Roseby (1895), and W.H. Wooster (1895) of Ballarat, Victoria (Bio Box 13.2).

**Bio Box 13.2: William H. Wooster**

William H. Wooster (Fig. 13.3) was born in London in 1844, and his family emigrated to Sydney in 1853. After completing his schooling he spent several years working on farms in New South Wales before obtaining a post as a teaching assistant in Sydney. He then trained formally as a teacher and after graduating in 1866 was assigned to Freeman's Reach, near Windsor (Wooster 1924). It was while there that his appetite for astronomy was whetted through visits to Windsor Observatory. Tebbutt also loaned him a book on astronomy (Wooster 1890).

**Fig. 13.3** W.H. Wooster at the age of 80 (after Wooster 1924)

(continued)

**Bio Box 13.2** (continued)

At the end of 1868, Wooster moved to a school in Melbourne and then to other Victorian centres, and it was only after he had settled in Ballarat in 1882 that he found the time and financial means to pursue an avid interest in astronomy.

He lived close to Oddie Observatory (Burk 1986), and in 1890 purchased an 8-in. (20.3-cm) reflector made by the Observatory's Director, Henry Evans Baker. This was later superseded first by a 12.25-in. (31.1-cm) reflector and later by a 15-in. (38.1-cm) instrument (Wooster 1890, 1892, 1924). Wooster was an active member of the Royal Society of Victoria; he also was an early member of the BAA and a contributor to its Jupiter and Star Colours Section (Wooster 1895), and he and published a couple of papers in the Association's *Journal.*

Wooster always was happy to supply local enquirers with astronomical information and, despite the superior facilities available at nearby Oddie Observatory, also ran popular public viewing nights at his own observatory. The *Ballarat Courier* reports a group visit on 8 February 1892 in order to view the Moon through his newly-acquired 12.25-in. reflector, ending with the flattering comment that he is always willing '... to entertain any number of the residents of Ballarat or elsewhere ...' (A glance at the Moon 1892). Two years later, Wooster (1894) reported that as a result of his newspaper articles in the *Ballarat Courier* '... very many strangers to me have requested permission to spend an evening at my telescope, and have always been astonished and delighted at what they saw; and the seeing was always accompanied by running comments descriptive and explanatory. So you see the good seed is being sown'. In 1924, he wrote that his 15-in. telescope was open to the public every Monday evening, up to fifty people being present at any one time (Wooster 1924, p. 8). By that time he was 80 years of age and in his twilight years.

Wooster was yet another Australian independent astronomer who successfully combined research and popular astronomy, and his biography deserves to be written.

Innes joined Gale in moving quickly to place the NSW Branch on a firm footing. They provided Tebbutt with a list of proposed office-bearers and councillors (Innes 1895b), publicly announced the inaugural meeting of the Branch via a circular letter (Gale 1895b) that was distributed to all NSW BAA members and others with an interest in astronomy (Fig. 13.4), and proceeded to get letterhead paper printed for both letters and memoranda.

The first meeting of the BAA NSW Branch went smoothly and after the elections those in attendance were treated to an 'inaugural speech' by Tebbutt that was predominantly about the observational possibilities open to the non-professional astronomer. He said, *inter alia*:

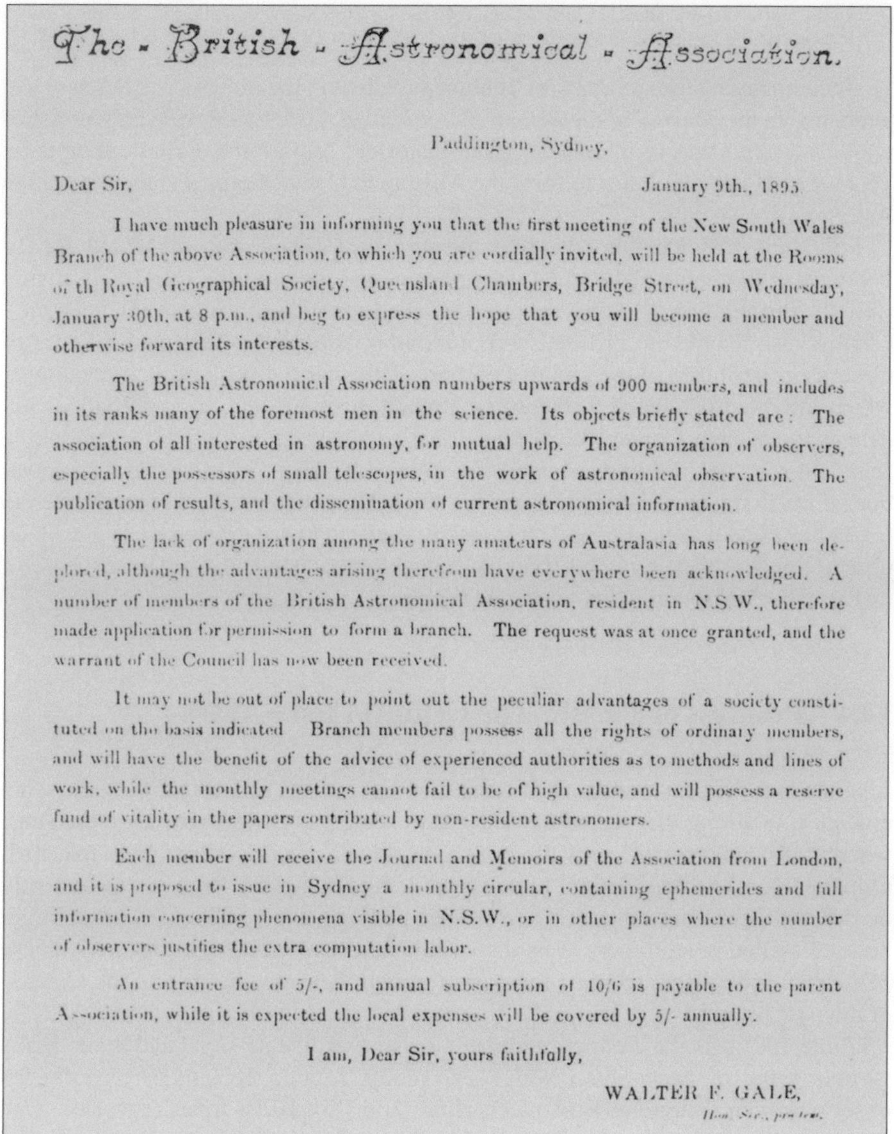

The - British - Astronomical - Association.

Paddington, Sydney,

Dear Sir, January 9th., 1895.

I have much pleasure in informing you that the first meeting of the New South Wales Branch of the above Association, to which you are cordially invited, will be held at the Rooms of th Royal Geographical Society, Queensland Chambers, Bridge Street, on Wednesday, January 30th, at 8 p.m., and beg to express the hope that you will become a member and otherwise forward its interests.

The British Astronomical Association numbers upwards of 900 members, and includes in its ranks many of the foremost men in the science. Its objects briefly stated are: The association of all interested in astronomy, for mutual help. The organization of observers, especially the possessors of small telescopes, in the work of astronomical observation. The publication of results, and the dissemination of current astronomical information.

The lack of organization among the many amateurs of Australasia has long been deplored, although the advantages arising therefrom have everywhere been acknowledged. A number of members of the British Astronomical Association, resident in N.S.W., therefore made application for permission to form a branch. The request was at once granted, and the warrant of the Council has now been received.

It may not be out of place to point out the peculiar advantages of a society constituted on the basis indicated Branch members possess all the rights of ordinary members, and will have the benefit of the advice of experienced authorities as to methods and lines of work, while the monthly meetings cannot fail to be of high value, and will possess a reserve fund of vitality in the papers contributed by non-resident astronomers.

Each member will receive the Journal and Memoirs of the Association from London, and it is proposed to issue in Sydney a monthly circular, containing ephemerides and full information concerning phenomena visible in N.S.W., or in other places where the number of observers justifies the extra computation labor.

An entrance fee of 5/-, and annual subscription of 10/6 is payable to the parent Association, while it is expected the local expenses will be covered by 5/- annually.

I am, Dear Sir, yours faithfully,

WALTER F. GALE,
*Hon. Sec., pro tem.*

**Fig. 13.4** Circular letter about the inaugural meeting of the New South Wales Branch of the British Astronomical Association (*Courtesy* Mitchell Library)

The field of observation in the southern hemisphere was a very wide one, and if properly and zealously worked, it would be a fertile one. It should, at the onset, be distinctly understood that the notion that very little could be done with a small instrument was an erroneous one … no one possessed of an instrument of moderate power need despair of achieving something valuable for science. As special objects for the observation of amateurs, there were the sun, the moon, the four brilliant planets—Venus, Mars, Jupiter, and Saturn—the double stars, variable stars, coloured stars, comets, meteors, the zodiacal light, and auroras. (Reports of the Branches 1895a, pp. 244–245).

Accompanying the summary of Tebbutt's address in the first NSW Branch report published in the *Journal of the British Astronomical Association* were an account of the business portion of the meeting and historical data on the earlier attempts by Tebbutt and Gale and Innes to form the Australian Comet Corps and the Australian Astronomical Society.

Following this meeting, the first meeting of Council was scheduled for 14 February at the home of Dr. A.M. Megginson, and amongst the matters discussed were the formulation of additional rules and the formation of a library (Gale 1895e). Council also decided to proceed with a monthly Branch Circular, and the first of these was distributed at the end of February. This was a 4-page printed document, which provided details of forthcoming Branch meetings and other activities, and data on the planets, Comet P/Encke and lunar occultations, as well as lists of double stars, clusters, nebulae and red stars. The title page (Fig. 13.5) listed the composition of the first Council and the Branch's rules. Note that membership was open to '... ladies as well as gentlemen'. It is also interesting to see the emphasis placed on observational astronomy (as outlined in 'Objects' and 'Methods'), at the expense of mathematical astronomy and telescope-making.

## 13.4 Progress or Perish: The Critical First 5 Years

Like the first 5 weeks of human life, the first 5 years of a new scientific society are crucial. It is during this period that the infant society either progresses to the stage where survival is ensured, or wallows in a succession of mini-catastrophes that spell ultimate doom and destruction. In this regard, the BAA NSW Branch was basically no different from any other newborn society, except that it had a somewhat larger body of skilled practitioners at its disposal bend on nurturing it to maturity. These gentlemen functioned, for the most part, through their membership of the Council of the new Branch.

Table 13.2 lists the Council members of the first five sessions and reads like a 'Who's Who' of New South Wales astronomy. Joining Tebbutt as experienced observers were Gale (Bio Box 12.1), Hirst (Bio Box 10.3), Innes (Bio Box 8.2), Lenehan (Bio Box 7.2), Macdonnell (Bio Box 9.2), Merfield (Bio Box 11.4) and Wright (Bio Box 2.1), while Tebbutt, Furber (Bio Box 8.4), Innes, Knibbs (Bio Box 13.1), Lenehan and Merfield (Fig. 13.6) were accomplished in mathematical astronomy (Orchiston 1989). Gale also had demonstrated considerable ability as a telescope-maker (see Orchiston and Bembrick 1995), while Tebbutt, Gale, Macdonnell and Roseby (Bio Box 12.4) were committed to the popularisation of astronomy (Orchiston 1997a).

By 1895, Tebbutt, Gale, Furber, Hirst, Innes, Knibbs, Lenehan, Macdonnell, Merfield, Roseby and Wright had all achieved national or international visibility in

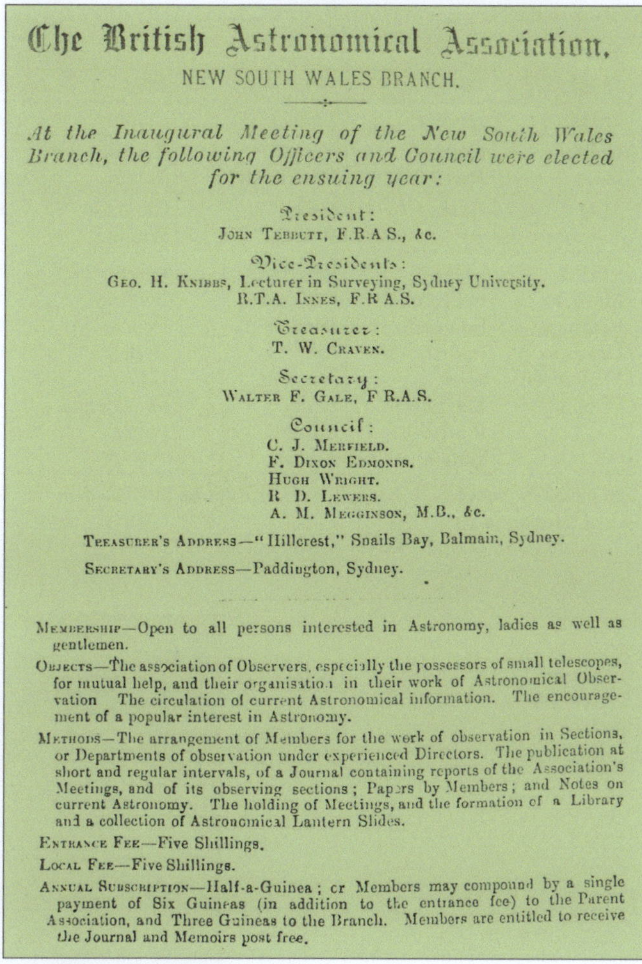

The British Astronomical Association.

NEW SOUTH WALES BRANCH.

*At the Inaugural Meeting of the New South Wales Branch, the following Officers and Council were elected for the ensuing year:*

President:
JOHN TEBBUTT, F.R.A.S., &c.

Vice-Presidents:
GEO. H. KNIBBS, Lecturer in Surveying, Sydney University.
R.T.A. INNES, F.R.A.S.

Treasurer:
T. W. CRAVEN.

Secretary:
WALTER F. GALE, F.R.A.S.

Council:
C. J. MERFIELD.
F. DIXON EDMONDS.
HUGH WRIGHT.
R. D. LEWERS.
A. M. MEGGINSON, M.B., &c.

TREASURER'S ADDRESS—"Hillcrest," Snails Bay, Balmain, Sydney.

SECRETARY'S ADDRESS—Paddington, Sydney.

MEMBERSHIP—Open to all persons interested in Astronomy, ladies as well as gentlemen.

OBJECTS—The association of Observers, especially the possessors of small telescopes, for mutual help, and their organisation in their work of Astronomical Observation The circulation of current Astronomical information. The encouragement of a popular interest in Astronomy.

METHODS—The arrangement of Members for the work of observation in Sections, or Departments of observation under experienced Directors. The publication at short and regular intervals, of a Journal containing reports of the Association's Meetings, and of its observing sections; Papers by Members; and Notes on current Astronomy. The holding of Meetings, and the formation of a Library and a collection of Astronomical Lantern Slides.

ENTRANCE FEE—Five Shillings.

LOCAL FEE—Five Shillings.

ANNUAL SUBSCRIPTION—Half-a-Guinea; or Members may compound by a single payment of Six Guineas (in addition to the entrance fee) to the Parent Association, and Three Guineas to the Branch. Members are entitled to receive the Journal and Memoirs post free.

**Fig. 13.5** Title page of the first Branch circular, of February 1895. Note that one guinea was equal to 21 shillings (*Courtesy* Mitchell Library)

their respective fields of astronomy, and by the end of the century all bar Wright were Fellows of the Royal Astronomical Society. Lenehan was the only one employed as a professional astronomer (he was Deputy-Director of Sydney Observatory at the time), although Innes and Merfield were later to make this transition, in 1896 and 1905, respectively (Orchiston 2001, 2003, 2015). The occupations of others on the early Councils varied considerably. Furber and Knibbs had trained as surveyors; Gale and Macdonnell were senior bank officers; Merfield was an engineer-draftsman; Innes and Hirst conducted successful liquor businesses; Roseby was a clergyman; and Wright was a librarian. Meanwhile, although ostensibly a farmer, Tebbutt worked almost full time as an astronomer. In life, as in astronomy,

**Table 13.2** Members of the New South Wales Branch Council, 1895–1899

| | Year | | | | |
|---|---|---|---|---|---|
| Office | 1895 | 1896 | 1897 | 1898 | 1899 |
| President | *Tebbutt* | *Tebbutt* | *Knibbs* | *Knibbs* | *Roseby* |
| Vice-President | Knibbs<br>*Innes* | Knibbs<br>*Roseby* | *Furber*<br>*Macdonnell* | *Furber*<br>*Roseby* | Darley<br>*Merfield* |
| Secretary | *Gale* | *Gale* | *Gale*<br>Wright | Close<br>Wright | Close<br>Wright |
| Treasurer | Craven | Craven | Batholomew | Batholomew | Halligan |
| Councillors | Merfield<br>Edmonds<br>Lewers<br>Megginson<br>Wright | Bedford<br>Edmonds<br>Furber<br>*Hirst*<br>Megginson<br>Porter<br>Wright | Edmonds<br>*Hirst*<br>*Lenehan*<br>*Merfield*<br>Porter<br>*Roseby*<br>*Tebbutt* | Bedford<br>Bell<br>Darley<br>*Gale*<br>*Macdonnell*<br>Porter<br>*Tebbutt* | Bedford<br>Bell<br>Craven<br>*Furber*<br>*Gale*<br>*Knibbs*<br>*Macdonnell* |

*Notes*: Entries are based on successive Branch Annual Reports published in the *Journal of the British Astronomical Association*
Italicised names indicate Fellows of the Royal Astronomical Society

**Fig. 13.6** A photograph showing Charles Merfield's appearance in 1895 when the BAA NSW Branch was formed (*Courtesy* Melbourne Camera Club)

all 11 were highly-intelligent, highly-successful, upwardly-mobile individuals, and the new BAA Branch was safe in their capable collective hands.

Only two Sydney astronomers of equivalent reputation were not actively involved with the new Branch, and they were Sydney Observatory's Henry Russell (for biographical details, see the preceding chapter) and the New South Wales Lands Department's Field Astronomer, Joseph Brooks (Bio Box 8.5). Despite his jaundiced view of Gale's attempt to form the NSW Branch of the BAA, Russell was elected to Branch membership at the inaugural meeting (Reports of the Branches 1895a), although he rarely attended meetings. For his part, Brooks (1895) was a member of the parent body, but declined to join the Branch '... owing to my strong objection to one of the members ...'. He never revealed who that person was.

Despite non-involvement by Brooks and Russell, hopes generally were high for the new Branch. Gale expressed this opinion in 1895:

> That the New South Wales Branch will be a success is already in a fair way of realisation. More than half of the present members are provided with instruments and are desirous of advancing the interest of Astronomy, either as computers or as observers. In the Southern Hemisphere the amateur has still a splendid chance of accomplishing much work of an original character. (Gale 1895a, p. 371).

To be a success, the Branch had to address itself to the objects listed in its initial circular (Fig. 13.5). The first of these was to encourage systematic astronomical observations, mainly by way of specialist observing sections, and this was quickly achieved thanks to the excellent examples set by the experienced observers on Council (but particularly Gale, Innes, Tebbutt and Wright). The leading active observers during the Branch's first 5 years are listed in Table 13.3, together with data on their telescopes and the primary targets of these instruments.

In addition to the more popular targets, some observers found time to engage in 'special projects': for example, Gale (1897a) studied the planet Mercury; Corbett made drawings of Saturn (Reports of the Branches 1895b); and Bayldon (1900), McDonall (1897b) and Gale (Reports of the Branches 1899c) observed the zodiacal light. Merfield observed a near-occultation of a star by Uranus (see Merfield 1897c; Tebbutt 1897a); Merfield (1896e; 1897a) reported on two partial eclipses (one solar, one lunar) and a number of members observed the total lunar eclipse of 23/24 June 1899 (Reports of the Branches 1899b); Gale (1896c; d) and Innes (1896b) observed

**Table 13.3** Main observational programs undertaken by NSW Branch members, 1895–1899[a]

| Observer | Telescope(s) (mm) | C | DS | JS | J | LO | M | MP | SC | SS | VS |
|---|---|---|---|---|---|---|---|---|---|---|---|
| J. Bell | ? | | | | | | | | | × | |
| T.H. Close | 64R | | | | | × | | | | | |
| T.W. Craven | 64R | | | × | × | | × | | | | |
| J.E. Davidson | 152R | × | | | | | | | | | |
| W.F. Gale | 127,152,254 R | × | × | | × | × | × | | | | |
| R.T.A. Innes | 419S | | × | | | | | | × | | × |
| F.K. McDonall | 54R | | | | | × | × | | × | × | |
| W.J. Macdonnell | 95R | | | × | | × | | | | | |
| Ms C. Maclellan | ? | | | | | | × | | | | |
| C.J. Merfield | 191S | | | | × | × | | | × | | |
| J. Tebbutt | 114, 203R | × | × | × | × | × | | × | | | × |
| H. Wright | 216S | × | × | | × | × | × | | × | | × |

*Key*: R = refractor, S = reflector, C = comets, DS = double stars, JS = Jovian satellites, J = Jupiter, LO = lunar occultations, M = meteors, MP = minor planets, SC = star colours, SS = sunspots, VS = variable stars

[a]Based on data in Davidson (1897a, b, c), Gale (1895f, g, 1897b, c), Innes (1896a, b, c), McDonall (1895, 1896a, b, c, d, 1897a), Macdonnell (1897), Merfield (1895, 1896a, d, e, 1897b), Reports of the Branches (1895b, c, d, 1896a, d, 1897b, c, d, 1898b, c, d, e, f, g), Tebbutt (1895, 1896a, b, 1897b, c, 1899), Wright (1896a, b, f, 1897a, b, c, 1899a, b)

star clusters and each discovered a new nebula; and Wright and Wilcox experimented with astrophotography (Reports of the Branches 1899a; Wright 1896d). Some use was made of the NSW observations by others. Thus, Gale (1896a) proudly reported in May 1896 that 'Denning has determined seven southern meteor radiants from our observations last year …'. He was referring to the famous British astronomer William Frederick Denning (1848–1931; Beech 1998).

Initially, all observations by Branch members were carried out on an individual basis, except for those determining star colours, but by the end of April 1897 four different observing sections had been formed as a means of formalising and regulating much of the work undertaken (Table 13.4). Of other popular observing targets listed in Table 13.3, only double stars, lunar occultations and variable stars did not come under the auspices of specific sections, although the question of a variable star section was raised as early as 1895 (Gale 1895a) and was strongly endorsed by Tebbutt (Wright 1896e). Tebbutt also advocated the establishment of a lunar section (Reports of the Branches 1897a)—which, presumably, would have accommodated lunar occultations, as well as lunar surface features.

Merfield proved an inspiring leader of the Star Colours Section which, in its first 2 years, was particularly active. Although he was joined by just three other members (Innes, McDonall and Wright), in 1896 they produced a 6-page paper reporting their initial results: 1125 observations of 104 stars (Merfield 1896f). This was followed in 1897 by an event larger paper, running to 13 pages and recording 3962 observations of 354 different southern stars (Merfield 1897b). After this supernova-like start, the Section became inactive as the demands of Merfield's post with the Department of Public Works and his quest for a position in professional astronomy began to take their toll (see details in Orchiston 2015).

The other observing Section which initially also was a noted success was the Comet Section, whose active members were Tebbutt (Director), Davidson, Gale and Wright. From the start, two different programs were pursued: monitoring of known comets and systematic searches for new comets. Unfortunately, the latter programme proved unsuccessful.

The overall observational output from the Branch during its first 5 years was phenomenal, and most of the 38 different research papers and notes published in the *Journal of the British Astronomical Association* between 1896 and 1899 (inclusive) related to observational astronomy (Table 13.5). In addition, between them Tebbutt, Innes and Merfield published more than 50 papers in other journals (but mainly *Astronomische Nachrichten* and *Monthly Notices of the Royal Astronomical Society*)

**Table 13.4** The first observing Sections of the NSW Branch

| Section | Founding year | Founding director |
|---|---|---|
| Comet | 1897 | Tebbutt |
| Solar and meteor | 1897 | McDonall |
| Star colours | 1895 | Merfield |
| Jupiter | 1897 | Gale |

**Table 13.5** Papers and observational notes published by NSW Branch members in the *BAA Journal*, 1896–1899 (inclusive)

| | Year | | | | |
|---|---|---|---|---|---|
| Astronomer | 1896 | 1897 | 1898 | 1899 | Totals |
| T.W. Craven | | | 1 | | 1 |
| W. Gale | 3 | 3 | | | 6 |
| R.T.A. Innes | 4 | 1 | – | – | 5 |
| F.K. McDonall | 5 | 3 | | | 8 |
| W.J. Macdonnell | | 1 | | | 1 |
| C.J. Merfield | 4 | 3 | | | 7 |
| J. Tebbutt | 2 | 3 | | 1 | 6 |
| H. Wright | 3 | | | 1 | 4 |
| Totals: | 21 | 14 | 1 | 2 | 38 |

during this period. Of all Branch members, Tebbutt published the greatest number of papers (in all journals) during this interval, with Merfield in obvious second place, followed by Innes, and then Gale and McDonall in equal fourth.

Although both Gale and Innes placed great emphasis on observational astronomy (Reports of the Branches 1897a), some highly-respected Branch members, as we have noted, gave vent to their celestial interests through mathematical astronomy. Innes and Merfield, in particular, quickly achieved international recognition in this branch of the discipline and published a number of important papers in *Astronomische Nachrichten* and *Monthly Notices of the Royal Astronomical Society* (Orchiston 2015). Innes (1896d) also published a mathematical paper on the telescope objective in the *Journal of the British Astronomical Association.* Given the Branch's strengths in mathematical astronomy—which Gale (1896b) was quick to recognise—it is a little surprising that this type of work is not mentioned in the 'Objects' statement.

Another aim of the new Branch was to circulate current astronomical information, and this was achieved through the monthly Circulars and by presentations at the monthly meetings. These latter arose as a result of the 21 March 1899 meeting when Roseby gave a '… resume of the most recent astronomical researches and observations …' (Reports of the Branches 1899a, p. 258). This was so successful that it became of feature of every meeting, thereby keeping members in touch with what is going on in the astronomical world (Macdonnell 1904, p. 80).

While the 'Monthly Notes' fell to Roseby, preparation of the Circulars was a collective effort, and it is heartening to see that there was still a healthy level of group-cooperation, even after 5 years. But perhaps this should be seen as no more than a feature of the 'honeymoon period'. On the topic of publications, it should be mentioned that Branch members obviously found the parent Association's journal adequate and decided not to produce their own local journal.

The last aim of the Branch was to encourage popular interest in astronomy, and this was achieved at the level of the existing membership through regular meetings, occasional visits to Sydney Observatory (see Reports of the Branches 1896b; 1898g;

1899a), and formation of a specialist loans library. Although the library was high on the Council's list of priorities—indeed it was discussed at the very first Council meeting—it only came into existence in 1899 (*B.A.A. N.S.W. Branch Circular* 1899).

Meetings of the Branch were held monthly and generally featured from one to three guest speakers, the presentation of observational papers by members, Roseby's report on recent astronomical developments, displays of astronomical photographs or books and announcements of a topical nature. From all accounts, meetings were long drawn-out affairs—even if packed with assorted astronomical 'treasures'—and automatically involved one in a late night. Given the collective expertise of the members, some meetings proved a trifle overwhelming when a less than judicious mix of topics was presented. For instance, at one meeting in 1898 (Reports of the Branches 1898a) the following succession of papers formed the main fare for the evening:

"Astronomy as applied to surveying" (by T.F. Furber)
"The mass and brightness of certain binary stars" (by the Reverend Dr Thomas Roseby)
"A study in gravitational astronomy" (by Dr A.M. Megginson)

Is it little wonder that some of the newer members, especially those not well-versed in mathematics, found meetings a trifle intimidating? Even T.W. Craven, one of the Branch founders and a member of Council for three of the first 5 years, expressed concerns in this regard: '… the meetings seem to me to be losing their interest to some of us—the subjects brought up while very good are a little deep for beginners like myself' (Craven 1898).

It is quite possible that the pseudo-professional level of papers at some monthly meetings was at least in part responsible for the notable decline in Branch membership that was experienced between 1897 and 1899 (see Fig. 13.6). This graph is based on data in successive Annual Reports published in the *Journal of the British Astronomical Association*, and it shows that after experiencing rapid initial growth Branch membership stood at 76 by the end of the first session (Reports of the Branches 1896b) and remained close to this figure at the end of the second session (Reports of the Branches 1897a). However, the apparent drop of nearly 30 % over the next 2 years may not have been as steep as it would appear, for the 1897 membership total included a small non-financial component (see Reports of the Branches 1898b). It is also likely that the straitened economic conditions which persisted in Australia during the 1890s led some members to leave the Branch, but membership eventually stabilised at just above 50 during 1899. As on September 1900, the official 'Membership List' of the Branch stood at 56, with 44 of those deriving from Sydney itself, eight (including Tebbutt and Gale) from elsewhere in New South Wales, two from Tasmania, and one each from Queensland and South Australia (Membership List 1900).

Figure 13.7 reveals another interesting statistic relating to Branch performance in that it includes data on the annual average attendance at monthly meetings (based on successive Annual Reports). Although the general trend mirrors the membership curve—as would be expected—the actual proportion of members attending meetings decreased significantly (from almost 40 % to just above 30 %) during the

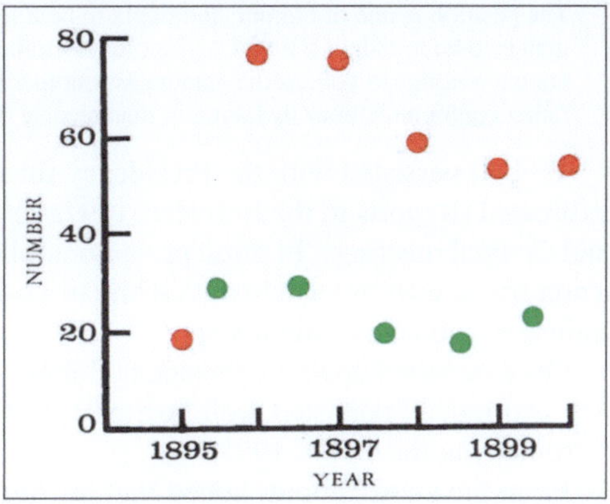


**Fig. 13.7** A plot showing Branch membership at the beginning of each session (red dots), and annual average attendance at meetings (*green dots*), 1895–1900 (*Plot* Wayne Orchiston)

decline in membership, only to return to initial levels during 1899 once membership had stabilised.

If we return to Table 13.2 once more, we note that there was a good deal of jockeying for power prior to the 1897 Annual General Meeting. Roseby, formerly a Vice-President, became an ordinary Councillor when his original place was taken by Council newcomer, Macdonnell. Interestingly, their positions were reversed just 1 year later. Furthermore, Tebbutt stood down as President at the end of the 1896 session and Gale was transferred by the bank to Newcastle and only agreed to continue as Secretary when Wright offered to assist. The following year he stood down altogether and simply served as an ordinary Council member. So, by the start of the 1898 session, two of the most respected members of Council, Tebbutt and Gale, no longer occupied positions of power, while the other leading figure in the formation of the Branch, Innes, had already moved to South Africa.

Despite the membership hiccup in 1897–1898 and a degree of political instability at about the same time, the Branch had a fruitful and dynamic first 5 years and established itself as a strong force in Australian astronomy. In his Presidential Address to the Australasian Association for the Advancement of Science in 1901, R.L.J. Ellery (Bio Box 7.3) was able to report that the NSW Branch of the BAA '… is now a very active and flourishing society, many of its members doing excellent observational work …' (Ellery 1901, p. 17).

## 13.5 Leading by Example: The Vital Role Played by Tebbutt

Tebbutt played a critical role in the early evolution of the Branch, initially in his capacity of President, and some were even of the view that this was part of the reason for the early success of the group. In 1896, when speaking of the Presidency, Roseby said this:

> The position is one of honour, and ought to be filled, as it is in England, by a man, if such a man is to be found, of the first Astronomical eminence. We all of us feel that … the place of honour belongs to you, as the veteran astronomer of Australia. And indeed we feel that you rather confer an honour by taking it, than receive one. (Roseby 1896; his underlining).

Tebbutt persisted with the Presidency for two sessions, despite concerns that he expressed (Reports to the Branches 1897a) over the difficulty of getting to monthly and Council meetings. In those pre-automobile days, each return trip to Sydney was a non-trivial exercise and involved him in a 6-km buggy ride to the station, a 110-km train trip, and an overnight stay.

Once he stood down as President, Tebbutt's influence on Branch affairs continued and indeed persisted well beyond his official 'retirement' from observational astronomy at the end of 1903.

From the start, Tebbutt hoped that his observational achievements would serve as an inspiration to others, and '… would make some amends for his shortcomings in the way of such attendance [at meetings]'. (Reports of the Branches 1897a). He was equally quick to dispel the myth that a large expensive telescope was required if serious work was contemplated. He believed that the skill and dedication of the observer were paramount, and he was always ready to assist any member who wished to embark on a career in observational astronomy. However, he was keen to emphasise the importance of systematic observing rather than general sky-browsing, and his attitude mirrored that of Dr. Sydney Samuel Hough (1870–1923), Director of the Royal Observatory at the Cape of Good Hope:

> We are often told that an astronomer pointing his telescope more or less at random to the sky and faithfully recording what he sees or photographs is bound to add something to our fund of knowledge. While this is perhaps true, promiscuous observing is certainly never to be encouraged in an observatory. The fact is that astronomy of to-day demands answers to definite questions and the astronomer who goes to his telescope without concentrating his whole energy on this particular problem is at least partly wasting his time. (Cape Astronomical Association 1912).

Tebbutt's desire to help newcomers went far beyond observational astronomy and was widely recognised and appreciated. Wright was one of the stalwarts on Council, and he voiced this in a letter dated 6 February 1896: 'As an enthusiastic amateur, I can assure you that I heartily appreciate your kindness, and for the interest of astronomy, I wish there were many others in the Colony who were so ready to help beginners as you are' (Wright 1896c). Despite this willingness, many new members were in awe of Tebbutt because of his international standing.

As Wright's letter also suggests, Tebbutt was always prepared to assist, encourage and inspire experienced members of the Branch (e.g. see Furber 1899), even though he was concerned that few were interested in undertaking 'systematic work' as opposed to casual observing of interesting events (see Merfield 1896c). One of the things that he made a point of doing was to provide experienced observers with 'alerts' on worthwhile observing projects. In 1895, he stressed the value of timing phenomena of Jupiter's fourth satellite (Reports of the Branches 1895b), which Macdonnell (1897) subsequently responded to. In January 1896, he observed two dark spots on Jupiter (Tebbutt 1896b), and soon had Gale, Merfield and Wright monitoring them (see Merfield 1896b). The following year he advised some Branch colleagues of a probable occultation of 41 Librae by Uranus, and Close, Merfield and Macdonnell succeeded in obtaining observations very soon after the event (see Merfield 1897c). He also made a point of notifying members of his Section of any new comets, and he encouraged them to undertake regular comet searches. Meanwhile, his catholic observational strategy particularly inspired Ernest H. Beattie (Bio Box 13.3) and James Nangle (Bio Box 13.4) during the first decade of the twentieth century. Both would go on to make their mark in New South Wales non-professional and professional astronomy, respectively.

**Bio Box 13.3: Ernest H. Beattie**

Ernest H. Beattie (Fig. 13.8) was born in Sydney on 6 December 1864 and trained as an architect (Baracchi 1914; Orchiston n.d.). For a number of years, he was employed as a draftsman by the Sydney City Council, before launching a private architectural practice.

**Fig. 13.8** Ernest Beattie with his 6-in. Grubb refractor (Orchiston collection)

A childhood interest in astronomy was rekindled in the 1890s, and Beattie purchased a 3.75-in. (9.5-cm) refractor and installed this in an observatory. This instrument was subsequently replaced by the 6-in. Grubb (15.2-cm) Grubb refractor shown above that once was the property of W.J. Macdonnell and then W. Gale (see Orchiston 1997b). Beattie used his telescopes to observe comets, planetary conjunctions, sunspots, Jovian satellite phenomena, lunar

(continued)

(continued)

occultations and double stars, and he published a number of papers on this work in the *Journal of the British Astronomical Association*. Beattie also was a member of the Australian expedition that went to Vavau in 1911 to observe the total solar eclipse.

In addition to his observational astronomy, Beattie maintained a time-service that was used by people living in the suburb of Mosman in Sydney.

Beattie was elected an FRAS, and for a time served as Secretary of the New South Wales Branch of the BAA.

Ernest Beattie died in Sydney on 1931.

In addition to observational efforts, Tebbutt actively encouraged those with an interest in mathematical astronomy, and he provided Innes and Merfield with micrometric cometary positions so that they could calculate the orbital elements.

When necessary, Tebbutt also placed his outstanding research and reference library at the disposal of Branch members. For example, when Knibbs was researching his paper on the Sun's motion in space (see Knibbs 1900) he was delighted to find that the Windsor Observatory library contained material that was not available at Sydney Observatory, Sydney University or in the Public Library of New South Wales.

**Bio Box 13.4: James Nangle**

James Nangle (Fig. 13.9) was born in Sydney on 28 December 1868, and after leaving school at the age of 11 studied architecture at the Sydney Technical College (Baracchi 1914; Cobb 1986; The late James Nangle 1941; Mr. James Nangle 1907; Obituary 1942; Orchiston 1988a; Wood 1958).

**Fig. 13.9** A photograph of James Nangle taken in 1928 for the in-house magazine of Australasian Footware (*Courtesy* Museum of Applied) Arts and Sciences, Sydney, P3549-112)

(continued)

(continued)

From 1890 he practised as an architect, and in 1905 was appointed Lecturer-in-Charge of the Architecture Department at his old *alma mater*. In 1913, Nangle became Superintendent of Technical Education for the State of New South Wales and proceeded to totally reorganize technical education.

Nangle had a long-standing passion for astronomy and maintained a private observatory with a 6.5-in. (16.5-cm) refractor (see Orchiston 1997b). This was used for a variety of observational programs, but with the emphasis on double stars. Nangle wrote a number of descriptive papers on his work, and these were published in the *Journal of the British Astronomical Association*.

After Cooke's forced resignation from Sydney Observatory in 1925 Nangle was appointed unpaid Director, and it is largely to his credit that the institution was able to survive the difficult years of the Great Depression. He also was interested in popularising astronomy, and while at Sydney Observatory wrote his well-known book, *Stars of the Southern Heavens* (Nangle 1929).

James Nangle ('Jimmy' to his friends) was an FRAS, and in 1920 received an O.B.E. He served one session as President of the Royal Society of New South Wales and as President of the Royal Australian Institute of Architects and was a Fellow of the Royal Institute of British Architects. He died in 1941 while still in office at Sydney Observatory, having exerted '… important influence in this State [New South Wales] in the fields of architecture, education and astronomy' (Obituary 1942).

Another way in which Tebbutt exerted a positive influence on Branch members was by inspiring them to emulate his example of writing on astronomical matters for the local newspapers. For example, in 1895 and 1896 articles penned by Gale, Innes, McDonall, Roseby and Wright appeared in the Sydney newspapers (see Orchiston 1997a).

As a result of his network of interstate contacts, Tebbutt was able to entice at least two non-local independent astronomers to join the Branch, namely A.B. Biggs of Launceston, Tasmania (Bio Box 10.1) and J. Ewen Davidson of Branscombe near Mackay in Queensland (Bio Box 9.4). Then in his twilight years, Biggs derived considerable pleasure from presenting material—*in absentia*—at meetings (e.g. see Orchiston 1985; Reports of the Branches 1896c, d), as did Davidson before his return to England in 1900 (Darlington and Orchiston 2017).

Finally, Tebbutt's ongoing correspondence with Melbourne's David Ross (Fig. 13.10; Bio Box 9.3) was probably instrumental in tempting Ross to occasionally attend Branch meetings in Sydney, where members were particularly interested in his telescope-making and astrophotography achievements (Reports of the Branches 1898e). Like Gale and Macdonnell, Ross was a bank officer. He was also one of the founders of the Victoria Branch of the BAA in 1897 (Orchiston and Perdrix 1990, 2002).

**Fig. 13.10** David Ross was one of the BAA Victoria Branch's most active members. Later, with the outbreak of WWI, he established 'The Peace Observatory' near Melbourne, in anticipation of a quick resolution of the conflict (Orchiston collection)

## 13.6 Concluding Remarks

Why, we may ask, did the New South Wales BAA Branch succeed when both the Australian Comet Corps and Section A of the Royal Society of New South Wales failed? Partly it was because the time was right. By 1894, there simply were more active astronomers—professional and non-professional—in Sydney, and under the collective guidance of Gale, Innes and Tebbutt a degree of group cohesion had begun to emerge.

But more than this, the Branch was the progeny of gradual development, which started back in 1891 when Gale and Innes first voiced the idea of forming the Australian Astronomical Society. As a result of this 3-year lead-in time, many people were consulted and lobbied, and by 1894 there was a strong body of support for a local Sydney-based (as opposed to national) astronomical group of some kind. Of all the options available, the formation of a local Branch of the BAA was seen as the most attractive option. Not only would London offer tangible support, but the kudos of a formal affiliation with this vital, young Association based in 'Mother England' was all too obvious.

When the Branch was formed, the groundswell of support it received in its infant years and collective skills of so distinguished a group as served on the early Councils made its future assured. After all, it is hard to envisage any society failing that could boast between six and nine Fellows of the Royal Astronomical Society on Council in any one year (in the days when FRAS had a somewhat different connotation than it has today).

Once operating, the Branch quickly served to formalise non-professional astronomy in Sydney, through organised observing programs, monthly meetings and the monthly circular. As well as attracting recognised experts in astronomy, the Branch also welcomed '… every lover of the science' (Macdonnell 1904, p. 81), and thus provided an ideal structured way for newcomers to enter the hobby. And although there was a strong observational bias amongst Councillors, new members were not ostracised if they chose to follow other lines of astronomical interest. In all, there was much enthusiasm, tolerance and guidance from senior members of the Branch, who collectively formed the most powerful and active localised group of independent astronomers in the nation at the time. Leading by example was their founding President, John Tebbutt.

Although the Branch experienced a meteoric rise to prominence during its first 5 years, it continued in good health after Innes's departure and the 'defections' of Gale and Tebbutt from key positions on Council. Early in the twentieth century, two newcomers to Council were the architects Beattie and Nangle, both of whom were inspired by Tebbutt's catholic observational tastes (see Chap. 11). Both were to prove stalwarts of NSW astronomy for many years, and teamed with Gale, Roseby and Wright to take the Branch safely through to the era of World War I. However, during this vital period they had to bid farewell to Merfield (who transferred to Melbourne Observatory in 1907), and then Macdonnell and Tebbutt, who passed away in 1910 and 1916, respectively.

John Tebbutt will always be remembered by Branch members as their illustrious founding President, from an era when it was possible for a talented independent astronomer to make a significant contribution to world astronomy.

**Acknowledgements** I wish to thank Iwona Hetherington (Museum of Applied Arts and Sciences, Sydney), Professor Nick Lomb (University of Southern Queensland, Toowoomba), staff at the Mitchell Library (Sydney), Royal Astronomical Society Archives (London) and Sydney Observatory for their assistance in various ways, and the Melbourne Camera Club, the Mitchell Library (Sydney) and the Museum of Applied Arts and Sciences for kindly supplying Figs. 13.2, 13.4–13.6 and 13.9.

## References[1]

A glance at the Moon. *Ballarat Couier*, 13 February1892. In TL.
*B.A.A. N.S.W. Branch Circular*, May-June (1899).

[1] The following abbreviation is used: TL = Letters to J. Tebbutt, bound manuscript letters in the Mitchell Library, Sydney.

Bambrick, S. (1983). Knibbs, Sir George Handley (1858–1929). In B. Nairn & G. Serle (Eds.), *Australian Dictionary of Biography. Volume 9: 1891–1939. Gil-Las* (pp. 620–621). Melbourne: Melbourne University Press.
Baracchi, P. (1914). Astronomy and geodesy in Australia. In G.H. Knibbs (Ed.), *Federal Handbook Prepared in Connection with the Eighty-fourth Meeting of the British Association for the Advancement of Science, Held in Australia, August, 1914* (pp. 326–390). Melbourne: Government Printer.
Bayldon, F.J. (1900). The Zodiacal lights and Gegenschein. *Journal of the British Astronomical Association, 10*, 260–263.
Beech, M. (1998). W.F. Denning—The doyen of amateur astronomers. *WNG, 26,* 19–24.
Berman, M. (1975). "Hegemony" and the amateur tradition in British science. *Journal of Social History, 8,* 30–50.
Bracher, K. (1989). A centennial history of the Astronomical Society of the Pacific. *Mercury, 18* (5), 3–43.
Branagan, D.F. (1972). Words, actions, people: 150 years of scientific societies in Australia. *Journal and Proceedings of the Royal Society of New South Wales, 104,* 123–141.
Brasch, F.E. (1931). The Royal Society of London and its influence upon scientific thought in the American colonies. *The Scientific Monthly, 33,* 336–355, 448–469.
Brooks, J. (1895). Letter to John Tebbutt, dated 21 December. In TL.
Burk, K. (1986) A history of James Oddie, Captain Henry Evans Baker and the Ballarat Municipal Observatory. *The Oddie-Baker Bulletin, 8,* 3–15.
Cape Astronomical Association. President's Inaugural Address. Science and Utility. Press clipping reporting the inaugural Meeting of the Association on 1912 November 8. In Skjellerup, J.F. (1912–48). Astronomy. Notes of Observations and Remarks Thereon. MS, Orchiston Collection.
Cobb, J. (1986). Nangle, James (1868–1941). In B. Nairn & G. Serle (Eds.), *Australian Dictionary of Biography. Volume 10: 1891–1939. Lat-Ner* (pp. 661–663). Melbourne: Melbourne University Press.
Craven, T.W. (1898). Letter to John Tebbutt, dated 20 April. In TL.
Daniels, G.H. (1967). The process of professionalization in American science: the emergent period, 1820–1860. *Isis, 58,* 151–166.
Darlington, W. & Orchiston, W. (2017). North Queensland and the 1882 transit of Venus: our final role in the quest for the Astronomical Unit. *Memoirs of the Queensland Museum*, in press.
Davidson, J.E. (1897a). Letter to John Tebbutt, dated 16 July. In TL.
Davidson, J.E. (1897b). Letter to John Tebbutt, dated 15 November. In TL.
Davidson, J.E. (1897c). Letter to John Tebbutt, dated 13 December. In TL.
Ellery, R.J. (1901). A brief history of the beginnings and growth of astronomy in Australasia. In *Report of the Eighth Meeting of the Australasian Association for the Advancement of Science* (pp. 1–17). Melbourne: AAAS.
Furber, T.F. (1899). Letter to John Tebbutt, dated 29 March. In TL.
Gale, W. (1894a). Letter to John Tebbutt, dated 6 April. In TL.
Gale, W. (1894b). Letter to John Tebbutt, dated 31 May. In TL.
Gale, W. (1894c). Letter to John Tebbutt, dated 11 July. In TL
Gale, W. (1894d). Letter to John Tebbutt, dated 23 July. In TL.
Gale, W. (1894e). Letter to John Tebbutt, dated 4 October. In TL.
Gale, W. (1894f). Letter to W.H. Wesley, dated 14 November. In RAS Letter Archives
Gale, W. (1895a). The British Astronomical Association in N.S.W. Branch. *Cosmos, 1,* 371–374.
Gale, W. (1895b). Circular letter, dated 9 January. Copy in the Mitchell Library (Q520.6/7).
Gale, W. (1895c). Letter to John Tebbutt, dated 14 January. In TL.
Gale, W. (1895d). Letter to John Tebbutt, dated 28 January. In TL.
Gale, W. (1895e). Letter to John Tebbutt, dated 6 February. In TL.
Gale, W. (1895f). Letter to John Tebbutt, dated 2 May. In TL.
Gale, W. (1895g). Occultation of Antares, 1895, May 10. *Journal of the British Astronomical Association, 5,* 513.

Gale, W. (1896a). Letter to John Tebbutt, dated 7 May. In TL.

Gale, W. (1896b). Letter to W.H. Wesley, dated 22 August. In RAS Letter Archives.

Gale, W. (1896c). Nebulae and star clusters. *Journal of the British Astronomical Association, 6,* 422.

Gale, W. (1896d). A new ring nebula. *Journal of the British Astronomical Association, 6,* 218–219.

Gale, W. (1897a). The brightness and colour of the planet Mercury. *Journal of the British Astronomical Association, 7,* 145–146.

Gale, W. (1897b). Interim report of the Jupiter Section of the New South Wales Branch. *Journal of the British Astronomical Association, 7,* 453–454.

Gale, W. (1897c). Present aspect of Jupiter. *Journal of the British Astronomical Association, 7,* 426.

Haynes, R., Haynes, R., Malin, D. & McGee, R. (1996). *Explorers of the Southern Sky. A History of Australian Astronomy.* Cambridge: Cambridge University Press.

Hoare, M. (1974). *Science and Scientific Societies in Eastern Australia 1820–1890.* Ph.D. Thesis, Australian National University, Canberra, Australia.

Hoare, M.E. (1976). Wright, Horatio George Anthony (1827–1901). In G. Serle & R. Ward (Eds.), *Australian Dictionary of Biography. Volume 6: 1851–1890. R-Z* (pp. 442–443). Melbourne: Melbourne University Press.

Inkster, I. & Todd, J. (1988). Support for the scientific enterprise, 1850–1900. In R. Home (Ed.), *Australian Science in the Making* (pp. 102–132). Cambridge: Cambridge University Press.

Innes, R.T.A. (1891a). Letter to John Tebbutt, dated 31 March. In TL.

Innes, R.T.A. (1891b). Letter to John Tebbutt, dated 26 August. In TL.

Innes, R.T.A. (1891c). Letter to John Tebbutt, dated 3 September. In TL.

Innes, R.T.A. (1891d). Letter to John Tebbutt, dated 18 November. In TL.

Innes, R.T.A. (1892a). Letter to John Tebbutt, dated 17 January. In TL.

Innes, R.T.A. (1892b). Letter to John Tebbutt, dated 3 February. In TL

Innes, R.T.A. (1895a). Letter to John Tebbutt, dated 25 January. In TL

Innes, R.T.A. (1895b). Letter to John Tebbutt, dated 28 January. In TL

Innes, R.T.A. (1896a). New double stars. *Journal of the British Astronomical Association, 6,* 261–262.

Innes, R.T.A. (1896b). Observations made in 1895. *Journal of the British Astronomical Association, 6,* 257–261.

Innes, R.T.A. (1896c). Observations of variable stars. *Journal of the British Astronomical Association, 6,* 486–489.

Innes, R.T.A. (1896d). The telescope objective. *Journal of the British Astronomical Association, 6,* 483–486.

Kelly, H.L. (1948). *The History of the British Astronomical Association. The First Fifty Years.* London: British Astronomical Association (Historical Section Memoir, Volume 36, Part 2).

Knibbs, G.H. (1894). Letter to John Tebbutt, dated 17 August In TL.

Knibbs, G.H. (1900). The Sun's motion in space. Part 1. history and bibliography. *Journal and Proceedings of the Royal Society of New South Wales, 34,* 148–232.

*List of Members of the British Astronomical Association, December 31, 1890.* London: British Astronomical Association (1891).

McDonall, F.K. (1895). Note on apparent umbral protrusion of sun-spots. *Journal of the British Astronomical Association, 6,* 29.

McDonall, F.K. (1896a). Daylight occultation of Antares, 1896, September 13. *Journal of the British Astronomical Association, 7,* 15.

McDonall, F.K. (1896b). Meteors, April 1986. *Journal of the British Astronomical Association, 6,* 451–452.

McDonall, F.K. (1896c). Note on solar phenomena. *Journal of the British Astronomical Association, 6,* 390–391.

McDonall, F.K. (1896d). Observations of meteors. *Journal of the British Astronomical Association, 6,* 496.

McDonall, F.K. (1897a). The umbral protrusion of sun-spots. *Journal of the British Astronomical Association, 7,* 458–460.

McDonall, F.K. (1897b). The zodiacal light. *Journal of the British Astronomical, 7,* 144.

Macdonnell, W.J. (1897). Note on a dark transit of Jupiter's IVth satellite. *Journal of the British Astronomical Association, 7,* 401.

Macdonnell, W.J. (1904). Presidential address. *Journal of the British Astronomical Association, 15,* 80–86.

McKim, R. (1990). E.W. Maunder and the formation of the British Astronomical Association. *Journal of the British Astronomical Association, 100,* 166–168.

MacLeod, R. (1988). *The Commonwealth of Science: ANZAAS and the Scientific Enterprise in Australasia, 1888–1988*. Melbourne: Oxford University Press.

Maiden, J.H. (1918). A contribution to the history of the Royal Society of New South Wales, with information in regard to other New South Wales societies. *Journal and Proceedings of the Royal Society of New South Wales, 52,* 215–361.

Maunder, T.F. (1891a). Letter to John Tebbutt, dated 7 January.

Maunder, T.F. (1891b). Letter to John Tebbutt, dated 9 October. In TL.

Membership List. *Journal of the British Astronomical Association*, 11 (1900).

Merfield, C.J. (1895). Occultation of Antares. *Journal of the British Astronomical Association, 5,* 512.

Merfield, C.J. (1896a). Conspicuous spots on Jupiter. *Journal of the British Astronomical Association, 6,* 387–388.

Merfield, C.J. (1896b). Letter to John Tebbutt, dated 2 February. In TL.

Merfield, C.J. (1896c). Letter to John Tebbutt, dated 16 September. In TL.

Merfield, C.J. (1896d). Occultations, June 1896. *Journal of the British Astronomical Association, 6,* 421.

Merfield, C.J. (1896e). Partial eclipse of the Moon, 1896, February 28. *Journal of the British Astronomical Association, 6,* 391.

Merfield, C.J. (1896f). Section for the observation of the colours of stars. *Journal of the British Astronomical Association, 6,* 416–421.

Merfield, C.J. (1897a). Eclipse of the Sun, February 1897. *Journal of the British Astronomical Association, 7,* 338–339.

Merfield, C.J. (1897b). Star Colour Section. *Journal of the British Astronomical Association, 7,* 417–418.

Merfield, C.J. (1897c). Uranus. Conjunction with the star 41 Librae. *Journal of the British Astronomical Association, 8,* 87–88.

Mr James Nangle. In *The Cyclopedia of New South Wales* (pp. 417–418). Sydney, McCarron, Stewart & Co. (1907).

Nangle, J. (1929). *Stars of the Southern Heavens*. Sydney: Angus and Robertson.

Notes of the Association. *Journal of the British Astronomical Association, 1,* 533–534 (1891).

Obituary: Sir George Knibbs. *Monthly Notices of the Royal Astronomical Society, 90,* 372–373 (1930).

Obituary: James Nangle. *Monthly Notices of the Royal Astronomical Society, 102,* 69–70 (1942).

Oleson, A. & Brown, S.C. (Eds.), 1976. *The Pursuit of Knowledge in the Early American Republic. American Scientific and Learned Societies from Colonial Times to the Civil War*. Baltimore: Johns Hopkins University Press.

Orchiston, W. (n.d.). Unpublished study of E.H. Beattie.

Orchiston, W. (1982). Illuminating incidents in Antipodean astronomy: John Tebbutt and the abortive Australian association of comet observers. *Journal of the Astronomical Society of Victoria, 35,* 70–83.

Orchiston, W. (1985). The contribution of Alfred Barrett Biggs to Australian and Tasmanian astronomy. *Records of the Queen Victoria Museum, 89,* 1–28.

Orchiston, W. (1987). Contribution of the Lands Department to the development of astronomy in New South Wales during the nineteenth century. *Australian Journal of Astronomy, 2,* 65–74.

Orchiston, W. (1988a). From research to recreation: the rise and fall of Sydney Observatory. *Vistas in Astronomy, 32,* 49–63.

Orchiston, W. (1988b). Illuminating incidents in Antipodean astronomy: formation of the New South Wales Branch of the British Astronomical Association. *Journal of the British Astronomical Association, 98,* 75–84.

Orchiston, W. (1989). The role of the amateur in early Australian astronomy. *Search, 20,* 16–23.

Orchiston, W. (1997a). The role of the amateur in popularising astronomy: an Australian case study. *Australian Journal of Astronomy, 7,* 33–66.

Orchiston, W. (1997b). The role of the large refracting telescope in Australian amateur astronomy: an historical perspective. *Australian Journal of Astronomy, 7,* 89–114.

Orchiston, W. (1998). Amateur-professional collaboration in Australian science: the earliest astronomical groups and societies. *Historical Records of Australian Science, 12,* 163–182.

Orchiston, W. (2001). From amateur astronomer to observatory director: the curious case of R.T.A. Innes. *Publications of the Astronomical Society of Australia, 18,* 317–327.

Orchiston, W. (2002). Tebbutt vs Russell: passion, power and politics in nineteenth century Australian astronomy. In Ansari, S.M.R. (ed.), *History of Oriental Astronomy* (pp. 169–201). Dordrecht: Kluwer.

Orchiston, W. (2003). Passion and patronage: the Afro-Australian astronomy of Robert Thorburn Ayton Innes. *Monthly Notices of the Astronomical Society of Southern Africa, 62,* 4–13.

Orchiston, W. (2015). The amateur-turned-professional syndrome: two Australian case studies. In W. Orchiston, D.A. Green & R. Strom (Eds.), *New Insights from Recent Studies in Historical Astronomy: Following in thew Footsteps of F. Richard Stephenson* (pp. 259–350). Cham: Springer.

Orchiston, W. & Bembrick, C. (1995). The role of the large reflecting telescope in amateur astronomy: an Australian case study. *Australian Journal of Astronomy, 6,* 53–74.

Orchiston, W. & Bhathal, R. (1984). Illuminating incidents in Antipodean astronomy: attempted formation of the Australian Astronomical Society in 1892. *Proceedings of the Astronomical Society of Australia, 5,* 422–425.

Orchiston, W. & Bhathal, R. (1991). Illuminating incidents in Antipodean astronomy: Section A of the Royal Society of New South Wales. *Australian Journal of Astronomy, 4,* 49–67.

Orchiston, W. & Perdrix, J. (1990). The British Astronomical Association in Australia: a centennial perspective. *Australian Journal of Astronomy, 3,* 163–179.

Orchiston, W. & Perdrix, J. (2002). A history of the British Astronomical Association in Australia: the fate of the Branches. *Journal of the British Astronomical Association, 112,* 68–77.

Pescott, R.T.M. (1961). The Royal Society of Victoria from then, 1854 to now, 1959. *Proceedings of the Royal Society of Victoria, 73,* 1–40.

Piesse, E.L. (1913). The foundation and early work of the Society; with some account of earlier institutions in Tasmania. *Proceedings of the Royal Society of Tasmania*, 117–166.

Reports of the Branches. New South Wales Branch. *Journal of the British Astronomical Association, 5:* 243–245 (1895a).

Reports of the Branches. New South Wales Branch. *Journal of the British Astronomical Association, 5,* 399–400 (1895b).

Reports of the Branches. New South Wales Branch. *Journal of the British Astronomical Association, 5,* 454–455 (1895c).

Reports of the Branches. New South Wales Branch. *Journal of the British Astronomical Association, 5,* 508 (1895d).

Reports of the Branches. New South Wales Branch. *Journal of the British Astronomical Association, 6,* 13 (1896a).

Reports of the Branches. New South Wales Branch. *Journal of the British Astronomical Association, 6,* 373–376 (1896b).

Reports of the Branches. New South Wales Branch. *Journal of the British Astronomical Association, 6,* 416 (1896c).

Reports of the Branches. New South Wales Branch. *Journal of the British Astronomical Association, 7,* 11–13 (1896d).

Reports of the Branches. New South Wales Branch. *Journal of the British Astronomical Association, 7,* 376–378 (1897a).
Reports of the Branches. New South Wales Branch. *Journal of the British Astronomical Association, 7,* 451–452 (1897b).
Reports of the Branches. New South Wales Branch. *Journal of the British Astronomical Association, 7,* 472 (1897c).
Reports of the Branches. New South Wales Branch. *Journal of the British Astronomical Association, 8,* 31 (1897d).
Reports of the Branches. New South Wales Branch. *Journal of the British Astronomical Association, 8,* 139 (1898a).
Reports of the Branches. New South Wales Branch. *Journal of the British Astronomical Association, 8,* 219–222 (1898b).
Reports of the Branches. New South Wales Branch. *Journal of the British Astronomical Association, 8,* 308 (1898c).
Reports of the Branches. New South Wales Branch. *Journal of the British Astronomical Association, 8,* 340–341 (1898d).
Reports of the Branches. New South Wales Branch. *Journal of the British Astronomical Association, 8,* 373 (1898e).
Reports of the Branches. New South Wales Branch. *Journal of the British Astronomical Association, 8,* 398–399 (1898f).
Reports of the Branches. New South Wales Branch. *Journal of the British Astronomical Association, 9,* 56–58 (1898g).
Reports of the Branches. New South Wales Branch. *Journal of the British Astronomical Association, 9,* 257–258 (1899a).
Reports of the Branches. New South Wales Branch. *Journal of the British Astronomical Association, 9,* 423 (1899b).
Reports of the Branches. New South Wales Branch. *Journal of the British Astronomical Association, 10,* 53–54 (1899c).
Reports of the Branches. New South Wales Branch. *Journal of the British Astronomical Association, 34,* 44–46 (1923).
Report of the Council on the work of the First Session. October 1890 to October 1891. *Journal of the British Astronomical Association, 1,* 535–546 (1891).
Roseby, T. (1895). Letter to John Tebbutt, dated 25 March. In TL.
Roseby, T. (1896). Letter to John Tebbutt, dated 25 March. In TL.
Rothenberg, M. (1981). Organisation and control: professionals and amateurs in American astronomy, 1899–1918. *Social Studies in Science, 11,* 305–325.
Schofield, R.E. (1963). Histories of scientific societies: needs and opportunities for research. *History of Science, 2,* 70–83.
Swindlehurst, W. (1891). Letter to John Tebbutt, dated 9 October. In TL.
Tebbutt, J. (1891). The colour of r Velorum. *Journal of the British Astronomical Association, 1,* 423.
Tebbutt, J. (1895). Occultation of Antares, 1895, May 10, as observed at Windsor, New South Wales. *Journal of the British Astronomical Association, 5,* 511–512.
Tebbutt, J. (1896a). Daylight occultation of B.A.C. 6127. *Journal of the British Astronomical Association, 6,* 219–220.
Tebbutt, J. (1896b). Observations of dark spots on the planet Jupiter, January to May, 1896. *Journal of the British Astronomical Association, 7,* 59–62.
Tebbutt, J. (1897a). On the occultation of 41 Librae by Uranus on 1897 September 8. *Journal of the British Astronomical Association, 8,* 86–87.
Tebbutt, J. (1897b). The variable star R Carinae. *Journal of the British Astronomical Association, 7,* 138–138.
Tebbutt, J. (1897c). The variable star R Carinae. *Journal of the British Astronomical Association, 8,* 85.

Tebbutt, J. (1899). Tempel's Comet 1873 II. *Journal of the British Astronomical Association, 10,* 71–72.
The late James Nangle. An outstanding figure passes. *The Australasian Engineer, 41* (299), 6 (1941).
Waters, B. (1980). *A Reference History of the Astronomical Society of South Australia Inc. Volume 1. The First Decade 1891–1901.* Adelaide: printed for the author.
Waters, B. (1981). *A Reference History of the Astronomical Society of South Australia Inc. Volume 2. The Second Decade 1902–1911.* Adelaide: printed for the author.
Wood, H. (1958). *Sydney Observatory 1858 to 1958.* Sydney: Government Printer (Sydney Observatory Papers No. 31).
Wooster, W.H. (1890). Letter to John Tebbutt, dated 12 July. In TL.
Wooster, W.H. (1892). Letter to John Tebbutt, dated 21 February. In TL.
Wooster, W.H. (1894). Letter to John Tebbutt, dated 1 October. In TL.
Wooster, W.H. (1895). Letter to John Tebbutt, dated 19 March. In TL.
Wooster, W.H. (1924). *Sketch of the Life and Work of W. H. Wooster.* Sydney: The Worker Trade Union Print.
Wright, H. (1896a). Daylight occultation of Antares, 1896, September 13. *Journal of the British Astronomical Association, 7,* 15.
Wright, H. (1896b). Double stars. *Journal of the British Astronomical Association, 6,* 422.
Wright, H. (1896c). Letter to John Tebbutt, dated 6 February. In TL.
Wright, H. (1896d) Letter to John Tebbutt, dated 3 September. In TL.
Wright, H. (1896e). Letter to John Tebbutt, dated 21 October. In TL.
Wright, H. (1896f). Observations of meteors. *Journal of the British Astronomical Association, 6,* 495.
Wright, H. (1897a). Letter to John Tebbutt, dated 17 May. In TL.
Wright, H. (1897b). Letter to John Tebbutt, dated 15 November. In TL.
Wright, H. (1897c). Letter to John Tebbutt, dated 3 December. In TL.
Wright, H. (1899a). Letter to John Tebbutt, dated 15 June. In TL.
Wright, H. (1899b). Occultations observed during the lunar eclipse, June 1899. *Journal of the British Astronomical Association, 10,* 73–74.
Wright, H. (1908). Presidential address. *Journal of the British Astronomical Association, 19,* 80–89.

# Chapter 14
# The End of an Era: A Welcome and Well-Earned Retirement

## 14.1 Introduction

All good things must come to an end, and so it was with John Tebbutt's active life as an observational astronomer. The problem for him was choosing when to retire, for he loved his astronomy—it was his life! But eventually an intolerable workload and recurrent bouts of illness took their toll, and in his annual report for 1903 he intimated that '… having attained my seventieth year, I felt compelled to give up systematic work …' (Tebbutt 1908a, p. 107). Instead, we find him pursuing what can only be described as a 'Clayton's retirement' (the retirement you have when you don't have a retirement)! Indeed, he continued to observe and to publish almost up until his death in 1916 (Fig. 14.1).

It was also after his retirement that the tributes and honours to which he was entitled openly flowed in from around Australia and from overseas, and when Tebbutt finally passed away towards the end of 1916 an era in Australian astronomy also ended. Professional astronomy worldwide was increasingly espousing astrophysics, and no longer would it be possible for an independent astronomer, working in comparative isolation and with modest instruments, to make so meaningful a contribution to international astronomy.

In this final chapter on Tebbutt's long and eventful life as an astronomer and meteorologist, we trace his inevitable path towards retirement and share his joy in receiving a succession of well-deserved honours, awards and tributes. We discuss his 'Clayton's retirement' from observational astronomy and witness his funeral. We also discuss the two most important posthumous honours bestowed upon him: the naming of a crater on the Moon after him and his appearance on the Australian $100 bank note in 1984. We end this chapter by examining Tebbutt's place in the battle between 'metropolitan' and 'colonial' interests in the evolution of Australian science. In so doing, we farewell one of Australia's most remarkable astronomers.



W. Orchiston, *John Tebbutt*, Historical & Cultural Astronomy,
DOI 10.1007/978-3-319-44521-2_14

**Fig. 14.1** An old and somewhat battered photograph of an equally old and battered-looking John Tebbutt and his 8-in. Grubb telescope taken in 1915, the year before he passed away (Orchiston collection)

## 14.2 An Inevitable Decision: The End of an Illustrious 'Career'

During the latter half of the 1890s, Tebbutt increasingly found the workload intolerable for someone in his 60s. Not only was he responsible for running the Peninsula Estate and overseeing his various business enterprises in Windsor (N.S.W. 1893), but he was obliged to conduct all of the astronomical and meteorological observations himself and to carry out most of the associated reductions. This was an inhuman burden when we consider the minimum level of staff deemed necessary for a professional observatory carrying out similar research (e.g. see Dewhirst 1985).

The computations were a continuing source of frustration to Tebbutt for some were particularly onerous, and in the days before the pocket calculator or personal computer they took considerable time and mental exertion. For example, the determination of the orbital elements of a comet typically involved many pages of calculations, all done by hand (e.g. see Tebbutt 1853–94). Consequently, Tebbutt was always on the lookout for competent 'computers' who could assist with the reductions, but suitable candidates proved to be rather elusive (see Tebbutt 1899, p. 17) for they needed to have a university-level knowledge of mathematics and to be quick and, above all, reliable.

After the reductions came the writing up of the associated research papers or notes, and the all-too-frequent reports for the local newspapers. Tebbutt also had to find time to prepare his *Annual Reports* of Windsor Observatory and his meteorological monographs. In addition, he was in demand as a speaker in both Windsor and Sydney and had to focus his energies on preparing his talks—they were always thoroughly researched. Finally, as we have seen in Chap. 13, during the second half of the decade he had to assign time for administrative tasks that he automatically inherited as founding President, and later a Committee member, of the newly-formed New South Wales Branch of the British Astronomical Association.

It was clear to any dispassionate observer that Tebbutt was hopelessly over-committed, but this realisation only began to permeate *his* consciousness when illness became an increasingly conspicuous factor (see Table 14.1).

Despite his habit of sunbathing in order to restore his 'nervous vitality' (Ramsay 1892), as early as 1892 matters had come to a head, as the *Annual Report* for that year clearly indicates. Under the heading 'Work Proposed for the Year 1893,' Tebbutt writes:

> Many suggestions have been received from eminent astronomers as to work desirable to be done, but the author much regrets that he is unable to comply with them. Indeed, the time is at hand, when he will have to seek some relaxation, for the work during the past few years has been sufficient to tax his powers to the utmost. In due course, however, he trusts to take up the work again with renewed energy. (Tebbutt 1893, p. 21; cf. Biggs 1893).

This is from a man fast approaching 60 years of age, who had carried an almost unbelievable workload for more than 30 years!

In fact, his astronomical work continued unabated (see Tebbutt 1894, p. 3), and it was only in mid-1897 that he lightened his load by abandoning most of his meteorological observing (Maynard 1897; Tebbutt 1898a). Even so, he continued to feel the pressure (e.g. see Tebbutt 1899), and this was reflected in his intended observational programme for 1900:

> During the current year it is his intention to devote himself mainly to the observation of double stars, the reductions of which are comparatively easy. At his age he finds the reduction of planet and comet work too great a burden when added to the work of observing and recording. Were he able to secure the occasional services of two competent computers, the case would be different … (Tebbutt 1900, p. 18).

**Table 14.1** Tebbutt's documented illnesses, 1890–1903

| Year | Time of Year | Complaint | Reference |
|---|---|---|---|
| 1890 | February | Unspecified | White (1890) |
| | November | Unspecified | Biggs (1890) |
| 1891 | January | Unspecified | Innes (1891a, b) |
| | June | Unspecified | White (1891) |
| | November | Flu | Innes (1891c) |
| 1893 | Winter | Flu | Young (1893) |
| 1894 | May | Flu | Gale (1894a) |
| | July | Flu | Gale (1894b) |
| 1896 | Late | Unspecified | Lenehan (1897) |
| 1897 | Late | Eyesight; digestion | Brooks (1897), Merfield (1897), Wright (1897) |
| 1898 | Winter | Unspecified | Merfield (1898a, b) |
| 1899 | Late | Flu | Merfield (1899a) |
| 1900 | March | Flu | Merfield (1900b) |
| 1901 | February | 'Serious illness' | Wright (1901) |
| | June | Flu | Merfield (1901a) |
| | August–Sept | Unspecified | Merfield (1901b) |
| | October | Deafness | Merfield (1901c, d) |
| 1902 | Year-long | Deafness | Merfield (1902a, 1903) |
| | October | Unspecified | Butterfield (1902), Merfield (1902b) |
| 1903 | Ongoing | Deafness | Merfield (1903) |

Fortunately, fellow British Astronomical Association New South Wales Branch stalwart C.J. Merfield (1899b, 1900a) was able to forestall this plan by offering to assist with the reductions.

An illness that was serious enough to be mentioned in the *Sydney Morning Herald* newspaper intervened in February 1901 (Table 14.1) and was followed by further prolonged bouts of illness later in the year. At this time, Lenehan (1901) was so concerned that he was moved to write: 'The astronomical world can ill afford to lose such an enthusiastic worker, and I trust you will long be spared to carry on your valued labors in the cause of science …'. This illness sapped his strength, and by the end of the year the observational load was once again too much and Tebbutt (1902, p. 13) announced that he anticipated '… a considerable relaxation in his efforts for the year 1902'. The *Annual Report* for 1902 confirms that this did indeed occur (Tebbutt 1903). Partly this was dictated by further illnesses during 1902 although one suspects that Tebbutt was somewhat relieved that it was deafness rather than eyesight problems that continued to afflict him. A dedicated observational astronomer could ill afford to contemplate this.

A bitter blow occurred on 7 February 1903 when Tebbutt lost his beloved wife, Jane, just 1 month short of her 65th birthday (see the genealogy in Tebbutt 1986, p. 103). Meanwhile, illness continued to plague him, and later in the year he decided to finally retire from observational astronomy. The formal announcement appeared in his 1903 *Annual Report*:

> The author, having now attained his seventieth year, finds that he is compelled to discontinue systematic work. His declining years will be devoted to business of a private nature, with occasional attention to astronomical observations. The present [annual] Report will therefore conclude the series emanating from his Observatory. (Tebbutt 1904, p. 11).

This difficult decision only became widely known during 1904 as copies of the Windsor Observatory *Annual Report* reached overseas friends and colleagues, and it brought responses from Agnes Clerke (Fig. 9.5), Sir David Gill (Fig. 14.2), R.T.A. Innes (Fig. 12.21), A.W. Roberts (Fig. 14.3) and Professor H.H. Turner, amongst others. In Australia, British Astronomical Association New South Wales Branch President, W.J. Macdonnell (Fig. 14.4), spoke of the '… distinct loss to Australian astronomy …' (Macdonnell 1904), while independent astronomer and Superintendent of Technical Education James Nangle (1904) was '… greatly inspired by the excellent example of untiring industry which you give so much evidence of'. C.J. Merfield (1904a), who finally was about to join the staff of Sydney Observatory (Orchiston 2015), was concerned that Australian astronomy '… will suffer in no small degree …' with Tebbutt's retirement, and believed that Tebbutt's career as an astronomer and gentleman was an excellent example for young Australians to follow (Merfield 1904b).

H.A. Lenehan (Fig. 7.12), from Sydney Observatory, was even more flattering:

**Fig. 14.2** Sir David Gill, 1843–1914 (Wikimedia Commons)

**Fig. 14.3** Alexander William Roberts (after Orchiston 2006, p. 149)

**Fig. 14.4** William John Macdonnell (adapted from Russell, 1892, Frontispiece)

> Australia and Australians will ever be proud of your work and for all time, searchers and investigators in the astronomical field will look to your complete work and know that they can rely on it absolutely. You certainly deserve a rest after so many years of useful work which you undertook solely for the love of Astronomical and Meteorological science. (Lenehan 1904).

Meanwhile, R.P. Sellors (ex-Sydney Observatory, and by then working in the Trigonometrical Branch of the New South Wales Lands Department) spoke of

> … the long and most excellent series of Astronomical observations which have justly won for the Windsor Observatory a foremost place among the leading astronomical institutions of the world.
>
> Your wonderfully fine series of cometary, planetary, and double-star observations are unrivalled as a record of work done in a private observatory, and unequalled by many of the public establishments both here and in other lands.
>
> I feel that I cannot let your retirement from active astronomical labour pass without expressing the esteem which I, for many years a humble worker in the same great field, have always regarded your untiring efforts to promote the advancement of this the Queen of the Sciences. (Sellors 1904).

Back in 1894 University of Oxford's Savilian Professor of Astronomy, Herbert Hall Turner invited John Tebbutt to visit England (Turner 1894), and although he found the idea appealing, his financial circumstances following the recent depression militated again this. However, by 1900 his situation had changed (Merfield 1900c) and he was looking forward to the trip. But floods and illness intervened, so his departure had to be postponed until 1901 (see Merfield 1900d, e). At the grand old age of 67, this would be Tebbutt's first trip outside of Australia, and he would tour observatories and university astronomy departments, visit the Royal Astronomical Society and British Astronomical Association headquarters and met people he had been corresponding with or reading about for years. But unfortunately further illness during 1901 put paid to these well-laid plans, and in the future there would never be an opportunity for him to reactivate them, something that Turner (1908) truly regretted. Tebbutt would die without ever leaving Australia although his name was widely known internationally throughout the astronomical world.

## 14.3 A 'Clayton's Retirement': Post-1903 Astronomical Activities

True to his forecast, Tebbutt from time to time was drawn to the telescope after his 'retirement', as Table 14.2 indicates (after Tebbutt 1908a, 1915b; Annual Reports 1905, 1906, 1907, 1908, 1909, 1910, 1911, 1912, 1913, inclusive). Merfield hoped for this, and on 22 September 1905 wrote Tebbutt that 'It is very gratifying to me to know that we will still have some occasional work from one who is so capable of doing it well and so completely'. Almost exactly 2 years later, Melbourne's David Ross was equally enthusiastic. After omitting to contact Tebbutt when he discovered his second comet (C/1906 F1), Ross (1907; my italics) explained:

> I would have sought your assistance to find it at the time but I considered that you were not able for the task at your time of life, and retired from the Stage of Duty. *Yet you seem to have considerable fire in you yet.*

Table 14.2 shows that Tebbutt did not come out of retirement in 1909 especially in order to observe Comet 1P/Halley, as has been suggested by Roser and White (1986). Instead, he continued to conduct intermittent astronomical observations from 1904 almost up to the time of his death in 1916.

Having said that, although his ongoing double star observations continued a series initiated in 1880, undoubtedly his most useful observations between 1904 and 1915 were those of Comets 1P/Halley and C/1912 R1 (Gale). His micrometric positions for Comet Halley (Tebbutt 1910c, 1911) were reduced by the European Space Agency and used, along with others, to determine the orbital elements of the comet for its 1985 return (see Roser and White 1986).

**Table 14.2** Tebbutt's post-'retirement' astronomical observations

| Year | Object(s) observed |
|---|---|
| 1904 | Comet Encke; Uranus; minor planets Ceres and Hebe |
| 1905 | Uranus; double stars α and γ Centauri, γ Coronae Australis, β Muscae and β 416 |
| 1906 | Lunar occultation of Saturn; double star p Eridani |
| 1907 | Double stars α and γ Centauri, γ Circini, γ Coronae Australis, p Eridani, Lacaille 2145, d Lupi, μ Lupi, π Lupi, β Muscae, β Phoenicis, β 416 and h 5014 |
| 1908 | Lunar occultations of stars; phenomena of Jupiter's satellites |
| 1909 | Comet 1P/Halley; double stars α and γ Centauri, γ Circini, γ Coronae Australis, π Lupi, β Muscae and β 416 |
| 1910 | Comet 1P/Halley |
| 1911 | Double stars α and γ Centauri, γ Coronae Australis, β Muscae and β 416 |
| 1912 | Comet Gale |
| 1913 | Double stars α Centauri and γ Coronae Australis |
| 1914 | Double star α Centauri |
| 1915 | Double star α Centauri |

Tebbutt's final cometary observations, in 1912, were at Gale's specific request. On 9 September 1912, Gale (Bio Box 12.1) wrote Tebbutt announcing his discovery the previous night while using binoculars to compare the magnitudes of some stars in Centaurus: 'I am well acquainted with this area of sky and am confident the object is a comet of about seventh magnitude or brighter. I had no telescope with which to examine it last night'. Gale (1912) then asked if Tebbutt would obtain positional measurements of the new comet (C/1912 R1). This was a considerable challenge for a 78-year-old, but Tebbutt coped admirably and subsequently published observations obtained over 9 nights in the *Journal of the British Astronomical Association* and in *Astronomische Nachrichten* (see Tebbutt 1912a, 1913). The latter was to be his final paper on cometary astronomy. For good measure, he also sent a brief report for publication in *Monthly Notices of the Royal Astronomical Society* (Fig. 14.5), but this was not accepted.

This was the first time that the Royal Astronomical Society refused to publish one of Tebbutt's papers or observing notes, which must have come as a shock but, as the Society pointed out, information in his report had already appeared in papers Tebbutt that had published in *Astronomische Nachrichten* and (in part) in *Journal of the British Astronomical Association* papers (Royal Astronomical Society 1913).

Comets had long been Tebbutt's forté (e.g. see Orchiston 1982, 2004; Proctor 1926), and in addition to Comets Halley and Gale he also attempted to observe Comet C/1907 G1 (Grigg-Mellish), but by the time he received the orbital elements from Merfield the Moon was too bright and the comet had moved too far north and become too faint for observation from Windsor. In a short article published in *The Observatory*, he complains about not being advised of this comet immediately after the discovery:

> I cannot understand why the information was suppressed so long at the State Observatories, or why the discoverer himself, who is a correspondent of mine and who knows that nearly all the comet observations in Australia for many years past have devolved upon me, ignores the existence of my Observatory. (Tebbutt 1907a).

Co-discoverer, Grigg (Bio Box 14.1) was quick to point out to Tebbutt that he was not troubled on this occasion because of his recent retirement. Instead, Grigg (1907) cabled Baracchi at Melbourne Observatory, who neglected to conduct or arrange the appropriate follow-up observations (see Orchiston 1999).

Apart from his cometary and double star papers, Tebbutt prepared a number of short papers on his earlier observations and on matters relating to Australian astronomy or astronomy in general (Tebbutt 1905b, 1906, 1907b, 1910a, b, 1915c). In 1913, 7 months after the Royal Astronomical Society had refused his Comet Gale note, Tebbutt wrote a letter about η Argus and submitted this to the *Journal of the British Astronomical Association*, but it also was refused publication (Levander 1913). This was a new scenario for Tebbutt and must have reinforced the perception that his observing days were all but over.

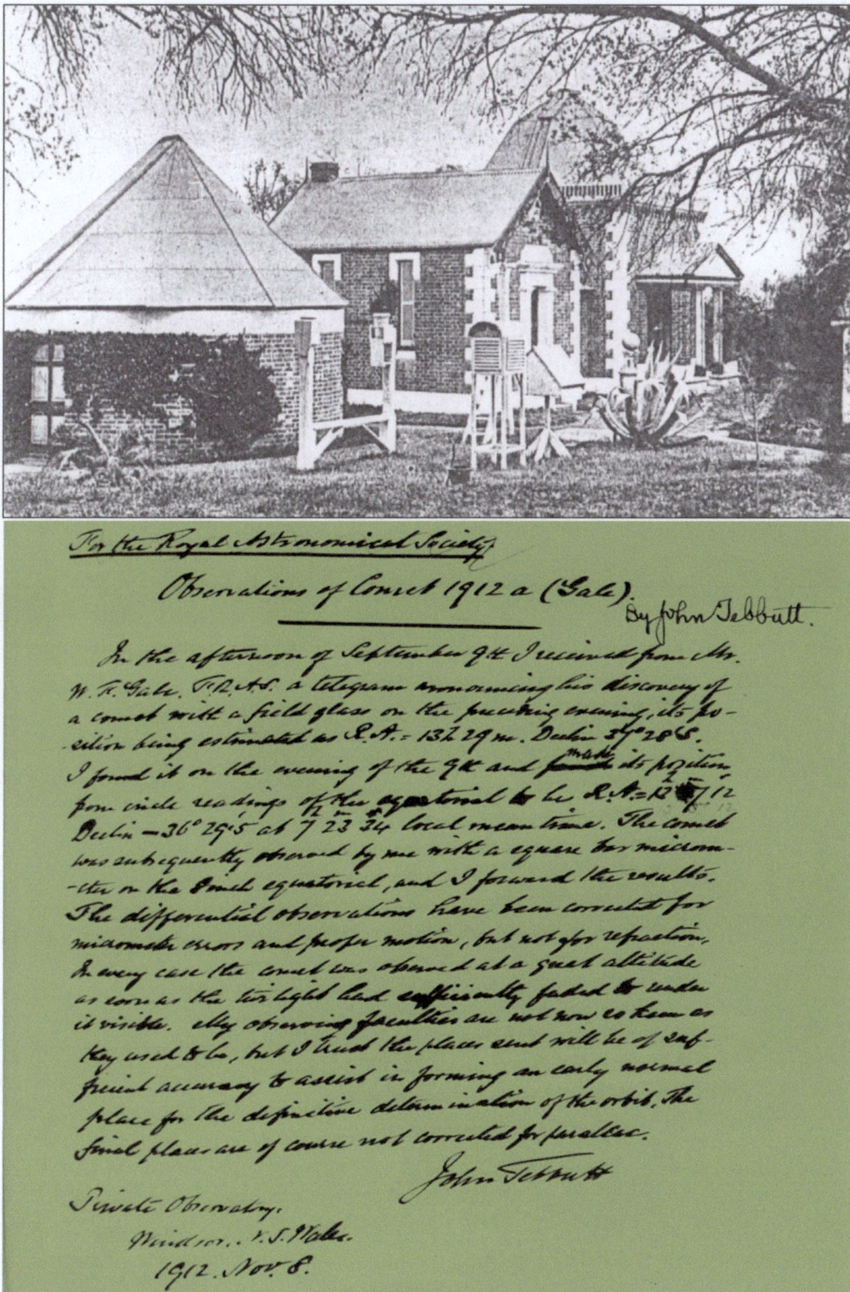

For the Royal Astronomical Society

Observations of Comet 1912 a (Gale). By John Tebbutt.

In the afternoon of September 9th I received from Mr. W. F. Gale, F.R.A.S. a telegram announcing his discovery of a comet with a field glass on the preceding evening, its position being estimated as R.A. = 13h 29m. Decln 57° 28 S. I found it on the evening of the 9th and made its position from circle readings of the equatorial to be R.A. = 13h [illegible] 12 Decln −36° 29'.5 at 7 23 34 local mean time. The comet was subsequently observed by me with a square bar micrometer on the 8inch equatorial, and I forward the results. The differential observations have been corrected for micrometer errors and proper motion, but not for refraction. In every case the comet was observed at a great altitude as soon as the twilight had sufficiently faded to render it visible. My observing faculties are not now so keen as they used to be, but I trust the places sent will be of sufficient accuracy to assist in forming an early normal place for the definitive determination of the orbit. The final places are of course not corrected for parallax.

John Tebbutt

Private Observatory.
Windsor. N.S. Wales.
1912. Nov. 8.

**Fig. 14.5** The short paper about C/1912 R1 (Gale) and 1906 photograph of Windsor Observatory that Tebbutt sent to the Royal Astronomical Society (*Courtesy* Mitchell Library)

**Bio Box 14.1: John Grigg**

John Grigg (Fig. 2.4) was born in London on 4 June 1838 to parents who were successful in business and made sure their son received a sound education, including a grounding in mathematics and 'the sciences'. In 1863, John Grigg and his wife emigrated to Auckland, New Zealand, and 5 years later, after the unexpected death of his wife, John Grigg settled in the booming North Island gold-mining town of Thames where he opened a furniture shop. Later, he added a music shop, and he also taught singing.

During the 1870s, Grigg's boyhood interest in astronomy returned, inspired in part by the 1874 transit of Venus, and in March 1885 he purchased a 3.5-in. (8.9-cm) Wray refractor. He then erected a small wooden observatory for it in the backyard of his business premises in the centre of Thames, and this facility included a transit annex with a small transit telescope by A.J. Frost of London (Fig. 14.6). In 1894, Grigg relocated the observatory to another part of Thames and enlarged it into a two-storey structure with an office and workshop below the dome room and transit annex (see Orchiston 2001).

**Fig. 14.6** John Grigg's original observatory (Orchiston Collection)

Grigg's primary observational interest was in comets (Orchiston 1993), and although he certainly would have noticed the Great Comets of 1880, 1881 and 1882, the first comet that he systematically observed was C/1885 X1 (Fabry) over 33 different evenings in May–June 1886. However, his favourite comet was 2P/Encke, which he observed on three different returns (1895, 1898 and 1904)—he and John Tebbutt shared the recovery of this comet in 1898.

(continued)

**Bio Box 14.1** (continued)

Apart from known comets, Grigg was interested in detecting new ones and in 1886 he began a systematic search programme, but it was only in the new century—after he had taken early retirement from business in 1894 so that he could devote more time to astronomy—that he met with success. Thus, he eventually discovered and was assigned credited for three comets: 26P/Grigg-Skjellerup (in 1902), C/1903 H1 (Grigg) and C/1907 G1 (Grigg-Mellish), and he also independently discovered C/1906 F1 (Ross). It is clear that 'John Grigg made a significant contribution to international cometary astronomy, and was the "founding father" of New Zealand cometary astronomy' (Orchiston 1998, p. 108). At a trans-Tasman level, he also was actively involved in an unpleasant confrontation about Australasian cometary discoveries between Melbourne Observatory Director Pietro Baracchi and a number of leading Sydney independent astronomers (see Orchiston 1999 for details).

Although 'addicted' to comets, Grigg also observed other astronomical objects and events, including the 1882 transit of Venus, sunspots, lunar occultations of stars and eclipses.

Another area of astronomy in which Grigg made an important local contribution was astronomical photography. Between 1882 and 1910 he used two home-made cameras attached to his Wray telescope to successfully photograph the 12 December 1890 partial solar eclipse, sunspots, the Moon, comets and star fields. Although none of these was published at the time, '… many played an important educational role, for Grigg used them as popular visual aids during his numerous public lectures' (Orchiston 2016, p. 622). John Grigg was one of New Zealand's earliest astrophotographers, and his work in this field is reviewed in Orchiston (1995, 2016, Chapter 22).

As the foregoing paragraph suggests, Grigg also was committed to the popularisation of astronomy, '… and he ran his observatory as a *de facto* "city observatory" for the citizens of and visitors to Thames. He conducted public viewing nights at the telescope, maintained a local time-service and gave public lectures on astronomy. At a national level, he occasionally lectured on astronomy in other centres, and he contributed an astronomical column to the *Otago Daily Witness* newspaper' (Orchiston 1998, p. 108).

By the end of the nineteenth century, John Grigg was New Zealand's leading non-professional astronomer. He died in Thames on 20 June 1920, not long after his 82nd birthday, and further details of his astronomical achievements are presented in Orchiston 2016 (Chapters 10, 17 and 22).

In early 1907, Tebbutt decided to write his astronomical autobiography (Merfield 1907), and he devoted considerable time and effort to this task during the year. *Astronomical Memoirs* (Tebbutt 1908a) appeared in mid-1908 as a 132-page soft cover illustrated book, complete with a year-by-year summary of the Observatory's principal achievements and a master list of research papers emanating from Windsor Observatory. Just 298 copies were printed (Tebbutt 1908b), and for those with an astronomical bent it is an enthralling tale of achievement in the face of adversity and intellectual isolation.

Tebbutt (1908a, p. 117) concludes the text of his *Astronomical Memoirs* with the following statement: 'I shall now bring these Memoirs to a close in the hope that if they do not prove interesting to the present generation of Australians they will, at all events, be so to those who follow'. This has certainly proved to be the case, as evidenced by this book and the various other publications about Tebbutt's astronomical activities which have appeared in the course of the last 35 years.

During his retirement, Tebbutt gradually wound down his astronomical work and activities, but he continued to take the following meteorological readings: daily rainfall, and monthly maximum and minimum shade temperature and solar radiation. In an Annual Report for 1912 submitted to the Royal Astronomical Society, he points out that 'The rainfall record now extends over fifty years, namely, from 1862 December 31 to 1912 December 31, the mean annual result being 30.613 inches' (Annual Reports 1913).

In fact, Tebbutt continued his meteorological observations virtually up until the time of his death and published these for the period 1898–1915 as an 18-page booklet in 1916 (Tebbutt 1916; Fig. 14.7). In addition to monthly rainfall and temperature data, this publication includes a 'Catalogue of Floods and Freshes in the Hawkesbury River and South Creek …'. The 1916 meteorological volume was the last in a series of eight (see Tebbutt 1868, 1874a, 1877, 1882a, 1886, 1891b, 1898b) that total 389 pages and between them provide a document of the weather and flooding at Windsor that rivals the records obtained by the much better-resourced Australian government observatories.

Tebbutt not only published and distributed the Windsor Observatory meteorological monographs at his own expense, but he wrote prolifically for the Sydney and Hawkesbury district newspapers (e.g. see the listing on pages 38–46 in Tebbutt 1887), published an important research paper on Australian storms (Tebbutt 1863c), and went to considerable pains to respond to public inquiries. One of his most interesting meteorological contributions was the proposal that there was a 3-year periodicity in New South Wales rainfall. This idea was promoted in a series of newspaper articles published during the 1870s and was contrary to a 19-year periodicity claimed by Russell (1876).

In the 2 March 1874 issue of the *Sydney Morning Herald*, Tebbutt wrote:

> It has been pointed out by my friend, the Government Astronomer, in his valuable abstract of meteorological observations up to the end of 1869, that a period of *nineteen* years obtains in the annual rainfall of Sydney; but it appears to me that a period of *three* years is much more distinctly marked. Floods have been prevalent in this district since 1857, and they seem to have a tendency to group themselves about every third year during the period which has elapsed since that time… If we may regard the valley of the Hawkesbury as a huge rain-gauge for the watershed of that river it is obvious that the annual rainfall over the source of the river reaches a maximum every *third* year. (Tebbutt 1874b).

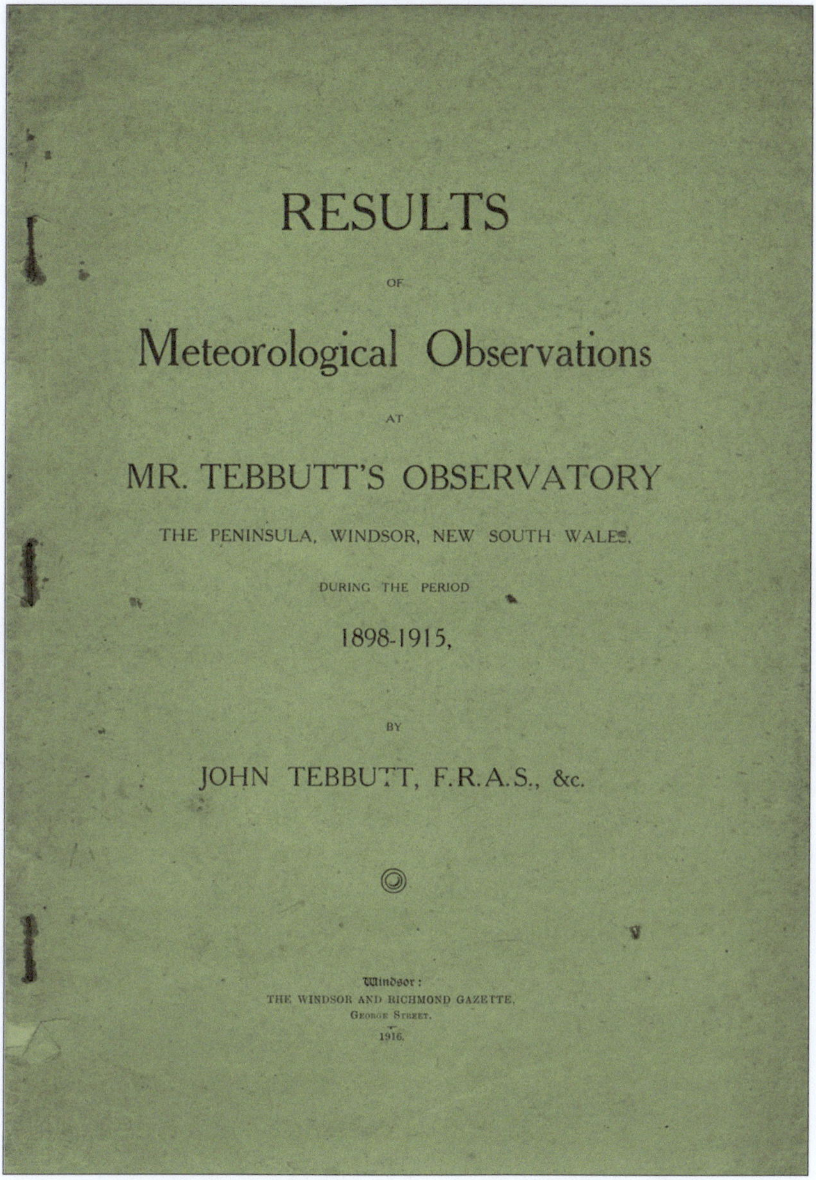


**Fig. 14.7** The front cover of Tebbutt's final meteorological monograph, which was published in the year in which he died (Orchiston Collection)

In a follow-up article, Tebbutt (1874c) found that some meteorological parameters (but particularly mean minimum air temperature, mean maximum terrestrial radiation and mean maximum solar radiation) provided '… striking confirmation of the three year period'. Further discussion of the 3-year period was presented in two later articles, published the following year (Tebbutt 1875a, b), and it was raised again in 1882 (Tebbutt 1882b).

In addition, during the 1860s Tebbutt agitated for the establishment of a network of weather reporters (including masters of ships entering Sydney Harbour) in order to facilitate more accurate weather forecasting (see Tebbutt 1863a, b), and in 1864 he published a long and critical analysis of Saxby's weather prediction model which '… assigns to the moon the power of influencing the meteorological phenomena of the earth …' on the basis of an assumed electrical connection between the Moon and the Earth's atmosphere (Tebbutt 1864).

It is apparent that in addition to his outstanding astronomical record, Tebbutt deserves to be recognised as a pioneer of Australian meteorology (e.g. see Hunt 1913; Todd 1893), and a separate detailed study of his overall contribution in this important scientific field is long overdue.

## 14.4 Honours at Home and Abroad: Rightful Recognition at Long Last

The Royal Astronomical Society had a long tradition of awarding Gold Medals for outstanding astronomical discoveries and achievements, and near the close of the nineteenth century it initiated the presentation of a second medal, the Jackson-Gwilt Medal and Gift (see Dreyer and Turner 1987). On 10 December 1904, the Council of the Society decided to award this Medal and Gift (Fig. 14.8) to Tebbutt in recognition of his long contribution to positional astronomy. President of the Society, Professor H.H. Turner, wrote Tebbutt the very next day announcing this honour, and added: 'I wish you could have heard the cordial expressions of appreciation of your work from various members of the Council'.

**Fig. 14.8** The obverse (*left*) and reverse (*right*) of the Jackson-Gwilt Medal awarded to John Tebbutt by the Royal Astronomical Society (after Tebbutt 1986)

At the Annual General Meeting of the Society, on 10 February 1905, Turner gave the following speech:

> The Jackson-Gwilt Gift and Medal have been awarded to Mr. John Tebbutt, of Windsor, N.S.W., for his important observations of comets and double stars and his long continued services to astronomy in Australia, extending over forty years.
>
> This gift has only twice been awarded previously—in 1902 to Dr. Anderson, and in 1897 to Mr. Lewis Swift… there can be no reason why I should not recall to your memory that Mr. Tebbutt began astronomical work in 1854, and is only now relinquishing it at the age of seventy; that amid surroundings which gave him little encouragement he has made regular and systematic observations during half a century, including, for instance, those of some 1400 occultations of stars by the Moon and valuable measures of double stars; that he has contributed over eighty papers to the *Monthly Notices*; and has, moreover, discovered several comets, including the notable ones of 1861 and 1881, the orbits of which he computed with his own observations.
>
> In handing this medal to the Secretary for transmission, with the gift, to Mr. Tebbutt, I will ask him to convey our hearty congratulations to the recipient on the accomplishment of half a century of single-handed astronomical work for which it would be difficult, if not impossible, to find a parallel, and our delight that Mr. Tebbutt should be entering upon the rare enjoyment of a thoroughly well-earned period of rest.

The Medal itself and accompanying Gift of £25 reached Tebbutt in April 1905, and he immediately replied and expressed his thanks and appreciation. He went on to elaborate:

> The Medal I shall carefully preserve as a memorial of the Council's appreciation of my services, but as I think the money part of the award should properly be devoted to the advancement of astronomy I have handed it over to our local Branch of the British Astronomical Association to be expended by them in the most suitable way for that object. (Tebbutt 1905a; cf. Reports of the Branches 1905).

Wright (1905a) penned the Branch's official letter of heartfelt thanks on 20 April, and at the June meeting of Council Tebbutt was elected a Life Member of the Branch (Wright 1905b). One year latter Wright (1906) was to report that some of the money had been spent on a lantern projector, and plans were afoot to spend some more improving the slide collection.

Meanwhile, as would be expected, once news of the Jackson-Gwilt Medal and Gift became known, Tebbutt received congratulatory letters from friends, relatives and colleagues. White (1905), from Melbourne Observatory, saw the award as '… a tardy recognition of the splendid work you have accomplished during the last half century, and all Australia is honoured by the award'. His letter succinctly summarises the sentiments expressed in all of the other letters.

In fact, this Medal was a long time coming, for the idea that Tebbutt deserved official recognition by the Royal Astronomical Society was first formally proposed more than 10 years earlier. In a letter to the Assistant Secretary of the Society W.H. Wesley dated 30 January 1893, Innes (Bio Box 8.1) writes: 'Mr. Tebbutt is a very earnest amateur & has done some good work. If the Society were ever to award him a medal he is if I may be permitted to say it the kind of gentleman who would appreciate the award' (Innes 1893). Wesley's reply was illuminating:

> I am sure there is no amateur observer of whom we think more highly than Mr. Tebbutt. With regard to the medal, however, it appears to be the practice to award it only for some distinctly original line of research. (Wesley 1893; his underlining).

What Wesley was referring to was the Society's Gold Medal, and a perusal of the list of recipients during the preceding 20 years confirms his emphasis on original research as opposed to programmes of systematic observations. They included such notables as A.A. Common, H.L. D'Arrest, David Gill, Asaph Hall, William Huggins, U.J.J. Le Verrier, Simon Newcomb and E.C. Pickering (see Dreyer and Turner 1987). However, this did not dissuade Innes, and he made two further pleas on Tebbutt's behalf (but unbeknown to Tebbutt) in 1895. This time, the response was more encouraging: '... it is thought that Mr. Tebbutt had scarcely done enough original work—I mean of an original kind. We shall be having another medal shortly the 'Jackson-Gwilt'. Perhaps this will give him a chance.' (Turner 1895). It took another 8 years before this actually occurred.

Further recognition of Tebbutt's status, after his retirement, is reflected in Lenehan's invitation in 1907 for him to serve on a committee to recommend a new site for Sydney Observatory (Lenehan 1907a). We must presume that it was a combination of age and a non-Sydney location which prompted Tebbutt to decline this offer (see Lenehan 1907b).

In 1908, Tebbutt's *Astronomical Memoirs* (Fig. 14.9) came off the press and were distributed widely to institutions with which Windsor Observatory maintained exchange relations and to selected acquaintances and astronomers.

As would be expected, this exercise generated an avalanche of acknowledgements. Wright (1908) thought the volume reflected Tebbutt's own temperament, 'Precise, exact, modest & not verbose'. while Macdonnell (1908) was somewhat more flattering:

> ... it is really a true memorial of the splendid work done by you for Astronomy, single handed and after surmounting difficulties and obstacles that would have discouraged any one less zealous than yourself. As a record of what can be done, given ability and enthusiasm, it should prove a guide and incentive to future workers perhaps yet unborn.

Amongst others, the South African astronomer A.W. Roberts and the noted British astronomers George Frederick Chambers and W.F. Denning wrote similar letters (Chambers 1909; Denning 1909; Roberts 1909), while Professor William Wallace Campbell (1862–1938; Fig. 14.10), the Director of Lick Observatory, read *Astronomical Memoirs* with great interest and admiration (Campbell 1909). As might be expected, the Commonwealth Meteorologist, Henry Ambrose Hunt (1866–1946), took a broader view of Tebbutt's overall contribution to science: '... in addition to the astronomical monumental work you have done for Australia, we in the Meteorological Service regard your climatological observations as amongst the very best we have from any part of Australia' (Hunt 1913). Following the widespread distribution of his *Astronomical Memoirs*, details of Tebbutt's achievements in astronomy, meteorology and tidal studies also became more widely known amongst the Australian public.

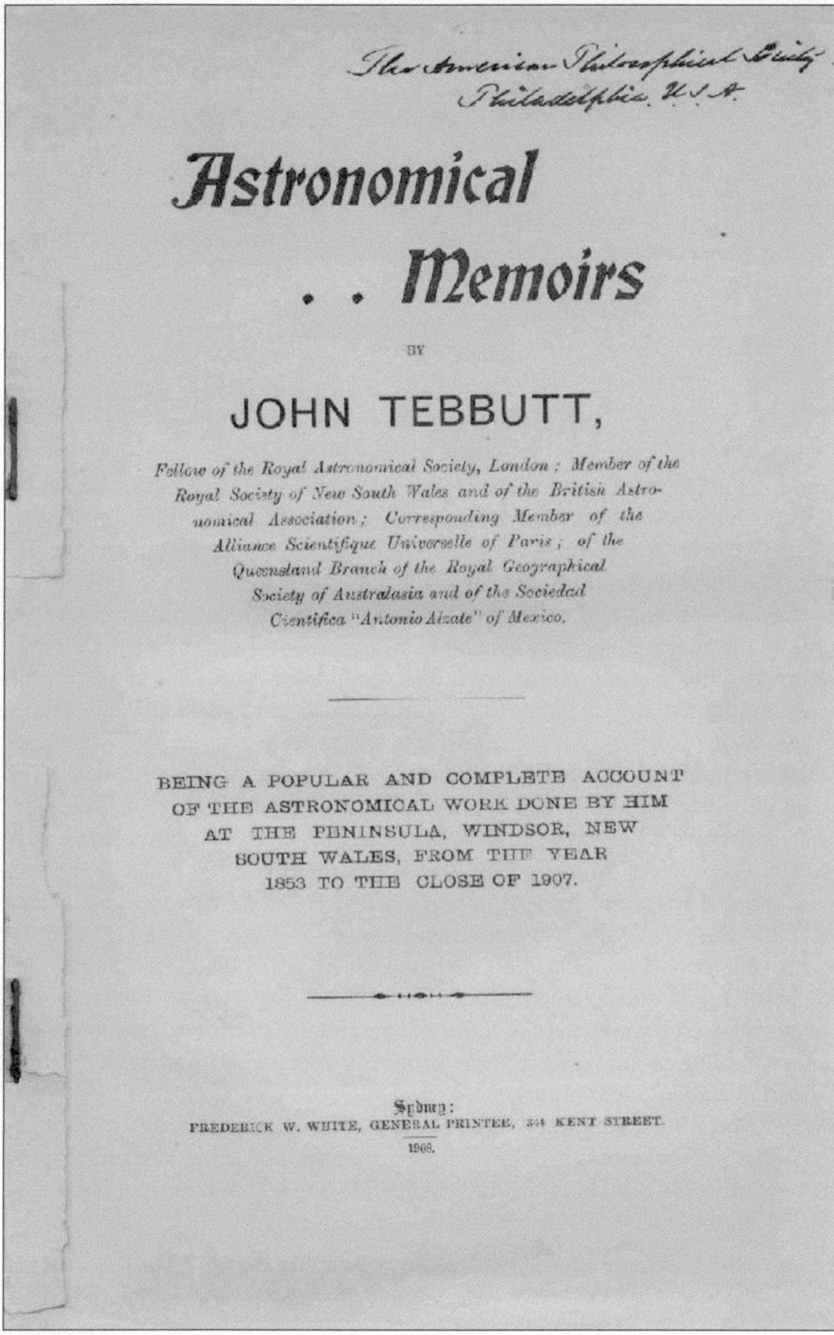
The American Philosophical Society
Philadelphia, U.S.A.

Astronomical
. . Memoirs

BY

JOHN TEBBUTT,

*Fellow of the Royal Astronomical Society, London; Member of the Royal Society of New South Wales and of the British Astronomical Association; Corresponding Member of the Alliance Scientifique Universelle of Paris; of the Queensland Branch of the Royal Geographical Society of Australasia and of the Sociedad Cientifica "Antonio Alzate" of Mexico.*

BEING A POPULAR AND COMPLETE ACCOUNT OF THE ASTRONOMICAL WORK DONE BY HIM AT THE PENINSULA, WINDSOR, NEW SOUTH WALES, FROM THE YEAR 1853 TO THE CLOSE OF 1907.

Sydney:
FREDERICK W. WHITE, GENERAL PRINTER, 344 KENT STREET.
1908.

**Fig. 14.9** The title page of Tebbutt's *Astronomical Memoirs*, which was published at his own expense in 1908 (Orchiston Collection)

**Fig. 14.10** W.W. Campbell (after Macpherson 1905, facing p. 240)

Another indication of Tebbutt's international status is the extent to which professional observatories around the world entered into exchange relationships with Windsor Observatory, thereby providing Tebbutt with a research and reference library. The institutions that contributed in this way as at 1908 are listed in his *Astronomical Memoirs* and run to 5 pages. Thirty-one different countries and virtually all of the major observatories and astronomical institutions of the world are represented. This international support allowed Tebbutt to assemble an absolutely outstanding library, and one that rivalled those found in many smaller professional observatories.

What is remarkable is that it was Tebbutt's own reputation which allowed him to build up his library, for all he had to offer in exchange were reprints of some of his published papers, a 75-page little book titled *History and Description of Mr. Tebbutt's Observatory, Windsor, New South Wales* published at his own expense in 1887 (Tebbutt 1887), and *Annual Reports of Mr. Tebbutt's Observatory, Windsor, New South Wales* … which he published as 19–33 page booklets (Table 14.3) between 1888 and 1903 inclusive (Tebbutt 1889, 1890, 1891a, 1892, 1893, 1894, 1895, 1896, 1897, 1898a, 1899, 1900, 1901, 1902, 1903, 1904). To a large extent, the length of each *Annual Report* reflected the amount of observational work conducted at the Observatory, which was primarily dependent upon Tebbutt's health and the frequency of clear nights.

Intermittent illness continued to pursue Tebbutt with the advancing of the years and, despite his best intentions, observing became an increasing burden. Moreover, there were other calls on his time and energy. He alludes to these in a letter sent to the Royal Astronomical Society in 1912:

> I cannot now do much for our science. Next May I shall have arrived at the 78th milestone of life, and I feel that the conduct of my private business is quite enough for me without the addition of scientific work. (Tebbutt 1912b).

In 1913, Tebbutt celebrated his 79th birthday, and amongst the letters he received was one from Lick Observatory's W.W. Campbell (1913):

**Table 14.3** The size of the Windsor Observatory *Annual Reports*

| Year | Pages | Year | Pages | Year | Pages | Year | Pages |
|---|---|---|---|---|---|---|---|
| 1888 | 33 | 1892 | 31 | 1896 | 31 | 1900 | 20 |
| 1889 | 26 | 1893 | 25 | 1897 | 24 | 1901 | 23 |
| 1890 | 28 | 1894 | 26 | 1898 | 24 | 1902 | 19 |
| 1891 | 24 | 1895 | 24 | 1899 | 26 | 1903 | 19 |

**Fig. 14.11** A photograph of W.J. Hussey taken by B.A. Hinsdale in 1906 (https://en.wikipedia.org)

> May I congratulate you most cordially in the name of myself and colleagues of the Lick Observatory staff? You have, indeed, accomplished a great deal for astronomy in the southern hemisphere during the past half century, and we hope that you will be permitted to have many additional years' use of your telescope.

Dr. William Joseph Hussey (1862–1926; Fig. 14.11) from La Plata Observatory was equally lavish in his praise of Tebbutt's work:

> It is sufficient that a paper should be printed in Windsor to make it of interest to any Observatory, for owing to your continuous and excellent work Windsor is known and esteemed wherever there is an astronomer pointing his telescope to the stars. (Hussey 1913).

By this time, Tebbutt's eyesight was failing (Tebbutt 1915b), so further observations were by no means a certainty.

In August 1914, just 3 months after Tebbutt's 80th birthday, the 84th Meeting of the British Association for the Advancement of Science (BAAS) was held in Sydney (see Robertson 1980), and amongst the distinguished British astronomers who attended were the Astronomer Royal Dr. Frank (from 1915 Sir Frank) Watson Dyson (1868–1939; Fig. 14.12) and the Oxford's Professor H.H. Turner (Fig. 14.13). Turner was well known to Tebbutt through their correspondence. On 20 August, a combined meeting of the BAAS and local astronomers was scheduled, and the organisers were particularly eager for Tebbutt to attend:

**Fig. 14.12** Sir Frank Dyson (https://en.wikipedia.org)

**Fig. 14.13** Professor H.H. Turner (http://galnet.wikia.com/wiki/Herbert_Hall_Turner)

> I am writing to you, personally, concerning the meeting of Astronomers from England … I am specially anxious that you should be one of the party to meet them on that unique occasion.
>
> Yours is the outstanding name among us all. They will feel more interest in meeting the "Nestor" of Australian Astronomy than in seeing any of us. Do make an effort to be there. They will feel a special delight in seeing the face of one so well known to them … (Roseby 1914; his underlining).

Because of his age and Windsor domicile Tebbutt had to refuse, and Turner (1914, his underlining) was moved to write: 'If I were to visit N.S.W. without seeing you I should never forgive myself …'. As a result, Turner and Dyson determined to visit Tebbutt that Sunday (22 August 1914), and James Nangle (Bio Box 13.5) drove them to Windsor, along with Cooke (the newly-appointed Government Astronomer of New South Wales and Director of Sydney Observatory). Nangle (1914) reported that Dyson and Turner '… had a very pleasant day & enjoyed meeting you very much'

**Fig. 14.14** John Tebbutt's meeting with Professor H.H. Turner and Dr. Frank Dyson at Tebbut's Homestead in 1914. From *left* to *right*: one of Tebbutt's daughters, Mr. and Mrs. Nangle, Professor Turner, John Tebbutt, Mr. Cooke and Professor Dyson (Orchiston collection)

(see Fig. 14.14). Turner (1918) later related that their party was made very welcome, and that they '… found the observatory still in excellent working order, and were proudly shown the sextant and other apparatus which had done such historic work'. As we have seen, 15 years earlier an unfortunate combination of circumstances had forced Tebbutt to abort an intended trip to England, so Turner was particularly delighted to be able to meet him at long last, and to explore Windsor Observatory in person (Evershed 1914; Turner 1915).

Another memorable feature of the 1914 BAAS meeting was the 618-page *Federal Handbook on Australia* … prepared for the delegates. This was edited by the Commonwealth Statistician, G.H. Knibbs (1914), formerly a prominent member of the non-professional astronomical fraternity in Sydney (see Bio Box 13.1), and includes maps of the Australian states and 14 different chapters on Australian life, landscape, science and inhabitants. One of these chapters is titled 'Astronomy and Geodesy in Australia' and was prepared by Baracchi from Melbourne Observatory. This runs to 64 pages (Baracchi 1914), 48 of which are devoted to astronomy. In all, a little over 10 pages deal exclusively with non-professional astronomers, and as the doyen of Australian astronomy Tebbutt is the first one to be discussed and his work is summarised in 3 pages. His name also features in the account of the NSW Branch of the BAA; in the summaries of the 1874 and 1882 transit of Venus programmes; and in the section dealing with the determination of Australian longitudes. Baracchi's original manuscript was much longer and contained far more information about Tebbutt, but unfortunately the manuscript had to be severely 'abbreviated' (see Baracchi 1914; Merfield 1914).

By this time, Tebbutt was recognised as one of Windsor's most valued 'living treasures', and in 1915 the Council decided to hang a large framed photograph of him and a smaller one of Windsor Observatory in the Council Chambers. The Town Clerk wrote requesting copies of these (Morgan 1915a, b), and Tebbutt was only too happy to oblige (Morgan 1915c).

In order to publicise John Tebbutt's 81st birthday, Sydney's *Daily Telegraph* newspaper ran a feature article and photograph of Tebbutt on 25 May 1915, and this brought further public recognition. Hugh Wright (1915b) was moved to write:

> It must be gratifying to you to find that at least among your own countrymen you are well appreciated: though for many years your good work was keenly appreciated by a very small band here, it gained its award of high distinction for great accuracy from the eminent mathematicians abroad.

Meanwhile, the international astronomical community was finally coming to terms with the fact that Tebbutt's observational days (and nights) were all but over. In February 1915, he wrote the Royal Astronomical Society that he had been a Fellow for 42 years, '… and I think I have fulfilled the promise which I made when I was elected, namely, to further the interests of the Society to the best of my ability' (Tebbutt 1915a). For its part, the Society was '… sorry to think we shall have no more reports from your observatory. But it would be hard to find a private observatory which has as fine a record of work as that of the Peninsula, New South Wales' (Wesley 1915). This reply came from William Henry Wesley (ca. 1841–1922; Crommelin 1922), the Society's long-serving Assistant Secretary, who got to know Tebbutt through their correspondence.

Meanwhile, Tebbutt could still promote astronomy in the Windsor-Sydney region, which he did by writing letters and reports for the local newspapers, in this way '… keeping alive a public interest in astronomy' (Wright 1915a).

Despite his advanced years and astronomical inactivity, Tebbutt was asked to join the Board of Visitors of Sydney Observatory when this was finally reactivated in November 1915 (see Board 1915). However, it is more than likely that the Department of Public Instruction merely issued this invitation as a courtesy, in the light of Tebbutt's reputation and his well-known earlier criticism of the Observatory's astronomical output—or lack of it (see Chap. 12).

Just 1 year later, on 29 November 1916, Tebbutt passed away at the age of 82, survived only by a son and three of his six daughters. His funeral was a major event for Windsor (see Death of Mr. John Tebbutt 1916), and in addition to family members and the local Mayor attracted the Minister of Mines and Assistant Treasurer (representing the New South Wales State Government), the Reverend Dr. Thomas Roseby (representing the Royal Astronomical Society), Mr. W.H. Hamlett (from the Royal Society of New South Wales), James Nangle (Superintendent of Technical Education) and Walter Gale (President of the New South Wales Branch of the

British Astronomical Association). Roseby, Nangle and Gale had all been close astronomical colleagues, and collectively they also represented the Sydney astronomical fraternity.

The service was held at St. Matthew's in Windsor (Fig. 1.14), the church which Tebbutt had supported throughout much of his life. Tebbutt was an ardent Christian, and for many years was an active member of the St. Matthew's congregation (see Bhathal 1993, pp. 34–39). He was eager to reconcile his religious and astronomical beliefs, and drew some inspiration from the book, *Discourses on the Christian Revelation, Viewed in Connection with the Modern Astronomy* (Chalmers n.d.), which was in his library. He even lectured on the topic in both Windsor and Sydney (e.g. see Tebbutt 1878) and had pamphlets published. However, in his twilight years, he vehemently objected to the way in which the Anglican religion was being presented at St. Matthew's and turned his back on his church. Amongst other places, this is well documented in a letter written by the Reverend Norman Jenkyn (Rector of St. Matthew's) to Tebbutt in 1915: '... some time ago we had a conversation together in which you gave me to understand that you would not attend the Church nor could you now subscribe to and accept her teachings'. In this light, at first sight it appears amazing that Tebbutt's funeral was held at St. Matthew's and presided over by the very same Reverend Jenkyn. The facts of the matter are revealed in the address that the Reverend S.G. Fielding, a long-time Tebbutt friend, delivered during the funeral:

> One of the most remarkable traits of the late John Tebbutt's character was his great reverence for Truth, whether scientific or religious. He had no tolerance for what he deemed to be a sham or make-believe. It was his great love for the Truth that caused him to be misunderstood from the religious point of view. He never really left the Church, as so many people thought. (Death of Mr. John Tebbutt 1916).

During his address, the Reverend Fielding also described Tebbutt as one of the greatest scientists Australia had produced, while in his speech, Dr. Roseby referred to Tebbutt's

> ... constant and uniform courtesy and his ready helpfulness to those occupied with kindred studies and engaged in the same work ... Mr. Tebbutt had the true scientific temper—a convention happily established among such workers—of rendering prompt, laborious, and ungrudging help to all fellow-workers. And there was one feature about this help that made it of special value—it was the veteran astronomer's passion for accuracy. (ibid.).

Tebbutt's coffin was placed in the family vault that had been constructed to his own design just 5 or 6 years earlier. This is an impressive structure, and the marble celestial spheres located at each of the four corners are a particularly appropriate astronomical touch (see Fig. 14.15). The names of others buried in the Tebbutt Family vault as at 6 August 1985 when I listed them are shown in Table 14.4, below. The birth and death dates are taken from the vault inscriptions or from the genealogy in Tebbutt (1986, p. 103); where there are differing dates, the former source has been favoured.

Since Tebbutt's death, streets in Windsor and in the Sydney suburb of Leichhardt have been named after him (White 1979), but in more recent years two far greater honours have been bestowed upon him.

**Fig. 14.15** The Tebbutt family vault at St. Matthew's, Windsor (*Photograph* Wayne Orchiston)

**Table 14.4** Names of those buried in the Tebbutt Family vault at St. Matthew's Cemetery, Windsor, as at 6 August 1985

| Full name | Relation to John Tebbutt | Birth date | Death date | Age at death |
|---|---|---|---|---|
| Jane | Wife | 21 Mar 1838 | 7 Feb 1903 | 65 |
| Mary Alice | Daughter | 2 Jan 1862 | 12 May 1910 | 48 |
| John, FRAS | – | 25 May 1834 | 29 Nov 1916 | 82 |
| James, Luke and Francis | (Triplets) | 6 Apr 1981 | 6 Apr 1981 | – |
| John Thomas | Grandson | 10 May 1900 | 20 Jul 1931 | 31 |
| Virginia Sarah | Daughter | 5 Jun 1858 | 27 Aug 1931 | 73 |
| John Thomas | Son | 2 Nov 1868 | 12 Aug 1936 | 67 |
| Jane Margaret | Daughter | 9 Aug 1866 | 5 May 1943 | 76 |
| Sybell Mary | Daughter-in-law | 25 Mar 1879 | 12 May 1950 | 71 |
| Edmund Halley | Grandson | 3 Jan 1910 | 8 Aug 1978 | 68 |
| Ann | Grandmother | 1765 | 16 Jan 1837 | 71 |
| Ann | Sister | 10 Mar 1838 | 26 Dec 1838 | <1 |
| John | Grandfather | 1762 or 1763 | 14 May 1844 | 81 |
| Robert Bruce | Grandson | 7 May 1904 | 27 Jul 1963 | 59 |
| Virginia | Mother | 4 Jun 1801 | 9 Oct 1865 | 64 |
| Thomas | Uncle | 1791 | 3 Feb 1866 | 74 |
| John | Father | 2 Dec 1794 | 20 Dec 1870 | 76 |
| Victoria Susannah | Daughter | 31 Jul 1871 | 22 Nov 1889 | 18 |
| Ann Elizabeth | Daughter | 23 Jan 1860 | 20 May 1891 | 31 |
| Jean | Granddaughter | 28 Mar 1916 | 27 Feb 1981 | 64 |

In 1973, the International Astronomical Union recognised Tebbutt's contribution to world astronomy by renaming a crater on the Moon after him. This is a very considerable honour, for these days it is rarely afforded to non-professional astronomers, let alone those who have lived in comparatively recent times. Crater Tebbutt was originally known as Picard G and is situated in mountainous terrain near the southern margin of the flat lunar plain known as Mare Crisium (see Fig. 14.16). It is a circular crater about 36 km in diameter, with an extension leading from a small break in the northern wall.

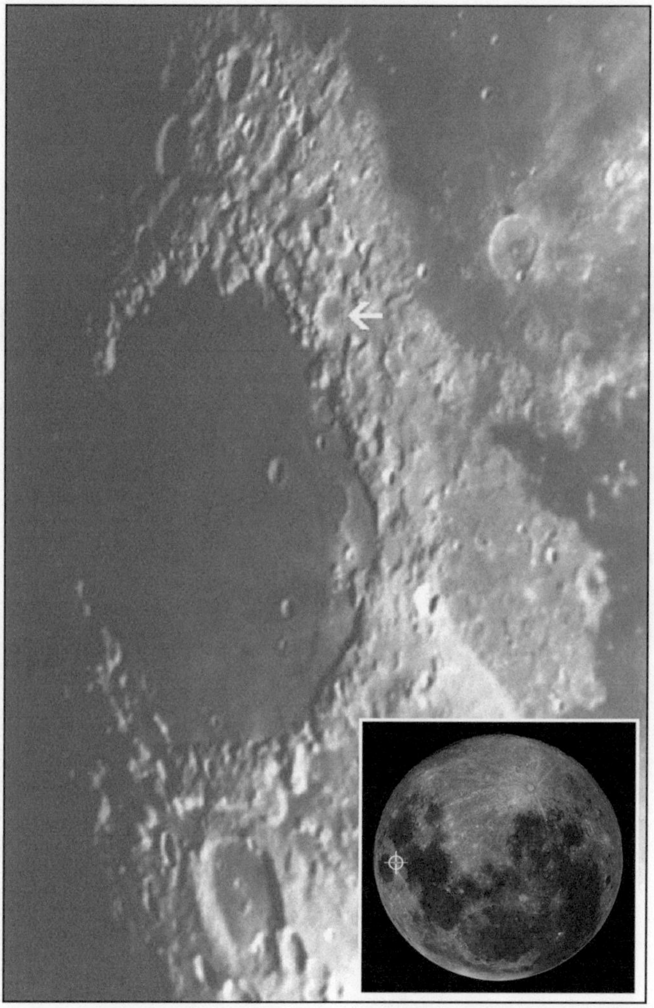

**Fig. 14.16** A close-up of the Mare Crisium region and lunar crater Tebbutt (marked by the *white arrow*). Sir Patrick Moore took this photograph with a 24-in. reflector (*Courtesy* Patrick Moore); the *inset* photograph shows the location of Mare Crisium on the Moon, with crater Tebbutt marked by the *white circle* and *cross*

According to Sir Patrick Moore (pers. comm., 1996), this crater

> … has very low, discontinuous walls; the floor has been flooded by Mare lava though there is no direct connection with the interior of the Mare Crisium itself. There is very little detail on the interior, which in colour is very similar to that of the main Mare.

Crater Tebbutt is readily visible in a small telescope, but it takes a larger instrument to reveal the two small craters on its floor, nestled up against the southern and south-eastern margins of the main crater wall.

As a matter of interest, the famous French astronomer Jean-Félix Picard (1620–1682) did not lose his own place on the Moon with the renaming of Picard G, for a little to the north of 'Tebbutt' there is another crater named simply 'Picard', and this stands out as the largest feature on the floor of Mare Crisium. Wilkins and Moore (1961, p. 197) describe Picard as a 'The largest ring-plain on the Mare … with bright walls, rising 5,000 feet above the depressed floor on which is a small central hill'. Picard is about 33.5 km in diameter.

The more recent great honour which was bestowed upon Tebbutt was the decision to feature him on the Australian $100 note that was issued in March 1984, almost exactly 150 years after his birth. As Fig. 14.17 indicates, this bank note shows Tebbutt, two of his three Windsor Observatory buildings, and background designs meant to '… symbolise elements to do with the sky and a comet's tail' (Williamson 1984).

The concept of a $100 note was first suggested in 1975. A $50 note had been in circulation for more than a year at the time and was accepted by the public, and with rising inflation it was opportune to consider the possibility of a higher value note. The Reserve Bank formed a steering panel, and they invited three designers to develop possible lines of approach (Stuart 1984). The theme 'Scientific Discovery' had already been decided on, as had the basic colour of the new $100 note.

**Fig. 14.17** The reverse side of the $100 Australian bank note, showing John Tebbut and two of his Windsor Observatory buildings (Orchiston collection)

**Fig. 14.18** Harry Williamson, designer of the 1984 $100 bank note (after Williamson 1984, p. 4)

According to Williamson (1984), over a 3-year period the three designers considered a number of designs and possible subjects, including George Bass (1771–1803), James Cook (1727–1779), Matthew Flinders (1774–1814), Phillip Parker King and Abel Tasman (1603–1659), but it was a design by the Sydney-based graphic designer Harry Williamson (Fig. 14.18) featuring Tebbutt and the distinguished Antarctic explorer, Sir Douglas Mawson, which eventually won the day (see Stuart 1984).

Tebbutt was perceived to be less eminent than Mawson by world standards, so his face was consigned to the back of the new bank note. It also was recognised that his name (and face) would be unknown to the public in general, even though he was '… a pioneer in Australian astronomy. [and] His work … is highly regarded by astronomers in Australia and overseas' (Australian $100 Note 1984). Williamson (1984) also saw Tebbutt's meteorological and astronomical observations '… as part of the foundation to Australian involvement in astronomy … the observatory that he built at Windsor was recognised internationally'.

Terri Williamson, Harry's wife, carried out much of the archival research on Tebbutt in the Mitchell Library and helped assemble the necessary information for the note design and supporting documentation.

Figure 14.19 shows the artist Lyal Dolan from the Reserve Bank's Note Printing Branch drawing Tebbutt's portrait prior to the printing of the $100 note (Alan Flint, pers. comm., 1985).

Following close on the heels of the $100 note was the 1985 Windsor Telephone Directory, which featured an image of Tebbutt on the front cover (Fig. 14.20).

**Fig. 14.19** Preparation of the art work for the $100 note prior to the engraving of the steel die (after Reserve Bank … 1984)

However, this Directory was soon replaced, and in May 1996 a new $100 bank note was issued. With these changes, Tebbutt's visage began to slip from the public gaze back into the very mists of antiquity from which it had so suddenly and mysteriously emerged in 1984.

**Fig. 14.20** The front cover of the 1985 Windsor Telephone Directory (Orchiston Collection)

## 14.5 Concluding Remarks

In retrospect, it is apparent that Tebbutt lived during a particularly interesting time in Australian history. It was a period characterised by fabulous gold rushes, a dramatic population increase, and economic boom and bust. To the power-brokers of Britain, Australia was one of the 'new frontiers', ripe for political manipulation and economic harvesting. It was also ripe for scientific exploitation.

The second half of the nineteenth century saw a gradual change as imperial (British-controlled) science gave way to colonial (Australian-based) science. Increasingly, Australian-born scientists sought to unravel the intricacies of the Australian environment, setting agendas that specifically addressed local issues rather than those formulated in the 'scientific metropolises' of the northern hemisphere. In many different fields of science and technology, Australia won its independence at this time (see Inkster and Todd 1988; MacLeod 1988c, 1993; MacLeod and Jarrell 1994; Todd 1993).

Back in 1967, Basalla published a seminal paper in the journal *Science* which presented a three-phase model for the development of science beyond Europe. These phases have been succinctly described by Moyal (1976, pp. 3–4):

> (i) the period when a newly opened territory provides a fresh source of data for European science; (ii) the period of colonial endeavour, still concerned primarily with natural history,

> but in which the indigenous scientists enlarge the spectrum of their interests and accept fuller responsibility for the investigation and interpretation of the scientific world; and (iii) the final phase of a struggle to attain an independent scientific tradition when the colonial scientists build self-supporting institutions, receive scientific training in their own country, develop independent societies, and formulate indigenous scientific attitudes and goals.

When tested in an Australian context, Inkster (1985) and MacLeod (1982) both found Basalla's model simplistic and rejected it, and although MacLeod (1982) developed an alternative model, a later evaluation suggested that '… the time may yet be premature for the application of elaborate developmental models and fine weightings of internal versus external influence as appropriate means for analysing the meaning of colonial or imperial science' (Stafford 1988, p. 69).

As an individual immersed in the nineteenth century Australian science, Tebbutt confirms this interpretation for he does not fit neatly into any of the expounded models. His is a strange blend of imperial and colonial science. He began his astronomical and meteorological career at a time when much of Australian science was prosecuted by transplanted Britons and Germans, yet he was locally born and bred. While it is true that he espoused observational programmes that were in vogue in Britain at the time, as an independent astronomer he was free to choose from amongst a range of options. In MacLeod's terms, the 'metropolis' set the overall scientific agenda in positional astronomy but Tebbutt exercised his autonomy by deciding his own specific agenda. Nor was he unique in this regard, for

> Gradually … [other] Australians recognised and made choices—drawing upon indigenous materials and wrestling significance from the common-place, or working on problems defined internationally, forming ranks in the army of science, if not always in its advance guard. (MacLeod 1988b, p. 7).

Tebbutt was no W.H. Bragg (see Home 1984), conducting important *original* research in the intellectual isolation of distant Australia. He was content to conduct observations and supply data for the benefit (mainly) of overseas professional astronomers although he did succeed in joining MacLeod's 'advance guard' on two specific occasions: when he discovered the two Great Comets and was able to offer these to northern hemisphere colleagues for more exhaustive analyses. Apart from these exceptions, he probably saw himself as part of an informal international research collaboration, best emphasised by the successive Vulcan-search programmes and the British 1874 transit of Venus initiative.

Nor were these collaborations always driven by British astronomers, in spite of Australia's historical affiliations. When we examine Tebbutt's wide-ranging observational programmes, we see that they pay as much allegiance to Germany as they do to England, and towards the turn of the nineteenth century Tebbutt made no secret of the fact that he believed Germany to be leading the world in positional astronomy (Elkington 1901). We see this reflected in Tebbutt's publications. When he commenced his astronomical career and was busy establishing his local scientific identity all of his publications appeared in the Sydney newspapers. Once he joined the local predecessor of the Royal Society he began publishing in their journal, but it did not take him long to recognise the international supremacy of *Astronomische Nachrichten* and *Monthly Notices of the Royal Astronomical Society*. Consequently, the majority of his very numerous publications appeared in these two journals.

Despite this, he continued to publish in the *Journal and Proceedings of the Royal Society of New South Wales*, but reserved his most significant contributions for the overseas journals and international audiences.

Perceived relative status also was implicit in the extent to which Tebbutt was actively associated with these and other scientific societies. While the Philosophical Society of New South Wales and its 'Royal' successor had their place, to be elected a Fellow of the Royal Astronomical Society was the pinnacle of achievement for an independent astronomer in far-flung Australia. Through this affiliation, one gained access to the metropolitan astronomy centres of Britain and Europe, and established a network of links with colleagues in the British colonies and beyond.

Yet there was a place for local scientific organisations, as we have seen, and Tebbutt was not backward in coming forward and forming the Australian Comet Corps at a time when there was agitation locally to found independent specialist scientific societies (see Branagan 1972; Hoare 1974). However, it is important to note that the concept of an Australian Comet Corps was not of indigenous origin; surprisingly, it derived from the United States, rather than from Britain or Germany. Clearly, Tebbutt was prepared to seriously consider trialing any new initiative in the Australian astronomical environment, so long as the anticipated outcome was deemed worthwhile.

This perceived pre-eminence of non-indigenous scientific bodies was to impact negatively on Tebbutt when inter-colonial developments led to the formation of the Australasian Association for the Advancement of Science in 1888 (see MacLeod 1988a). Modelled on the highly successful British counterpart, the Australian body '… played a central role in forging the disparate groups of scientists in the various colonies into a truly Australian (or rather, Australasian) scientific community' (Home 1984, p. 21). Section A was specifically established to accommodate the astronomers, mathematicians and physicists (see Home 1988b), and although it received strong support from Australia's professional astronomers and some leading non-professional astronomers, Tebbutt virtually boycotted the group on account of his strained relations with Russell at this time. As a consequence, his strongest societal allegiances remained with England.

This situation was exacerbated at the end of 1894 when he was one of these who formed the New South Wales Branch of the British Astronomical Association. For reasons which appeared sound at the time, the alternative scenarios (an autonomous society or an Astronomical Section of the local Royal Society) were passed over in favour of establishing closer ties with the (largely non-professional) astronomical fraternity in the mother country.

Yet by this time, Tebbutt was able to influence the intercolonial movement in astronomy through his own ranking as arguably Australia's leading astronomer and a world figure in the area of positional astronomy. Although preferring to work in isolation (one gets the impression that he was not, by choice, a 'team player'), he did attract disciples (such as Biggs and Merfield) and served as an excellent role model. He also was an adviser to and confidant of independent astronomers from throughout Australia. Likewise, he enjoyed close and friendly working relations with all of the nation's professional astronomers, that is, until the development of the Russell-Tebbutt feud. His publications and the 'Tebbutt Letters' in the Mitchell Library clearly demonstrate that he also supported imperial science at this time by maintaining close links with noted professional and non-professional astronomers from around the world.

This dual support for imperial and colonial astronomy is also reflected in Tebbutt's choice of instruments for his burgeoning Windsor Observatory. While his first transit telescope and first two micrometres were of local origin, later micrometres and his second transit telescope derived from England. When it came to non-meridian work, all three equatorial telescopes were imported English refractors, even though locally-made research-quality reflectors of substantially larger aperture were becoming available locally at the time he purchased the Grubb instrument (see Orchiston and Bembrick 1995). Of his remaining scientific equipment, both chronometers and most of the meteorological instruments were of British origin, although some of the latter, and the tide gauge on South Creek, were manufactured in the colony.

This same mix typifies the architecture of the four different observatory buildings that Tebbutt designed. Although the basic design of his earliest observatory resembled that of various overseas observatories, the specific design was his own. In contrast, his second observatory building (for the 4.5-in. Cooke telescope) was inspired by a Cambridge University Observatory design, but his two later observatories were again local initiatives.

Finally, there is the matter of honours and recognition. While Tebbutt wanted—and received—recognition at local, national and international levels, the only major tangible honour he received during his lifetime came from the Royal Astronomical Society. Because of local politics, he never received any honours from the local Royal Society, or from the University of Sydney, despite his long and important contribution to New South Wales astronomy and meteorology and his ranking as one of the world's foremost independent astronomers. In this respect, it could be argued that imperial science gave him due recognition while colonial science abdicated its rightful responsibility. As we have seen (Tebbutt 1908a), in later life this was one of John Tebbutt's abiding regrets.

**Acknowledgements** I wish to thank Alan Flint (formerly Reserve Bank, Note Printing Branch), Professor Nick Lomb (University of Southern Queensland, Toowoomba) and staff at the Mitchell Library (Sydney) for their assistance in various ways, and the Mitchell Library and the late Sir Patrick Moore (Selsey, England) for kindly providing Figs. 14.5 and 14.16.

## References[1]

Annual Reports: Mr. Tebbutt's Observatory … *Monthly Notices of the Royal Astronomical Society, 65,* 380–381 (1905).

Annual Reports: Mr. Tebbutt's Observatory … *Monthly Notices of the Royal Astronomical Society, 66,* 215 (1906).

Annual Reports: Mr. Tebbutt's Observatory … *Monthly Notices of the Royal Astronomical Society, 67,* 268 (1907).

[1] The following abbreviation is used:
TL = Letters to J. Tebbutt, bound manuscript letters in the Mitchell Library, Sydney.

Annual Reports: Mr. Tebbutt's Observatory … *Monthly Notices of the Royal Astronomical Society, 68,* 287 (1908).
Annual Reports: Mr. Tebbutt's Observatory … *Monthly Notices of the Royal Astronomical Society, 69,* 291 (1909).
Annual Reports: Mr. Tebbutt's Observatory … *Monthly Notices of the Royal Astronomical Society, 70,* 481 (1910).
Annual Reports: Mr. Tebbutt's Observatory … *Monthly Notices of the Royal Astronomical Society, 71,* 315 (1911).
Annual Reports: Mr. Tebbutt's Observatory … *Monthly Notices of the Royal Astronomical Society, 72,* 293 (1912).
Annual Reports: Mr. Tebbutt's Observatory … *Monthly Notices of the Royal Astronomical Society, 73,* 263 (1913).
Australian $100 Note. Press release issued by the Reserve Bank of Australia, 20 March 1984.
Baracchi, P. (1914). Astronomy and geodesy in Australia. In Knibbs, 326–390.
Bhathal, R. (1993). *Australian Astronomer John Tebbutt. The Life and World of the Man on the $100 Note*. Sydney: Kangaroo Press.
Biggs, A.B. (1890). Letter to John Tebbutt, dated 18 November. In TL.
Biggs, A.B. (1893). Letter to John Tebbutt, dated 23 March. In TL.
Board, R. (1915). Letter to John Tebbutt, dated 5 November. In TL.
Branagan, D.F. (1972). Words, actions, people: 150 years of scientific societies in Australia. *Journal and Proceedings of the Royal Society of New South Wales, 104,* 123–141.
Brooks, J. (1897). Letter to John Tebbutt, dated 7 November. In TL.
Butterfield, G. (1902). Letter to John Tebbutt, dated 31 October. In TL.
Campbell, W.W. (1909). Letter to John Tebbutt, dated 9 September. In TL.
Campbell, W.W. (1913). Letter to John Tebbutt, dated 20 June. In TL.
Chambers, G.F. (1908). Letter to John Tebbutt, dated 15 December. In TL.
Chalmers, T. (n.d.). *Discourses on the Christian Revelation, Viewed in Connection with the Modern Astronomy*. Glasgow: Collins.
Crommelin, A.C.D. (1922). Obituary: W.H. Wesley. *Nature, 110,* 509–610.
Death of Mr. John Tebbutt. Famous Australian Astronomer. *Windsor and Richmond Gazette,* 8 December 1916 (reproduced in Tebbutt 1986, 143–149).
Denning, W.F. (1909). Letter to John Tebbutt, dated 27 December. In TL.
Dewhirst, D.W. (1985). Meridian astronomy in the private and university observatories of the United Kingdom: rise and fall. *Vistas in Astronomy, 28,* 147–158.
Dreyer, J.L.E. & Turner, H.H. (1987). *History of the Royal Astronomical Society. Volume 1 1820–1920*. London: Royal Astronomical Society (Blackwell Scientific Publications reprint).
Elkington, H. (1901). Mr. John Tebbutt, F.R.A.S., interviewed. *The Speaker,* 2 (2), 40–42.
Evershed, M.A. (1914). Letter to John Tebbutt, dated 13 October. In TL.
Gale, W. (1894a). Letter to John Tebbutt, dated 31 May. In TL.
Gale, W. (1894b). Letter to John Tebbutt, dated 23 July. In TL.
Gale, W. (1912). Letter to John Tebbutt, dated 9 September. In TL.
Grigg, J. (1907). Letter to John Tebbutt, dated 22 May. In TL.
Hoare, M. (1974). *Science and Scientific Societies in Eastern Australia 1820–1890*. Ph.D. Thesis, Australian National University, Canberra, Australia.
Home, R. (1984). The problem of intellectual isolation in scientific life: W.H. Bragg and the Australian scientific community, 1886–1909. *Historical Records of Australian Science, 6,* 19–30.
Hunt, H.A. (1913). Letter to John Tebbutt, dated 2 May. In TL.
Hussey, W.J. (1913). Letter to John Tebbutt, dated 5 July. In TL.
Inkster, I. (1985). Scientific enterprise and the colonial "model": observations on the Australian experience in historical context. *Social Studies of Science, 15,* 677–704.
Inkster, I. & Todd, J. (1988). Support for the scientific enterprise, 1850–1900. In R. Home (Ed), *Australian Science in the Making* (pp. 102–132). Cambridge: Cambridge University Press.

Innes, R.T.A. (1891a). Letter to John Tebbutt, dated 29 January. In TL.
Innes, R.T.A. (1891b). Letter to John Tebbutt, dated 6 June. In TL.
Innes, R.T.A. (1891c). Letter to John Tebbutt, dated 18 November. In TL.
Innes, R.T.A. (1893). Letter to W.H. Wesley, dated 30 January. In RAS Letter Archives.
Jenkyn, N. (1915). Letter to John Tebbutt, dated 9 July. In TL.
Knibbs, G.H. (Ed.) (1914). *Federal Handbook Prepared in Connection with the Eighty-fourth Meeting of the British Association for the Advancement of Science, Held in Australia, August, 1914*. Melbourne: Government Printer.
Lenehan, H.A. (1897). Letter to John Tebbutt, dated 27 January. In TL.
Lenehan, H.A. (1901). Letter to John Tebbutt, dated 19 February. In TL.
Lenehan, H.A. (1904). Letter to John Tebbutt, dated 13 September. In TL.
Lenehan, H.A. (1907a). Letter to John Tebbutt, dated 4 April. In TL.
Lenehan, H.A. (1907b). Letter to John Tebbutt, dated 10 April. In TL.
Levander, F.W. (1913). Letter to John Tebbutt, dated 21 October. In TL.
Macdonnell, W.J. (1904). Letter to John Tebbutt, dated 30 September. In TL.
Macdonnell, W.J. (1908). Letter to John Tebbutt, dated 25 October. In TL.
MacLeod, R. (1982). On visiting the 'moving metropolis': reflections on the architecture of imperial science. *Historical Records of Australian Science, 5* (3), 1–16.
MacLeod, R. (1988a). *The Commonwealth of Science: ANZAAS and the Scientific Enterprise in Australasia, 1888–1988*. Melbourne: Oxford University Press.
MacLeod, R. (1988b). Introduction. In MacLeod 1988a, 1–16.
MacLeod, R. (1988c). Organising science under the southern cross. In MacLeod 1988a, 19–39.
MacLeod, R. (1993). Passages in imperial science: from empire to commonwealth. *Journal of World History, 4*, 117–150.
MacLeod, R. & Jarrell, R. (Eds.) (1994). *Dominions Apart: Reflections on the Culture of Science and Technology in Canada and Australia, 1850–1945*. Ontario: Canadian Science & Technology Historical Association.
Macpherson, H. (1905). *Astronomers of Today*. London: Gall & Inglis.
Maynard, J. (1897). Letter to John Tebbutt, dated 28 October. In TL.
Merfield, C.J. (1897). Letter to John Tebbutt, dated 14 November. In TL.
Merfield, C.J. (1898a). Letter to John Tebbutt, dated 25 June. In TL.
Merfield, C.J. (1898b). Letter to John Tebbutt, dated 3 September. In TL.
Merfield, C.J. (1899a). Letter to John Tebbutt, dated 13 December. In TL.
Merfield, C.J. (1899b). Letter to John Tebbutt, dated 28 December. In TL.
Merfield, C.J. (1900a). Letter to John Tebbutt, dated 11 February. In TL.
Merfield, C.J. (1900b). Letter to John Tebbutt, dated 30 March. In TL.
Merfield, C.J. (1900c). Letter to John Tebbutt, dated 5 May. In TL.
Merfield, C.J. (1900d). Letter to John Tebbutt, dated 22 September. In TL.
Merfield, C.J. (1900e). Letter to John Tebbutt, dated 20 December. In TL.
Merfield, C.J. (1901a). Letter to John Tebbutt, dated 28 June. In TL.
Merfield, C.J. (1901b). Letter to John Tebbutt, dated 13 August. In TL.
Merfield, C.J. (1901c). Letter to John Tebbutt, dated 6 October. In TL.
Merfield, C.J. (1901d). Letter to John Tebbutt, dated 31 October. In TL.
Merfield, C.J. (1902a). Letter to John Tebbutt, dated 18 May. In TL.
Merfield, C.J. (1902b). Letter to John Tebbutt, dated 9 November. In TL.
Merfield, C.J. (1903). Letter to John Tebbutt, dated 17 May. In TL.
Merfield, C.J. (1904a). Letter to John Tebbutt, dated 30 April. In TL.
Merfield, C.J. (1904b). Letter to John Tebbutt, dated 9 October. In TL.
Merfield, C.J. (1907). Letter to John Tebbutt, dated 2 February. In TL.
Merfield, C.J. (1914). Letter to John Tebbutt, dated 22 October. In TL.
Morgan, S. (1915a). Letter to John Tebbutt, dated 16 April. In TL.
Morgan, S. (1915b). Letter to John Tebbutt, dated 27 April. In TL.
Morgan, S. (1915c). Letter to John Tebbutt, dated 15 July. In TL.

Moyal, A. (1976). *Scientists in Nineteenth Century Australia: A Documentary History*. Sydney: Cassell Australia.
Nangle, J. (1904). Letter to John Tebbutt, dated 11 November. In TL.
Nangle, J. (1914). Letter to John Tebbutt, dated 8 September. In TL.
N.S.W. (1893). A self-taught astronomer. *The Australasian,* 26 August.
Orchiston, W. (1982). John Tebbutt and the Whakatane Eight Inch Refractor: a review of "The Australian Connection". *Southern Stars, 29,* 215–244.
Orchiston, W. (1993). John Grigg, and the genesis of cometary astronomy in New Zealand. *Journal of the British Astronomical Association, 103,* 67–76.
Orchiston, W. (1995). John Grigg, and the development of astrophotography in New Zealand. *Australian Journal of Astronomy, 6,* 1–14.
Orchiston, W. (1998). *Nautical Astronomy in New Zealand: The Voyages of James Cook.* Wellington: Carter Observatory.
Orchiston, W. (1999). Comets and communication: amateur-professional tension in Australian astronomy. *Publications of the Astronomical Society of Australia, 16,* 212–221.
Orchiston, W. (2001). The Thames observatories of John Grigg. *Southern Stars, 40* (3), 14–22.
Orchiston, W. (2004). John Tebbutt and observational astronomy at Windsor Observatory. *Journal of the British Astronomical Association, 114,* 141–154.
Orchiston, W. (2006). Amateurs in the Antipodes: the common denominators of nineteenth and early twentieth century South African, Australian and New Zealand astronomy. *Monthly Notices of the Astronomical Society of Southern Africa, 65,* 148–159; 194–208.
Orchiston, W. (2015). The amateur-turned-professional syndrome: two Australian case studies. In W. Orchiston, D.A. Green & R. Strom (Eds), *New Insights from Recent Studies in Historical Astronomy: Following in the Footsteps of F. Richard Stephenson* (pp. 259–350). Cham: Springer.
Orchiston, W. (2016). *Exploring the History of New Zealand Astronomy: Trials, Tribulations, Telescopes and Transits*. Cham: Springer.
Orchiston, W. & Bembrick, C. (1995). The role of the large reflecting telescope in amateur astronomy: an Australian case study. *Australian Journal of Astronomy, 6,* 53–74.
Proctor, M. (1926). *The Romance of Comets*. New York: Harper.
Ramsay, J. (1892). Letter to John Tebbutt, dated 4 April. In TL.
Reports of the Branches. New South Wales Branch (Sydney). *Journal of the British Astronomical Association, 15*, 367–368 (1905).
*Reserve Bank of Australia Report and Financial Statements 30 June 1984*. Sydney: Reserve Bank of Australia (1984).
Roberts, A.W. (1908). Letter to John Tebbutt, dated 12 December. In TL.
Robertson, P. (1980). Coming of age: the British Association in Australia, 1914. *Australian Physicist, 17* (2), 23–27.
Roseby, T. (1914). Letter to John Tebbutt, dated 15 August. In TL.
Roser, S. & White, G.L. (1986). John Tebbutt's observations of Halley's Comet in 1910. *Proceedings of the Astronomical Society of Australia, 6,* 354–358.
Royal Astronomical Society (1913). Letter to John Tebbutt, dated 1 March. In TL.
Russell, H.C., 1876). Meteorological periodicity. *Journal and Proceedings of the Royal Society of New South Wales, 10,* 151–177.
Sellors, R.P. (1904). Letter to John Tebbutt, dated 5 October. In TL.
Stafford, R.A. (1988). The long arm of London: Sir Roderick Murchison and imperial science in Australia. In Home 1988a, 69–101.
Stuart, R. (1984). The Bank's baby. *Currency, 25* (4), 1–3.
Tebbutt, J. (1853–94). *Reductions of Observations, Windsor. Calculation of Comet Orbits 1853–94*. MSS, Mitchell Library, Sydney (A3729).
Tebbutt, J. (1863a). The gales of February, 1863. *Sydney Morning Herald,* 11 June.
Tebbutt, J. (1863b). Meteorology. *Sydney Morning Herald,* 13 October.

Tebbutt, J. (1863c). On Australian storms. *Transactions of the Philosophical Society of New South Wales,* 153–164.

Tebbutt, J. (1864). Mr. Saxby's weather predictions. *Sydney Morning Herald,* 13 June.

Tebbutt, J. (1868). *Meteorological Observation Made at the Private Observatory of John Tebbutt, Junr., The Peninsula, Windsor, New South Wales, in the Years 1863, 1864, 1865, and 1866.* Sydney: printed for the author.

Tebbutt, J. 1874a. *Meteorological Observations Made at the Private Observatory of John Tebbutt, The Peninsula, Windsor, New South Wales, in the Years 1867, 1868, 1869, and 1870.* Sydney: printed for the author.

Tebbutt, J. (1874b). The periodic occurrence of heavy rains and floods. *Sydney Morning Herald,* 2 March.

Tebbutt, J. (1874c). Periodicity of the meteorological elements. *Sydney Morning Herald,* 7 March.

Tebbutt, J. (1875a). The drought of 1876. *Sydney Morning Herald,* 29 December.

Tebbutt, J. (1875b). Rainfall at Sydney and Windsor. *Sydney Morning Herald,* 15 February.

Tebbutt, J. (1877). *Results of Meteorological Observations Made at the Private Observatory of John Tebbutt, The Peninsula, Windsor, New South Wales, in the Years 1871, 1872, 1873, 1874, 1875 & 1875.* Sydney: printed for the author.

Tebbutt, J. (1878). *The Testimony Which Astronomy Furnishes to the Attributes of The Creator.* Sydney: Young Men's Christian Association.

Tebbutt, J. (1882a). *Results of Meteorological Observations Made at the Private Observatory of John Tebbutt, The Peninsula, Windsor, New South Wales, in the Years 1877, 1878, 1879, 1880, & 1881.* Parramatta & Sydney: printed for the author.

Tebbutt, J. (1882b). Temperature and rainfall. *Sydney Morning Herald*, 18 February.

Tebbutt, J. 1886). *Results of Meteorological Observations Made at the Private Observatory of John Tebbutt, The Peninsula, Windsor, New South Wales, in the Years 1882, 1883, 1884, & 1885.* Parramatta & Sydney: printed for the author.

Tebbutt, J. (1887). *History and Description of Mr. Tebbutt's Observatory, Windsor, New South Wales.* Sydney: printed for the author.

Tebbutt, J. (1889). *Report of Mr. Tebbutt's Observatory, The Peninsula, Windsor, New South Wales, For the Year 1888.* Sydney: printed for the author.

Tebbutt, J. (1890). *Report of Mr. Tebbutt's Observatory, The Peninsula, Windsor, New South Wales, For the Year 1889.* Sydney: printed for the author.

Tebbutt, J. (1891a). *Report of Mr. Tebbutt's Observatory, The Peninsula, Windsor, New South Wales, For the Year 1890.* Sydney: printed for the author.

Tebbutt, J. (1891b). *Results of Meteorological Observations Made at the Private Observatory of John Tebbutt, The Peninsula, Windsor, N.S. Wales, in the Years 1886, 1887, 1888, 1889, & 1890.* Sydney: printed for the author.

Tebbutt, J. (1892). *Report of Mr Tebbutt's Observatory, The Peninsula, Windsor, New South Wales, For the Year 1891.* Sydney: printed for the author.

Tebbutt, J. (1893). *Report of Mr. Tebbutt's Observatory, The Peninsula, Windsor, New South Wales, For the Year 1892.* Sydney: printed for the author.

Tebbutt, J. (1894). *Report of Mr. Tebbutt's Observatory, The Peninsula, Windsor, New South Wales, For the Year 1893.* Sydney: printed for the author.

Tebbutt, J. (1895). *Report of Mr. Tebbutt's Observatory, The Peninsula, Windsor, New South Wales, For the Year 1894.* Sydney: printed for the author.

Tebbutt, J. (1896). *Report of Mr. Tebbutt's Observatory, The Peninsula, Windsor, New South Wales, For the Year 1895.* Sydney: printed for the author.

Tebbutt, J. (1897). *Report of Mr. Tebbutt's Observatory, The Peninsula, Windsor, New South Wales, For the Year 1896.* Sydney: printed for the author.

Tebbutt, J. (1898a). *Report of Mr. Tebbutt's Observatory, The Peninsula, Windsor, New South Wales, For the Year 1897.* Sydney: printed for the author.

Tebbutt, J. (1898b). *Results of Meteorological Observations Made at the Private Observatory of John Tebbutt, The Peninsula, New South Wales, in the Years 1891, 1892, 1893, 1894, 1895, 1896, and 1897.* Sydney: printed for the author.

Tebbutt, J. (1899). *Report of Mr. Tebbutt's Observatory, The Peninsula, Windsor, New South Wales, For the Year 1898*. Sydney: printed for the author.

Tebbutt, J. (1900). *Report of Mr Tebbutt's Observatory, The Peninsula, Windsor, New South Wales, For the Year 1899*. Sydney: printed for the author.

Tebbutt, J. (1901). *Report of Mr Tebbutt's Observatory, The Peninsula, Windsor, New South Wales, For the Year 1900*. Sydney: printed for the author.

Tebbutt, J. (1902). *Report of Mr. Tebbutt's Observatory, The Peninsula, Windsor, New South Wales, For the Year 1901*. Sydney: printed for the author.

Tebbutt, J. (1903). *Report of Mr. Tebbutt's Observatory, The Peninsula, Windsor, New South Wales for the Year 1902*. Sydney: printed for the author.

Tebbutt, J. (1904). *Report of Mr. Tebbutt's Observatory, The Peninsula, Windsor, New South Wales, For the Year 1903*. Sydney: printed for the author.

Tebbutt, J. (1905a). Letter to F.W. Dyson, dated 24 April. In RAS Letter Archives.

Tebbutt, J. (1905b). Observations of phenomena of Jupiter's satellites at Windsor, New South Wales, in the years 1900 and 1902. *Monthly Notices of the Royal Astronomical Society, 66,* 14–17.

Tebbutt, J. (1906). Star-occultations observed at Windsor, N.S. Wales, 1900-1-2. *Astronomische Nachrichten, 170,* 91–92.

Tebbutt, J. (1907a). Comet Grigg, 1907, April. *The Observatory, 30,* 285–286.

Tebbutt, J. (1907b). Mr Crommelin's lecture on Halley's Comet. *Journal of the British Astronomical Association, 17,* 361.

Tebbutt, J. (1908a). *Astronomical Memoirs*. Sydney: printed for the author.

Tebbutt, J. (1908b). Memorandum, dated 3 October (Orchiston Collection).

Tebbutt, J. (1910a). The first naked-eye observation of the Great Comet of 1882 in full daylight. *Journal of the British Astronomical Association, 20,* 438.

Tebbutt, J. (1910b). The Great Comet of 1882. *The Observatory, 33,* 252.

Tebbutt, J. (1910c). Note on Halley's Comet. *Journal of the British Astronomical Association, 20,* 199–200.

Tebbutt, J. (1911). Observations of Halley's Comet at Windsor, N.S. Wales. *Monthly Notices of the Royal Astronomical Society, 71,* 224–226.

Tebbutt, J. (1912a). Gale's Comet. *Journal of the British Astronomical Association, 23,* 139–140.

Tebbutt, J. (1912b). Letter to W.H. Wesley, dated 28 March. In RAS Letter Archives.

Tebbutt, J. (1913). Komet 1912a (Gale). Observations at Windsor, N.S. Wales. *Astronomische Nachrichten, 193,* 313–314.

Tebbutt, J. (1915a). Letter to the Secretaries, dated xx February. In RAS Letter Archives.

Tebbutt, J. (1915b). Measures of southern binary stars. *Monthly Notices of the Royal Astronomical Society, 76,* 36.

Tebbutt, J. (1915c). Sir Thomas Makdougall Brisbane. *The Observatory, 38,* 221.

Tebbutt, J. (1916). *Results of Meteorological Observations at Mr. Tebbutt's Observatory The Peninsula, Windsor, New South Wales, During the Period 1898-1915*. Windsor, printed for the author.

Tebbutt, J. (1986). *Astronomical Memoirs*. Windsor: Hawkesbury Shire Council (Reprinted edition). (Although the reprint date is given as 1986, in fact this book was in fact printed in 1987 and was officially launched on 9 April 1987).

Todd, J. (1993). Science at the periphery: an interpretation of Australian scientific and technological dependency and development prior to 1914. *Annals of Science, 50,* 33–58.

Turner, H.H. (1894). Letter to John Tebbutt, dated 19 January. In TL.

Turner, H.H. (1895). Letter to R.T.A. Innes, dated 9 November. In RAS Letter Archives.

Turner, H.H. (1905). Letter to R.T.A. Innes, dated 9 November. In RAS Letter Archives.

Turner, H.H. (1908). Letter to John Tebbutt, dated 17 November. In TL.

Turner, H.H. (1914). Letter to John Tebbutt, dated [21] August. In TL.

Turner, H.H. (1915). Letter to John Tebbutt, dated 19 February. In TL.

Turner, H.H. (1918). Obituary: John Tebbutt. *Monthly Notices of the Royal Astronomical Society, 78,* 252–255.
Wesley, W.H. (1893). Letter to R.T.A. Innes, dated 4 March. In RAS Letter Archives.
Wesley, W.H. (1915). Letter to John Tebbutt, dated 26 March. In TL.
White, E.J. (1890). Letter to John Tebbutt, dated 11 February. In TL.
White, E.J. (1891). Letter to John Tebbutt, dated 7 June. In TL.
White, E.J. (1905). Letter to John Tebbutt, dated 13 February. In TL.
White, G. (1979). John Tebbutt and the astronomy at the Windsor Observatory. *Proceedings of the Astronomical Society of Australia, 3,* 408–411.
Wilkins, H.P. & Moore, P. (1961). *The Moon.* London: Faber and Faber.
Williamson, H. (1984). My designs on money. *Currency, 25* (4), 4–6.
Wright, H. (1897). Letter to John Tebbutt, dated 3 December. In TL.
Wright, H. (1901). Letter to John Tebbutt, dated 12 February. In TL.
Wright, H. (1905a). Letter to John Tebbutt, dated 20 April. In TL.
Wright, H. (1905b). Letter to John Tebbutt, dated 24 June. In TL.
Wright, H. (1906). Letter to John Tebbutt, dated 21 March. In TL.
Wright, H. (1908). Letter to John Tebbutt, dated 25 October. In TL.
Wright, H. (1915a). Letter to John Tebbutt, dated 10 May. In TL.
Wright, H. (1915b). Letter to John Tebbutt, dated 1 June. In TL.
Young, J.S. (1893). Letter to John Tebbutt, dated 10 July. In TL.

# Chapter 15
# The Sequel: From Scientific Observatory to Museum of Astronomy

## 15.1 Introduction

It is always interesting to see the chain of events that follows the demise of a notable individual. Sometimes, their effects are dispersed or destroyed and all but lost to later scholars, while in other instances the estate is kept intact and developed as a museum of some sort. What was the case with Windsor Observatory, Tebbutt's astronomical and meteorological records, and his outstanding research and reference library?

Since none of John Tebbutt's children was interested in astronomy, dispersal of the collection was inevitable. As we saw in Chap. 2, most of his records and library went to the Mitchell Library in Sydney, while his astronomical instruments, for the most part, were acquired in the first instance by that well-known Sydney independent astronomer, Walter Gale (Bio Box 12.1). The current whereabouts of some are unknown, but after a series of international moves the 8-in. Grubb telescope eventually returned to Windsor, and it now graces its original dome in a refurbished Windsor Observatory (Fig. 15.1) that has been developed by John Halley Tebbutt—the great-grandson of the famous astronomer—as a unique museum of astronomy and local tourist attraction.

This chapter draws on some material in Orchiston (1982, 2001) and Orchiston and Bembrick (1997). In it, we follow the successive chains of ownership of Tebbutt's 4.5- and 8-in. telescopes, noting how the New Zealand astronomer, Frank Bateson, was able to use the latter instrument to continue the observational tradition that Tebbutt had established many years earlier; and we document the changing fortunes of the three observatory buildings that existed at the time of Tebbutt's death in 1916, which culminated in the refurbishment of two of these as one of the world's most interesting little museums of the nineteenth century astronomy.



W. Orchiston, *John Tebbutt*, Historical & Cultural Astronomy,
DOI 10.1007/978-3-319-44521-2_15

**Fig 15.1** A November 1996 photograph of John Halley Tebbutt, the great-grandson of the astronomer, with the 8-in. telescope, which is now back in its original dome at Windsor Observatory (*Photograph* Wayne Orchiston)

## 15.2 A Tale of Loose Ends: Dispersal of the Windsor Observatory Scientific Instruments

As we saw in Chap. 2, identifying the current whereabouts of Tebbutt's library and the majority of his astronomical records is straightforward, but this is generally not the case when it comes to his scientific instruments.

Clearly, as early as 1908 the fate of Windsor Observatory and its scientific instruments weighed heavily on the mind of W.E. Cooke, Director of Perth Observatory, and on 23 October he felt compelled to write Tebbutt: did Tebbutt know of any young enthusiasts who could carry on the 'Tebbutt tradition' and use the telescopes at Windsor Observatory? Cooke (1908) felt that it would be a great pity to donate them to a Government institution or to bequeath the Observatory to a relative who would fail to appreciate (and use) it. We should note that Cooke would be in an ideal position to pursue this line of thought further in the not-too-distant future upon his appointment as the Director of Sydney Observatory, but the fact that he did not do so—even after visiting Tebbutt and Windsor Observatory in 1914 (as outlined in the previous chapter)—suggests that his 1908 letter did not bring the desired response.

However, maybe Tebbutt deliberated on Cooke's suggestions, for in December 1911 he intimated to an inquirer that in a year or two his astronomical instruments would be offered for sale (Foley 1912a), but by February 1912 he had had a change of heart:

> Your reply re disposal of instruments is what I expected. I heartily approve of your family's determination to preserve them as heirlooms. Besides they have an exceptional value thro' being used by him who will go down in history as Australia's first astronomer. (Foley 1912b).

So which of the 'family heirlooms' is still owned by the Tebbutt Family? Two of the meteorological instruments are: the Negretti and Zambra minimum shade thermometer (No. 2358) and a barometer made by Tornaghi (see Fig. 15.2). The thermometer was one of the suite of Negretti and Zambra instruments purchased towards the end of 1862, while the barometer would have been constructed by Tornaghi at about the same time because a letter dated 8 January 1863 from Henry Russell (who was then Officer-in-Charge at Sydney Observatory) states that it will be a great pleasure to compare the readings of the Sydney Observatory and Windsor Observatory barometers. These, then, were the only two meteorological 'heirlooms' that the family ended up retaining, and over the years all of the other meteorological instruments were disposed of and can no longer be traced.

The two chronometers suffered a similar fate. Meanwhile, the ever-dependable grandfather clock went to one of Tebbutt's daughters and then to her daughter, but inquiries by John Halley Tebbutt (pers. comm., 1996) revealed that it had been left under a house and rotted and rusted away and that nothing even was left of the mechanical mechanism.

Meanwhile, the marine sextant with which Tebbutt carried out many of his earliest astronomical observations has disappeared without trace. Likewise, neither transit telescope can now be located. Small transit telescopes formed part of the instrument stable of serious non-professional astronomers during the nineteenth century and were used to maintain a reliable local time-service, something critical for those involved in positional astronomy (see Stott and Hughes 1987), but with the advent of accessible radio time signals during the twentieth century

**Fig. 15.2** John Halley Tebbutt photographed in July 1973 with the Tornaghi barometer (*Photograph* Wayne Orchiston)

they quickly became redundant. By the time Tebbutt's astronomical equipment came on the market, the transit telescope was nearing the end of its useful life, and apart from their curiosity value and historical associations, the two Windsor Observatory instruments would have been of little interest to practising astronomers, professional or non-professional. It is ironic, therefore, that scientific instruments made by Tornaghi—the manufacturer of Tebbutt's first transit telescope—are now highly-prized collectors' items and attract good prices at auction (see Maguire 1985).

Despite numerous inquiries, the current whereabouts of the 3.25-in. Jones refractor also is unknown. Apparently, this telescope was retained by the Tebbutt family at Windsor, and during the 1940s passed from one of the grandsons, Edmund Halley Tebbutt, to his brother, Bruce Tebbutt (J.H. Tebbutt, pers. comm., 1981). At some time prior to 1959 (when I first met Bruce Tebbutt), this telescope was disposed of.

Fortunately, the little marine telescope with which Tebbutt conducted his first recorded astronomical observations in 1853 (Fig. 8.1) is still with the family.

The 8-in. Grubb telescope also is there, but it has enjoyed a peripatetic existence since Tebbutt's death in 1916 (see Orchiston 1997a). At some undetermined date it was acquired by Walter Gale (see Bio Box 12.1) and mounted in a roll-off roof observatory in the Sydney suburb of Waverley. A 1923 report of the NSW Branch of the British Astronomical published in the Association's *Journal* (Reports of the Branches … 1923) mentions that Gale was involved in astrophotography using a camera mounted on the 8-in. Grubb, while a short paper by Gale (1925) about spectroscopic observations of Nova Pictoris 1925 again implicates the ex-Tebbutt telescope, and a photograph of it appears in a Sydney newspaper article about Mars which was published in 1924 (Biggest not best … 1924). It would appear, therefore, that Gale purchased this instrument in the early 1920s, and although he was a prolific planetary observer, apart from the afore-mentioned observations there is no evidence that he made very much use of this telescope. Possibly, this is because he also was the proud owner at this time of two different 18-in. (46-cm) reflectors (Orchiston and Bembrick 1995), and the increased light grasp made one of these the preferred observing instrument.

Whatever the circumstances, in the late 1920s Gale sold the telescope to the well-known Sydney independent astronomer, George Hoskins (Bio Box 15.1), and up until 1940 it was housed in a roll-off roof observatory at his country property at Brewongle, 10 miles (16.1 km) east of Bathurst and at an altitude of 2250 ft (686 m) (Hoskins 1946). This observatory is shown in Fig. 15.4. Hoskins used the Grubb refractor mainly for solar and planetary observing and remembers that the opposition of Mars of 1939

> … gave me 30 or 40 of the most interesting nights I have ever had with a Telescope. I remember one evening when looking at the S. Pole with its serrated edge noticing that it was as clear white as porcelain, when I saw a gleam on the point of one of the serrations—it lasted half an hour or so and then faded away. In my library I found a record of a similar observation in Lowell's first book; he says it is a reflection from the icy slopes of the Mitchell Mountains. (ibid.).

Meanwhile, the solar eyepiece gave '… good details of the willow leaf convolutions of some of the [sun]spots' (ibid.).

**Bio Box 15.1: George Herbert Hoskins**

George Herbert Hoskins (1883–1953; Fig. 15.3) was born in Sydney and educated at Newington College before joining the family firm of G. & C. Hoskins Ltd. and eventually become a Director (Orchiston and Bembrick 1997). The founders of G. & C. Hoskins Ltd. were George John Hoskins (1847–1926) and his brother Charles Henry Hoskins (1851–1926); George Herbert was George John's son. The Hoskins were an affluent dynasty, and pioneers of the Australian steel industry (see Hoskins 1969), so G.H. Hoskins had the financial means to indulge his avocational passion for astronomy.

(continued)

**Bio Box 15.1** (continued)

**Fig. 15.3** G.H. Hoskins (*Courtesy* Colin Bembrick)

For many years, Hoskins was a prominent member of the New South Wales Branch of the British Astronomical Association (see Orchiston and Bembrick 1997). He joined the Branch in 1902 when still a teenager and was elected to the Council 12 years later.

By 1903, Hoskins had acquired a 12-in. (30.5-cm) reflector (see Hoskins 1903), and in 1917 he replaced this with an 18-in. (46-cm) f/8.2 Newtonian fork-mounted equatorial reflector with a primary mirror by Robert W. Wigmore of Melbourne, a drive manufactured by the Sydney scientific instrument-maker and importer William Edward Esdaile (1858/59–1947; Drawing compass … 1877), and two spectroscopes—one by Zollner and the other made by local astronomer James Nangle (Orchiston and Bembrick 1995). This fine telescope was installed in Hoskins' observatory in Beecroft, at that time a far north-western suburb of Sydney, but just two years later he donated it to the NSW Branch of the British Astronomical Association, and it was housed in a 15-ft (4.6-m) diameter domed observatory at the Sydney Technical College (where James Nangle was ensconced as Director and the Superintendent of Technical Education for New South Wales). In 1921, it acquired a 6-in. (15.2-cm) Cooke refractor, which served as a guide scope. Records show that the telescope was to be used twice-monthly for public viewing nights and was available to BAA Branch members for their observing programmes. The subsequent history of this telescope is beyond the scope of this book, but is recounted in a paper by Orchiston and Bembrick (ibid.) and in the records of the Museum of Applied Arts and Sciences in Sydney (which now has custody of the telescope).

George Herbert Hoskins died on 3 August 1953, at the age of 70. At that time, he was recognised as the longest-serving member of the New South Wales Branch of the British Astronomical Association, having been a member for more than 50 years (Orchiston and Bembrick 1997).

**Fig. 15.4** The simple roll-off roof observatory at Brewongle near Bathurst, New South Wales, that Hoskins erected for the 8-in. Grubb refractor (*Courtesy* Colin Bembrick)

However, when it came to the history of this telescope, Hoskins (1946) had a far from accurate story to tell:

It had been imported by Mr. Tebbutt [it was actually Dr William Bone] …

When Mr. Tebbutt died, about 17 or 18 years ago [actually it was 30 years ago] the Telescope was acquired by Mr. W.F. Gale, who held it for a year or so [more likely, about 5 years], when I purchased it

In 1940, Hoskins (ibid.) sold his country property, and '… having no immediate prospect of using it, I lent the instrument [i.e. the 8-in Grubb] to Mr. Howarth of Newcastle …' Mark Howarth (Bio Box 15.2) was a well-known independent astronomer, and he installed the telescope in a roll-off roof observatory (Orchiston and Bembrick 1997) and used it mainly for double star and variable star observations (Howarth 1946).

**Bio Box 15.2: Mark Howarth**

Mark Howarth (Fig. 15.5) was born in Warrington, England, on 26 November 1884, and after completing his schooling and receiving the requisite technical training served in the Royal Navy.

(continued)

**Bio Box 15.2** (continued)

**Fig. 15.5** Mark Howarth (*Courtesy* Newcastle Region Public Library: 0104-005241)

In 1921, he emigrated to Australia and settled in Newcastle, New South Wales' second-largest city (after Sydney), 100 miles (161 km) north of Sydney, where he founded the engineering firm Ryland Bros. Pty. Ltd. (Burke 1997).

Howarth was a Fellow of the Royal Astronomical Society and a prominent member of the New South Wales Branch of the British Astronomical Association (Orchiston and Bembrick 1997; Orchiston and Perdrix 1990), and from 1933 until 1964 he maintained the Grange Mount Observatory at his home in the suburb of Mayfield (Obituary 1971). The Observatory comprised two roll-off roof buildings, which were later converted into a little cottage and a storage shed; these have survived through to the present day.

In 1958, Mark Howarth was awarded an Order of the British Empire for his services to the community, and he died in 1971.

As we have just seen, Grange Mount Observatory comprised two adjacent roll-off roof buildings (see Fig. 15.6), and the larger one, measuring 14-ft (4.3-m) square and with 8-ft (2.4-m) high walls, housed the 8-in. Grubb telescope (Esdaile, 1945b). This is building '1' in Fig. 15.6. Meanwhile, 'The roof is divided into two parts which separate and run out away from the Observatory over extended rails. I operate them by means of two small winches' (ibid.).

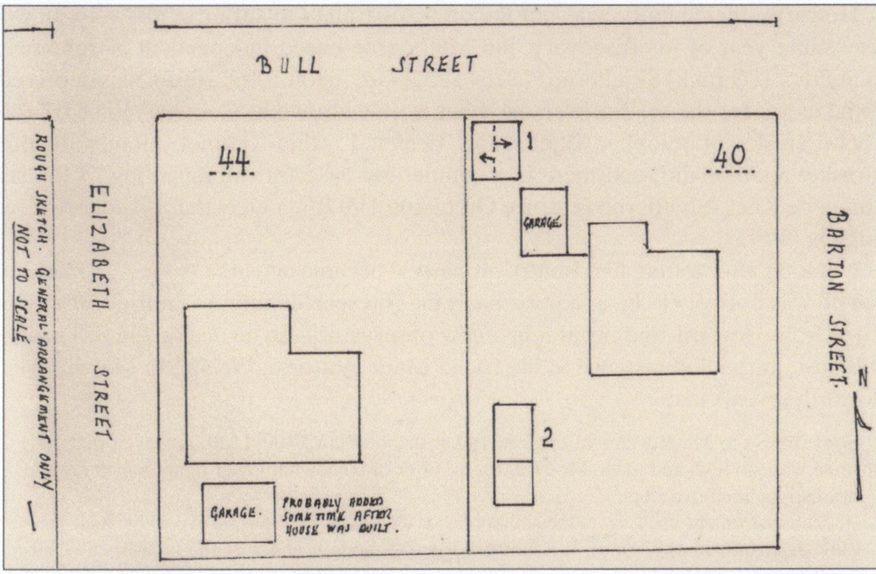


**Fig. 15.6** A schematic plan (not to scale) showing Howarth's two adjacent properties and houses in Newcastle. The two roll-off roof buildings comprising his Grange Mount Observatory were at 40 Bull Street and are marked '1' and '2'. The Grubb telescope was housed in observatory '1' (*Courtesy* Pat Burke)

Although Howarth used the Grubb telescope for some solar observing (including photography), his main preoccupation was variable stars, and he was one of the BAA New South Wales Branch's most prolific observers. Apart from variables, he also used the Grubb telescope to observe double stars (Howarth 1946).

However, Howarth's main claim to fame is his independent discovery of Comet C/1941 K1 (van Gent) (Comet Howarth 1941). On 16 June 1941, while observing variables, he noticed a '... diffuse patch 1 min. of arc in diameter with a central nucleus, and ... of the tenth magnitude'. (Another new comet ... 1941), which proved to be a new comet. The Dutch-born South African professional astronomer, Dr. Hendrik van Gent (1899–1947), actually discovered this object on 27 May (Comet van Gent 1941; Marsden 1982, p. 37) but, because of telecommunication restrictions during wartime, news of this did not reach Australia until some time after Howarth's independent discovery. Since Howarth was only the second independent discoverer, there is a legitimate case for the official renaming of this object as Comet C/1941 K1 (van Gent-Howarth)—see Orchiston 1997b. The fact that Howarth was awarded both the Donovan Medal (from Sydney) and the Donohoe Medal by the Astronomical Society of the Pacific (Mr. Mark Howarth honoured 1942) adds further weight to this co-discovery claim.

Howarth, incidentally, was not the only Australian to suffer in this way. In the very same year of his discovery, the Melbourne-based independent astronomer, John Francis (Frank) Skjellerup (1875–1952; Orchiston 1999, 2003), was deprived of the credit for the co-discovery of what is now known as Comet C/1941 B2 (de Kock-Paraskevopoulos). Skjellerup detected this comet long before Paraskevopoulos did, and there is a compelling case for the renaming of this as Comet de Kock-Skjellerup (e.g. see Orchiston 1997b), a view that is supported by Hughes (1991).

Not long after selling his country property it became apparent to George Hoskins that he was unlikely to be in a position in the foreseeable future to retrieve the telescope from Howarth and instal it in a new observatory. So he decided to sell it, and offered it in the first instance to his friend Mark Howarth. Writing in March 1946, Howarth reveals that

> Some time ago Mr. Hoskins offered me this instrument for £400. I felt, however, that this price was too high and made Mr. Hoskins an offer of somewhat lower figure which he did not feel inclined to accept.
>
> The instrument itself, from the standpoint of its optical qualities, is excellent. As it was built, however, in 1884 [*sic*], it is not particularly modern as far as the adjustments and controls are concerned …
>
> [However] It is, as I have already mentioned, a splendid Telescope … (Howarth 1946).

It then was left to E. Esdaile & Sons in Sydney to arrange the sale of the telescope, one of the principals of the firm, Edward William Esdaile Jnr (better known as Ted Esdaile; 1882–1967), being well known to George Hoskins through the New South Wales Branch of the British Astronomical Association.

Esdaile (1945a) heard that the Cook Islands-based New Zealand astronomer Frank Maine Bateson (Bio Box 15.3) was interested in setting up an 'tropical observatory' (Bateson 1945a), and in October 1945 he offered him the 8-in. Grubb telescope—then still domiciled in Newcastle—for an all-up price of £485. This figure included £25 for a filar micrometre and £10 for a Dawes solar eyepiece. Esdaile (ibid.) pointed out that '… the original cost was £1200 in 1882 … [so] it will be seen that the price for this Grubb instrument is very low' (ibid.).

In a letter addressed to British-born Jesuit priest Dr. Daniel Joseph Kelly O'Connell (1896–1982), the Director of Riverview Observatory in Sydney, Bateson (1945b) indicated his initial interest in this telescope: 'It is a lot of money, but is by no means an excessive charge for a telescope of the kind'.

**Bio Box 15.3: Frank Maine Bateson**

Frank Maine Bateson (Fig. 15.7; Bateson 1989; Christie 2014) was born in Wellington, the New Zealand capital, in 1909, and after early schooling in Wanganui, New Zealand, moved with his family to Sydney, Australia, when he was a teenager. Although destined for a career in accountancy and business administration, his Sydney sojourn allowed him to pursue his passion for astronomy, mainly through the New South Wales Branch of the BAA.

(continued)

**Bio Box 15.3** (continued)

**Fig. 15.7** Frank Maine Bateson in July 1990 (*Photograph* Wayne Orchiston)

While in Sydney, Bateson made his first observations of meteors and variable stars. In 1927, he returned to New Zealand, and founded the Variable Star Section of the New Zealand Astronomical Society (NZAS, later the Royal Astronomical Society of New Zealand), at the age of 18. Over the next seven decades, he built this into a vibrant Southern Hemisphere group that published reports and papers, and supplied visual, and later photoelectric, variable star data that were used by professional astronomers in their astrophysical research projects (see Christie 2014).

Once back in New Zealand, Bateson also was able to continue his observational interest in meteors and was a founding member and very keen participant in the NZAS's Meteor Section, which was established by Ronald Alexander McIntosh (1900–1977; Orchiston 2016, Chap. 19) in February 1928 (see McIntosh 1946).

Bateson's third main area of observational interest was Jupiter, and after purchasing the ex-Tebbutt 8-in. Grubb refractor he was able to time many Jovian satellite phenomena and record the never-ending changes that occurred in the system of belts on the planet.

After residing in the Cook Islands from 1946 to 1960, Bateson returned to New Zealand with the dream of establishing a high altitude professional observatory that could contribute to astrophysical research (and especially of variable stars). Through US contacts, he was funded to conduct a national site survey, which culminated in the establishment of Mt John University Observatory, overlooking Lake Tekapo, in 1963. This was a joint venture between the University of Canterbury (in Christchurch) and the University of Pennsylvania, and Bateson was appointed officer-in-charge (Hearnshaw and Gilmore 2015). The first two telescopes on the mountain were personally owned by Bateson; one was the 8-in. Grubb refractor.

(continued)

**Bio Box 15.3** (continued)

Upon retiring from mountain life in 1969 Bateson sold both telescopes, and he settled in Tauranga where he was able to devote much of his time to the successful running of the Variable Star Section.

Over the years, Frank Bateson received many national and international honours for his variable star work and his contributions to New Zealand astronomy, including an honorary doctorate from Waikato University in 1979. Meanwhile, in 1970 he was awarded a high civilian honour, an Order of the British Empire, for his services to New Zealand and international astronomy, and in 1981 minor planet 2434 was named 'Bateson'.

Bateson's story (1989) is one filled with poignant lessons for 'young players' of how it was still possible in twentieth century New Zealand astronomy for a dedicated, talented non-professional astronomer with the necessary political acumen and observational prowess to cross the boundary and join the ranks of the professional astronomer. Indeed, Bateson is the pre-eminent—but by no means the only—example of a New Zealand ATP, an Amateur-Turned-Professional (and in his case also a PTA = Professional-Turned-Amateur), as defined and discussed (primarily in a trans-Tasman context) by Orchiston (2015).

Frank Main Bateson died in Tauranga on 16 April 2007 at the age of 97.

Two weeks later, Bateson (1945c) told Esdaile he was interested in the Grubb telescope and he asked for details of the observatory in which it currently was housed. He also stated that £485 was more than he was hoping to pay, given the additional costs associated with freighting the instrument from Newcastle to the Cook Islands and constructing an observatory for it (Bateson 1945b). On the same day, Bateson (1945d) also sent a further letter to Father O'Connell, where he repeated that he was interested in the telescope, but it was more than he was expecting to pay. He wondered, therefore, whether Hoskins might consider a time-payment plan: £100 on the date of sale (say 1 March 1946) and the balance in three successive annual instalments. O'Connell (1945) obviously did see merit in the time-payment option for he suggested that Bateson offer Hoskin a lower price—after all, this was a 60-year-old telescope. Furthermore, he suggested that Bateson write to Howarth, as 'I have just heard that he himself wants a telescope, and it is likely that he considers the price too high'. As we have already seen, Howarth (1946) subsequently confirmed this suspicion.

Bateson (1946a) then wrote directly to Hoskins, stating that he wanted to purchase the Grubb telescope in order to

> … further the work on which I have been engaged for a number of years, i.e. variable star observations and planetary work. Dr. O'Connell … can certify, I think, that the work I have been doing warrants a larger instrument and that my work has been a definite contribution of value to Astronomy.

Dr. O'Connell certainly was in a good position to judge as he also was involved in variable star work—Riverview Observatory's forté—albeit through photography rather than visual observations (Orchiston 1985). After appealing to Hoskins' good nature in supporting his variable star programme, Bateson mentioned that the price was a hurdle, and he therefore proposed (1) that Hoskins loan him the telescope for one year, from 1 March 1946; and (2) that the purchase price be paid in four equal annual instalments, starting on 1 March 1947. Bateson was not only an experienced astronomer but he also was an astute businessman, so while waiting for a reply from Hoskins he wrote to the Resident Commissioner of the Cook Islands, seeking his financial assistance, on the basis that '… I am not financially affluent and the erection of an observatory here will keep me poor for years' (Bateson 1946b).

This ploy paid off, and Bateson was promised free transportation of the telescope from Auckland to Rarotonga, but Hoskin was not interested in Bateson's proposal: he wanted a definite sale, not a trial period, with an immediate payment of £150, and the balance in three yearly instalments, and *with interest* (Esdaile 1946a; my italics). On 16 February 1946, Bateson (1946c) wrote accepting these conditions, and less than four months after he first heard that the Grubb telescope was available the sale was finalised. On the very same day that Bateson contacted Esdaile, he also wrote to the head office of his employer, A.B. Donald Ltd., in Auckland, with the mysterious news that the purchase of the telescope '… is being financed from grants that are being made by interested Observatories overseas' (Bateson 1946d). At no time does he elaborate on this tantalising claim, and it would be very interesting indeed to know which particular overseas observatories were willing to fund Bateson's astronomy as early as 1946.

With the sale finalised events moved along apace. On 18 March, Esdaile (1946b) advised Bateson that he was going to Newcastle at the end of the week to begin dismantling the telescope. He also mentioned Howarth's reaction: 'Mr. Howarth is very sorry to see it go but as he had his opportunity to buy it he has no complaints. He has found it a very efficient instrument'.

By 22 May, the telescope was dismantled and in Sydney ready to be overhauled by E. Esdaile & Sons before they were shipping it to New Zealand. At this time, Esdaile (1946c) was impressed with the quality of the telescope, and told Bateson so:

> The instrument I found in very good condition … The Optical parts are not damaged in the slightest and all mechanical parts are as good as original condition … I think you are very lucky to be the possessor of such a beautiful and efficient instrument and expect that you will have pleasure in using it.

Two weeks later, the telescope was on board the S.S. *Kiaranga*, en route to Auckland, and Esdaile (1946d) sent Bateson an explanation of the invoice: the total of £544.1.6 comprised the original figure of £485 plus 'sundry expenses' associated with dismantling the telescope in Newcastle, packing and transferring it to Sydney, packing and wharf costs in Sydney and a 50 % share of the cost of overhauling the telescope. This certainly was more than Bateson had bargained for, but towards the end of July the telescope arrived in the Cook Islands, and once it was installed in a

**Fig. 15.8** Frank Bateson and the 8-in. Grubb telescope in the Cook Islands (after Bateson 1957, p. 574)

new observatory (see Fig. 15.8) there was a chance to test it on Jupiter and some of his variable star fields and Bateson (1946f) had no regrets.

In November 1947, Bateson wrote Esdaile:

> The Telescope is most satisfactory and I am very pleased indeed with its performance. It is in constant use, and has enabled me to resume Planetary work. During the 1947 opposition of Jupiter nearly 400 transits of markings across the C.M. were recorded; a large number of drawings and many other observations.
>
> In addition it is used very regularly for Variable star work and has enabled the variables to be followed down to a much fainter magnitude. Over a thousand observations have been secured since last April. On occasions it has been possible to observe as faint as magnitude 14.2. (Bateson 1947).

Upon reflecting on the history of the telescope, Bateson (1946e) recalled the pleasure he had '... using the instrument in 1924 when I was in Sydney and it was in Mr. Gale's possession'. He also retained '... a vivid recollection of views of Mars through this particular instrument [in 1925, in Sydney]. Little did he dream that one day it would pass into his possession' (Bateson 1948, pp. 45–46). Moreover, Bateson was well aware of the telescope's reputation:

> The writer only hopes that he is able at Rarotonga to attain in some measure the skill with this telescope that was reached by Tebbutt and Gale. For not only has the instrument a high standard of optical perfection, but also a record of being in the hands of highly skilled observers. Their work is an inspiration to an unworthy successor. (Bateson 1948, p. 48).

As it turned out, Bateson proved an admirable and totally worthy successor. From 1947, the Grubb refractor was used extensively for visual observations of variable stars and Jupiter. We saw in Bio Box 15.3 how in 1960 Bateson returned to New Zealand and undertook the ambitious site-testing programme which led, eventually, to the founding of Mt John University Observatory (Hearnshaw and Gilmore 2015). Bateson served as Officer-in-Charge and used the 8-in. Grubb refractor '...

**Fig. 15.9** The historic 8-in. Grubb refractor at Hillcrest Observatory in Whakatane (*Courtesy* Whakatane Astronomical Society)

for the observation of Variable Stars and for the checking out of several series of CHARTS FOR SOUTHERN VARIABLES' (Bateson 1979).

After his retirement from Mt John Bateson realised that he could not set up the Grubb telescope near his home in light-polluted Tauranga, and 'My age precluded the idea of establishing it out in the country away from city lights, so I sold the instrument to the Whakatane Astronomical Society in 1969' (ibid.)—see Fig. 15.9.

Jim Duthie (1985) recalls that the price was NZ$2000, and 'The purchase was described locally as a milestone or alternatively millstone but we managed to scrape up the dough over an agreed term of 5 years'. In January 1971, the 8-in. Grubb finally was installed in the Society's Hillcrest Observatory with Frank Bateson's assistance (see The first quarter century 1985), where it was used mainly for public viewing nights (Bateson 1989), although Bateson (1979) notes that he also '… used it on a number of occasions for … the Courses that I have given throughout the Bay of Plenty for the University of Waikato'. After the long 'research' record of Tebbutt and Bateson, education proved a new but rather effective role for this venerable old telescope, and most visitors would have been totally unaware of its remarkable history.

The final leg of the telescope's travels saw it return to Windsor and its original observatory building in 1986, the culmination of years of repatriation attempts. Jim Duthie (1985) elaborates. Almost as soon as the 8-in. was mounted at Hillcrest Observatory

> … we were plagued with enquiries for it to be returned to NSW for museum purposes—first by the Tebbutt family who after 3/4 enquiries did nothing more. Then a Melbourne museum (University) enquired; it was at this stage that we gave the proposal real credence and offered to trade it in exchange for a C-14—no reply. Wayne Orchiston of Melbourne was next—same answer but it was too tough for his organisation and he had to opt out. Then a movie mogul of Sydney wanted to produce an epic of p/Halley—same offer, no response. Finally, the Hawkesbury Shire Council of Windsor came into the picture—same offer (I nearly got a swag of printed letters prepared) but this time, they accepted …

Some hard trans-Tasman bargaining (Kiwi stargazers … 1986) followed between the Hawkesbury Shire Council and the Whakatane Astronomical Society, then a feature article titled 'Tebbutt telescope to return' on the front page of the 7 February 1985 issue of the *Hawkesbury Courier* formally announced that agreement had been reached by the two parties, and that the Council for its part would provide the New Zealand society with a substitute telescope costing about $A12,000. The 8-in. Grubb then returned to Windsor (Fig. 15.10) and a 14-in. (35-cm) Celestron (C14) Schmidt-Cassegrain telescope went to New Zealand. As it turned out, members of

**Fig. 15.10** A newspaper clipping showing Eric Cleaver, Bruce Thompson and Pearce Atkins (*left* to *right*) dismantling the Grubb telescope in Whakatane so that it could be air-freighted to Australia (Orchiston collection)

the Whakatane Astronomical Society had no regrets because the C14 was a much more compact instrument and therefore proved far superior for public viewing (J. Duthie, pers. comm., 1986).

However, back in Australia the imminent return of the telescope to Windsor was merely the start of the yet another saga, for there was considerable debate as to where the telescope should be erected. Some wished to consign it to the local historical museum:

> The Tebbutt Telescope should be placed on exhibition at the Hawkesbury Museum which is owned by the Hawkesbury Shire Council—having been purchased by the council—the Hawkesbury Historical Society believes.
>
> According to the Historical Society, as the telescope is now community property it can best be displayed there in the interests of everybody … (Museum display … 1986).

Others (including Dr. Graeme White and I) actively lobbied for a return of the telescope to its original dome at Windsor Observatory. Fortunately, the latter option eventually prevailed, as we shall see in Sect. 15.3 below.

Unfortunately, we cannot report such a satisfactory outcome in the case of Tebbutt's 4.5-in. (11.4-cm) Cooke refractor, nor is the chain of ownership so clear. According to an inscription on the reverse side of a watercolour of Windsor Observatory painted by M.J. MacNally in 1931 (a copy of which is in the Mitchell Library), some time after Tebbutt's death Walter Gale also acquired this telescope. We also know that he then sold it to Mark Howarth in 1933, so for a time Howarth was the proud custodian of Tebbutt's two largest telescopes, the Cooke refractor occupying his second, smaller, observatory. In 1941, he published a short paper in the *Journal of the British Astronomical Association* about a transit of Mercury which he observed with this telescope (Howarth 1941). The roll-off roof building where this telescope was located (number '2' in Fig. 15.6) was still in existence in 2001 and had been converted into a comfortable little cottage—see Fig. 15.11.

With advancing years, Howarth eventually found observational astronomy a challenge and in 1964 moved to Wollongong (Obituary 1971) to live next door to his daughter. It was at this time that he decided to sell the Cooke refractor (P. Burke, pers. comm., 1983). The new owner was a man named Mashman, from Sydney (I. Debono, pers. comm., 1985), who has since died. The telescope remains with the Mashman family, and various attempts by John Halley Tebbutt (pers. comm., 1996) and others to purchase it or to borrow it and return it to Windsor Observatory have proved unsuccessful.

## 15.3 Refurbished as a Museum of Astronomy: A New Role for an Historic Observatory

After Tebbutt died and most of the Windsor Observatory instruments were disposed of, the buildings themselves underwent a transformation. The original wooden observatory built in 1863/1867 continued to nourish the white ants and was in so

**Fig. 15.11** By 2001, Howarth's roll-off roof building at Grange Mount Observatory, Newcastle, that once housed the 4.5-in. Cooke refractor was a comfortable little cottage (*Photograph* Wayne Orchiston)

poor a condition that it had to be demolished (Fig. 15.12). This was accomplished by Keith Tebbutt, one of John Tebbutt's grandsons, most probably during the 1920s, but well before the birth of John Halley Tebbutt in 1942 (J.H. Tebbutt, pers. comm., 1996). The Mitchell Library's copy of MacNally's 1931 painting of Windsor Observatory only includes the 1863 observatory building as an inset, which would suggest that it had already been demolished by 1931.

In about 1938, the 1879 brick observatory building was substantially modified (ibid.). After acquiring a raised transit room ceiling, altered doorways, and kitchen and bathroom additions, it became a cosy two-bedroom flat and was rented out (see Fig. 15.13). Only the 1894 circular brick observatory building survived intact, and as we noted in Chap. 2 it was used as a storeroom.

Meanwhile, Tebbutt's descendants continued to live in the original Homestead, up until Bruce Tebbutt assumed ownership. He then moved into one of the farm buildings set up as a cottage, and rented out rooms in the Homestead. So when I used to regularly visit the site, during 1959–1961—before we had to abandon the plan to install an 18-in. reflector in the 1894 circular observatory and turn it into a functioning astronomical centre once more—the Peninsula Estate was like a little residential village, with tenants' mailboxes lining the entrance to the complex (e.g. see Fig. 15.14).

**Fig. 15.12** The 1863/1867 original observatory building, shown here on the left, is thought to have been demolished during the 1920s (after Tebbutt 1908)

**Fig. 15.13** A rear view of the 1879 observatory-cottage in October 1981 (*Photograph* Wayne Orchiston)

**Fig. 15.14** A later photograph, taken in July 1973, with the mail boxes at the entrance to the Peninsula Estate and the picturesque 1879 observatory-cottage in the background (*Photograph* Wayne Orchiston)

This situation only changed in about 1977 when John Halley Tebbutt (pers. comm., 1996) took over the farm and moved his family into the Homestead. He also took an interest in the scientific achievements of his famous great-grandfather, and during the 1980s he and a number of Australian astronomers began thinking about the future of the observatory buildings. In this way, the concept of a museum of astronomy at Windsor Observatory was born.

In 1982, I formally developed this concept in two research papers (see Orchiston 1982, 1983) which were presented at conferences in New Zealand and Brisbane, respectively, and subsequently published. In the latter paper, I presented a case for the establishment of a 'Museum of Australian Astronomy' at a time when no museums of astronomy existed in Australia (for the decision by the New South Wales Government to convert Sydney Observatory into a museum only occurred later—see Orchiston 1988). After critically reviewing the alternatives, I came up with Windsor Observatory as the obvious site:

> At Windsor, near Sydney, is a more promising prospect, namely John Tebbutt's old observatory. Two of the three observatory buildings that were extant at the time of Tebbutt's death in 1916 survive, basically in sound condition. One of these originally housed the 8 inch refractor … Preliminary negotiations [by the author] have already taken place with the present owners of the instrument, the Whakatane Astronomical Society, in New Zealand [see Anonymous, 1985, p. 8] regarding the possible repatriation of this historical telescope.
>
> Tebbutt's Observatory has three other major advantages as a potential Museum of Australian Astronomy. Firstly, associated with the observatory buildings is a well-preserved 2-storey mansion of 1845 vintage, which could serve as an excellent museum building and interpretation centre, thereby allowing the two observatory buildings to be

devoted entirely to Tebbutt, one as a functioning observatory, complete with the original telescope … Secondly, the surrounding grounds … could easily accommodate large items of radio-astronomical hardware, and finally, since Windsor, as one of New South Wales' earliest towns, is already an established tourist mecca, an astronomy museum here would undoubtedly attract more than just the budding astronomer!

Given the crucial contribution of the Sydney area during astronomy's embryonic days in Australia, and the total dominance of the Sydney-based amateur astronomers of the late nineteenth century, Windsor is clearly the ideal place for Australia's first museum of astronomy. (Orchiston 1983, pp. 9–10).

These conclusions were reached in my then capacity as an active independent astronomer and professional museologist involved in the planning and running of post-graduate museum studies courses in Melbourne. One of my special research interests was science centres, which—from my perspective—encompassed both observatories and museums of astronomy.

In the other paper (Orchiston 1982), which was presented on my behalf by Dr. Frank Bateson (as a former owner of the Tebbutt 8-in. Grubb telescope), I suggested that the mooted Museum of Australian Astronomy at Windsor should have three primary functions:

1. To collect and preserve significant items of Australian astronomical hardware and software;
2. To serve as a 'public information centre', through displays, educational programs, publications, and the like; and
3. To carry out some form of research in optical astronomy, within the context of the original function of the Windsor Observatory (Orchiston 1982, p. 240).

Having a valid concept is one thing, but transforming it into reality is often something else again, and this proved to be the case at Windsor. Who should be responsible for the Windsor Observatory transformation? After arranging for the repatriation of the Grubb telescope, the Hawkesbury Shire Council was for a time keen on purchasing the Peninsula Homestead and remaining observatory buildings and developing them as a museum and tourist attraction (e.g. see Tebbutt telescope … 1985), but during 1985 this plan was abandoned (J.H. Tebbutt, pers. comm., 1985). Instead, in late 1986 John Halley Tebbutt decided to embark on this ambitious project himself. His starting point had to be the observatory buildings as they existed at that time (see Figs. 15.15 and 15.16), and the task was enormous. For example, John Halley Tebbutt pointed out that in the case the 1879 brick observatory,

The biggest problem is finding the money. I suppose it might cost at least $50,000 to completely restore.

Although the building is structurally sound, the staircase which once led into the dome is gone. The wheels on which the dome once rotated have rusted and only one of the dome flaps can be opened. (Tebbutt's observatory … 1984).

Because of the historic importance of Windsor Observatory and the heritage status of the buildings, restoration proved to be a time-consuming and extremely expensive exercise. It even included archaeological excavations in May and July 1987 by staff and students from the University of Sydney (Cremin 1987), carried

**Fig. 15.15** A photograph showing the general appearance and condition of the 1879 Windsor Observatory building in October 1981, prior to the start of restoration (*Photograph* Wayne Orchiston)

**Fig. 15.16** A photograph showing the general appearance and condition of the 1894 Windsor Observatory building in October 1981, prior to the start of restoration (*Photograph* Wayne Orchiston)

**Fig. 15.17** A view of the renovated John Tebbutt Observatories in November 1996, looking towards the front gate of the Peninsula Estate (*Photograph* Wayne Orchiston)

out primarily to investigate the original 1863 observatory building (even though there was no plan to construct a replica of this).

The accurate reconstruction of the 1879 observatory building proved a major conservation challenge, despite the wealth of historical information available and the existence of detailed plans and elevations prepared by University of New South Wales architects in about 1980 (J.H. Tebbutt, pers. comm., 1981). After various distractions, the refurbished 'John Tebbutt Observatories' complex was opened to the public in 1989 following three years of constructional work and a total financial outlay of around $400,000 (J.H. Tebbutt, pers. comm., 1992).

The John Tebbutt Observatories (see Figs. 1.22 and 15.17 and 15.18) does not quite mirror the functional image I presented in my 1982 paper, given that it is a privately-owned and operated museum, and given that the government-funded transformation of Sydney Observatory had occurred in the interim (Orchiston 1988). Only the two observatory buildings are involved in the museum development, and the new facility focusses solely on the astronomical and other scientific achievements of John Tebbutt. The transit and prime vertical rooms in the main observatory building and the former office beneath the dome room are all devoted to historical displays, while the former library serves as a lecture room.

After considerable negotiation, the Shire Council finally agreed to lease the repatriated Grubb telescope to the Observatory (for an annual fee), and this is now back in its original building (Fig. 15.19) and is used for public viewing nights. But before this could happen, it was necessary to carry out an extensive restoration of the tele-

**Fig. 15.18** A 25 April 1992 photograph of the renovated 1879 building at the John Tebbutt Observatories, showing the present-day John Tebbutt (*left*) and my brother Peter Orchiston (*Photograph* Wayne Orchiston)

scope, but particularly the rolled steel tube, and all of the optics (S. Williams, pers. comm., 1989). Meanwhile, a modern 8-in. (20-cm) Celestron (C8) telescope occupies the dome in the main observatory building, pending the possible return of the original Cooke inhabitant or an equatorially-mounted refractor of similar vintage and aperture.

While Windsor Observatory is currently addressing its potential educational and informational roles and serving as a unique Windsor tourist attraction (see Fig. 15.20), the prospect of carrying out meaningful research is remote. Not only is the Grubb telescope far too small by today's standards, but the types of observational programmes that were pursued by Tebbutt in the nineteenth century during the positional astronomy era have been well and truly bypassed by most contemporary professional observatories where the emphasis is now firmly on astrophysics (e.g. see Haynes et al. 1996 for an excellent overview of contemporary professional astronomy in Australia).

Therefore, if Windsor Observatory is to make any contribution to observational astronomy it will be as a non-professional observatory. The world of astronomy has moved on, and so long as it remains a monument to John Tebbutt and his type of astronomy, Windsor Observatory will never be able to rejoin the ranks of 'notable world observatories'. That it was able to achieve this status back in the nineteenth

**Fig. 15.19** A November 1996 photograph of the 8-in. Grubb refractor back in its original dome at the renovated John Tebbutt Observatories (*Photograph* Wayne Orchiston)

century is a tribute to John Tebbutt's genius (see Tebbutt 1986). In today's world, no independent astronomer could aspire to such a goal!

Perhaps, this was one of the reasons that eventually prompted John Halley Tebbutt to try and sell the old Windsor Observatory. In June 1994, the Tebbutt Homestead, Observatory buildings, associated outbuildings and the remaining 54 acres of land comprising the Peninsula Estate were put up for auction (J.H. Tebbutt, pers. comm., 1994 and 1996). This generated considerable interest and a price in

**Fig. 15.20** The front of the A4-size twofold brochure about the refurbished John Tebbutt Observatories (Orchiston collection)

excess of $1 million was anticipated (Edwards 1994), but the reserve figure was not reached. Ownership of the Estate thus remains with the Tebbutt family, preserving a tradition that has prevailed for more than 170 years, but the observatory complex is now the responsibility of those who run a restaurant at this historic site.

Windsor Observatory may live no more as an internationally-recognised research institution, but it survives in its present form as a memorial to one of Australia's most remarkable non-professional scientists.

**Acknowledgements** I wish to thank the following for their assistance in various ways: staff at the Alexander Turnbull Library (Wellington, New Zealand), the late Dr. Frank Bateson (Tauranga, New Zealand), Colin Bembrick (Walang, New South Wales), the late Pat Burke (Newcastle), Dr. Aedeen Cremin (University of Sydney), the late Ignace Debono (Sydney), George Odey, Bruce Thompson and the late Jim Duthie (Whakatane Astronomical Society), Professor Nick Lomb (University of Southern Queensland, Toowoomba), Peter Orchiston (Port Douglas, Australia), the late Jackie St. George (Sydney), Dr. Graeme White (Wagga Wagga), Steve Williams (formerly at the John Tebbutt Observatories) and especially John Halley Tebbutt. I also am grateful to Colin Bembrick, the late Pat Burke, the Newcastle Region Public Library and the Whakatane Astronomical Society for kindly supplying Figs. 15.3–15.6 and 15.9.

## References[1]

Anonymous (1985). *Whakatane Astronomical Society. The First Quarter Century*. Whakatane: Whakatane Astronomical Society.
Another new comet (1941). *Southern Stars, 7,* 71 (1941).
Bateson, F. (1945a). A tropical observatory. *Southern Stars, 11* (4), 9–12.
Bateson, F.M. (1945b). Letter to D. O'Connell, dated 29 October. In BR.
Bateson, F.M. (1945c). Letter to E. Esdaile, dated 4 November. In BR.
Bateson, F.M. (1945d). Letter to D. O'Connell, dated 4 November. In BR.
Bateson, F.M. (1946a). Letter to G. Hoskins, dated 27 January. In BR.
Bateson, F.M. (1946b). Letter to The Resident High Commissioner, dated 13 February. In BR.
Bateson, F.M. (1946c). Letter to E. Esdaile, dated 16 February. In BR.
Bateson, F.M. (1946d). Letter to A.B. Donald Ltd., Auckland, dated 16 February. In BR.
Bateson, F.M. (1946e). Letter to M. Howarth, dated 21 April. In BR.
Bateson, F.M. (1946f). Letter to E. Esdaile, dated 28 July. In BR.
Bateson, F.M. (1947). Letter to E. Esdaile, dated 16 November. In BR.
Bateson, F.M. (1948). A tropical observatory. *Southern Stars, 14* (3), 45–48.
Bateson, F.M. (1979). Letter to G. White, dated 31 January. In Orchiston Collection.
Bateson, F.M. (1989). *Paradise Beckons*. Waikanae: Heritage Press.
Biggest not best. Telescopes for Mars. Atmospheric obstacles. *The Sun,* 26 August 1924.
Burke, P., 1997. Two Newcastle amateur astronomers. Part 1—Mark Howarth. *The Astronomers' Bulletin, 2* (3), 6–7.
Christie, G. (2014). Bateson, Frank Maine. In T. Hockey et al. (Eds.), *Biographical Encyclopedia of Astronomers. 2nd Edition* (pp. 171–172). New York: Springer.
Comet Howarth. *British Astronomical Association, New South Wales Branch Bulletin, 211,* 1 (1941).
Comet van Gent. *Southern Stars, 7,* 97 (1941).
Cremin, A., 1987. Report on the Excavation of Tebbutt's First Observatory at Windsor, NSW May and July 1987. Unpublished Report, Department of History, University of Sydney.
Drawing compass in case, 1877. Power house Museum Collection web site/: www.powerhouse-museum.com/mob/collection/database/?m=289282&img=22647
Duthie, J (1985). Letter to J. St. George, dated 23 October. In Orchiston Collection.
Edwards, G., 1994. Mansion of note goes on the market. *Telegraph Mirror,* 14 May.
Esdaile, W. (1945a). Letter to F. Bateson, dated 22 October. In BR.
Esdaile, W. (1945b). Letter to F. Bateson, dated 24 December. In BR.

[1]The following abbreviations are used: BR = Bateson Records, Alexander Turnbull Library, Wellington, New Zealand (MS-Group-0826). TL = Letters to John Tebbutt, Mitchell Library, Sydney

Esdaile, W. (1946a). Letter to F. Bateson, dated 8 February. In BR.
Esdaile, W. (1946b). Letter to F. Bateson, dated 18 March. In BR.
Esdaile, E. (1946c). Letter to F. Bateson, dated 22 May. In BR.
Esdaile, W. (1946d). Letter to F. Bateson, dated 5 June. In BR.
Foley. M. (1912a). Letter to John Tebbutt, dated 9 February. In TL.
Foley, M. (1912b). Letter to John Tebbutt, dated 20 February. In TL.
Gale, W. (1925). Nova Pictoris (1925). *Journal of the British Astronomical Association, 36,* 30–31.
Haynes, R., Haynes, R., Malin, D. & McGee, R., 1996. *Explorers of the Southern Sky. A History of Australian Astronomy*. Cambridge: Cambridge University Press.
Hearnshaw, J.B. & Gilmore, A.C. (2015). *Mt John—The First Fifty Years. A Celebration of Half a Century of Optical Astronomy at the University of Canterbury*. Christchurch: Canterbury University Press.
Hoskins, Sir Cecil, 1969. *The Hoskins Saga*. Sydney: Halstead.
Hoskins, G.H. (1903). Letter to John Tebbutt, dated 29 November. In TL.
Hoskins, G.H. (1946). Letter to Frank Bateson, dated 8 March. In BR.
Howarth, M. (1941). Observation of the transit of Mercury from New South Wales. *Journal of the British Astronomical Association, 51,* 178–179.
Howarth, M. (1946). Letter to Frank Bateson, dated 29 March. In BR.
Hughes, D.W. (1991). John Grigg, J.F. Skjellerup and their comet. *Vistas in Astronomy, 34,* 1–10.
Kiwi stargazers seal a celestial deal. *Australasian Post,* 8 May 1986, page 30.
McIntosh, R.A. (1946). A history of the Meteor Section. *Southern Stars, 12,* 10–15.
Macey. R. (1984). Tebbutt's observatory awaits days of a brighter note. *Sydney Morning Herald*, 4 April, page 11.
Maguire, R. (1985). Angelo Tornaghi. An inventive Italian of the nineteenth century. *The Australian Antique Collector, 29,* 42–46
Marsden, B.G. (1982). *Catalogue of Cometary Orbits*. Cambridge (Mass.): Minor Planet Center.
Mr. Mark Howarth honoured. *Newcastle Morning Herald,* 12 August 1942.
Museum display would give public greatest access. *Hawkesbury Gazette,* 4 June 1986.
O'Connell, D. (1945). Letter to F. Bateson, dated 5 December. In BR.
Obituary. Mr. Mark Howarth. *Newcastle Morning Herald,* 30 October 1971.
Orchiston, W. (1982). John Tebbutt and the Whakatane Eight Inch Refractor: a review of "The Australian Connection". *Southern Stars, 29,* 215–244.
Orchiston, W. (1983). Preserving our astronomical heritage: towards a history of astronomy in Australia. In *Supplement to the Proceedings. Tenth National Australian Convention of Amateur Astronomers* (pp. 12–24). Brisbane: NACAA.
Orchiston, W. (1985). Life-history of a telescope: the 17.8-cm refractor at the Riverview Observatory in Sydney. *Australian Journal of Astronomy, 1,* 57–80.
Orchiston, W. (1988). From research to recreation: the rise and fall of Sydney Observatory. *Vistas in Astronomy, 32,* 49–63.
Orchiston, W. (1997a). The role of the large refracting telescope in Australian amateur astronomy: an historical perspective. *Australian Journal of Astronomy, 7,* 89–114.
Orchiston, W. (1997b). The "tyranny of distance" and Antipodean cometary astronomy. *Australian Journal of Astronomy, 7,* 115–126.
Orchiston, W. (1999). Of comets and variable stars: the Afro-Australian astronomical activities of J.F. Skjellerup. *Journal of the British Astronomical Association, 109,* 328–338.
Orchiston, W. (2001). 'Sentinel of the Southern Heavens': The Windsor Observatory of John Tebbutt. *Journal of the Antique Telescope Society, 21,* 11–23.
Orchiston, W. (2003). J.F. Skjellerup: a forgotten name in South African cometary astronomy. *Monthly Notices of the Astronomical Society of Southern Africa, 62,* 54–73.
Orchiston, W. (2015). The amateur-turned-professional syndrome: two Australian case studies. In W. Orchiston, D.A. Green & R. Strom (Eds.), *New Insights from Recent Studies in Historical Astronomy: Following in the Footsteps of F. Richard Stephenson* (pp. 259–350). Cham: Springer.

Orchiston, W. & Bembrick, C. (1995). The role of the large reflecting telescope in amateur astronomy: an Australian case study. *Australian Journal of Astronomy, 6,* 53–74.

Orchiston, W. & Bembrick, C. (1997). The potency of patronage: George Hoskins and the New South Wales Branch of the British Astronomical Association. *Australian Journal of Astronomy, 7,* 1–15.

Orchiston, W. & Perdrix, J. (1990). The British Astronomical Association in Australia: a centennial perspective. *Australian Journal of Astronomy, 3,* 163–179.

Reports of the Branches, New South Wales Branch. *Journal of the British Astronomical Association, 33,* 370–372 (1923).

Stott, C., and Hughes, D.W., 1987. The amateur's small transit telescope in the nineteenth century. *Quarterly Journal of the Royal Astronomical Society, 28,* 30–42.

Tebbutt, J. (1986). *Astronomical Memoirs*. Windsor: Hawkesbury Shire Council (Reprinted edition). (Although the reprint date is given as 1986, in fact this book was in fact printed in 1987 and was officially launched on 9 April 1987).

Tebbutt telescope to return. *Hawkesbury Courier*, 7 February 1985, page 1.

*The First Quarter Century*. Whakatane: Whakatane Astronomical Society (1985).

# Index




W. Orchiston, *John Tebbutt*, Historical & Cultural Astronomy,
DOI 10.1007/978-3-319-44521-2

**D**



**E**

MIX
Papier aus verantwortungsvollen Quellen
Paper from responsible sources
FSC® C105338

If you have any concerns about our products,
you can contact us on
**ProductSafety@springernature.com**

In case Publisher is established outside the EU,
the EU authorized representative is:
**Springer Nature Customer Service Center GmbH**
**Europaplatz 3, 69115 Heidelberg, Germany**

Printed by Libri Plureos GmbH
in Hamburg, Germany